Poisson Geometry in Mathematics and Physics

CONTEMPORARY MATHEMATICS

450

Poisson Geometry in Mathematics and Physics

International Conference
June 5–9, 2006
Tokyo, Japan

Giuseppe Dito
Jiang-Hua Lu
Yoshiaki Maeda
Alan Weinstein
Editors

American Mathematical Society
Providence, Rhode Island

2000 *Mathematics Subject Classification.* Primary 53D17, 53D55, 19L10, 58B34, 81T70, 14A22, 22A22, 47G30.

Library of Congress Cataloging-in-Publication Data

Poisson 2006 (2006 : Tokyo, Japan)
Poisson geometry : Poisson 2006, June 5–9, 2006, Tokyo, Japan / G. Dito... [et al.], editors.
p. cm. — (Contemporary mathematics, ISSN 0271-4132 ; v. 450)
Includes bibliographical references.
ISBN 978-0-8218-4423-6 (alk. paper)
1. Poisson algebras—Congresses. 2. Poisson processes—Congresses. 3. Geometric quantization—Congresses. I. Dito, Giuseppe. II. Title.

QA251.5.P64 2006
512′.482—dc22 2007060575

∞ The paper used in this book is acid-free and falls within the guidelines
established to ensure permanence and durability.
Visit the AMS home page at http://www.ams.org/

10 9 8 7 6 5 4 3 2 1   13 12 11 10 09 08

Contents

Foreword

The conference "Poisson 2006: Poisson Geometry in Mathematics and Physics" was held from June 5 through 9, 2006 at the National Olympics Memorial Youth Center in Tokyo. There were about 150 participants, including 25 invited speakers, and 20 presenters at a poster session.

The speakers were chosen by a Scientific Committee of ten members, chaired by Alan Weinstein, while local organization was handled by a separate committee headed by Yoshiaki Maeda and Giuseppe Dito.

The meeting was preceded by a school of about three days, organized by Giuseppe Dito, Yoshiaki Maeda and Alan Weinstein, consisting of a lecture series designed to provide background for the conference talks, as well as invited topical lectures by young participants.

Sponsoring organizations for the conference and school included the Mathematical Society of Japan, the European Mathematical Society, the American Mathematical Society, the Bernoulli Center at EPFL Lausanne, and the 21st Century Center of Excellence (COE) at Keio University. The COE provided the majority of funding, with additional support from the US National Science Foundation.

Poisson 2006 was the fifth in a series of international conferences on Poisson geometry, held every two years. The first, in 1998, took place at the Banach Centre in Warsaw, with subsequent meetings at CIRM in Luminy, IST in Lisbon, and the University of Luxembourg. Further information about all these meetings, as well as the one to be held in 2008 at EPFL in Lausanne, may be found on the Poisson Geometry Home Page at `poissongeometry.org`, which links to the videos of all the talks of the conference Poisson 2006 and principal lectures of the school.

The aim of these meetings has been to bring together mathematicians and mathematical physicists who work in diverse areas and share a common interest in Poisson geometry. With roots in classical mechanics from 200 years ago and the work of Sophus Lie from a century ago, the subject of Poisson geometry crystallized through the work of Kirillov and Lichnerowicz in the 1970's and has been particularly driven by the program of "deformation quantization", in which Poisson structures appear as the first deviation from commutativity in families of associative algebras. Subjects where Poisson geometry plays an essential role include symplectic geometry and topology, deformation theory, representation theory, hamiltonian dynamics, and field theory.

In preparing the program for Poisson 2006, the Scientific Committee made a special effort to include speakers from "outside" areas which were relevant to Poisson geometry and its applications. The program of Poisson 2006 (conference and school) was remarkable for the overlap of topics, some intentional, some fortuitous, between the lectures. Here are some examples:

- Generalized complex stuctures and other geometry on $TM+T^*M$ (Bursztyn, Gualtieri, Meinrenken, Uchino, Yoshimura)
- Stacks and twisting by a three-form (Gomi, Schapira, Tsygan, Van den Bergh)
- Orbifolds and other singular spaces as differentiable stacks (Crainic, Holm, Weinstein, Xu, Zhu)
- Normal forms of Poisson structures in the neighborhood of points and symplectic leaves (Dufour, Fernandes, Ratiu, Zung)
- Deformation of Poisson structures (Ikeda, Zhang)
- Reduction of systems with symmetry (Bursztyn, Cardona, Cattaneo, Holm, Ratiu, Yoshimura)
- Kontsevich formality and its variants (Alekseev, Kontsevich, Merkulov, Park, Tsygan, Van den Bergh, Waldmann)
- Log-canonical coordinates (Gekhtman, Kontsevich, Lu)
- Group-valued momentum maps (Meinrenken, Schaffhauser)
- Strict quantization of spaces via group actions (Rieffel, Voglaire, Waldmann)
- Quantization of canonical transformations via their graphs (Kontsevich, Schapira, Van den Bergh, Weinstein)

The present volume consists of refereed papers by many of the invited speakers at the conference and by the principal lecturers at the school. Papers by presenters at the poster session and by other speakers at the school will be appearing in *Travaux Mathématiques.*

Giuseppe Dito, Jiang-Hua Lu, Yoshiaki Maeda, Alan Weinstein

List of Participants

Alekseev, Anton
Université de Genève, Switzerland

Androulidakis, Iakovos
Universität Zürich, Switzerland

Antunes, Paulo
École Polytechnique, France

Arias Abad, Camilo
Universiteit Utrecht, The Netherlands

Baird, Tom
University of Toronto, Canada

Bialecki, Mariusz
University of Tokyo, Japan and
Academy of Sciences, Poland

Blohmann, Christian
University of California at Berkeley,
USA

Bonneau, Philippe
Université de Bourgogne, France

Brahami, Renaud
Université de Bourgogne, France

Brahic, Olivier
Instituto Superior Técnico, Portugal

Bursztyn, Henrique
Instituto Nacional de Matemática Pura
e Aplicada, Brazil

Cardona, Alexander
Universidad de Los Andes, Colombia

Cattaneo, Alberto
Universität Zürich, Switzerland

Chen, Zhuo
Capital Normal University, China

Chiang, River
National Cheng Kung University,
Taiwan

Claessens, Laurent
Université catholique de Louvain,
Belgium

Crainic, Marius
Universiteit Utrecht, The Netherlands

Dherin, Benoit
Université de Genève, Switzerland

Dinar, Yassir
International School for Advanced
Studies (SISSA), Italy

Dito, Giuseppe
Université de Bourgogne, France

Dragulete, Oana
Ecole Polytechnique Fédérale de
Lausanne, Switzerland

Dubrovskiy, Stanislav
Keio University, Japan

Dufour, Jean-Paul
Université de Montpellier 2, France

Fernandes, Rui
Instituto Superior Técnico, Portugal

Fregier, Yael
University of Luxembourg,
Grand-Duchy of Luxembourg

Fuchs, Shay
University of Toronto, Canada

Fujita, Daisuke
Keio University, Japan

Futaki, Akito
Tokyo Institute of Technology, Japan

Futaki, Masahiro
University of Tokyo, Japan

Gay-Balmaz, François
Ecole Polytechnique Fédérale de Lausanne, Switzerland

Gekhtman, Michael
University of Notre Dame, USA

George, Nathan
University of California at Berkeley, USA

Ginzburg, Viktor
University of California at Santa Cruz, USA

Gomi, Kiyonori
University of Tokyo, Japan

Grabowski, Janusz
Polish Academy of Sciences, Poland

Gualtieri, Marco
Massachusetts Institute of Technology, USA

Hamachi, Kentaro
Kyoto Sangyo University, Japan

Hamilton, Mark
University of Calgary, Canada

Harada, Megumi
University of Toronto, Canada

Haruyama, Daisuke
Keio University, Japan

Hayakawa, Yohei
Keio University, Japan

He, Long-Guang
Capital Normal University, China

Hirota, Yuji
Keio University, Japan

Ho, Nan-Kuo
The Fields Institute, Canada

Hofer, Laurent
Université de Haute-Alsace, France

Holm, Tara
University of Connecticut, USA

Iglesias Ponte, David
Instituto de Matemáticas y Física Fundamental, Spain

Ikeda, Kaoru
Keio University, Japan

Ikeda, Noriaki
Ritsumeikan University, Japan

Indelicato, Davide
Universität Zürich, Switzerland

Inoue, Rei
University of Tokyo, Japan

Ito, Yuji
Keio University, Japan

Iwata, Etsuji
Waseda University,Japan

Jane, James
University of Cambridge, United Kingdom

Kajiura, Hiroshige
Kyoto University, Japan

Kameta, Keisei
Institute of Physics Publishing, Japan

Kato, Daisuke
Keio University, Japan

Keller, Frank
Université du Luxembourg, Grand-Duchy of Luxembourg

Kieserman, Noah
University of Winsconsin, USA

Kim, Donghui
Yonsei University, Korea

Kim, Hoil
Kyungpook National University, Korea

Kimura Takashi
Boston University, USA

Kirillov, Anatol
RIMS, Kyoto University, Japan

Koda, Yuya
Keio University, Japan

Kohmoto, Daichi
Nagoya University, Japan

Konishi, Yukiko
University of Tokyo, Japan

Kontsevich, Maxim
Institut des Hautes études Scientifiques, France

Kori, Tosiaki
Waseda University, Japan

Kosmann-Schwarzbach, Yvette
Ecole Polytechnique, France

Kubarski, Jan
Technical University of Lodz, Poland

Kubo, Fujio
Hiroshima University, Japan

Larrain Hubach, Andres
Universidad de los Andes, Colombia

Le, Thanh Hieu
Quy Nhon University, Vietnam

Lee, Sungyun
Korean Advanced Institute of Science and Technology, Korea

Li, Hui
Instituto Superior Técnico, Portugal

Liu, Zhangju
Beijing University, China

Lu, Jiang-Hua
University of Hong Kong, Hong Kong

Lyapina, Oleksandra
University of Notre Dame, USA

Machida, Yoshinori
Numazu National college of Technology, Japan

Maeda, Yoshiaki
Keio University, Japan

Martens, Johan
University of Toronto, Canada

Martinez Torres, David
Utrecht University, The Netherlands

Mehta, Rajan
University of California at Berkeley, USA

Meinrenken, Eckhard
University of Toronto, Canada

Merkulov, Sergei
Stockholms Universitet, Sweden

Mikami, Kentaro
Akita University, Japan

Miranda, Eva
Université de Toulouse 3, France

Miyake, Ken
Keio University, Japan

Miyaoka, Reiko
Kyushu University, Japan

Miyazaki, Naoya
Keio University, Japan

Mizutani, Tadayoshi
Saitama University, Japan

Monnier, Philippe
Université de Toulouse 3, France

Moreau, Anne
Université Paris 7, France

Moriyoshi, Hitoshi
Keio University, Japan

Nakanishi, Nobutada
Gifu-Keizai University, Japan

Nest, Ryszard
Copenhagen University, Denmark

Nguyen, Minh Cong
University of Education of Hanoi, Vietnam

Nguyen, Tien Zung
Université de Toulouse 3, France

Nunes da Costa, Joana
Universidade de Coimbra, Portugal

Oguni, Shin-ichi
Kyoto University, Japan

Oikonomides, Catherine
Keio University, Japan

Omori, Hideki
Tokyo University of Science, Japan

Ono, Kaoru
Hokkaido University, Japan

Otgonbayar, Uuye
Pennsylvania State University, USA

Park, Jae-Suk
Yonsei University, Korea

Park, Seo-Ree
Seoul National Univesity, Korea

Peterka, Mira
University of California at Berkeley, USA

Pflaum, Markus
Johann Wolfgang Goethe-Universität, Germany

Pichereau, Anne
Université de Poitiers, France

Poncin, Norbert
University of Luxembourg, Grand-Duchy of Luxembourg

Ratiu, Tudor
École polytechnique fédérale de Lausanne, Switzerland

Raugas, Mark
RABA Technologies, USA

Rieffel, Marc
University of California at Berkeley, USA

Sako, Akifumi
Keio University, Japan

Sato, Nobuya
Rikkyo University, Japan

Schaetz, Florian
Universität Zürich, Switzerland

Schaffhauser, Florent
Keio University, Japan

Schapira, Pierre
Université de Paris 6, France

Schlichenmaier, Martin
University of Luxembourg, Grand-Duchy of Luxembourg

Siby, Hassène
Université de Montpellier 2, France

Signori, Daniele
Pennsylvania State University, USA

Sjamaar, Reyer
Cornell University, USA

Stefanini, Luca
Universität Zürich, Switzerland

Sternheimer, Daniel
Université de Bourgogne, France and Keio University, Japan

Stienon, Mathieu
Pennsylvania State University, USA

Suzuki, Haruo
Hokkaido University, Japan

Takhtajan, Leon
SUNY at Stony Brook, USA

Tanaka, Yuuji
Nagoya University, Japan

Terashima, Yuji
Tokyo Institute of Technology, Japan

Terilla, John
CUNY - Queens College, USA

Tomoda, Atsushi
Keio University, Japan

Tose, Nobuyuki
Keio University, Japan

Tsygan, Boris
Northwestern University, USA

Tudoran, Razvan Micu
Universitatea de Vest din Timisoara, Romania

Uchino, Kyosuke
Tokyo University of Science, Japan

Urbanski, Pawel
University of Warsaw, Poland

Van den Bergh, Michel
Universiteit Hasselt, Belgium

Voglaire, Yannick
Université catholique de Louvain, Belgium

Wade, Aissa
Pennsylvania State University, USA

Waldmann, Stefan
Albert-Ludwigs-Universität Freiburg, Germany

Watamura, Satoshi
Tohoku University, Japan

Weinstein, Alan
University of California at Berkeley, USA

Xu, Ping
Pennsylvania State University, USA

Yoshida, Takahiko
University of Tokyo, Japan

Yoshimura, Hiroaki
Waseda University, Japan

Yoshioka, Akira
Tokyo University of Science, Japan

Zambon, Marco
Universität Zürich, Switzerland

Zhang, Youjin
Tsinghua University, China

Zhu, Chenchang
ETH Zürich, Switzerland

Contemporary Mathematics
Volume **450**, 2008

Quantized Anti de Sitter spaces and non-formal deformation quantizations of symplectic symmetric spaces

Pierre Bieliavsky, Laurent Claessens, Daniel Sternheimer, and Yannick Voglaire

ABSTRACT. We realize quantized anti de Sitter space black holes, building Connes spectral triples, similar to those used for quantized spheres but based on Universal Deformation Quantization Formulas (UDF) obtained from an oscillatory integral kernel on an appropriate symplectic symmetric space. More precisely we first obtain a UDF for Lie subgroups acting on a symplectic symmetric space M in a locally simply transitive manner. Then, observing that a curvature contraction canonically relates anti de Sitter geometry to the geometry of symplectic symmetric spaces, we use that UDF to define what we call Dirac-isospectral noncommutative deformations of the spectral triples of locally anti de Sitter black holes. The study is motivated by physical and cosmological considerations.

1. Introduction

1.1. Physical and cosmological motivations. This paper, of independent interest in itself, can also be seen as a small part in a number of long haul programs developed by many in the past decades, with a variety of motivations. The references that follow are minimal and chosen mostly so as to be a convenient starting point for further reading, that includes the original articles quoted therein.

An obvious fact (almost a century old) is that anti de Sitter (AdS) space-time can be obtained from usual Minkowski space-time, deforming it by allowing a (small) non-zero negative curvature. The Poincaré group symmetry of special relativity is then deformed (in the sense of [**Ger64**]) to the AdS group $SO(2,3)$. In $n+1$ space-time dimensions ($n \geq 2$) the corresponding AdS_n groups are $SO(2,n)$. Interestingly these are the conformal groups of flat (or AdS) n space-times. The *deformation philosophy* [**Fla82**] makes it then natural, in the spirit of deformation quantization [**DS02**], to deform these further [**St05, St07**], i.e. quantizing them, which many are doing for Minkowski space-time.

2000 *Mathematics Subject Classification.* 81S10; 53D50, 58B34, 81V25, 83C57.

Key words and phrases. Non commutative geometry, BTZ-spaces, Deformation quantization, Symplectic symmetric spaces.

The first author thanks Victor Gayral for enlightening discussions on noncommutative spectral triples. This research is partially supported by the IAP grant "NOSY" at UCLouvain.

Laurent Claessens is a FRIA-fellow.

These deformations have important consequences. Introducing a small negative curvature ρ permits to consider [**AFFS**] massless particles as composites of more fundamental objects, the Dirac *singletons*, so called because they are associated with unitary irreducible representations (UIR) of $SO(2,3)$, discovered by Dirac in 1963, so poor in states that the weight diagram fits on a single line. These have been called Di and Rac and are in fact massless UIR of the Poincaré group in one space dimension less, uniquely extendible to the corresponding conformal group $SO(2,3)$ ($\mathrm{AdS}_4/\mathrm{CFT}_3$ symmetry, a manifestation of 't Hooft's holography). That kinematical fact was made dynamical [**FF88**] in a manner consistent with quantum electrodynamics (QED), the photons being considered as 2-Rac states and the creation and annihilation operators of the naturally confined Rac having unusual commutation relations (a kind of "square root" of the canonical commutation relations for the photon).

Later [**FFS**] this phenomenon has been linked with the (then very recently) observed oscillations of neutrinos (see below, neutrino mixing). Shortly afterwards, making use of flavor symmetry, Frønsdal was able to modify the electroweak model [**Frø00**], obtaining initially massless leptons (see below) that are massified by Yukawa interaction with Higgs particles. (In this model, 5 pairs of Higgs are needed and it predicts the existence of new mesons, parallel to the W and Z of the $U(2)$-invariant electroweak theory, associated with a $U(2)$ flavor symmetry.)

Quantum groups can be viewed [**BGGS**] as an avatar of deformation quantization when dealing with Hopf algebras. Of particular interest here are the quantized AdS groups [**FHT, Sta98**], especially at even root of unity since they have some finite dimensional UIR, a fact generally associated with compact groups and groups of transformations of compact spaces. It is then tempting to consider quantized AdS spaces at even root of unity $q = e^{i\theta}$ as "small black holes" in an ambient Minkowski space that can be obtained as a limit when $\rho q \to -0$. Note that, following e.g. 't Hooft (see e.g. [**Hoo06**] but his approach started around 1980) that some form of communication is possible with quantum black holes by interaction at their surface.

At present, conventional wisdom has it that our universe is made up mostly of "dark energy" (74% according to a recent Wilkinson Microwave Anisotropy Probe, WMAP), then of "dark matter" (22% according to WMAP), and only 4% of "our" ordinary matter, which we can more directly observe. Dark matter is "matter", not directly observed and of unknown composition, that does not emit nor reflect enough electromagnetic radiation to be detected directly, but whose presence can be inferred from gravitational effects on visible matter. According to the Standard Model, dark matter accounts for the vast majority of mass in the observable universe. Dark energy is a hypothetical form of energy that permeates all of space. It is currently the most popular method for explaining recent observations that the universe appears to be expanding at an accelerating rate, as well as accounting for a significant portion of the missing mass in the universe.

The Standard Model of particle physics is a model which incorporates three of the four known fundamental interactions between the elementary particles that make up all matter (the fourth one, weakest but long range, being gravity). It came after the electroweak theory that incorporated QED (electromagnetic interactions) associated with the photon and the so-called weak interactions, associated with the *leptons* that now exist in three generations (flavors): electron, muon and tau, and their neutrinos. The Standard Model, of phenomenological origin, encompasses also the so-called strong interactions, associated with (generally) heavier particles called baryons (the proton and neutron, and many more), now commonly assumed to be bound states of "confined" quarks with gluons, in three "colors". It contains 19 free parameters, plus 10 more in extensions needed to account for the recently observed *neutrino mixing* phenomena, which require nonzero masses for the neutrinos that are traditionally massless in the Standard Model.

Very recently Connes [**CCM**] developed an effective unified theory based on noncommutative geometry (space-time being the product of a Riemannian compact spin 4-manifold and a finite noncommutative geometry) for the Standard Model with neutrino mixing, minimally coupled to gravity. It has 4 parameters less and predicts for the yet elusive Higgs particle (responsible for giving mass to initially massless leptons) a mass that is slightly different (it is at the upper end of the expected mass range) from what is usually predicted. See also [**Co06**], and [**Ba06**] in a Lorentzian framework.

Previously, in part aiming at a possible description of quantum gravity, but mainly in order to study nontrivial examples of noncommutative manifolds, Fröhlich (in a supersymmetric context), then Connes and coworkers had studied quantum spheres in 3 and 4 dimensions [**FGR, CL01, CDV**]. The basic tool there is a spectral triple introduced by Connes [**Co94**]. In the present paper we are developing a similar approach, but for hyperbolic spheres and using integral universal deformation formulas in the deformation quantization approach.

The distant hope is that these quantized AdS spaces (at even root of unity) can be shown to be a kind of "small" black holes at the edge of our Universe in accelerated expansion, from which matter would emerge, possibly created as (quantized) 2-singleton states emerging from them and massified by interaction with ambient dark energy (or dark matter), in a process similar to those of the creation of photons as 2-Rac states and of leptons from 2-singleton states, mentioned above.

As fringe benefits that might explain the acceleration of expansion of our Universe, and the problems of baryogenesis and leptogenesis (see e.g. [**Cl06, SS07**]). Physicists love symmetries and even more to break them (at least at our level). One of the riddles that physics has to face is that, while symmetry considerations suggest that there should be as much matter as antimatter, one observes a huge imbalance in our region of the Universe. In a seminal paper published in 1967 that went largely unnoticed for about 13 years but has now well over a thousand citations (we won't quote it here), Andrei Sakharov addressed that problem, now called baryogenesis. If and when a mechanism along the lines hinted at above can be developed for creating baryons and other particles, it could solve that riddle.

Roughly speaking the idea is that there is no reason, except theological, why everything (whatever that means) would be created "in the beginning", or as conventional wisdom has it now, in a Big Bang. There could very well be "stem cells" of the primordial singularity that would be spread out, like shrapnel, mostly at the edge of the Universe. Our proposal is that these could be described mathematically as quantized AdS black holes. We shall now concentrate our study on them.

1.2. Mathematical introduction. Roughly speaking, a universal deformation formula (briefly UDF) for a given symmetry $\mathcal{G}$ is a procedure that, for every, say, topological algebra $\mathbb{A}$ admitting the symmetry $\mathcal{G}$, produces a deformation $\mathbb{A}_\theta$ of $\mathbb{A}$ within the *same* category of topological algebras. Such a UDF is called *formal* when the category it applies to is that of formal power series in a formal parameter with coefficients in associative algebras.

For instance, Drinfel'd twisting elements in elementary quantum group theory constitute examples of formal UDF's (see e.g. [**CP95**]). Other formal examples in the Hopf algebraic context have been given by Giaquinto and Zhang [**GZ98**]. In [**Zag94**], Zagier produced a formal example from the theory of modular forms. The latter has been used and generalized by Connes and Moscovici in their work on codimension-one foliations [**CM04**].

In [**Rie93**], Rieffel proves that von Neumann's oscillatory integral formula [**vN31**] for the composition of symbols in Weyl's operator calculus actually constitutes an example of a non-formal UDF for the actions of $\mathbb{R}^d$ on associative Fréchet algebras. The latter has been extensively used for constructing large classes of examples of noncommutative manifolds (in the framework of Connes' spectral triples [**Co94**]) via Dirac isospectral deformations[1] of compact spin Riemannian manifolds [**CL01**] (see also [**CDV**]). Some Lorentzian examples have been investigated in [**BDSR**] and [**PS06**]. Other very interesting related approaches can be found in [**HNW, Ga05**] and references quoted therein.

Oscillatory integral UDF's for proper actions of non-Abelian Lie groups have been given in [**Bie02, BiMs, BiMa, BBM**]. Several of them were obtained through geometrical considerations on solvable symplectic symmetric spaces. Nevertheless, the geometry underlying the one in [**BiMs**] remained unclear.

In the present work we build on these works in the AdS context, with when needed reminders of their main features so as to remain largely self-contained. First we show that the latter geometry is that of a solvable symplectic symmetric space which can be viewed as a curvature deformation of the rank one non-compact Hermitian symmetric space. [It can also be viewed as a curvature deformation of the AdS space-time, as we shall see in the last section of the article.]

Next, we develop some generalities on UDF's for groups which act strictly transitively on a symplectic symmetric space. We give some precise criteria. We end the section by providing new examples with exact symplectic forms such as UDF's for solvable one-dimensional extensions of Heisenberg groups, as well as examples with non-exact symplectic forms.

In the last section we apply these developments to noncommutative Lorentzian geometry. In anti de Sitter space $\mathrm{AdS}_{n\geq 3}$, every open orbit $\mathcal{M}_o$ of the Iwasawa component $\mathcal{AN}$ of $SO(2,n)$ is canonically endowed with a causal black hole structure [**CD07**] (generalizing the BTZ-construction in dimension $n=3$). We define the analog of a Dirac-isospectral noncommutative deformation for a triple built on $\mathcal{M}_o$. The deformation is maximal in the sense that its underlying Poisson structure is symplectic on the open $\mathcal{AN}$-orbit $\mathcal{M}_o$. In particular, it does not come from an application of Rieffel's deformation machinery for isometric actions of Abelian Lie groups. Moreover, via the group action, the black hole structure is encoded in the deformed spectral triple, with no other additional geometrical data, in contradistinction with the commutative level[2].

[1] A deformation triple $(\mathcal{A}_\theta, \mathcal{H}_\theta, D_\theta)$ is said **isospectral** when $\mathcal{H}_\theta$ and D_θ are the same for all values of θ.

[2] An interesting challenge would be to analyze which operator algebraic notions attached to the triple are responsible for the singular causality. That is not investigated in the present article.

2. Curvature deformations of rank one Hermitian symmetric spaces and their associated UDF's

2.1. Preliminary set up and reminder. In [**BiMs**], a formal UDF for the actions of the Iwasawa component $\mathcal{R}_0 := \mathcal{AN}$ of $SU(1,n)$ is given in oscillatory integral form. It has been observed in [**BBM**] that this type of UDF is actually non-formal for proper actions on topological spaces. The precise framework and statement are as follows. The group $\mathcal{R}_0$ is a one dimensional extension of the Heisenberg group $\mathcal{N}_0 := H_n$. Through the natural identification $\mathcal{R}_0 = SU(1,n)/U(n)$ induced by the Iwasawa decomposition of $SU(1,n)$, the group $\mathcal{R}_0$ is endowed with a (family of) left-invariant symplectic structure(s) ω. Denoting by $\mathfrak{r}_0 := \mathfrak{a}_0 \times \mathfrak{n}_0$ its Lie algebra, the map

$$\mathfrak{r}_0 \longrightarrow \mathcal{R}_0 : (a,n) \mapsto \exp(a)\exp(n) \tag{2.1}$$

turns out to be a global Darboux chart on $(\mathcal{R}_0, \omega)$. Setting $\mathfrak{n}_0 = V \times \mathbb{R}.Z$ with table $[(x,z)\,,\,(x',z')] = \Omega_V(x,x')\,Z$, and $\mathfrak{r}_0 = \{(a,x,z)\,|\,,a,z\in\mathbb{R}; x\in V\}$, one has

THEOREM 2.1. *For all non-zero $\theta \in \mathbb{R}$, there exists a Fréchet function space $\mathcal{E}_\theta$, $C^\infty_c(\mathcal{R}_0) \subset \mathcal{E}_\theta \subset C^\infty(\mathcal{R}_0)$, such that, defining for all $u, v \in C^\infty_c(\mathcal{R}_0)$*

(2.2)

$$\begin{aligned}
&(u \star_\theta v)(a_0,x_0,z_0)\\
&:= \frac{1}{\theta^{\dim \mathcal{R}_0}} \int_{\mathcal{R}_0\times\mathcal{R}_0} \cosh(2(a_1-a_2))\,[\cosh(a_2-a_0)\cosh(a_0-a_1)\,]^{\dim\mathcal{R}_0 - 2}\\
&\times \exp\Big(\frac{2i}{\theta}\Big\{S_V\big(\cosh(a_1-a_2)x_0, \cosh(a_2-a_0)x_1, \cosh(a_0-a_1)x_2\big)\\
&- \bigoplus_{0,1,2} \sinh(2(a_0-a_1))z_2\Big\}\Big)\\
&\times u(a_1,x_1,z_1)\,v(a_2,x_2,z_2)\,da_1da_2dx_1dx_2dz_1dz_2\;;
\end{aligned}$$

where $S_V(x_0,x_1,x_2) := \Omega_V(x_0,x_1) + \Omega_V(x_1,x_2) + \Omega_V(x_2,x_0)$ is the phase for the Weyl product on $C^\infty_c(V)$ and $\bigoplus_{0,1,2}$ stands for cyclic summation[3]*, one has:*

(i) *$u\star_\theta v$ is smooth and the map $C^\infty_c(\mathcal{R}_0)\times C^\infty_c(\mathcal{R}_0) \to C^\infty(\mathcal{R}_0)$ extends to an associative product on $\mathcal{E}_\theta$. The pair $(\mathcal{E}_\theta, \star_\theta)$ is a (pre-$C^\star$) Fréchet algebra.*

(ii) *In coordinates (a,x,z) the group multiplication law reads*

$$L_{(a,x,z)}(a',x',z') = \left(a+a', e^{-a'}x+x', e^{-2a'}z+z'+\frac{1}{2}\Omega_V(x,x')e^{-a'}\right).$$

The phase and amplitude occurring in formula (2.2) are both invariant under the left action $L : \mathcal{R}_0\times\mathcal{R}_0 \to \mathcal{R}_0$.

(iii) *Formula (2.2) admits a formal asymptotic expansion of the form:*

$$u\star_\theta v \sim uv + \frac{\theta}{2i}\{u,v\} + O(\theta^2)\;;$$

where $\{\,,\,\}$ denotes the symplectic Poisson bracket on $C^\infty(\mathcal{R}_0)$ associated with ω. The full series yields an associative formal star product on $(\mathcal{R}_0,\omega)$ denoted by $\tilde{\star}_\theta$.

---

[3]In [**BiMs**], the exponent $\dim\mathcal{R}_0 - 2$ was forgotten in the expression of the amplitude of the oscillating kernel.

The setting and (i-ii) may be found in [**BiMs**], while (iii) is a straightforward adaptation to $\mathcal{R}_0$ of [**BBM**].

2.2. Geometry underlying the product formula. We start with preliminary material concerning the symmetric spaces.

2.2.1. *Symmetric spaces.* A *symplectic symmetric space* [**Bie95, BCG**] is a triple $(\mathcal{M}, \omega, s)$ where $\mathcal{M}$ is a connected smooth manifold, ω is a non-degenerate two-form on $\mathcal{M}$ and $s : \mathcal{M} \times \mathcal{M} \to \mathcal{M} : (x, y) \mapsto s_x(y)$ is a smooth map such that $\forall x \in \mathcal{M}$ the map $s_x : \mathcal{M} \to \mathcal{M}$ is an involutive diffeomorphism of $\mathcal{M}$ preserving ω and admitting x as an isolated fixed point. Moreover, one requires that the identity $s_x \circ s_y \circ s_x = s_{s_x(y)}$ holds for all $x, y \in \mathcal{M}$ [**Loo69**]. In this situation, if $x \in \mathcal{M}$ and X, Y and Z are smooth tangent vector fields on $\mathcal{M}$,

$$\omega_x(\nabla_X Y, Z) := \frac{1}{2} X_x.\omega(Y + s_{x_\star} Y\,,\, Z) \tag{2.3}$$

defines an affine connection ∇ on $\mathcal{M}$, the unique affine connection on $\mathcal{M}$ which is invariant under the symmetries $\{s_x\}_{x \in \mathcal{M}}$. It is moreover torsion-free and such that $\nabla\omega = 0$. In particular, the two-form ω is necessarily symplectic.

It then follows that the group $\mathcal{G} = \mathcal{G}(\mathcal{M}, s)$ generated by the compositions $\{s_x \circ s_y\}_{x,y \in \mathcal{M}}$ is a Lie group of transformations acting transitively on $\mathcal{M}$. The group $\mathcal{G}$ is called the **transvection** group of $\mathcal{M}$. Given a base point o in $\mathcal{M}$, the conjugation by the symmetry s_o defines an involutive automorphism $\tilde{\sigma}$ of $\mathcal{G}$. Its differential at the unit element $\sigma := \tilde{\sigma}_{\star_e}$ induces a decomposition into ± 1-eigenspaces of the Lie algebra $\mathfrak{g}$ of $\mathcal{G}$: $\mathfrak{g} = \mathfrak{k} \oplus \mathfrak{p}$. The subspace $\mathfrak{k}$ of fixed vectors turns out to be a Lie subalgebra which acts faithfully on the subspace $\mathfrak{p}$ of "anti-fixed" vectors ($\sigma x = -x$ for $x \in \mathfrak{p}$). Moreover $[\mathfrak{p}, \mathfrak{p}] = \mathfrak{k}$. A pair $(\mathfrak{g}, \sigma)$ as above is called a *transvection pair*. The subalgebra $\mathfrak{k}$ corresponds to the Lie algebra of the stabilizer of o in $\mathcal{G}$, while the vector space $\mathfrak{p}$ is naturally identified with the tangent space $T_o(\mathcal{M})$ to $\mathcal{M}$ at point o. In particular the symplectic form at o, ω_o, induces a $\mathfrak{k}$-invariant symplectic bilinear form on $\mathfrak{p}$. Extending the latter by 0 on $\mathfrak{k}$ yields a Chevalley 2-cocycle Ω on $\mathfrak{g}$ with respect to the trivial representation of $\mathfrak{g}$ on $\mathbb{R}$. A triple $(\mathfrak{g}, \sigma, \Omega)$ as above is called a *symplectic transvection triple*. It it said to be *exact* when there exists an element ξ in $\mathfrak{g}^\star$ such that $\delta\xi = \Omega$, where δ denotes the Chevalley coboundary operator. Up to coverings, the correspondence which associates a symplectic transvection triple to a symplectic symmetric space is bijective. More precisely *there is an equivalence of categories between the category of connected simply connected symplectic symmetric spaces and that of symplectic transvection triples* – the notion of morphism being the natural one in both cases. In the above setting, exactness corresponds to the fact that the transvection group acts on $(\mathcal{M}, \omega)$ in a strongly Hamiltonian manner.

The above considerations can be adapted in a natural manner to the Riemannian or pseudo-Riemannian setting[4], essentially by replacing *mutatis-mutandis* the symplectic structure by a metric tensor. The canonical connection (cf. Formula (2.3) above) corresponds in this case to the Levi-Civita connection.

2.2.2. *Symmetric spaces of group type.* We observe that in coordinates (a, x, z) the map $\phi : \mathcal{R}_0 \to \mathcal{R}_0 : (a, x, z) \mapsto (-a, -x, -z)$ preserves the symplectic form ω (because the coordinates are Darboux coordinates), is involutive and admits the unit element $e = (0, 0, 0)$ as an isolated fixed point. It may be therefore called

[4]See [**CP80**] for an excellent reference.

"centered symmetry" of the associative kernel (2.2). Since the kernel is left-invariant as well, such a centered symmetry $s_g : \mathcal{R}_0 \to \mathcal{R}_0 : g' \mapsto s_g(g') := L_g \circ \phi \circ L_{g^{-1}}(g')$ is attached to every point g in $\mathcal{R}_0$. It turns out that endowed with the above family of symmetries, the manifold $\mathcal{R}_0$ becomes a symplectic symmetric space. More precisely:

PROPOSITION 2.2. *Let $s : \mathcal{R}_0 \times \mathcal{R}_0 \to \mathcal{R}_0$ be the map $(g, g') \mapsto s_g(g')$. Then the triple $(\mathcal{R}_0, \omega, s)$ is a symplectic symmetric space whose transvection group is solvable. The underlying affine connection is the unique affine symplectic connection which is invariant under the group of symmetries of the oscillatory kernel (2.2).*

PROOF. In coordinates (a, x, z), the symmetry map is expressed as

$$\begin{aligned} s_{(a,x,z)}(a', x', z') = (&2a - a', 2\cosh(a - a')x - x', \\ &2\cosh(2(a - a'))z + \Omega_V(x, x')\sinh(a - a') - z'). \end{aligned} \tag{2.4}$$

One then verifies that it satisfies the defining identity: $s_g \circ s_{g'} \circ s_g = s_{s_g(g')}$. Concerning the solvability of the transvection group, four-dimensional symplectic transvection triples have been classified in [**Bie95, Bie98**]. The one we are concerned with here is given by Table (1) ($\varepsilon = 1$) in Proposition 2.3 of [**Bie98**]; denoting here $\mathfrak{k} = \text{Span}\{u_1, u_2, u_3\}$ and $\mathfrak{p} = \text{Span}\{e_1, e_2, f_1, f_2\}$, it writes:

$$\begin{aligned} &[u_2, u_3] = u_1; \quad [u_2, e_2] = e_1; \quad [u_2, f_1] = -f_2; \quad [u_3, f_2] = e_1; \quad [u_1, f_1] = 2e_1; \\ &\qquad [e_1, f_1] = 2u_1; \quad [e_2, f_1] = u_3; \quad [e_2, f_2] = u_1; \quad [f_1, f_2] = u_2. \end{aligned}$$

In these notations the Lie algebra $\mathfrak{r}_0 = \mathfrak{a}_0 \oplus V \oplus \mathbb{R}Z$ of $\mathcal{R}_0$ (with Darboux chart (2.1)) is generated by $\mathfrak{a}_0 = \mathbb{R}.(-f_1)$, $V = \text{Span}\{u_2 - f_2, u_3 + e_2\}$, $Z = 2(u_1 + e_1)$. We now verify (by a short computation) that for all $a, z \in \mathbb{R}$ and $x \in V$, one has

$$\tilde{\sigma}(\exp(-af_1)\exp(x + 2z(u_1 + e_1))K = (\exp(af_1)\exp(-x - 2z(u_1 + e_1))K,$$

where K denotes the analytic subgroup associated with $\mathfrak{k}$. Thus $\tilde{\sigma}$ gives $s_e = \phi$ on $\mathcal{R}_0$. The triple considered here is therefore precisely the triple that induces on $\mathcal{R}_0$ the present symmetric space structure, hence the transvection group is solvable. The higher dimensional case is similar to the 4-dimensional one.

Note that the above symmetric space structure on $\mathcal{R}_0$ is canonically associated with the data of the oscillatory kernel (2.2). Indeed, coming from the stationary phase expansion of an oscillatory integral, the formal star product $\tilde{\star}_\theta$ mentioned in item (iii) of Theorem 2.1 is *natural* in the sense that for all positive integer r the r-th cochain of the star product is a bidifferential operator of order r. To every such natural star product is uniquely attached a symplectic connection [**Lic82, GR03**]. In our case the latter, being invariant under the symmetries, must coincide with the canonical connection associated with the symmetries $\{s_g\}_{g \in \mathcal{R}_0}$. □

The above considerations lead to the following definitions.

DEFINITION 2.3. Let $\mathcal{R}$ be a Lie group. A **symmetric structure** on $\mathcal{R}$ is a diffeomorphism $\phi : \mathcal{R} \to \mathcal{R}$ such that $\phi^2 = \text{id}_\mathcal{R}$; $\phi(e) = e$; $\phi_{\star_e} = -\text{id}_{T_e(\mathcal{R})}$; and setting, for all $g \in \mathcal{R}$, $s_g := L_g \circ \phi \circ L_{g^{-1}}$ then, for all g', we define $s_g \circ s_{g'} \circ s_g = s_{s_g(g')}$.

Or equivalently:

DEFINITION 2.4. A (symplectic) symmetric space, or more generally a homogeneous space, $\mathcal{M}$ of dimension m is **locally of group type** if there exists a m-dimensional (symplectic) Lie subgroup $\mathcal{R}$ of its automorphism group which acts freely on one of its orbits in $\mathcal{M}$. One says that it is **globally** of group type if it is locally and if $\mathcal{R}$ has only one orbit[5].

Lie groups are themselves examples of symmetric spaces (globally) of group type. In the symplectic situation, however, a symplectic symmetric Lie group must be Abelian [**Bie95**]. We will see in what follows other non-Abelian examples.

2.2.3. *Strategy.* Our strategy for constructing UDF's for certain Lie groups may now be easily described: starting with a symplectic symmetric space of group type admitting an invariant deformation quantization — obtained by geometric considerations at the level of the symmetric space structure — one deduces a UDF for every Lie subgroup $\mathcal{R}$ as above by either identifying $\mathcal{M}$ to $\mathcal{R}$, or, in the formal case, by restricting the deformation quantization to an open $\mathcal{R}$-orbit. This type of strategy, in contexts other than symmetric spaces, has already been proposed within a formal framework, see for example [**Xu93, CP95, Huy82, CM04, GZ98**].

3. Construction of UDF's for symmetric spaces of group type

The first example presented in the preceding section as well as the *elementary solvable exact triples* (briefly "ESET") of [**Bie02**] are special cases of the following situation.

3.1. Weakly nilpotent solvable symmetric spaces. In this section $(\mathfrak{g},\sigma)$ denotes a complex solvable involutive Lie algebra (in short, iLa) such that if $\mathfrak{g}=\mathfrak{k}\oplus\mathfrak{p}$ is the decomposition into eigenspaces of σ, the action of $\mathfrak{k}$ on $\mathfrak{p}$ is nilpotent[6]. We denote by $\mathrm{pr}_{\mathfrak{p}}:\mathfrak{g}\to\mathfrak{p}$ the projection onto $\mathfrak{p}$ parallel to $\mathfrak{k}$.

DEFINITION 3.1. A **good Abelian subalgebra** (in short, gAs) of $\mathfrak{g}$ is an Abelian subalgebra $\mathfrak{a}$ of $\mathfrak{g}$, contained in $\mathfrak{p}$, supplementary to a σ-stable ideal $\mathfrak{b}$ in $\mathfrak{g}$ and such that the homomorphism $\rho:\mathfrak{a}\to\mathtt{Der}(\mathfrak{b})$ associated with the split extension $0\to\mathfrak{b}\to\mathfrak{g}\to\mathfrak{a}\to 0$ is injective.

LEMMA 3.2. *If $(\mathfrak{g},\sigma)$ is not flat (i.e. $[\mathfrak{p},\mathfrak{p}]$ acts nontrivially on $\mathfrak{p}$) then a gAs always exists.*

PROOF. Since $\mathfrak{k}$ is nilpotent, $[\mathfrak{k},\mathfrak{p}]\neq\mathfrak{p}$. Moreover by non-flatness there exists $X\in\mathfrak{p}\backslash[\mathfrak{k},\mathfrak{p}]$ not central in $\mathfrak{g}$. Hence for every choice of a subspace V supplementary to $\mathfrak{a}:=\mathbb{R}X$ in $\mathfrak{p}$ and containing $[\mathfrak{k},\mathfrak{p}]$, one has that $\mathfrak{b}:=\mathfrak{k}\oplus V$ is an ideal of $\mathfrak{g}$ supplementary to $\mathfrak{a}$ and on which X acts nontrivially. □

Note that the centralizer $\mathfrak{z}_{\mathfrak{b}}(\mathfrak{a})$ of $\mathfrak{a}$ in $\mathfrak{b}$ is stable by the involution σ; indeed, for all $a\in\mathfrak{a}$ and $X\in\mathfrak{z}_{\mathfrak{b}}(\mathfrak{a})$, one has $[a,\sigma X]=-\sigma[a,X]=0$. Moreover, the map $\rho:\mathfrak{a}\to\mathtt{End}(\mathfrak{b})$ being injective, we may identify $\mathfrak{a}$ with its image: $\mathfrak{a}=\rho(\mathfrak{a})$. Let $\Sigma:\mathtt{End}(\mathfrak{b})\to\mathtt{End}(\mathfrak{b})$ be the conjugation with respect to the involution $\sigma|_{\mathfrak{b}}\in GL(\mathfrak{b})$, i.e. $\Sigma=Ad(\sigma|_{\mathfrak{b}})$. The automorphism Σ is involutive and preserves the canonical Levi decomposition $\mathtt{End}(\mathfrak{b})=\mathcal{Z}\oplus\mathfrak{sl}(\mathfrak{b})$, where $\mathcal{Z}$ denotes the center of $\mathtt{End}(\mathfrak{b})$. Writing the element $a=\rho(a)\in\mathfrak{a}$ as $a=a_Z+a_0$ within this decomposition, one has: $\Sigma(a)=a_Z+\Sigma(a_0)=-a=-a_Z-a_0$, because the endomorphisms a and $\sigma|_{\mathfrak{b}}$ anticommute. Hence $\Sigma(a_0)=-2a_Z-a_0$ and therefore $a_Z=0$. So, $\mathfrak{a}$

[5]In this case, for every choice of a base point o in $\mathcal{M}$, the map $\mathcal{R}\to\mathcal{M}:g\mapsto g.o$ is a diffeomorphism.

[6]This condition is automatic for a transvection algebra, but not in general. Indeed, consider the 2-dimensional non-Abelian (solvable) algebra $\mathfrak{g}=\mathrm{Span}\{k,p\}$ with table $[k,p]=p$.

actually lies in the semisimple part $\mathfrak{sl}(\mathfrak{b})$. For any $x \in \mathfrak{sl}(\mathfrak{b})$, we denote by $x = x^S + x^N, \quad x^S, x^N \in \mathfrak{sl}(\mathfrak{b})$, its abstract Jordan-Chevalley decomposition. Observe that, denoting by $\mathfrak{sl}(\mathfrak{b}) = \mathfrak{sl}_+ \oplus \mathfrak{sl}_-$ the decomposition in (± 1)-Σ- eigenspaces, one has that $\mathfrak{a} \subset \mathfrak{sl}_-$. Also, the algebra $\mathfrak{a}_N := \{a^N\}_{a \in \mathfrak{a}}$ is an Abelian subalgebra in $\mathfrak{sl}_-$ commuting with $\mathfrak{a}$. Setting $\mathfrak{a}_S := \{a^S\}_{a \in \mathfrak{a}}$, we have

DEFINITION 3.3. A gAs is called **weakly nilpotent** if $\mathfrak{z}_{\mathfrak{b}}(\mathfrak{a}_S) \subset \mathfrak{z}_{\mathfrak{b}}(\mathfrak{a}_N)$ and[7] $\mathfrak{a}_N \subset \mathtt{Der}(\mathfrak{b})$.

Let $\mathfrak{b}^c =: \bigoplus_{\alpha \in \Phi} \mathfrak{b}_\alpha$ be the weight space decomposition with respect to the action of $\mathfrak{a}_S$. Note that for all α, one has $\mathfrak{a}_N.\mathfrak{b}_\alpha \subset \mathfrak{b}_\alpha$. Moreover, for all $X_\alpha \in \mathfrak{b}_\alpha$ and $a^S \in \mathfrak{a}_S$, one has

$$\sigma(a^S.X_\alpha) = \alpha(a^S)\sigma(X_\alpha) = \sigma a^S \sigma^{-1} \sigma X_\alpha = \Sigma(a^S).\sigma(X_\alpha) = -a^S.\sigma(X_\alpha).$$

Therefore, $-\alpha \in \Phi$ and $\sigma\mathfrak{b}_\alpha = \mathfrak{b}_{-\alpha}$. Note in particular that $\sigma\mathfrak{b}_0 = \mathfrak{b}_0$.

DEFINITION 3.4. An involutive Lie algebra $(\mathfrak{g}, \sigma)$ is called **weakly nilpotent** if there exists a sequence of subalgebras $\{\mathfrak{a}_i\}_{0 \le i \le r}$ of $\mathfrak{g}$ such that

(i) $\mathfrak{a}_0$ is a weakly nilpotent gAs of $\mathfrak{g}$ with associated supplementary ideal $\mathfrak{b}^{(0)}$.

(ii) $\mathfrak{a}_{i+1}$ is a weakly nilpotent gAs of $\mathfrak{z}_{\mathfrak{b}^{(i)}}(\mathfrak{a}_i)$ $\quad (0 \le i \le r-1)$ where, for $i \ge 1$, $\mathfrak{b}^{(i)}$ denotes the σ-stable ideal of $\mathfrak{z}_{\mathfrak{b}^{(i-1)}}(\mathfrak{a}_{i-1})$ associated with $\mathfrak{a}_i$.

(iii) $\mathfrak{z}_{\mathfrak{b}^{(r)}}(\mathfrak{a}_r)$ is Abelian.

PROPOSITION 3.5. *Assume the iLa $(\mathfrak{g}, \sigma)$ to be weakly nilpotent. Then there exists a (complex) subalgebra $\mathfrak{s}$ of $\mathfrak{g}$ such that the restriction $pr_{\mathfrak{p}}|_{\mathfrak{s}} : \mathfrak{s} \to \mathfrak{p}$ is a linear isomorphism.*

PROOF. Let $\mathfrak{a}$ be a weakly nilpotent gAs of $\mathfrak{g}$ and set $V_\alpha := \mathfrak{b}_\alpha \oplus \mathfrak{b}_{-\alpha}$ for every $\alpha \in \Phi$. Note that $V_0 = \mathfrak{b}_0$ and that each subspace V_α is then σ-stable and one sets $V_\alpha = \mathfrak{k}_\alpha \oplus \mathfrak{p}_\alpha$ for the corresponding eigenspace decomposition. Choose a partition[8] of $\Phi \backslash \{0\}$ as $\Phi \backslash \{0\} =: \Phi^+ \cup \Phi^-$ with the properties that $-\Phi^+ = \Phi^-$ and that if $\alpha, \beta \in \Phi^+$ with $\alpha + \beta \in \Phi$ then $\alpha + \beta \in \Phi^+$. One has $\mathfrak{b} = \oplus_{\alpha \in \Phi^+} V_\alpha \oplus \mathfrak{b}_0$. We set $\mathfrak{b}^+ := \oplus_{\alpha \in \Phi^+} \mathfrak{b}_\alpha$ and $\mathfrak{p}^+ := \mathrm{pr}_{\mathfrak{p}}(\mathfrak{b}^+)$. It turns out that the restriction map $\mathrm{pr}_{\mathfrak{p}}|_{\mathfrak{b}^+} : \mathfrak{b}^+ \to \mathfrak{p}^+$ is a linear isomorphism. Indeed for $X \in \mathfrak{b}_{\mathfrak{k}} := \mathfrak{b} \cap \mathfrak{k}$ and $a \in \mathfrak{a}_S$ one has $\sigma(a.X) = \sigma a \sigma \sigma X = \Sigma(a).X = -a.X$; hence $\mathfrak{a}_S.\mathfrak{b}_{\mathfrak{k}} \subset \mathfrak{p}$. Therefore, for all $X \in \mathfrak{b}_\alpha \cap \mathfrak{k} \quad \alpha \ne 0$, one can find $a \in \mathfrak{a}_S$ such that $a.X = X \in \mathfrak{p} \cap \mathfrak{k}$; thus $\mathfrak{b}_\alpha \cap \mathfrak{k} = 0$ as soon as $\alpha \ne 0$, yielding $\ker(\mathrm{pr}_{\mathfrak{p}}|_{\mathfrak{b}^+}) = 0$. The last condition of Definition 3.3 implies that $\mathfrak{a}_S$ acts by derivations on $\mathfrak{b}$, hence the usual argument yields that $\mathfrak{b}^+$ is a subalgebra normalized by $\mathfrak{a} \oplus \mathfrak{b}_0$. Moreover the first condition

[7]The last condition is automatic when $\mathfrak{b}$ is Abelian. Observe also that it is satisfied when $\rho(\mathfrak{a})$ is contained in a Levi factor of the derivation algebra $\mathtt{Der}(\mathfrak{b})$.

[8]Such a partition can be defined as follows. Let $\mathfrak{h}$ be a Cartan subalgebra of $\mathfrak{sl}(\mathfrak{b})$ containing $\mathfrak{a}_S$ and let $\Lambda \in \mathfrak{h}^\star$ denote the set of weights of the representation of $\mathfrak{sl}(\mathfrak{b})$ on $\mathfrak{b}$. Note that the restriction map $\lambda \mapsto \lambda|_{\mathfrak{a}_S}$ from Λ to $\mathfrak{a}_S^\star$ is surjective onto Φ. Let $\mathfrak{h} = \mathfrak{h}_{\mathbb{R}} \oplus i\mathfrak{h}_{\mathbb{R}}$ be a real decomposition such that the restriction of the Killing form to $\mathfrak{h}_{\mathbb{R}}$ is positive definite. Every weight in Λ is then real valued when restricted to $\mathfrak{h}_{\mathbb{R}}$ [**Kna01**]. Now, any choice of a basis of $\mathfrak{h}_{\mathbb{R}}$ defines a partial ordering on Λ with the desired properties. To pass to the set of weights Φ, consider the $\mathfrak{h}_{\mathbb{R}}$-components, $\mathfrak{a}_S^{\mathbb{R}}$, of $\mathfrak{a}_S$ viewed as a vector subspace of $\mathfrak{h}$. The restriction map $\rho : \Phi \to \Phi|_{\mathfrak{a}_S^{\mathbb{R}}}$ is then a bijection. Indeed, for $\lambda \in \ker(\rho)$, one has, by $\mathbb{C}$-linearity, $\lambda(a+ia') = \lambda(a) + i\lambda(a') = 0 \, \forall a, a' \in \mathfrak{a}_S^{\mathbb{R}}$. Hence $\lambda = 0$ as an element of Φ. Therefore, an order on Λ induces on Φ an order having the same properties.

of the same definition implies $[\mathfrak{a}, \mathfrak{b}_0] = 0$. The proposition follows by induction. One starts applying the above considerations to $\mathfrak{a} = \mathfrak{a}_0$ and $\mathfrak{b} = \mathfrak{b}^{(0)}$. This yields a subalgebra $\mathfrak{s}_0$ supplementary to $\mathfrak{z}_{\mathfrak{b}^{(0)}}(\mathfrak{a}_0)$ and normalized by it. One then sets $\mathfrak{g}_1 := \mathfrak{z}_{\mathfrak{b}^{(0)}}(\mathfrak{a}_0)$, considers a weakly nilpotent $\mathfrak{a}_1$ in $\mathfrak{g}_1$ and gets a subalgebra $\mathfrak{s}_1$ that is now supplementary to and normalized by $\mathfrak{z}_{\mathfrak{b}^{(1)}}(\mathfrak{a}_1)$. Applying this procedure inductively, one gets a sequence a subalgebras $\mathfrak{s}_0, ..., \mathfrak{s}_{r-1}, \mathfrak{s}_r := \mathfrak{z}_{\mathfrak{b}_r}(\mathfrak{a}_r) \cap \mathfrak{p}$ such that $\mathfrak{s}_{i+1}$ normalizes $\mathfrak{s}_i$ and $\mathfrak{s}_i \cap \mathfrak{s}_j = 0$, and defines $\mathfrak{s} := \oplus_i \mathfrak{s}_i$. □

REMARK 3.6. In the exact symplectic case, one has $\mathfrak{p}_0 \perp \mathfrak{a}$ since $\xi[\mathfrak{a}, \mathfrak{b}_0] = 0$.

3.2. Darboux charts and kernels: an extension lemma. Let $(\mathfrak{s}_i, \Omega_i)$ with $i = 1, 2$ be symplectic Lie algebras, and denote by $(\mathbf{S}_i, \omega_i)$ the corresponding simply connected symplectic Lie groups (ω_i being left-invariant). Given a homomorphism $\rho : \mathfrak{s}_1 \to \mathtt{Der}(\mathfrak{s}_2) \cap \mathfrak{sp}(\Omega_2)$, form the corresponding semi-direct product $\mathfrak{s} := \mathfrak{s}_1 \times_\rho \mathfrak{s}_2$ and consider the associated simply connected Lie group $\mathbf{S}$ endowed with the left-invariant symplectic structure ω defined by $\omega_e := \Omega_1 \oplus \Omega_2$ (e denotes the unit element in $\mathbf{S}$).

LEMMA 3.7. *If $\phi_i : (\mathfrak{s}_i, \Omega_i) \to (\mathbf{S}_i, \omega_i)$ $(i = 1, 2)$ are Darboux charts, the map $\phi : (\mathfrak{s}, \Omega := \Omega_1 \oplus \Omega_2) \longrightarrow (\mathbf{S}, \omega) : (X_1, X_2) \mapsto \phi_2(X_2).\phi_1(X_1)$ is Darboux.*

PROOF. For $X \in \mathfrak{s}_1$ and $Y \in \mathfrak{s}_2$ one has:

$$L_{\phi^{-1}_{\star_\phi}} \phi_\star(X) = L_{\phi^{-1}_{\star_\phi}} \left(R_{\phi_1{}_{\star_{\phi_2}}} (\phi_{2_\star} X) \right) = L_{\phi_1^{-1}{}_\star} L_{\phi_2^{-1}{}_\star} R_{\phi_1{}_\star} \phi_{2\star} X$$

$$= \mathtt{Ad}\phi_1^{-1} \left(L_{\phi_2^{-1}{}_{\star_{\phi_2}}} \phi_{2\star} X \right)$$

while for $Y \in \mathfrak{s}_1$ one has

$$L_{\phi^{-1}{}_{\star_\phi}} \phi_\star(Y) = L_{\phi_1^{-1}{}_\star} L_{\phi_2^{-1}{}_\star} L_{\phi_2{}_\star} \phi_{1\star} Y = L_{\phi_1^{-1}{}_\star} \phi_{1\star} Y.$$

Hence

$$\omega_\phi(X, Y) = \Omega \Big(\mathtt{Ad}\phi_1^{-1} (L_{\phi_2^{-1}{}_{\star_{\phi_2}}} \phi_{2\star} X) \, , \, L_{\phi_1^{-1}{}_\star} \phi_{1\star} Y \Big) = 0,$$

because the first (resp. the second) argument belongs to $\mathfrak{s}_2$ (resp. to $\mathfrak{s}_1$), and for $X, X' \in \mathfrak{s}_2$,

$$\begin{aligned}
\omega_\phi(X, X') &= \Omega \Big(\mathtt{Ad}\phi_1^{-1} (L_{\phi_2^{-1}{}_{\star_{\phi_2}}} \phi_{2\star} X) \, , \, \mathtt{Ad}\phi_1^{-1} (L_{\phi_2^{-1}{}_{\star_{\phi_2}}} \phi_{2\star} X') \Big) \\
&= \mathtt{Ad}\phi_1^{-1\star} \Omega_2 \Big(L_{\phi_2^{-1}{}_{\star_{\phi_2}}} \phi_{2\star} X \, , \, L_{\phi_2^{-1}{}_{\star_{\phi_2}}} \phi_{2\star} X' \Big) \\
&= \Omega_2 \Big(L_{\phi_2^{-1}{}_{\star_{\phi_2}}} \phi_{2\star} X \, , \, L_{\phi_2^{-1}{}_{\star_{\phi_2}}} \phi_{2\star} X' \Big) \\
&= \omega_{2\star_{\phi_2}}(X, X') \ = \Omega_2(X, X').
\end{aligned}$$

A similar (and simpler) computation applies for two elements of $\mathfrak{s}_1$. □

A direct computation shows

LEMMA 3.8. *Let $K_i \in Fun((\mathbf{S}_i)^3)$ be a left-invariant three point kernel on $\mathbf{S}_i$ $(i = 1, 2)$. Assume $K_2 \otimes 1 \in Fun((\mathbf{S})^3)$ is invariant under conjugation by elements of $\mathbf{S}_1$. Then $K := K_1 \otimes K_2 \in Fun((\mathbf{S})^3)$ is left-invariant (under $\mathbf{S}$).*

In particular, given associative kernels satisfying the above hypotheses, their tensor product defines an associative invariant kernel on the semidirect product $\mathbf{S}$. We will call it **extension product** of K_2 by K_1.

3.3. Examples.

3.3.1. *One dimensional split extensions of Heisenberg algebras.* The strategy is as follows. First we observe that (the connected simply connected group associated to) every such extension acts simply transitively on the symmetric space $\mathcal{R}_0$, and is therefore diffeomorphic to it. Then, we remark that this diffeomorphism can be chosen to be symplectic. The latter is almost a homomorphism, up to an action of a subgroup of the automorphism group of the kernel on $\mathcal{R}_0$ (this is done by embedding the extension in a subgroup of the automorphism group of the kernel containing $\mathcal{R}_0$). Finally we check that the pullback of that kernel by the diffeomorphism gives an invariant kernel on the extension.

So let us start with a Heisenberg Lie algebra $\mathfrak{h}_n = V \oplus \mathbb{R}Z$, where (V, Ω_V) is a $2n$-dimensional symplectic vector space, Z is central, and $[v, v'] = \Omega_V(v, v')Z$. One easily sees that a split extension (with $i \circ p = \mathrm{Id}$) $\mathfrak{r}_n$ of $\mathfrak{h}_n$ by $\mathfrak{a} = \mathbb{R}A$:

$$0 \to \mathfrak{h}_n \to \mathfrak{r}_n \underset{i}{\overset{p}{\rightleftarrows}} \mathfrak{a} \to 0,$$

must be such that, for all $v \in V$ and $z \in \mathbb{R}$, $[A, v + zZ] = X.v + \mu(v)Z + 2dzZ$ with $X \in \mathrm{End}(V)$, $\mu \in V^*$ and $d \in \mathbb{R}$ such that $X - d\mathrm{Id} \in \mathfrak{sp}(V, \Omega_V)$. As symplectic form on the associated group, we choose the left-invariant 2-form whose value at the identity is the Chevalley 2-coboundary $\Omega = -\delta Z^* = Z^*[\cdot, \cdot]$, which is a natural generalization of that on $\mathcal{R}_0$ in [**BiMs**]. This 2-coboundary is nondegenerate if and only if $d \neq 0$, which we will assume from now on[9], and consider extensions with parameters dX, $d\mu$ and $2d$, which we will denote by $(dX, d\mu, 2d)$.

We have the following symplectic Lie algebra isomorphisms:

(i) $(dX, d\mu, 2d) \cong (X, \mu, 2)$ for all $d \in \mathbb{R}_0$, through the map $L(a, v, z) = (da, v, z)$.

(ii) $(X, \mu, d) \cong (X, 0, 2)$ for all $\mu \in V^*$, through the map $L(a, v, z) = (a, v + au, z)$ with $i_u\Omega_V = \mu$.

(iii) $(X, 0, 2) \cong (X', 0, 2)$ through the map $L(a, v, z) = (a, M.v, z)$, if and only if $M \in Sp(V, \Omega_V)$ is such that $MXM^{-1} = X'$, i.e. if $X - \mathrm{Id}$ and $X' - \mathrm{Id}$ belong to the same adjoint orbit of $Sp(V, \Omega_V)$.

Thus we concentrate on algebras of type $(X, 0, 2)$ from which we can recover the quantization of the others.

Now let $\mathfrak{r}_0 = (I, 0, 2)$, $\mathfrak{r}' = (X, 0, 2)$, and $\mathcal{R}_0$, $\mathcal{R}'$ the associated groups. Elements of these algebras will be denoted respectively by $aA + v + zZ$ and $aA' + v + zZ$. The difference between the actions of A' and A on V is $\bar{X} := X - \mathrm{Id}$ which lies in $\mathfrak{sp}(V, \Omega_V)$. Extending the action $\bar{X}$ on $\mathfrak{r}_0$ by $[\bar{X}, aA + v + zZ] := \bar{X}.v$, we can therefore view $\mathfrak{r}_0$ and $\mathfrak{r}'$ as subalgebras of the semidirect product $\mathfrak{g} = \mathfrak{r}_0 \times \mathfrak{s}$ of $\mathfrak{r}_0$ by $\mathfrak{s} = \mathrm{span}(\bar{X}) \subset \mathfrak{sp}(V, \Omega_V)$. At the level of the groups ($\mathcal{S}$ corresponding to $\mathfrak{s}$), on the one hand we can identify $\mathcal{R}_0$ with $\mathcal{G}/\mathcal{S}$ as manifolds, and on the other hand as a subgroup of $\mathcal{G}$, $\mathcal{R}'$ acts on the quotient. That action is simply transitive. Indeed, on $\mathcal{R}_0$ and $\mathcal{R}'$, we have global coordinate maps $I(aA + v + zZ) = \exp(aA)\exp(v + zZ)$ and $I'(aA' + v + zZ) = \exp(aA')\exp(v + zZ)$ such that $I'(aA' + v + zZ) = \exp(a\bar{X}).I(aA + v + zZ)$, giving a decomposition $g = sr \in \mathcal{R}'$ with $s \in \mathcal{S}$ and $r \in \mathcal{R}_0$. Thus, acting on $e\mathcal{S} \in \mathcal{G}/\mathcal{S}$, such an element g gives

[9] Quantizing extensions with $d = 0$ requires a non exact 2-form. Since our method needs an exact one, we shall have to apply it on central extensions of our algebras. An example of that procedure is given in the next section.

$g \cdot e\mathcal{S} = C_s(r)\mathcal{S} = I(aA + e^{a\bar{X}}v + zZ)\mathcal{S}$, where C_s denotes the action by conjugation, $C_s(g) = sgs^{-1}$. The map $\phi : aA' + v + zZ \mapsto aA + e^{a\bar{X}}v + zZ$ is a diffeomorphism from $\mathfrak{r}'$ to $\mathfrak{r}_0$, and the corresponding map $\Phi = I \circ \phi \circ I'^{-1}$, or $sr \mapsto C_s(r)$, is a diffeomorphism from $\mathcal{R}'$ to $\mathcal{R}_0$. We now observe

PROPOSITION 3.9. (i) *For $g = sr \in \mathcal{R}'$, we have $\Phi \circ L_g = L_{\Phi(g)} \circ C_s \circ \Phi$.*
(ii) *The kernel K on $\mathcal{R}_0$ is invariant under the conjugations by $\mathcal{S}$.*
(iii) *Denoting by ω and ω' the left-invariant 2-forms with value $-\delta Z^*$ at the identity on $\mathcal{R}_0$ and $\mathcal{R}'$ respectively, we have $\Phi^*\omega = \omega'$.*

PROOF. (i) For all $g = sr$, $g' = s'r' \in \mathcal{R}'$, we have

$$\Phi \circ L_g(g') = \Phi(ss'C_{s'^{-1}}(r)r') = C_{ss'}(C_{s'^{-1}}(r)r') = L_{\Phi(g)} \circ C_s \circ \Phi(g').$$

(ii) Recall that for $s = e^{a'\bar{X}}$, $C_s(I(aA + v + zZ)) = I(aA + e^{a'\bar{X}}v + zZ)$ and that $\bar{X} \in Sp(V, \Omega_V)$. Therefore in the kernel (2.2) the amplitude is independent of v, and in its phase, the function S_V is invariant under the symplectomorphisms of V. The kernel as a whole is thus also invariant under $Sp(V, \Omega_V)$.
(iii) By left-invariance, the condition is $\omega_e \circ \left(L_{\Phi(g)^{-1}} \circ \Phi \circ L_g\right)_{*e} = \omega'_e$. Using the first property of Φ above and the invariance of ω_e under $Sp(V, \Omega_V)$, we get $\omega_e \circ \left(L_{\Phi(g)^{-1}} \circ \Phi \circ L_g\right)_{*e} = \omega_e \circ (C_s \circ \Phi)_{*e} = \omega_e \circ \phi_{*0}$, which is readily seen to be equal to ω'_e.

□

Defining a kernel $K' : \mathcal{R}' \times \mathcal{R}' \times \mathcal{R}' \to \mathbb{C}$ on $\mathcal{R}'$ by $K' = \Phi^* K$, we now have

PROPOSITION 3.10. *The kernel K' is*
(i) *invariant under the diagonal left action of $\mathcal{R}'$,*
(ii) *associative.*

Together with the functional space $\Phi^*\mathcal{E}_\theta$, it thus defines a WKB quantization of $\mathcal{R}'$.

A quantization of these groups can be obtained by the same method as in [**BiMs**]. So let us first quickly review that method. On a connected simply connected symplectic solvable Lie group $(\mathcal{R}, \omega)$, with ω a left-invariant exact symplectic form (so that the action by left translations is strongly Hamiltonian), one chooses a global Darboux chart $I : \mathfrak{r} \to \mathcal{R}$ for which the Moyal star product is *covariant*, i.e. if for $X \in \nabla$, $\lambda_X \in C^\infty(\mathfrak{r})$ denotes the (dual) moment map in these coordinates, and $\star^M_\theta$ the Moyal product on $\mathfrak{r}$, one has $[\lambda_X, \lambda_Y]_{\star^M_\theta} = 2\theta\{\lambda_X, \lambda_Y\}$. Such charts always exist on these groups (see [**Puk90**], [**AC90**]).

Covariance of the Moyal product implies that $\rho_\theta : \mathfrak{r} \to \mathrm{End}(C^\infty(\mathfrak{r})[[\theta]]) : X \mapsto [\lambda_X, \cdot]_{\star^M_\theta}$ is a representation of $\mathfrak{r}$ by derivations of the algebra $(C^\infty(\mathfrak{r})[[\theta]], \star^M_\theta)$. In order to find an invariant product on $\mathfrak{r}$, one then tries to find an invertible operator T_θ which intertwines this action and that by fundamental vector fields, i.e. such that $T_\theta^{-1} \circ \rho_\theta(X) \circ T_\theta = X^*$. Those found up to now were all integral operators of the type $\mathcal{F}^{-1} \circ \phi_\theta^* \circ \mathcal{F}$, where $\mathcal{F}$ is a partial Fourier transform and ϕ_θ a diffeomorphism (see the proof of Theorem 4.6 for the precise form of the one in [**BiMs**]). An invariant product $\star_\theta$ on $\mathfrak{r}$ is then defined by $u \star_\theta v = T_\theta^{-1}(T_\theta u \star^W_\theta T_\theta v)$, where $\star^W_\theta$ is the Weyl product on $\mathfrak{r}$. Modulo some work on the function spaces, this gives a WKB invariant quantization of $\mathcal{R}$. In our case, choosing as Darboux chart the map $I(aA + v + zZ) = \exp(aA)\exp(e^{-a\bar{X}}v + zZ)$, one checks that the same integral operator T_θ as in [**BiMs**] works here, giving thus rise to the same kernel (2.2). This reflects again the fact that our groups are all subgroups of the automorphism group of a symplectic symmetric space on which they act simply transitively.

3.3.2. *Non exact example.* As mentioned before, all the examples of symplectic symmetric Lie groups shown up to now were endowed with an exact symplectic form, as our method requires exactness in order for the left translations to be strongly Hamiltonian. We present here an example with a non exact symplectic form showing that, as expected, considering a central extension allows to apply the same method.

Let $\mathfrak{r} = \text{span}\,\langle A, V_1, V_2, W_1, W_2, Z\rangle$ be the Lie algebra defined by $[A, V_i] = V_i$, $[A, W_i] = -W_i$, $[A, Z] = 2Z$, $[V_1, V_2] = Z$, and choose the non exact[10] Chevalley 2-cocycle Ω: $\Omega(A, Z) = 1$, $\Omega(V_i, W_i) = 1$, $\Omega(V_1, V_2) = 1/2$. We define $\mathfrak{g}$ as the central extension of $\mathfrak{r}$ by the element E with commutators $[X, Y]_{\mathfrak{g}} = [X, Y]_{\mathfrak{r}} + \Omega(X, Y).E$ for all $X, Y \in \mathfrak{g}$, where we extended $[\cdot,\cdot]_{\mathfrak{r}}$ and Ω by zero on E. Then $\Omega = -\delta E^*$ is a 2-coboundary.

Now the connected simply connected Lie group $\mathcal{R}$ whose Lie algebra is $\mathfrak{r}$ can be realized as the coadjoint orbit $\mathcal{O}$ of E^* in $\mathfrak{g}^*$, and a global Darboux chart J from $\mathbb{R}^6$ to $\mathcal{O}$ is given by:

$$\begin{aligned} J(q_1, p_1, q_2, p_2, q_3, p_3) \quad = \quad & \exp(q_3 H) \exp((p_3 + q_1 p_1 + q_2 p_2 - p_1 p_2/2) Z) \\ & \exp((q_2 + p_1/4) W_1) \exp(p_2 V_1) \\ & \exp((q1 - p_2/4) W_2) \exp(p_1 V_2) \cdot E^*. \end{aligned}$$

In this chart, the (dual) moment maps are linear in the p_i, so that the Moyal product is covariant. From now on, the method outlined above can be carried on the same way as in [**BiMs**], with the same integral operator T_θ as before.

3.3.3. *The Iwasawa factor of* $\mathfrak{sp}(2, \mathbb{R}) \simeq \mathfrak{so}(2, 3)$. The Lie algebra $\mathfrak{g}_0 := \mathfrak{sp}(n, \mathbb{R})$ of the group $\mathcal{G}_0 := Sp(n, \mathbb{R})$ is defined as the set of $2n \times 2n$ real matrices X such that ${}^\tau XF + FX = 0$ where $F := \begin{pmatrix} 0 & I_n \\ -I_n & 0 \end{pmatrix}$. One has

$$\mathfrak{g}_0 = \{ \begin{pmatrix} A & S_1 \\ S_2 & -{}^\tau A \end{pmatrix} \text{ where } A \in \text{Mat}(n \times n, \mathbb{R}) \text{ and } S_i = {}^\tau S_i\,; i = 1, 2\}. \tag{3.1}$$

In particular, $F \in \mathcal{G}_0$ and a Cartan involution of $\mathfrak{g}_0$ is given by $\theta := \mathtt{Ad}(F)$. The corresponding Cartan decomposition $\mathfrak{g}_0 = \mathfrak{k}_0 \oplus \mathfrak{p}_0$ is then given by

$$\mathfrak{k}_0 \simeq \mathfrak{u}(n) \text{ and } \mathfrak{p}_0 = \{ \begin{pmatrix} S & S' \\ S' & -S \end{pmatrix} \}, \tag{3.2}$$

where the matrices S and S' are symmetric. For $n = 2$, a maximal Abelian subalgebra $\mathfrak{a}$ in $\mathfrak{p}_0$ is generated by $H_1 = E_{11} - E_{33}$ and $H_2 = E_{22} - E_{44}$ where as usual E_{ij} denotes the matrix whose component are zero except the element ij which is one. The restricted roots Φ w.r.t $\mathfrak{a}$ are then given by

$$\Phi = \{\alpha_0,\ \alpha_1,\ \alpha_2 := \alpha_0 + \alpha_1,\ \alpha_3 := \alpha_0 + 2\alpha_1\} \tag{3.3}$$

with $\alpha_0 := 2H_2^*$, $\alpha_1 := H_1^* - H_2^*$, hence $\alpha_2 := H_1^* + H_2^*$ and $\alpha_3 := 2H_1^*$, where $H_i^*(H_j) := \delta_{ij}$. The corresponding root spaces $\mathfrak{g}_{\alpha_i}$ $(i = 0, ..., 3)$ are one-dimensional, generated respectively by $N_0 = E_{24}$, $N_1 = E_{12} - E_{43}$, $N_2 = E_{14} + E_{23}$, $N_3 = E_{13}$.

---

[10] One can actually show that every symplectic 2-cocycle on $\mathfrak{r}$ is non exact and that what follows can be applied to any one of them.

With this choice of generators, the minimal parabolic subalgebra $\mathfrak{s} := \mathfrak{a} \oplus \mathfrak{n}$ with $\mathfrak{n} := \bigoplus_{i=0}^{3} \mathfrak{g}_{\alpha_i}$ has the following multiplication table:

$$[H_1, N_1] = N_1, \qquad [H_1, N_2] = N_2, \tag{3.4a}$$

$$[H_2, N_1] = -N_1, \qquad [H_2, N_2] = N_2, \tag{3.4b}$$

$$[H_1, N_3] = 2N_3, \qquad [H_2, N_0] = 2N_0, \tag{3.4c}$$

$$[N_0, N_1] = -N_2, \qquad [N_1, N_2] = 2N_3; \tag{3.4d}$$

the other brackets being zero. Setting

$$\mathfrak{s}_1 := \mathrm{Span}\{H_2, N_0\} \text{ and } \mathfrak{s}_2 := \mathrm{Span}\{H_1, N_1, N_2, N_3\}, \tag{3.5}$$

one observes that $\mathfrak{s}$ is a split extension of $\mathfrak{s}_2$ by $\mathfrak{s}_1$:

$$0 \longrightarrow \mathfrak{s}_2 \longrightarrow \mathfrak{s} \longrightarrow \mathfrak{s}_1 \longrightarrow 0. \tag{3.6}$$

Note that $\mathfrak{s}_1$ is a minimal parabolic subalgebra of $\mathfrak{su}(1,1)$ while $\mathfrak{s}_2$ is a minimal parabolic subalgebra of $\mathfrak{su}(1,n)$. In particular, the Lie algebra $\mathfrak{s}$ is exact symplectic w.r.t. the element $\xi := \xi_1 \oplus \xi_2$ of $\mathfrak{s}^\star$ with $\xi_i \in \mathfrak{s}_i^\star \quad (i = 1, 2)$ defined as $\xi_1 := N_0^*$ and $\xi_2 := N_3^*$.

One therefore obtains a UDF for proper actions of $\mathcal{AN}$ by direct application of the above extension lemma 3.8.

4. Isospectral deformations of anti de Sitter black holes

4.1. Anti de Sitter black holes. Anti de Sitter (AdS) black holes have been introduced by Bañados, Teitelbaum, Zannelli and Henneaux [**BTZ, BHTZ**] as connected locally AdS space-times $\mathcal{M}$ (possibly with boundary and corners) admitting a *singular* causal structure in the following sense:

CONDITION BH. There exists a closed subset $\mathcal{S}$ in $\mathcal{M}$ called the *singularity* such that the subset $\mathcal{M}_{\text{bh}}$ constituted by all the points x such that every light like geodesic issued from x ends in $\mathcal{S}$ within a finite time is a proper open subset of $\mathcal{M}_{\text{phys}} := \mathcal{M} \backslash \mathcal{S}$.

Originally such solutions were constructed in space-time dimension 3, but they exist in arbitrary dimension $n \geq 3$ (see [**BDSR, CD07**]). More precisely the structure may be described as follows. Take $\mathcal{G} := SO(2, n-1)$ (the AdS group), fix a Cartan involution θ and a θ-commuting involutive automorphism σ of $\mathcal{G}$ such that the subgroup $\mathcal{H}$ of $\mathcal{G}$ of the elements fixed by σ is locally isomorphic to $SO(1, n-1)$. The quotient space $\mathcal{M} := \mathcal{G}/\mathcal{H}$ is an n-dimensional Lorentzian symmetric space, the *anti de Sitter space-time*. It is a solution of the Einstein equations without source. Let $\mathfrak{g}$ denote the Lie algebra of $\mathcal{G}$ and denote by $\mathfrak{g} = \mathfrak{h} \oplus \mathfrak{q}$ the ± 1-eigenspace decomposition with respect to the differential at e of σ that we denote again by σ. Denote by $\mathfrak{g} = \mathfrak{k} \oplus \mathfrak{p}$ the Cartan decomposition induced by θ, consider a σ-stable maximally Abelian subalgebra $\mathfrak{a}$ in $\mathfrak{p}$ and choose accordingly a positive system of roots. Denote by $\mathfrak{n}$ the corresponding nilpotent subalgebra. Set $\overline{\mathfrak{n}} := \theta(\mathfrak{n})$, $\mathfrak{r} := \mathfrak{a} \oplus \mathfrak{n}$ and $\overline{\mathfrak{r}} := \mathfrak{a} \oplus \overline{\mathfrak{n}}$. Finally denote by $\mathcal{R} := \mathcal{AN}$ and $\overline{\mathcal{R}} := A\overline{N}$ the corresponding analytic subgroups of $\mathcal{G}$. One then has

PROPOSITION 4.1. [**BDSR, CD07**] *The groups $\mathcal{R}$ and $\overline{\mathcal{R}}$ admit open orbits and finitely many closed orbits in the AdS space $\mathcal{M}$. Prescribing as singular the union of all closed orbits (of $\mathcal{R}$ and $\overline{\mathcal{R}}$) defines a structure of causal black hole on an open subset $\mathcal{M}_{phys}$ in $\mathcal{M}$ (in the sense of the above condition* (**BH**)*). In* *particular, every open orbit $\mathcal{M}_o$ of $\mathcal{R}$ in $\mathcal{M}$ containing $\mathcal{M}_{phys}$ is itself endowed with a black hole structure.*

Recall that if J denotes an element of $Z(\mathcal{K})$ whose associated conjugation coincides with the Cartan involution θ then the $\mathcal{R}$-orbit $\mathcal{M}_o$ in $\mathcal{G}/\mathcal{H}$ of an element $u\mathcal{H}$ with $u^2 = J$ is open and contains $\mathcal{M}_{phys}$, see [**CD07**]. Remark that the extension lemma 3.8 yields an oscillatory integral UDF for proper actions of $\mathcal{R}$. But here the situation is simplified by the following observation — for convenience of the presentation we write it below for $n = 4$ but the results are valid for any $n \geq 3$.

PROPOSITION 4.2. *The $\mathcal{R}$-homogeneous space $\mathcal{M}_o$ admits a unique structure of globally group type symplectic symmetric space. The latter is isomorphic to $(\mathcal{R}_0, \omega, s)$ described in section 2.*

For the proof, recall first (see [**CD07**]) that the solvable part of the Iwasawa decomposition of $\mathfrak{so}(2,3)$ may be realized with as nilpotent part $\mathfrak{n} = \{W, V, M, L\}$ and Abelian $\mathfrak{a} = \{J_1, J_2\}$ with the commutator table $[V, W] = M$, $[V, L] = 2W$, $[J_1, W] = W$, $[J_2, V] = V$, $[J_1, L] = L$, $[J_2, L] = -L$, $[J_1, M] = M$, $[J_2, M] = M$. Notice that $W, J_1 \in \mathfrak{h}$, and $J_2 \in \mathfrak{q}$. This decomposition is related to the one given in (3.4) by

$$N_0 = L \qquad N_2 = 2W \qquad H_1 = J_1 + J_2$$
$$N_1 = V \qquad N_3 = M \qquad H_2 = J_1 - J_2.$$

We choose to study the orbit of the element $\vartheta = u\mathcal{H}$ with

$$u = \begin{pmatrix} 0 & 1 & \\ -1 & 0 & \\ & & I_{3\times 3} \end{pmatrix}.$$

We denote by $\mathcal{R}_\theta$ its stabilizer group in $\mathcal{R}$, by $\mathfrak{r}_\theta$ the Lie algebra of $\mathcal{R}_\theta$; $\mathfrak{r}'$ is the subalgebra of $\mathfrak{r}$ generated by elements of $\mathfrak{r}$ minus the generator of $\mathfrak{r}_\theta$ and $\mathcal{R}'$ is the analytic subgroup of $\mathcal{R}$ whose algebra is $\mathfrak{r}'$.

LEMMA 4.3. *The action of $\mathcal{R}'$ on $\mathcal{U}$ is simply transitive, i.e. $\mathcal{R}'u\mathcal{H} = \mathcal{R}u\mathcal{H}$.*

PROOF. The first step is to prove that $\mathcal{R}_\theta$ is connected and $\mathcal{U}$ simply connected in order to prevent any double covering problem. The stabilizer of $u\mathcal{H}$ is

$$\mathcal{R}_\theta = \{r \in \mathcal{R} \mid r \cdot u\mathcal{H} = u\mathcal{H}\} = \{r \in \mathcal{R} \mid C_{u^{-1}}(r) \in \mathcal{H}\}. \tag{4.1}$$

Since $\mathcal{R}$ is an exponential group, we have $\mathfrak{r}_\theta = \{X \in \mathcal{R} \mid \mathtt{Ad}(u^{-1})X \in \mathfrak{h}\}$ with $\mathcal{R}_\theta = \exp \mathfrak{r}_\theta$. The set $\mathfrak{r}_\theta$ being connected, $\mathcal{R}_\theta$ is connected too. A long exact sequence argument using the fibration $\mathcal{R}_\theta \to \mathcal{R} \to \mathcal{U}$ shows that $H^0(\mathcal{R}_\theta) \simeq H^1(\mathcal{U})$, which proves that $\mathcal{U}$ is simply connected.

As an algebra, $\mathfrak{r}$ is a split extension $\mathfrak{r} = \mathfrak{r}_\theta \oplus_{\mathsf{ad}} \mathfrak{r}'$. Hence, as group, $\mathcal{R} = \mathcal{R}_\theta\mathcal{R}'$, or equivalently $\mathcal{R} = \mathcal{R}'\mathcal{R}_\theta$. This proves that the action is transitive. The action is even simply transitive because $\mathcal{U}$ is simply connected. □

Let us now find the algebra $\mathfrak{r}_\theta$. The Cartan involution $X \mapsto -X^t$ is implemented as C_J with

$$J = \begin{pmatrix} -I_{2\times 2} & \\ & I_{3\times 3} \end{pmatrix}.$$

Using the relations $u^2 = J$ and $\sigma(u) = u^{-1}$, one sees that $C_{u^{-1}}(r) \in \mathcal{H}$ if and only if $\sigma\Big(C_{u^{-1}}r\Big) = C_{u^{-1}}r$. This condition is equivalent to $\theta\sigma(r) = r$. The involution σ

splits $\mathfrak{a}$ into two parts: $\mathfrak{a} = \mathfrak{a}^+ \oplus \mathfrak{a}^-$ with $J_1 \in \mathfrak{a}^+ = \mathfrak{a} \cap \mathfrak{h}$ and $J_2 \in \mathfrak{a}^- \cap \mathfrak{q}$. Let $\beta_1, \beta_2 \in \mathfrak{a}^*$ be the dual basis; we have $W \in \mathfrak{g}_{\beta_1}$, $V \in \mathfrak{g}_{\beta_2}$, $L \in \mathfrak{g}_{\beta_1-\beta_2}$, $M \in \mathfrak{g}_{\beta_1+\beta_2}$, and, in terms of positive roots, the space $\mathfrak{n}$ is given by $W \in \mathfrak{g}_{\alpha+\beta}$, $V \in \mathfrak{g}_{\beta}$, $L \in \mathfrak{g}_{\alpha}$, $M \in \mathfrak{g}_{\alpha+2\beta}$. We are now able to compute the vectors $X \in \mathfrak{r}$ such that $\sigma\theta(X) = X$. Let us take $X \in \mathfrak{r} = \mathfrak{a} \oplus \mathfrak{n}$ and apply $\sigma\theta$:

$$\begin{aligned} X &= X_{J_1} + X_{J_2} + X_\alpha + X_\beta + X_{\alpha+\beta} + X_{\alpha+2\beta}, \\ \sigma\theta X &= -X_{J_1} + X_{J_2} + Z_{-(\alpha+2\beta)} + Z_\beta + Z_{-(\alpha+\beta)} + Z_{-\alpha} \end{aligned} \tag{4.2}$$

where X_φ and Z_φ denote elements of $\mathfrak{g}_\varphi$. It is directly apparent that $X = J_2$ belongs to $\mathfrak{r}_\theta$. The only other component common to X and $\sigma\theta X$ is in $\mathfrak{g}_\beta$, but it is *a priori* not clear that $X_\beta = Z_\beta$. The dimension of $\mathcal{U}$ is 4 and that of $\mathcal{R}$ is 6, hence $\mathcal{R}_\theta$ is at least 2-dimensional; it is generated by J_2 and $\mathfrak{g}_\beta = \mathbb{R}V$, i.e. $\mathfrak{r}_\theta = \text{Span}\{J_2, V\}$. This proves that the orbit of $u\mathcal{H}$ is open.

The fact that $\mathcal{R}'$ acts freely on $\mathcal{U} = \mathcal{R}/\mathcal{R}_\theta$ proves that $\mathcal{U}$ is locally of group type and since, by definition, $\mathcal{U}$ is only one orbit of $\mathcal{R}$, the space $\mathcal{U}$ is globally of group type. From now on, $\mathcal{M}_o = \mathcal{R}/\mathcal{R}_\theta$ will be identified with $\mathcal{U}$ as homogeneous space, so what we have to find is a group $\tilde{\mathcal{R}}$ which

- acts transitively on $\mathcal{U}$, i.e. $\tilde{\mathcal{R}}u\mathcal{H} = \mathcal{R}u\mathcal{H}$,
- admits a symplectic structure.

It is immediate to see that the algebra $\mathfrak{r}'$ fails to fulfil the symplectic condition. The algebra $\tilde{\mathfrak{r}} = \text{Span}\{A, B, C, D\}$ of a group which fulfils the first condition must at least act transitively on a small neighborhood of $u\mathcal{H}$ and thus be of the form

$$A = J_1 + aJ_2 + a'V \tag{4.3a}$$

$$B = W + bJ_2 + b'V \tag{4.3b}$$

$$C = M + cJ_2 + c'V \tag{4.3c}$$

$$D = L + dJ_2 + d'V. \tag{4.3d}$$

The problem is now to fix the parameters $a, a', b, b', c, c', d, d'$ in such a way that $\text{Span}\{A, B, C, D\}$ is a Lie algebra (i.e. it is closed under the Lie bracket) which admits a symplectic structure and whose group acts transitively on $\mathcal{U}$. We will begin by proving that the surjectivity condition imposes $b = c = d = 0$. Then the remaining conditions for $\tilde{\mathfrak{r}}$ to be an algebra are easy to solve by hand.

First, remark that A acts on the algebra $\text{Span}\{B, C, D\}$ because J_1 does not appears in $[\mathfrak{r}, \mathfrak{r}]$. We can write $\tilde{\mathfrak{r}} = \mathbb{R}A \oplus_{\text{ad}} \text{Span}\{B, C, D\}$ and therefore a general element of the group $\tilde{\mathcal{R}}$ reads $\tilde{r}(\alpha, \beta, \gamma, \delta) = e^{\alpha A} e^{\beta B + \gamma C + \delta D}$ because a subalgebra of a solvable exponential Lie algebra is solvable exponential. Our strategy will be to split this expression in order to get a product $\mathcal{S}\mathcal{R}'$ (which is equivalent to a product $\mathcal{R}'\mathcal{S}$). As Lie algebras, $\text{Span}\{B, C, D\} \subset \mathbb{R}J_2 \oplus_{\text{ad}} \{W, M, L, V\}$. Hence there exist functions w, m, l, v and x of $(\alpha, \beta, \gamma, \delta)$ such that

$$e^{\beta B + \gamma C + \delta D} = e^{xJ_2} e^{wW + mM + lL + vV}. \tag{4.4}$$

We are now going to determine $l(\alpha, \beta, \gamma, \delta)$ and study the conditions needed in order for l to be surjective on $\mathbb{R}$. Since J_2 does not appear in any commutator, the Campbell-Baker-Hausdorff formula yields $x = \beta b + \gamma c + \delta d$. From the fact that $[J_2, L] = -L$, we see that the coefficient of L in the left hand side of (4.4) is $-l(1 - e^{-x})/x$. The V-component in the exponential can also get out without changing the coefficient of L. We are left with $\tilde{r}(\alpha, \beta, \gamma, \delta) = e^{\alpha A} e^{xJ_2} e^{yV} e^{w'W + m'M + lL}$ where

w' and m' are complicated functions of (β,γ,δ) and l is given by

$$l(\beta,\gamma,\delta)=\frac{-\delta(\beta b+\gamma c+\delta d)}{1-e^{-\beta b-\gamma c-\delta d}}, \tag{4.5}$$

which is not surjective except when $b=c=d=0$. Taking the inverse a general element of $\tilde{\mathcal{R}}u\mathcal{H}$ reads $\left[e^{-wW-mM-lM}e^{j_1J_1}u\right]$, where the range of l is not the whole $\mathbb{R}$. Since the action of R' is *simply* transitive, $\tilde{\mathcal{R}}$ is not surjective on $\mathcal{R}u\mathcal{H}$.

When $b=c=d=0$, the conditions for (4.3) to be an algebra are easy to solve, leaving only two *a priori* possible two-parameter families of algebras:

Algebra 1.

$$\begin{aligned} A&=J_1+\frac{1}{2}J_2+sV & [A,B]&=B+sC\\ B&=W & [A,C]&=\frac{3}{2}C\\ C&=M & [A,D]&=2sB+\frac{1}{2}D\\ D&=L+rV & [B,D]&=-rC. \end{aligned}$$

with $r\neq 0$. The general symplectic form on that algebra is given by

$$\omega_1=\begin{pmatrix}0 & -\alpha & -\beta & -\gamma\\ \alpha & 0 & 0 & \frac{2\beta r}{3}\\ \beta & 0 & 0 & 0\\ \gamma & -\frac{2\beta r}{3} & 0 & 0\end{pmatrix}, \tag{4.7}$$

Since $\det\omega=\left(\frac{2\beta r}{3}\right)^2$ we must have $\beta\neq 0$, $r\neq 0$. That algebra will be denoted by $\mathfrak{r}_1$. The analytic subgroup of $\mathcal{R}$ whose Lie algebra is $\mathfrak{r}_1$ is denoted by $\mathcal{R}_1$.

Algebra 2.

$$\begin{aligned} A&=J_1+rJ_2+sV & [A,B]&=B+sC\\ B&=W & [A,C]&=(r+1)C\\ C&=M & [A,D]&=2sB+(1-r)D\\ D&=L. \end{aligned}$$

There is no way to get a non-degenerate symplectic form on that algebra.

REMARK 4.4. One can eliminate the two parameters in algebra $\mathfrak{r}_1$ by the isomorphism

$$\phi=\begin{pmatrix}1 & 0 & 0 & 0\\ 0 & 1 & 0 & 4s\\ 0 & 2sr & 1/r & 4s^2/r\\ 0 & 0 & 0 & 1\end{pmatrix} \tag{4.8}$$

which fixes $s=0$ and $r=1$ and transforms $\mathfrak{r}_1$ into the algebra defined by $[A',B']=B'$, $[A',C']=\frac{3}{2}C'$, $[A',D']=\frac{1}{2}D'$, $[B',D']=-C'$.

It is now easy to prove that

PROPOSITION 4.5. *The group* $\mathcal{R}_1$ *of algebra* $\mathfrak{r}_1$ *acts transitively on* $\mathcal{U}$, *i.e.* $\mathcal{R}u\mathcal{H}=\mathcal{R}_1u\mathcal{H}$.

PROOF. The algebra $\mathfrak{r}_1$ can be written $\mathfrak{r}_1 = \mathbb{R}A \oplus_{\text{ad}} \mathbb{R}D \oplus_{\text{ad}} \text{Span}\{B, C\}$, a split extension, hence a general element reads $r_1(\alpha,\beta,\gamma,\delta) = e^{\alpha A}e^{\delta D}e^{\beta W+\gamma M}$. One can use Campbell-Baker-Hausdorff formula to split it into a factor in $\mathcal{R}_\theta$ and one in $\mathcal{R}'$ (where f and g are some auxiliary functions):

$$r_1(\alpha,\beta,\gamma,\delta) = \underbrace{e^{\alpha sV+\frac{\alpha}{2}J_2}e^{\delta rV}}_{\in\mathcal{R}_\theta}\underbrace{e^{\alpha J_1}e^{\left(f(\delta)+\beta\right)W\left(g(\delta)+\gamma\right)M}}_{\text{surjective on }\mathcal{R}'} \tag{4.9}$$

□

The conclusion is that $\mathcal{R}_1$ is the group $\tilde{\mathcal{R}}$ that we were searching for. To summarize, the structure is as follows.

(i) The AdS space is decomposed into a family of cells: the orbits of a symplectic solvable Lie group $\tilde{\mathcal{R}}$ as in Proposition 4.2 above. Note that these cells may be viewed as the symplectic leaves of the Poisson generalized foliation associated with the left-invariant symplectic structure on $\tilde{\mathcal{R}}$.

(ii) The open $\tilde{\mathcal{R}}$-orbit $\mathcal{M}_o$, endowed with a black hole structure, identifies with the group manifold $\tilde{\mathcal{R}}$.

4.2. Deformation triples for $\mathcal{M}_o$.

4.2.1. *Left-invariant Hilbert function algebras on $\mathcal{R}_0$.* In this section, we present a modified version of the oscillatory integral product (2.2) leading to a left-invariant associative algebra structure on the space of square integrable functions on $\mathcal{R}_0$.

THEOREM 4.6. *Let u and v be smooth compactly supported functions on $\mathcal{R}_0$. Define the following three-point functions:*

$$\begin{aligned} S := &S_V\big(\cosh(a_1-a_2)x_0, \cosh(a_2-a_0)x_1, \cosh(a_0-a_1)x_2\big) \\ &- \bigoplus_{0,1,2} \sinh\big(2(a_0-a_1)\big)z_2\ ; \end{aligned} \tag{4.10}$$

and

$$A := \Big[\cosh\big(2(a_1-a_2)\big)\cosh(2(a_2-a_0))\cosh(2(a_0-a_1)) \\ \big[\cosh(a_1-a_2)\cosh(a_2-a_0)\cosh(a_0-a_1)\big]^{\dim\mathcal{R}_0-2}\Big]^{\frac{1}{2}}.$$

Then the formula

$$u \star_\theta^{(2)} v \ := \ \frac{1}{\theta^{\dim\mathcal{R}_0}}\int_{\mathcal{R}_0\times\mathcal{R}_0} A\, e^{\frac{2i}{\theta}S}u\otimes v \tag{4.11}$$

extends to $L^2(\mathcal{R}_0)$ as a left-invariant associative Hilbert algebra structure. In particular, one has the strong closedness[11] *property:*

$$\int u \star_\theta^{(2)} v \ = \ \int uv\ .$$

PROOF. The oscillatory integral product (2.2) may be obtained by intertwining the Weyl product on the Schwartz space $\mathcal{S}$ (in the Darboux global coordinates (2.1)) by the following integral operator [**BiMs**]:

$$\tau := \mathcal{F}^{-1}\circ(\phi_\theta^{-1})^\star\circ\mathcal{F}\ ,$$

[11]The notion of strongly closed star product was introduced in [**CFS**] in the formal context.

$\mathcal{F}$ being the partial Fourier transform with respect to the central variable z:

$$\mathcal{F}(u)(a,x,\xi) := \int e^{-i\xi z} u(a,x,z)\mathrm{d}z \ ;$$

and ϕ_θ the one parameter family of diffeomorphism(s):

$$\phi_\theta(a,x,\xi) = (a, \frac{1}{\cosh(\frac{\theta}{2}\xi)}x, \frac{1}{\theta}\sinh(\theta\xi)).$$

Set $\mathrm{J} := |(\phi^{-1})^\star \mathrm{Jac}_\phi|^{-\frac{1}{2}}$ and observe that for all $u \in C^\infty \cap L^2$, the function $\mathrm{J}\,(\phi^{-1})^\star u$ belongs to L^2. Indeed, one has

$$\int |\mathrm{J}\,(\phi^{-1})^\star u|^2 = \int |\phi^\star \mathrm{J}|^2\,|\mathrm{Jac}_\phi|\,|u|^2 = \int |u|^2 \,.$$

Therefore, a standard density argument yields the following isometry:

$$T_\theta : L^2(\mathcal{R}_0) \longrightarrow L^2(\mathcal{R}_0) : u \mapsto \mathcal{F}^{-1} \circ m_{\mathrm{J}} \circ (\phi^{-1})^\star \circ \mathcal{F}(u) \,,$$

where m_{J} denotes the multiplication by J. Observing that $T_\theta = \mathcal{F}^{-1} \circ m_{\mathrm{J}} \circ \mathcal{F} \circ \tau$, one has $\star_\theta^{(2)} = \mathcal{F}^{-1} \circ m_{\mathrm{J}} \circ \mathcal{F}(\star_\theta)$. A straightforward computation (similar to the one in [**Bie02**]) then yields the announced formula. □

REMARK 4.7. Let us point out two facts with respect to the above formulas:

(i) Note the cyclic symmetry of the oscillating three-point kernel $A\, e^{\frac{2i}{\theta} S}$.

(ii) The above oscillating integral formula gives rise to a strongly closed, symmetry invariant, formal star product on the symplectic symmetric space $(\mathcal{R}_0, \omega, s)$.

PROPOSITION 4.8. *The space $L^2(\mathcal{R}_0)^\infty$ of smooth vectors in $L^2(\mathcal{R}_0)$ of the left regular representation closes as a subalgebra of $(L^2(\mathcal{R}_0), \star_\theta^{(2)})$.*

PROOF. First, observe that the space of smooth vectors may be described as the intersection of the spaces $\{V_n\}$ where $V_{n+1} := (V_n)_1$, with $V_0 := L^2(\mathcal{R}_0)$ and $(V_n)_1$ is defined as the space of elements a of V_n such that, for all $X \in \mathfrak{r}_0$, $X.a$ exists as an element of V_n (we endow it with the projective limit topology).

Let thus $a, b \in V_1$. Then, $(X.a) \star b + a \star (X.b)$ belongs to V_0. Observing that $\mathcal{D} \subset V_1$ and approximating a and b by sequences $\{a_n\}$ and $\{b_n\}$ in $\mathcal{D}$, one gets (by continuity of $\star$): $(X.a)\star b + a\star(X.b) = \lim(X.a_n \star b_n + a_n \star (X.b_n)) = \lim X.(a_n \star b_n) = X.(a \star b)$. Hence $a \star b$ belongs to V_1. One then proceeds by induction. □

4.2.2. *Twisted L^2-spinors and deformations of the Dirac operator.* We now follow in our four dimensional setting the deformation scheme presented in [**BDSR**] in the three-dimensional BTZ context.

At the level of the (topologically trivial) open $\mathcal{R}$-orbit, the spin structure over $\mathcal{M}_o$ and the associated spinor $\mathbb{C}^2$-bundle – restriction of the spinor bundle on AdS_n – are trivial. The space of (smooth) spinor fields may then be viewed as $\mathbf{S} := C^\infty(\mathcal{R}_0, \mathbb{C}^2)$, on which the (isometry) group $\mathcal{R}_0$ acts on via the left regular representation. In this setting, the restriction to $\mathcal{M}_o$ of the Dirac operator D on AdS_n may be written as $D = \sum_i \gamma^i\,(\tilde{X}_i + \Gamma_i)$, where

- $\{X_i\}$ denotes an orthonormal basis of $\mathfrak{r}_0 = T_\vartheta(\mathcal{M}_o)$ (w.r.t. the adS-metric at the base point ϑ of $\mathcal{M}_o$);
- for $X \in \mathfrak{r}_0$, $\tilde{X}$ denotes the associated left-invariant vector field on $\mathcal{R}_0$;

- γ^i and Γ_i denote respectively the Dirac γ-endomorphism and the spin-connection element associated with X_i.

In that expression the elements γ^i's and Γ_i's are constant. However, already at the formal level, a left-invariant vector field $\tilde{X}$ as infinitesimal generator of the right regular representation does not in general act on the deformed algebra. In order to cure this problem, we twist the spinor module in the following way.

DEFINITION 4.9. Let $\mathrm{d}^r g$ be a right-invariant Haar measure on $\mathcal{R}_0$ and consider the associated space of square integrable functions $L^2_{\mathrm{right}}(\mathcal{R}_0)$. Set

$$\mathcal{H} := L^2_{\mathrm{right}}(\mathcal{R}_0) \otimes \mathbb{C}^2 ;$$

and denote by $\mathcal{H}^\infty$ the space of smooth vectors in $\mathcal{H}$ of the natural right representation of $\mathcal{R}_0$ on $\mathcal{H}$. Then intertwining $\star^{(2)}_\theta$ by the inverse mapping

$$\iota : L^2_{\mathrm{right}}(\mathcal{R}_0) \to L^2(\mathcal{R}_0) : \iota(u)(g) := u(g^{-1})$$

yields a right invariant noncommutative $L^2_{\mathrm{right}}(\mathcal{R}_0)$-bi-module structure (respectively a $(L^2_{\mathrm{right}}(\mathcal{R}_0))^\infty$-bi-module structure) on $\mathcal{H}$ (resp. $\mathcal{H}^\infty$). The latter will be denoted $\mathcal{H}_\theta$ (resp. $\mathcal{H}^\infty_\theta$).

We summarize the main results of this paper in the following:

THEOREM 4.10. *The Dirac operator D acts in $\mathcal{H}^\infty_\theta$ as a derivation of the noncommutative bi-module structure. In particular, for all $a \in (L^2_{\mathrm{right}}(\mathcal{R}_0))^\infty$, the commutator $[D, a]$ extends to $\mathcal{H}$ as a bounded operator. In other words, the triple $(L^2_{\mathrm{right}}(\mathcal{R}_0))^\infty, \mathcal{H}^\infty_\theta, D)$ induces on $\mathcal{M}_o$ a pseudo-Riemannian deformation triple.*

5. Conclusions, remarks and further perspectives

To the AdS space we associated a symplectic symmetric space (M, ω, s). That association is *natural* by virtue of the uniqueness property mentioned in Proposition 4.2. The data of any invariant (formal or not) deformation quantization on (M, ω, s) yields then canonically a UDF for the actions of a non-Abelian solvable Lie group. Using it we defined the noncommutative Lorentzian spectral triple $(\mathcal{A}^\infty, \mathcal{H}, D)$ where $\mathcal{A}^\infty := (L^2_{\mathrm{right}}(\mathcal{R}_0))^\infty$ is a noncommutative Fréchet algebra modelled on the space of smooth vectors of the regular representation on square integrable functions on the group $\mathcal{R}_0$. The underlying commutative limit is endowed with a causal black hole structure encoded in the $\mathcal{R}_0$-group action. A first question that this construction raises is that of defining within the present Lorentzian context the notion of causality at the operator algebraic level.

Another direction of research is to analyze the relation between the present geometrical situation and the corresponding one within the quantum group context. Indeed, our universal deformation formulas can be used at the algebraic level to produce nonstandard quantum groups $SO(2, n-1)_q$ via Drinfeld twists. An interesting challenge would then be to study the behaviour of the representation theory under the deformation process.

More generally the somewhat elliptic sentence with which we started the paper may now be better understood if we remark that the physical motivation section and the quantum group framework suggest to study a number of questions related to (noncommutative) singleton physics, in particular:

(i) Since [**FHT, Sta98**] we know that for q even root of unity, there are unitary irreducible finite dimensional representations of the Anti de Sitter groups. Interestingly (cf. [**FGR**] p.122) the "fuzzy 3-sphere" is related to the Wess-Zumino-Witten models and is conjectured to be related to the non-commutative geometry of the quantum group $U_q(\mathrm{sl}_2)$ for $q = e^{2\pi i/(k+2)}$, $k > 0$, a root of unity.

(ii) The last remark suggests to look more closely at the phenomenon of dimensional reduction which appears in a variety of related problems. In this paper we considered only $n \geq 3$. The reason is that for $n = 2$ the context is in part different: the conformal group of 1+1 dimensional space-time is infinite dimensional, and there are no black holes [**CD07**]. But many considerations remain true, and furthermore many group-theoretical properties find their origin at the 1+1 dimensional level, e.g. the uniqueness of the extension to conformal group [**AFFS**]. Another exemple of dimensional reduction is the fact that the massless UIR of the 2+1 dimensional Poincaré group Di and Rac satisfy Di $\oplus$ Rac $=$ D(HO)$\oplus$D(HO) where D(HO) is the representation D(1/4)$\oplus$D(3/4) of the metaplectic group (double covering of $SL(2,\mathbb{R})$) which is the symmetry of the harmonic oscillator in the deformation quantization approach (see e.g. Section (2.2.4) in [**DS02**]).

(iii) What do the degenerate representations Di and Rac become under deformation? Furthermore there may appear, for our nonstandard quantum group $SO(2,n)_q$, new representations that have no equivalent at the undeformed level (e.g. in a way similar to the supercuspidal representations in the p-adic context). These may have interesting physical interpretations.

(iv) We have seen that for q even root of unity $SO(2,n)_q$ has some properties of a compact Lie group. Our cosmological Ansatz suggests that the qAdS black holes are "small." It is therefore natural to try and find a kind of generalized trace that permits to give a finite volume for qAdS. Note that, in contradistinction with infinite dimensional Hilbert spaces, the notions of boundedness and compactness are the same for closed sets in Montel spaces, and that our context is in fact more Fréchet nuclear than Hilbertian. This raises the more general question to define in an appropriate manner the notion of "q-compactness" (or "q-boundedness") for noncommutative manifolds.

(v) Possibly in relation with the preceding question, one should perhaps consider deformation triples in which the Hilbert space is replaced by a suitable locally convex topological vector space (TVS), on which D could be continuous.

(vi) The latter should yield a natural framework for implementing quantum symmetries in deformation triples, since Fréchet nuclear spaces and their duals are at the basis of the topological quantum groups (and their duals) introduced in the 90's, especially in the semi-simple case with preferred deformations (see the review [**BGGS**]). We would thus in fact have quadruples $(\mathcal{A}, \mathcal{E}, D, \mathcal{G})$ where $\mathcal{A}$ is some topological algebra, $\mathcal{E}$ an appropriate TVS, D some (bounded on $\mathcal{E}$) "Dirac" operator and $\mathcal{G}$ some symmetry. [Being in a Lorentzian noncompact framework, we did not address here questions such as the resolvent of D when $\mathcal{E}$ is a Hilbert space, which we did not need at this stage; eventually one may however have to deal with the reasons that motivated the additional requirements on triples in the Riemannian compact context; note that here the restriction to an open orbit was needed in order to have bounded commutators $[D, a]$ in the Hilbertian context, but a good choice of $\mathcal{E}$ could lift the restriction.] That framework should be naturally extendible to the supersymmetric context, which is the one considered in [**FGR**] with modified spectral triples and is natural also for the problems considered here since e.g. Di⊕Rac and D(HO) are UIR of the corresponding supersymmetries.

(vii) If we want to incorporate "everything," the (external) symmetry $\mathcal{G}$ should be the Poincaré group $SO(1,3)\cdot\mathbb{R}^4$ in the ambient Minkowski space (possibly modified by the presence of matter) and $SO(2,3)_q$ in the $q\mathrm{AdS}_4$ black holes, or possibly some supersymmetric extension. The unified (external) symmetry could therefore be something like a groupoid. The latter should be combined in a subtle way (as hinted e.g. in [**St07**]) with the "internal" symmetry associated with the various generations, colors and flavors of (composite) "elementary" particles in a generalized Standard Model, possibly in a noncommutative geometry framework analogous to what is done in [**Co06, CCM, Ba06**]. There would of course remain the formidable task to develop quantized field theories on that background, incorporating composite QED for photons on AdS as in [**FF88**] and some analog construction for the electroweak model (touched in part in [**Frø00**]) and for QCD, possibly making use of some formalism coming from string theory.

(viii) The Gelfand isomorphism theorem permits to realize commutative involutive algebras as algebras of functions on their "spectrum." Finding a noncommutative analog of it has certainly been in the back of the mind of many, since quite some time (see e.g. [**St05**]). We now have theories and many examples of deformed algebras, quantum groups and noncommutative manifolds. The above mentioned quadruples could provide a better understanding of that situation.

References

[AFFS] E. Angelopoulos, M. Flato, C. Fronsdal, and D. Sternheimer, *Massless particles, conformal group, and de Sitter universe*, Phys. Rev. D (3) **23** (1981), no. 6, 1278–1289.

[AC90] D. Arnal and J.-C. Cortet, *Représentations ∗ des groupes exponentiels*, J. Funct. Anal. **92** (1990), no. 1, 103–135.

[BHTZ] M. Bañados, M. Henneaux, C. Teitelboim, and J. Zanelli, *Geometry of the* $2+1$ *black hole*, Phys. Rev. D (3) **48** (1993), no. 4, 1506–1525 (`gr-qc/9302012`).

[BTZ] M. Bañados, C. Teitelboim, and J. Zanelli, *Black hole in three-dimensional space-time*, Phys. Rev. Lett. **69** (1992), no. 13, 1849–1851 (`hep-th/9204099`).

[Ba06] J.W. Barrett, *A Lorentzian version of the non-commutative geometry of the standard model of particle physics*, J. Math. Phys. **48** (2007) 012303 (`hep-th/0608221`).

[Bie95] P. Bieliavsky, *Espaces symétriques symplectiques*, Thesis, Université Libre de Bruxelles 1995 (`math.DG/0703358`).

[Bie98] ———, *Four-dimensional simply connected symplectic symmetric spaces*, Geom. Dedicata **69** (1998), no. 3, 291–316.

[Bie02] ———, *Strict quantization of solvable symmetric spaces*, J. Symplectic Geom. **1** (2002), no. 2, 269–320 (`math.QA/0010004`).

[BBM] P. Bieliavsky, P. Bonneau, and Y. Maeda, *Universal deformation formulae for three-dimensional solvable Lie groups*, Quantum field theory and noncommutative geometry, Lecture Notes in Phys., vol. 662, Springer, Berlin, 2005, pp. 127–141 (`math.QA/0308188`).

[BCG] P. Bieliavsky, M. Cahen, and S. Gutt, *Symmetric symplectic manifolds and deformation quantization*, Modern group theoretical methods in physics (Paris, 1995), Math. Phys. Stud., vol. 18, Kluwer Acad. Publ., Dordrecht, 1995, pp. 63–73.

[BDSR] P. Bieliavsky, S. Detournay, P. Spindel, and M. Rooman, *Star products on extended massive non-rotating BTZ black holes*, J. High Energy Phys. (2004), no. 6, 031, 25 pp. (electronic) (`hep-th/0403257`).

[BiMa] P. Bieliavsky and Y. Maeda, *Convergent star product algebras on "$ax+b$"*, Lett. Math. Phys. **62** (2002), no. 3, 233–243 (`math.QA/0209295`).

[BiMs] P. Bieliavsky and M. Massar, *Oscillatory integral formulae for left-invariant star products on a class of Lie groups*, Lett. Math. Phys. **58** (2001), no. 2, 115–128.

[BGGS] P. Bonneau, M. Gerstenhaber, A. Giaquinto, and D. Sternheimer, *Quantum groups and deformation quantization: explicit approaches and implicit aspects*, J. Math. Phys. **45** (2004), no. 10, 3703–3741.

[CP80] M. Cahen and M. Parker, *Pseudo-Riemannian symmetric spaces*, Mem. Amer. Math. Soc. **24** (1980), no. 229, iv+108.

[CP95] V. Chari and A. Pressley, *A guide to quantum groups*, Cambridge University Press, Cambridge, 1995, Corrected reprint of the 1994 original.

[CD07] L. Claessens and S. Detournay, *Solvable symmetric black hole in anti-de Sitter spaces*, J. Geom. Phys. **57** (2007), 991–998 (`math.DG/0510442`).

[Cl06] J.M. Cline, *Baryogenesis*, Lectures at Les Houches Summer School Session 86: Particle Physics and Cosmology: the Fabric of Spacetime, 7-11 Aug. 2006, `hep-ph/0609145`.

[Co94] A. Connes, *Noncommutative geometry.* Academic Press, Inc., San Diego, CA, 1994.

[Co06] A. Connes, *Noncommutative Geometry and the standard model with neutrino mixing*, JHEP 0611 (2006) 081 (`hep-th/0608226`).

[CCM] A. Connes, A.H. Chamseddine, and M. Marcolli, *Gravity and the standard model with neutrino mixing*, `hep-th/0610241`.

[CDV] A. Connes and M. Dubois-Violette, *Moduli space and structure of noncommutative 3-spheres*, Lett. Math. Phys. **66** (2003), no. 1-2, 91–121 (`math.QA/0511337`).

[CFS] A. Connes, M. Flato, and D. Sternheimer, *Closed star products and cyclic cohomology*, Lett. Math. Phys. **24** (1992), no. 1, 1–12.

[CL01] A. Connes and G. Landi, *Noncommutative manifolds, the instanton algebra and isospectral deformations*, Comm. Math. Phys. **221** (2001), no. 1, 141–159 (`math.QA/0011194`).

[CM04] A. Connes and H. Moscovici, *Rankin-Cohen brackets and the Hopf algebra of transverse geometry*, Mosc. Math. J. **4** (2004), no. 1, 111–130, 311 (`math.QA/0304316`).

[DS02] G. Dito and D. Sternheimer, *Deformation quantization: genesis, developments and metamorphoses*, Deformation quantization (Strasbourg, 2001), IRMA Lect. Math. Theor. Phys., vol. 1, de Gruyter, Berlin, 2002, pp. 9–54 (`math.QA/0201168`).

[Fla82] M. Flato, *Deformation view of physical theories*, Czechoslovak J. Phys. B **32** (1982), no. 4, 472–475.

[FF88] M. Flato and C. Fronsdal, *Composite electrodynamics*, J. Geom. Phys. **5** (1988), no. 1, 37–61.

[FFS] M. Flato, C. Frønsdal, and D. Sternheimer, *Singletons, physics in AdS universe and oscillations of composite neutrinos*, Lett. Math. Phys. **48** (1999), no. 1, 109–119, Moshé Flato (1937–1998). See also `hep-th/9901043`.

[FHT] M. Flato, L. K. Hadjiivanov, and I. T. Todorov, *Quantum deformations of singletons and of free zero-mass fields*, Found. Phys. **23** (1993), no. 4, 571–586.

[FGR] J. Fröhlich, O. Grandjean and A. Recknagel, *Supersymmetric quantum theory and non-commutative geometry.* Comm. Math. Phys. **203** (1999), no. 1, 119–184 (`math-ph/9807006`).

[Frø00] C. Frønsdal, *Singletons and neutrinos*, Lett. Math. Phys. **52** (2000), no. 1, 51–59, Conference Moshé Flato 1999 (Dijon) (`hep-th/9911241`).

[Ga05] V. Gayral, *Déformations isospectrales non compactes et théorie quantique des champs*, Thesis CPT Aix-Marseille I, 2005 (`hep-th/0507208`).

[Ger64] M. Gerstenhaber, *On the deformation of rings and algebras*, Ann. of Math. (2) **79** (1964), 59–103.

[GZ98] A. Giaquinto and J.J. Zhang, *Bialgebra actions, twists, and universal deformation formulas*, J. Pure Appl. Algebra **128** (1998), no. 2, 133–151 (hep-th/9411140).

[GR03] S. Gutt and J. Rawnsley, *Natural star products on symplectic manifolds and quantum moment maps*, Lett. Math. Phys. **66** (2003), no. 1-2, 123–139 (math.SG/0304498).

[HNW] J.G. Heller, N. Neumaier, and S. Waldmann, *A C^*-Algebraic Model for Locally Noncommutative Spacetimes*, math.QA/0609850.

[Hoo06] G. 't Hooft, *The black hole horizon as a dynamical system*, Internat. J. Modern Phys. D **15** (2006), no. 10, 1587–1602 (gr-qc/0606026).

[Huy82] T. V. Huynh, *Invariant-quantization associated with the affine group*, J. Math. Phys. **23** (1982), no. 6, 1082–1087.

[Kna01] A. W. Knapp, *Representation theory of semisimple groups*, Princeton Landmarks in Mathematics, Princeton University Press, Princeton, NJ, 2001, An overview based on examples, Reprint of the 1986 original.

[Lic82] A. Lichnerowicz, *Déformations d'algèbres associées à une variété symplectique (les $*_\nu$-produits)*, Ann. Inst. Fourier, Grenoble **32** (1982), 157–209.

[Loo69] O. Loos, *Symmetric spaces. I: General theory*, W. A. Benjamin, Inc., New York-Amsterdam, 1969.

[PS06] M. Paschke and A. Sitarz, *Equivariant Lorentzian Spectral Triples*, math-ph/0611029.

[Puk90] L. Pukanszky, *On a property of the quantization map for the coadjoint orbits of connected Lie groups*, The orbit method in representation theory (Copenhagen, 1988), Progr. Math., vol. 82, Birkhäuser Boston, Boston, MA, 1990, pp. 187–211.

[Rie93] M. Rieffel, *Deformation quantization for actions of* $\mathbf{R}^d$, Mem. Amer. Math. Soc. **106** (1993), no. 506, x+93.

[SS07] N. Sahu and U. Sarkar, *Predictive model for dark matter, dark energy, neutrino masses and leptogenesis at the TeV scale*, hep-ph/0701062 (in press in Phys. Rev. D).

[Sta98] H. Steinacker, *Finite-dimensional unitary representations of quantum anti-de Sitter groups at roots of unity*, Comm. Math. Phys. **192** (1998), no. 3, 687–706 (q-alg/9611009).

[St05] D. Sternheimer, *Quantization is deformation.* Noncommutative geometry and representation theory in mathematical physics, 331–352, Contemp. Math. **391**, Amer. Math. Soc., Providence, RI, 2005.

[St07] D. Sternheimer, *The geometry of space-time and its deformations, from a physical perspective*, From Geometry to Quantum Mechanics – in honor of Hideki Omori (Y. Maeda, P. Michor, T. Ochiai, and A. Yoshioka, eds.), Progress in Mathematics series, no. 252, 2007, pp. 287–301.

[vN31] J. von Neumann, *Die Eindeutigkeit der Schödingerschen Operatoren*, Math. Ann **104** (1931), 570—-578.

[Xu93] P. Xu, *Poisson manifolds associated with group actions and classical triangular r-matrices*, J. Funct. Anal. **112** (1993), no. 1, 218–240.

[Zag94] D. Zagier, *Modular forms and differential operators*, Proc. Indian Acad. Sci. Math. Sci. **104** (1994), no. 1, 57–75, K. G. Ramanathan memorial issue.

Université catholique de Louvain, département de mathématiques, Chemin du Cyclotron 2, B-1348 Louvain-la-Neuve, Belgium

E-mail address: bieliavsky claessens voglaire @math.ucl.ac.be

Institut de Mathématiques de Bourgogne, Université de Bourgogne, BP 47870, F-21078 Dijon Cedex, France

Current address: Department of Mathematics, Keio University,, 3-14-1 Hiyoshi, Kohoku-ku, Yokohama, 223-8522 Japan

E-mail address: Daniel.Sternheimer@u-bourgogne.fr

Contemporary Mathematics
Volume **450**, 2008

Group-like objects in Poisson geometry and algebra

Christian Blohmann and Alan Weinstein

Abstract. A group, defined as a set with an associative multiplication and an inverse, is a natural structure describing the symmetry of a space. The concept of group generalizes to group objects internal to other categories than sets. But there are yet more general objects that can still be thought of as groups in many ways, such as quantum groups. We explain some of the generalizations of groups which arise in Poisson geometry and quantization: the germ of a topological group, Poisson Lie groups, rigid monoidal structures on symplectic realizations, groupoids, 2-groups, stacky Lie groups, and hopfish algebras.

1. Introduction

Every mathematician learns that a group is a set with an associative multiplication admitting an identity and inverses. But there are other objects, such as group germs, Lie groups, Poisson groups, and quantum groups, which qualify as groups in many senses, but are either more or less than simply sets with operations satisfying the group axioms. These notes will describe some of these objects, with an emphasis on the role of groupoids, both as examples of group-like objects and as a tool for describing the objects themselves. A goal toward which we strive (but which we will not reach) is to give a unified "categorical" notion which encompasses all of our examples.

Some of our examples will be explicitly geometric. Others will be algebraic, but may still be considered as geometric from the viewpoint which identifies geometric objects with suitable algebras of functions on them and, more generally, considers algebras, even noncommutative ones, as if they were the functions on a space. Even more abstract is the view of spaces as represented by categories, such as the category of representations of an algebra, or the derived category of coherent sheaves on an algebraic variety.

2000 *Mathematics Subject Classification.* 20L05 (Primary); 16W30, 53D17 (Secondary).

Key words and phrases. group object, Poisson Lie group, symplectic realization, groupoid, 2-group, stacky Lie group, hopfish algebra.

These notes are loosely based on the three lectures given by Weinstein at the School on Poisson Geometry and Related Topics, Keio University, May 31–June 2, 2006. We would like to thank Nathan George for the use of his preliminary notes. Christian Blohmann was supported by a Marie Curie Fellowship of the European Union; Alan Weinstein was partially supported by NSF grant DMS-0204100.

2. Symmetry groups

The (global) symmetry of a space is described by a set of transformations closed under composition and inversion. Conversely, Cayley's theorem states that every abstract group is also a group of transformations, acting on itself by left multiplication.

At first, a group was just a set, and its structure morphisms were maps between sets. But it is useful to consider groups which themselves have some additional structure, for example a differentiable structure. The product, unit, and inversion morphisms of the group are required to respect this structure, in which case they have to be smooth maps. This leads us naturally to the concept of a Lie group, which is in turn an example of a notion of group internal to a category, meaning that the group is given by an object and group structure morphisms in that category. In this sense, an ordinary group is a group object in the category of sets, a topological group in the category of topological spaces, a Lie group in the category of manifolds, an algebraic group in the category of algebraic varieties, and so on. This notion works well for many categories in which the objects are spaces with geometric or topological structure.

On the other hand, in categories of spaces with algebraic structure, the group objects often turn out to be surprisingly rare. For example, the group objects in the category of vector spaces are the vector spaces themselves with the underlying abelian group structure, while in the category of groups they are the abelian groups. In the category of rings the only group object is the trivial ring with one element.

But there are other kinds of objects which we can naturally associate to symmetries. For example, to the action of a finite group G on a finite set S, the Gelfand "algebraization" functor associates the commutative algebras of functions $\mathcal{A}(G)$ and $\mathcal{A}(S)$ to both the group and the space. This is a contravariant functor, so the group product, unit, and inverse in the group become the comultiplication, counit, and coinverse (antipode) of a Hopf structure on $\mathcal{A}(G)$. The action of the group on the set becomes a coaction of the Hopf algebra $\mathcal{A}(G)$ on the algebra $\mathcal{A}(S)$. Since we can recover the group G from $\mathcal{A}(G)$ as the set of group-like elements and the space S from $\mathcal{A}(S)$ as the set of characters, the description of the symmetry of S in terms of the Hopf algebra $\mathcal{A}(G)$ is completely equivalent to the description in terms of the group. This is why such Hopf algebras, even noncommutative ones, have been termed quantum groups. But quantum groups are not group objects, at least not in any of the underlying categories of vector spaces, of algebras, or of coalgebras. For instance, the coproduct is a map from $\mathcal{A}(G)$ to $\mathcal{A}(G) \otimes \mathcal{A}(G)$, but the tensor product is not a product in the category of algebras (nor in the dual, with arrows reversed). Our conclusion is that there are yet more general structures which we may associate with the concept of a symmetry group.

In fact, there is an ample collection of structures which are considered to be "groups" in the sense that they encode symmetries: groupoids, inverse semi-groups, hypergroups, n-groups, Lie algebras, Hopf algebras, etc. — just to name a few. Our goal in these lectures is to identify some of such group-like structures that arise in Poisson geometry and quantization.

3. Group objects

A group germ is an example of a group object in a category which is not (at least in its usual presentation) a subcategory of the category of sets.

We define a **topological germ** to be the collection of all pointed topological spaces (X, x) modulo the equivalence relation in which two spaces (X, x) and (Y, y) are identified if $x = y$ and if this common point has open neighborhoods in X and Y which are equal as sets and homeomorphic with the induced topologies. A morphism between topological germs is an equivalence class of continuous maps between representatives of the germs, where two maps are considered equivalent if they agree on some neighborhood of the basepoint. This category admits products defined as products of representatives, and a terminal object 1 consisting of a single point. In any such category, a **group object** is defined to be an object $\mathcal{U}$ together with the structure morphisms of multiplication $m : \mathcal{U} \times \mathcal{U} \to \mathcal{U}$, unit $e : 1 \to \mathcal{U}$, and inverse $\mathrm{inv} : \mathcal{U} \to \mathcal{U}$, such that the following diagrams are commutative. The diagrams for associativity and the unit are

(1)
$$
\begin{array}{ccc}
\mathcal{U}\times\mathcal{U}\times\mathcal{U} & \xrightarrow{m\times\mathrm{id}} & \mathcal{U}\times\mathcal{U} \\
\downarrow{\scriptstyle \mathrm{id}\times m} & & \downarrow{\scriptstyle m} \\
\mathcal{U}\times\mathcal{U} & \xrightarrow{m} & \mathcal{U}
\end{array}
\qquad
\begin{array}{ccccc}
1\times\mathcal{U} & \xleftarrow{\cong} & \mathcal{U} & \xrightarrow{\cong} & \mathcal{U}\times 1 \\
\downarrow{\scriptstyle e\times\mathrm{id}} & & \downarrow{\scriptstyle \mathrm{id}} & & \downarrow{\scriptstyle \mathrm{id}\times e} \\
\mathcal{U}\times\mathcal{U} & \xrightarrow{m} & \mathcal{U} & \xleftarrow{m} & \mathcal{U}\times\mathcal{U}
\end{array}
$$

while that for the inverse is

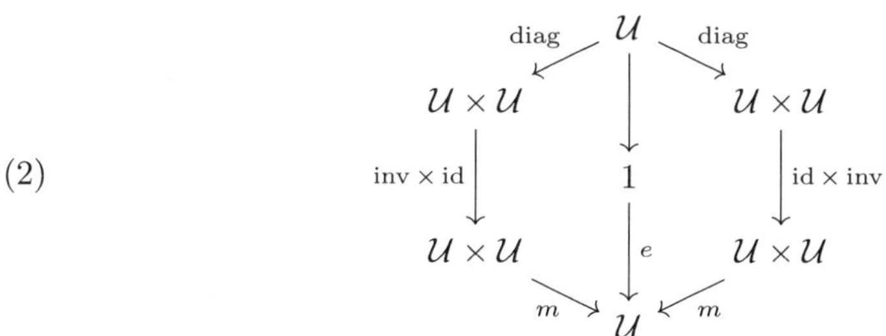

where diag is the diagonal morphism, that is, the unique morphism which lifts the identity on $\mathcal{U}$ to the product $\mathcal{U} \times \mathcal{U}$:

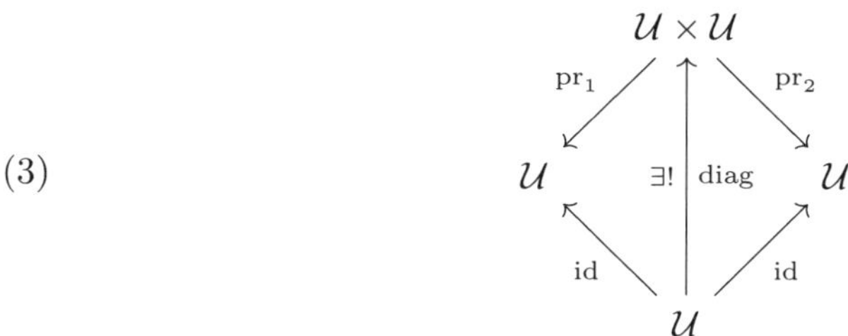

Restricting the group structure of a topological group to its germ at the unit element e yields such a group object in the category of topological germs which is called the germ of the group at e. This is no longer a group in the usual sense, because "it has only one point". In a similar way, one can define manifold germs and group objects in the category thereof. This is useful, for instance, when one tries to integrate a Banach Lie algebra to a group. The global object does not always exist [**9, 10**], but its germ does (and is unique up to isomorphism).

4. Poisson Lie groups

What is a Poisson Lie group? It is usually defined as a Lie group G with a Poisson structure such that the multiplication morphism $G \times G \to G$ is a Poisson map. It follows from this that the unit map from a point to G is a Poisson map, and the inversion map is anti-Poisson.

If we try to define a Poisson Lie group as a group object in the category of Poisson manifolds, a first problem is that the category of Poisson manifolds does not admit categorical products. The cartesian product $X \times Y$ does not work, since Poisson maps $A \to X$ and $A \to Y$ yield a Poisson map $A \to X \times Y$ only when the images in $C^\infty(A)$ of $C^\infty(X)$ and $C^\infty(Y)$ Poisson commute.

But let's forget this for a moment and admit $X \times X$ as some kind of product, if not a categorical one, and assume that X is a Poisson Lie semigroup, i.e. a Poisson manifold with an associative multiplication map $m : X \times X \to X$, which is a Poisson map, and a Poisson unit map, $e : 1 \to X$. If X is a group, we know that inversion is an anti-Poisson map, i.e. a Poisson map $\mathrm{inv} : X^{\mathrm{op}} \to X$. The categorical defining property of inversion is the commutativity of the diagram (2), but we must put in the opposite Poisson structure to get

(4)

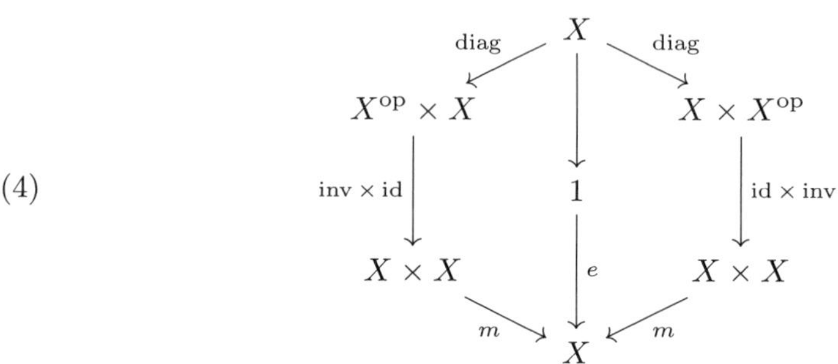

But the diagonal map is not a Poisson map from X to $X^{\mathrm{op}} \times X$ or $X \times X^{\mathrm{op}}$ unless the Poisson bivector is zero. So the inverse axiom does not have an evident interpretation in the Poisson category.

Another approach is to analyze the inversion map via its graph, which is

$$\mathrm{graph}(\mathrm{inv}) = \{(x, x^{-1}) \,|\, x \in X\} = \{(x,y) \in X \times X \,|\, xy = e\},$$

i.e., the pull-back, $(X \times X) \times_{m,e} 1$. This can be expressed in terms of the graph of multiplication and the opposite of the graph of the unit, $\mathrm{graph}(e)^{\mathrm{op}} = \{(e,1)\}$, which are coisotropic submanifolds,

$$\mathrm{graph}(m) \in \mathrm{Cois}((X \times X) \times X^{\mathrm{op}}), \quad \mathrm{graph}(e)^{\mathrm{op}} \in \mathrm{Cois}(X \times 1^{\mathrm{op}}).$$

The composition of these two coisotropic submanifolds, viewed as Poisson relations, is again a coisotropic submanifold,

$$\mathrm{graph}(\mathrm{inv}) = \mathrm{graph}(m) \times_X \mathrm{graph}(e)^{\mathrm{op}} \in \mathrm{Cois}(X \times X),$$

where we have used that $\mathrm{Cois}((X \times X) \times 1^{\mathrm{op}}) \cong \mathrm{Cois}(X \times X)$.

This makes it possible to define a Poisson Lie group in the following way. Starting with any Poisson semigroup, we may first require that m be transverse to e, so that the fiber product $\mathrm{graph}(m) \times_X \mathrm{graph}(e)^{\mathrm{op}} \in \mathrm{Cois}(X \times X)$ is a coisotropic submanifold, and then require that the projection of the manifold to one of the factors of $X \times X$ be a diffeomorphism. This is still not completely "categorical," but we will see below that it has a useful algebraic analogue.

5. Poisson Lie groups and symplectic realizations

It is sometimes possible to describe a group structure on a mathematical object as an extra structure on a category of "representations" of the object.

For a Poisson manifold X, one such category is that of the symplectic realizations, in which the objects are the Poisson maps from symplectic manifolds to X [**20**], and the morphisms are symplectic maps forming commutative diagrams with these. We may think of these as symplectic "points" of X, or as geometric representations of X in the sense that a Poisson map $J : S \to X$ induces a representation of the Poisson Lie algebra of functions on X by that of S, or by hamiltonian vector fields:

$$C^\infty(X) \longrightarrow \mathcal{X}(S)\,, \qquad f \longmapsto -X_{J^*(f)}\,.$$

There always exists a symplectic realization which is a surjective submersion [**8, 11**], for which (up to locally constant functions) this representation is faithful. Therefore, the collection of symplectic realizations encodes all the structural information of X.

Given two symplectic realizations $J_1 : S_1 \to X$ and $J_2 : S_2 \to X$ we can use a Poisson Lie structure on X to construct a product realization $J_1 \otimes J_2 : S_1 \otimes S_2 := S_1 \times S_2 :\to X$ by

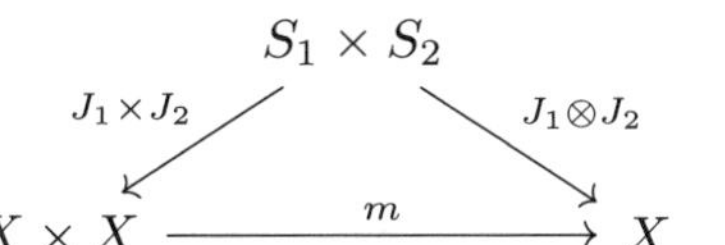

This multiplication of symplectic realizations is associative because the multiplication on X is. Furthermore, viewing the terminal object in the category of Poisson manifolds $1 = \{\mathrm{pt}\}$ as a zero-dimensional symplectic manifold, the unit $e : 1 \to X$ can also be viewed as symplectic realization. It is the identity for the product of realizations $e \otimes J = J = J \otimes e$, where we identify $1 \times S = S = S \times 1$. In this way, the monoidal structure on X naturally equips the category of symplectic realizations with a monoidal structure.

What structure is induced on the category of symplectic realizations by the inverse on X? From the analogous algebraic situation, we might expect that the inverse leads to a *rigid* monoidal structure [**19**]. We can try to define a dual symplectic realization by

$$J^\vee : S^\vee \equiv S^{\mathrm{op}} \xrightarrow{J} X^{\mathrm{op}} \xrightarrow{\mathrm{inv}} X\,.$$

But in the category of symplectic realizations of X the evaluation map would have to make the diagram

$$\begin{array}{ccc} S \otimes S^\vee & \xrightarrow{\mathrm{ev}} & 1 \\ & {\scriptstyle J\otimes J^\vee}\searrow \quad \swarrow{\scriptstyle e} & \\ & X & \end{array} \qquad (5)$$

,

commutative, which is not possible unless J maps all of S to a single point in X. If we want to equip the category of symplectic realizations with a rigid structure, we will need a more general notion of morphism.

6. Generalized morphisms

In many categories, morphisms are given by set theoretic maps, but we may allow relations instead of just maps, especially when maps of a certain type are characterized by properties of their graphs. For instance, a smooth map $f : A \to B$ between symplectic manifolds is symplectic if and only if its graph is a lagrangian submanifold of $A \times B^{\mathrm{op}}$. Moreover, for the graph of the composition of f with another map $g : B \to C$ we have

$$\operatorname{graph}(g \circ f) = \operatorname{graph}(f) \circ \operatorname{graph}(g) := \operatorname{graph}(f) \times_B \operatorname{graph}(g) . \tag{6}$$

This suggests allowing arbitrary lagrangian submanifolds of products as morphisms, i.e., defining

$$\operatorname{Hom}(A, B) := \operatorname{Lag}(A \times B^{\mathrm{op}}) ,$$

the symplectic A-B relations. Note that, for the morphisms in $\operatorname{Lag}(A \times B^{\mathrm{op}})$, there is no natural distinction between source and target, as there is for maps. A lagrangian submanifold L of $A \times B^{\mathrm{op}}$ is the same as a lagrangian submanifold of $(A \times B^{\mathrm{op}})^{\mathrm{op}} = A^{\mathrm{op}} \times B \cong B \times A^{\mathrm{op}}$. So L can be equivalently viewed as a morphism from B to A. This is why we prefer to denote the source and target maps of a category by l and r, because everyone agrees what is left and right.

Using this generalized notion of morphisms we return to the symplectic realizations of a Poisson Lie group X. A generalized morphism between two symplectic realizations $J_1 : S_1 \to X$ and $J_2 : S_2 \to X$ is given by a lagrangian submanifold $L \in \operatorname{Lag}(S_1 \times S_2^{\mathrm{op}})$ such that the following diagram commutes:

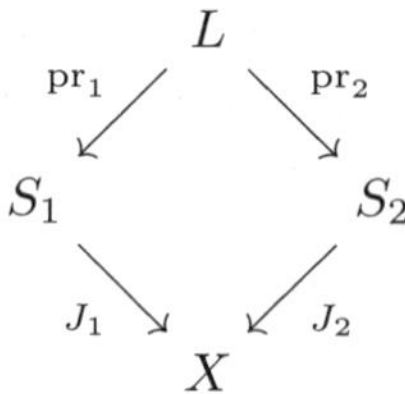

Now we can try again to find an evaluation morphism from $S \otimes S^{\vee}$ to 1 as in Eq. (5). What we need is a Lagrangian submanifold $L_{\mathrm{ev}} \in \operatorname{Lag}(S \times S^{\mathrm{op}}) \cong \operatorname{Lag}((S \times S^{\mathrm{op}}) \times \{\mathrm{pt}\})$ such that for all $(s, s') \in L_{\mathrm{ev}}$ we have $J(s)J(s')^{-1} = e$. The natural lagrangian submanifold satisfying these requirements is the diagonal $L_{\mathrm{ev}} := \Delta_S = \{(s, s) \mid s \in S\}$. The same reasoning leads us to define the coevaluation morphism also by $L_{\mathrm{cv}} := \Delta_S$. Moreover, we have the same evaluation and coevaluation morphisms for $S^{\vee} \otimes S$. It is easy to see that the morphism of symplectic realizations

$$S \xrightarrow{\cong} S \otimes 1 \xrightarrow{\mathrm{Id}_S \otimes L_{\mathrm{cv}}} S \otimes (S^{\vee} \otimes S) \xrightarrow{\cong} (S \otimes S^{\vee}) \otimes S \xrightarrow{L_{\mathrm{ev}} \otimes \mathrm{Id}_S} 1 \otimes S \xrightarrow{\cong} S$$

is the identity morphism, which is also given by the diagonal $\mathrm{Id}_S = \Delta_S$. Going in an analogous way from $S^{\vee}$ via $S^{\vee} \otimes S \otimes S^{\vee}$ to $S^{\vee}$ we obtain the identity on $S^{\vee}$ which is also given by the diagonal. We are tempted to conclude that the category of symplectic realizations and generalized morphisms is rigid monoidal. However, there is a catch:

Unfortunately, the symplectic relations are not really the morphisms of a category; when the projections to B of submanifolds in $\mathrm{Lag}(A\times B^{\mathrm{op}})$ and $\mathrm{Lag}(B\times C^{\mathrm{op}})$ intersect badly, their composition as defined in Eq. (6) is not a manifold. To avoid this, we can define $\mathrm{Hom}(A, B)$ to be $A\times B^{\mathrm{op}}$ itself, rather than the set of lagrangian submanifolds therein. The price we pay is that the composition operation

$$\mathrm{Hom}(A,B)\times\mathrm{Hom}(B,C) = (A\times B^{\mathrm{op}})\times(B\times C^{\mathrm{op}}) \longrightarrow A\times C^{\mathrm{op}} = \mathrm{Hom}(A,C)$$

is not a mapping of sets, but a relation, namely the lagrangian submanifold of

$$((A\times B^{\mathrm{op}})\times(B\times C^{\mathrm{op}}))\times\left(A\times C^{\mathrm{op}}\right)^{\mathrm{op}}$$

consisting of the product $\{(a,b,b,c,a,c)|a\in A, b\in B, c\in C\}$ of three diagonals. The result is what we have called in Section 5.2 of [**6**] a "symplectic category," i.e. a category internal to the "category" of symplectic relations.

7. Groupoids and stacks

Even more general than a relation between the sets X and Y is a "multi-relation", i.e. a map from a set M to $X\times Y$, which might not be injective. The best theory of such generalized morphisms comes about when M is acted upon by groupoids over X and Y.[1] The result is the theory of stacks, which we will describe in its smooth version [**2, 3, 18**]. (See [**14**] for the topological case.)

Roughly speaking, a stack is a device to describe a "bad" quotient. Here is a simple example. Let $\mathbb{Z}_2$ act on the open disc of unit radius in $\mathbb{R}^2$ by reflection at the origin, $1\cdot(x,y)=(-x,-y)$. The $\mathbb{Z}_2$ action is not free, because the origin is a fixed point. Taking the quotient amounts to cutting the disc along, say, the positive x-axis and rolling it up until you have two layers at every point with the exception of the origin.

The result is the surface of a cone, which is no longer a smooth manifold. You can do this for other finite cyclic groups, say, $\mathbb{Z}_3$ where now $1\in\mathbb{Z}_3$ acts by rotation by a third of the full circle. Now you obtain a cone with a smaller opening angle.

We can smoothly glue together two of these cones along an outer annulus of each disk to obtain a "Christmas tree ornament":

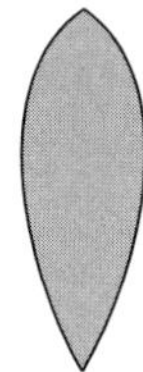

[1] It may not always be necessary to bring in the groupoids from the beginning; see for instance the use of "bisubmersions" in [**1**].

This is an example for an orbifold, a manifold with lower dimensional singularities which look locally like the quotient of $\mathbb{R}^n$ by a finite group. (The example above is still a manifold, topologically, but this is not true for more general orbifolds.) Like a manifold, an orbifold can be described by charts with these quotient spaces as local models. A drawback of this description is that there is no natural tangent bundle of an orbifold which is itself an orbifold, and the definition of morphism is rather complicated and unnatural-looking.

A more effective way to describe a bad quotient space is to remember all the gluings. This leads us to the concept of a groupoid. As a set, the gluing groupoid for $D^2/\mathbb{Z}_2$ is $G_1 := \mathbb{Z}_2 \times D^2$. There are two maps to the base $G_0 = D^2$, which map each element g of the groupoid to the two points which g glues together. We denote these maps by l and r,

$$\begin{aligned} l\big(0,(x,y)\big) &= (x,y)\,, & r\big(0,(x,y)\big) &= (x,y)\,,\\ l\big(1,(x,y)\big) &= (-x,-y)\,, & r\big(1,(x,y)\big) &= (x,y)\,, \end{aligned}$$

that is, $l(a,p) = a \cdot p$, $r(a,p) = p$ for $a \in \mathbb{Z}_2$ and $p \in D^2$. We can compose two elements of the groupoid $(a,p)(a',p') = (a+a',p')$ whenever $r(a,p) = p = a' \cdot p' = l(a',p')$. This construction can be extended to arbitrary group actions, but there are more general groupoids. For example, the groupoid presenting the Christmas ornament orbifold looks like this.

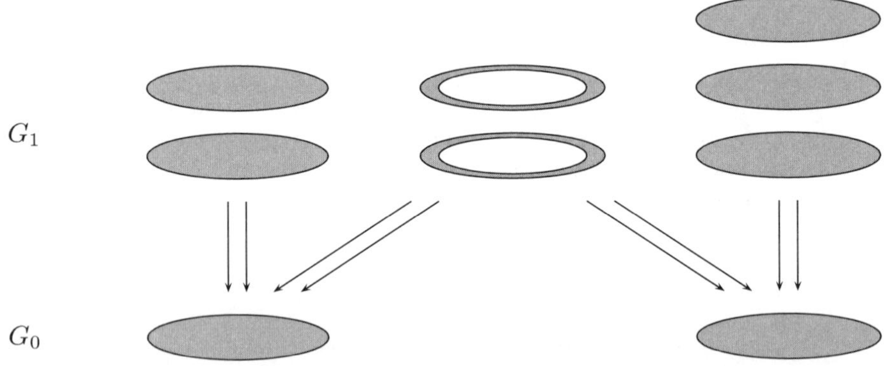

The l and r maps on the annuli are given by the embeddings into the disks.

All structure maps of the groupoid are smooth (and l and r submersions) so we have a Lie groupoid. In order to see what is special about the tips of the Christmas ornament, we have to look at the isotropy group $\mathrm{Iso}(p) := l^{-1}(p) \cap r^{-1}(p)$ for $p \in G_0$, which for an action groupoid is the stabilizer of p. The origins of the two disks of the base $G_0 = D^2 \cup D^2$ are the only points which have non-trivial isotropy, $\mathbb{Z}_2$ for the left disk and $\mathbb{Z}_3$ for the right disk. For orbifolds, the stabilizers are by definition finite groups. For the groupoid it means that it is proper étale, i.e., the anchor map $(l,r) : G_1 \to G_0 \times G_0$, $g \mapsto (l(g), r(g))$ is proper étale. Indeed, we can give the following definition of an orbifold [**13**]:

DEFINITION 1. *An orbifold is a differentiable stack presented by a proper étale Lie groupoid.*

Now we have to explain how we can associate a stack to a Lie groupoid. We can think of a Lie groupoid as a generalized equivalence relation describing the quotient space of equivalence classes. As we have seen, the actual quotient space

G_0/G_1 of G_1-orbits in G_0 is usually not a nice space, so we write $G_0//G_1$ for the as-if quotient the groupoid is thought to describe. The usual notion of isomorphism of Lie groupoids is that of a diffeomorphism which is compatible with all the structure maps. However, now two groupoids should be considered to be equivalent if they present the same quotient.

Given two groupoids G and H we must find a smooth way to associate the G_1-orbits on G_0 with the H_1-orbits on H_0. This cannot be just a map $G_0/G_1 \to H_0/H_1$, because these quotient spaces are in general not manifolds, and so there is no notion of smoothness. Our description of a bad quotient as a groupoid which is a generalized equivalence relation suggests defining morphisms between them in an analogous manner. Now the two spaces we want to relate are the bases G_0 and H_0 of the groupoids. A generalized relation consists of triples $(x \stackrel{m}{\to} y)$ where we say that $x \in G_0$ is related via m to $y \in H_0$. We denote the set of all such triples by M. The projections of the elements of M on the elements of the groupoid bases they relate, $l_M(x \stackrel{m}{\to} y) := x$ and $r_M(x \stackrel{m}{\to} y) := y$, are called the moment maps of M. When no confusion can arise, we will drop the subscripts of the moment maps.

We do not require M to be a map from G_0 to H_0. For example, a single pair x and y can be related by several elements of M. But we want the relation M to descend to a map on the set of orbits. For notational reasons it is convenient to use left orbits in G_0 and right orbits in H_0. (Note that the left orbits for a groupoids acting on its base are the same as the right orbits.) For the generalized relation M between elements $x \in G_0$ and $y \in H_0$, to descend to a well-defined relation on the orbits $G_1 \cdot x$ and $y \cdot H_1$ we have to require that for all g acting on x and h acting on y we have elements $g \cdot m$ and $m \cdot h$ of M such that

$$x \stackrel{m}{\to} y \;\Rightarrow\; (g \cdot x) \stackrel{g \cdot m}{\longrightarrow} y \qquad \text{and} \qquad x \stackrel{m}{\to} y \;\Rightarrow\; x \stackrel{m \cdot h}{\longrightarrow} (y \cdot h)\,.$$

We want to be able to choose $g \cdot m$ and $m \cdot h$ in a consistent way, such that we get maps $m \mapsto g \cdot m$ and $m \mapsto m \cdot h$ which are compatible with the groupoid structures, $g \cdot (g' \cdot m) = gg' \cdot m$ and $(m \cdot h) \cdot h' = m \cdot hh'$, as well as, $(g \cdot m) \cdot h = g \cdot (m \cdot h)$ whenever defined, i.e., we have two commuting groupoid actions on M. Such an object M is called a groupoid bibundle.

So far, M only descends to a relation on $G_0/G_1 \times H_0/H$. To obtain a function from G_0/G_1 to H_0/H_1, we first have to require that l_M is surjective, so that the function will be defined on all of G_0/G_1. Second, for the relation to be a function, a given $x \in G_0$ has to be related only to elements of one orbit in H_0;

$$x \stackrel{m}{\to} y \;\text{ and }\; x \stackrel{m'}{\to} y' \quad\Rightarrow\quad y \cdot h = y' \quad\Rightarrow\quad x \stackrel{m \cdot h}{\longrightarrow} y'$$

for some groupoid element $h \in H_1$. Again, we want to be able to choose h in a nice way, requiring that there is a unique h such that $m \cdot h = m'$. This gives us a right principal bibundle [**4, 14**]. Finally, we require all the structures to be smooth, that is, M is a manifold, the moment maps are smooth, and the groupoid actions are smooth.

We can depict the situation by the diagram

(7)

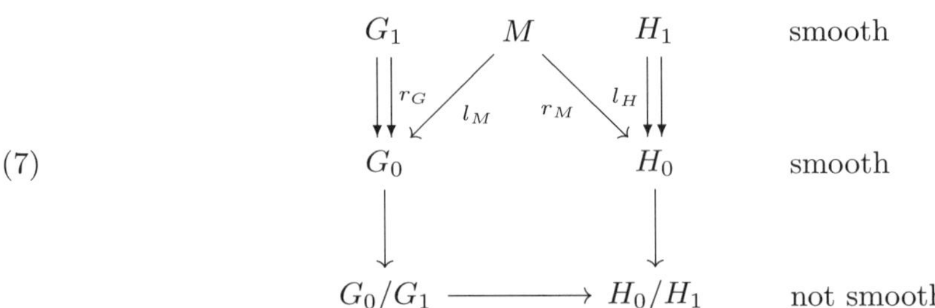

Now let G, H, K be Lie groupoids, M a smooth right principal G-H bibundle and N a smooth right principal H-K bibundle. The composition of the induced functions on the quotient spaces can be lifted to a smooth composition of the bibundles:

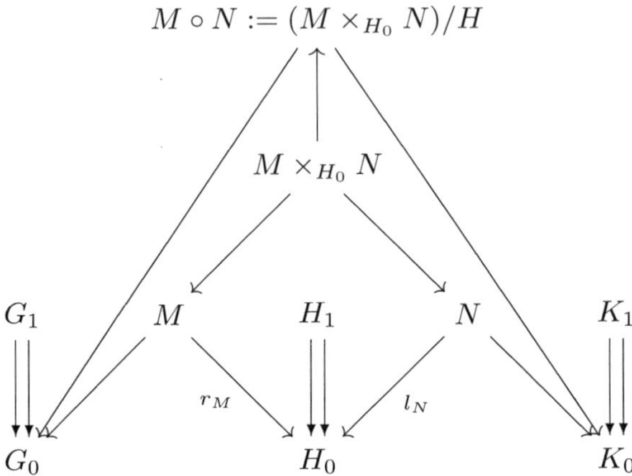

Here, the right H-action on $M \times_{H_0} N$ is given by $(m, n) \cdot h = (m \cdot h, h^{-1} \cdot n)$. The G-action and the K-action descend to actions on the quotient, since they both commute with the H-action. The conclusion is that $M \circ N$ is a G-K bibundle, which we call the composition of M and N. This composition is only associative up to a biequivariant diffeomorphism of bibundles. This means that we should really be working in the weak 2-category having Lie groupoids as objects, smooth right principal bibundles as 1-morphisms, and smooth biequivariant maps of bibundles as 2-morphism. We denote this category by LieGrpdPrBibu.

A good way to study the generalized space described by a groupoid G is the Grothendieck approach of considering all the morphisms from ordinary manifolds to G, where a manifold X is described by the groupoid $X_1 = X_0 = X$. The collection of all such morphisms to G, more concretely described as

$$\{M \mid M \text{ a right principal } X\text{-}G \text{ bibundle, } X \text{ a manifold}\}$$

is denoted by BG and called the classifying space of the groupoid G. This is a smooth stack in the usual sense which is presented by the groupoid G. Two stacks BG and BH are isomorphic if and only if the groupoids G and H are Morita equivalent (Theorem 2.24 in [**3**]). We thus get a 1-to-1 correspondence of isomorphism classes of presentable stacks and Morita equivalence classes of groupoids. This is often stated as "a stack is a groupoid up to Morita equivalence". But beware that the actual functor between the category of stacks and the weak 2-category of groupoids and bibundles is a weak 2-equivalence of 2-categories.

8. Stacky Lie groups

It can be shown that, in the weak 2-category LieGrpdPrBibu of Lie groupoids and smooth right principal bibundles, all products exist and the one-element groupoid 1 is a terminal object. (Note that, if the bibundles are not required to be principal, this is no longer true.) Since we are dealing with a weak 2-category, the categorical product is associative only up to weak 1-isomorphisms, that is, up to Morita equivalence of groupoids.

If we have products and a terminal objects we also have the notion of group objects, which we will call stacky Lie groups [**4, 22**]. The question is whether in LieGrpdPrBibu the notion of group objects is useful, as for differentiable spaces, or uninteresting as for groups or algebras. This is not easy to see because, on the one hand we think of a groupoid as a generalized quotient space, but on the other hand a Lie groupoid is itself an algebraic structure internal to the category of manifolds. It turns out that while not many examples of truly stacky groups have been studied until now, there are some particularly interesting ones.

Consider the group S^1 and the dense subgroup which is the image of the embedding $\mathbb{Z} \to S^1 \cong U(1)$, $k \mapsto e^{i\lambda k}$ where, $\lambda/2\pi$ is an irrational number. By abuse of notation, we will also denote the subgroup itself by $\mathbb{Z}$. The quotient $S^1/\mathbb{Z}$ is an abelian group, in which we will denote the multiplication by $m : S^1/\mathbb{Z} \times S^1/\mathbb{Z} \to S^1/\mathbb{Z}$. Because the subgroup $\mathbb{Z}$ is dense, the quotient topology is trivial. This suggests that we should work with the stack $S^1//\mathbb{Z}$ rather than with the actual quotient. We then try to lift the multiplication map m on the bottom level of the diagram of the form (7) to the smooth top levels:

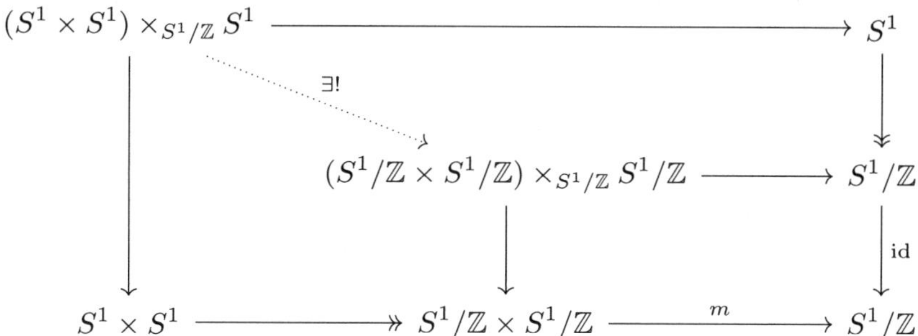

The inner pull-back $(S^1/\mathbb{Z} \times S^1/\mathbb{Z}) \times_{S^1/\mathbb{Z}} S^1/\mathbb{Z}$ is the graph of the group multiplication m, the pull-back projections being the range and image maps. Any object in the top left upper corner which makes the diagram commute can be viewed as a lift of the graph of m to S^1, the pull-back being the universal lift. The diagonal arrow is the unique map which exists by the universal property of the inner pull-back. Explicitly, the pull-back is

$$(S^1 \times S^1) \times_{S^1/\mathbb{Z}} S^1 = \{(\theta_1, \theta_2, \theta) \in (S^1 \times S^1) \times S^1 \mid \theta_1 + \theta_2 = \theta \bmod 2\pi\lambda\}$$

where θ is the angle representing $e^{i\theta} \in U(1) \cong S^1$. This set can be identified with $S^1 \times S^1 \times \mathbb{Z}$ and viewed as the graph of a multi-valued multiplication on S^1. It has the structure of a smooth manifold and inherits smooth actions of the action groupoid $G := S^1 \rtimes \mathbb{Z} \rightrightarrows S^1$ presenting the stack $S^1//\mathbb{Z}$. We thus obtain the smooth right principal $(G \times G)$-G bibundle E_m of multiplication. In an analogous manner we construct the 1-G bibundle E_e of the identity element, and the G-G bibundle of the inverse E_{inv}. It can be checked that the bibundles E_m, E_e, E_{inv} equip the groupoid G with the structure of a stacky Lie group.

9. Hopfish algebras

The Gelfand Theorem tells us that a locally compact topological space and its commutative algebra of continuous functions vanishing at infinity contain the same structural information. In our example, $S^1/\mathbb{Z}$, the quotient topology is trivial, so the only continuous functions are constant. But we are working instead with the groupoid presenting the stack $S^1//\mathbb{Z}$. What is the algebra of continuous "functions" on $S^1//\mathbb{Z}$? It is one of the main ideas of noncommutative geometry that we should consider the convolution algebra of the groupoid $G := S \rtimes \mathbb{Z}$ to be the analogue of the algebra of functions [**7, 15**]. For two compactly supported functions a and b on G, the convolution product is

$$(a * b)(g) = \int a(h)b(h^{-1}g)dh\,,$$

which looks just like the convolution algebra of a group. The difference is that on a groupoid the product $h^{-1}g$ is only defined if $l(h) = l(g)$, so for a given g we have to integrate over the left groupoid fiber of $r(g) = x$. Since we have to do this for all g, we need a whole family of Haar measures $dh = d_x h$. Alternatively, we can work with half-densities instead of functions, as in [**7**].

Recalling that Morita equivalent groupoids have Morita equivalent (but generally not isomorphic) convolution algebras, we also conclude that Morita equivalent algebras should be thought of as representing the "same noncommutative space".[2]

In our example, the Lie groupoid $G = S^1 \rtimes \mathbb{Z}$ is étale, so the Haar measures are merely counting measures. Thus,

$$(a * b)(\theta, k) := \sum_{k' \in \mathbb{Z}} a(\theta + \lambda k', k - k')b(\theta, k')\,, \tag{8}$$

for all compactly supported functions a and b on $S^1 \times \mathbb{Z}$. Among such functions are the standard Fourier basis functions on the circle times the characteristic functions of integers, $a_{nl}(k, \theta) := e^{in\theta}\delta_{lk}$, for which we obtain

$$a_{n_1 l_1} * a_{n_2 l_2} = e^{i\lambda n_1 l_2} a_{n_1+n_2, l_1+l_2}\,.$$

Let us denote the algebra of finite linear combinations of these functions by $\mathcal{A}$. The closure of $\mathcal{A}$ with respect to a suitable norm is known as the algebra continuous functions on a noncommutative torus,[3] which is a deformation of the usual algebra of continuous functions on the 2-torus [**16**].

---

[2] One must use this identification with some care. For instance, there are plenty of examples of nonisomorphic groups with isomorphic group algebras (e.g. pairs of finite abelian groups with the same number of elements); thinking of these groups as representing stacks of the form BG, one finds different stacks with the same "algebra of functions". Even more different, it seems to us, are the stacks presented by the group $\mathbb{Z}_2$ and the trivial groupoid over a set with two elements, yet again they have isomorphic group algebras These examples suggest that it takes more than an algebra to make a "noncommutative space," but it is not clear to us exactly what that "more" should be.

[3] By abuse of terminology, the algebra itself is often known as a noncommutative torus.

Now the quotient space $S^1/\mathbb{Z}$ is also a quotient group. Since $\mathcal{A}$ plays the role of an algebra of functions on this quotient, we might expect the group structure to translate into a Hopf structure on $\mathcal{A}$. But this is not the case. In fact, $\mathcal{A}$ is simple, so it does not even possess a counit. So what happened to the group structure? The answer lies with the stacky group structure.

The space of functions on every G-H bibundle naturally acquires the structure of an $\mathcal{A}(G)$-$\mathcal{A}(H)$ bimodule. Composition of bibundles corresponds to tensor product of bimodules, so, under suitable technical assumptions, we get a functor from LieGrpdPrBibu to a category in which the objects are algebras and the morphisms are bimodules. By analogy with the usual Gelfand functor, we view it as contravariant. Since this functor takes categorical products of groupoids to tensor products of algebras, which are not categorical products, it certainly does not take group objects to group objects. In fact, it does not even take them to Hopf algebras, as we saw already for the example of the noncommutative torus. (There, the functor takes the unit of the group to a $\mathbb{C}$-$\mathcal{A}$ bimodule rather than to a homomorphism $\mathbb{C} \leftarrow \mathcal{A}$.) Instead, the image of a stacky group is another structure which we call a **hopfish algebra** [**5**, **17**].

A space of functions on the bibundle E_m of multiplication for a stacky group with function algebra $\mathcal{A}$ becomes an $(\mathcal{A} \otimes \mathcal{A})$-$\mathcal{A}$ bimodule, which we denote by $\boldsymbol{\Delta}$ and view as the bimodule of comultiplication. Analogously, the bibundle E_e of the unit element, already mentioned above, becomes the $\mathbb{C}$-$\mathcal{A}$ bimodule $\boldsymbol{\epsilon}$ of the counit. By functoriality these bibundles satisfy coassociativity and counitality relations

$$(\mathcal{A} \otimes \boldsymbol{\Delta}) \otimes_{\mathcal{A}\otimes\mathcal{A}} \boldsymbol{\Delta} \cong (\boldsymbol{\Delta} \otimes \mathcal{A}) \otimes_{\mathcal{A}\otimes\mathcal{A}} \boldsymbol{\Delta}\,,$$
$$(\boldsymbol{\epsilon} \otimes \mathcal{A}) \otimes_{\mathcal{A}\otimes\mathcal{A}} \boldsymbol{\Delta} \cong \mathcal{A} \cong (\mathcal{A} \otimes \boldsymbol{\epsilon}) \otimes_{\mathcal{A}\otimes\mathcal{A}} \boldsymbol{\Delta}\,,$$

up to isomorphism of bimodules. We thus obtain a weak comonoidal object in the weak 2-category of algebras and bimodules, which is also called a sesquilinear sesquialgebra.

The algebraic image of the bibundle E_{inv} of the inverse is more difficult to interpret. Because E_{inv} is a G-G bibundle, the space of functions on E_{inv} is an $\mathcal{A}$-$\mathcal{A}$ bimodule, which we would expect to become the bimodule of an antipode. But a Hopf antipode is an algebra antihomomorphism, so the hopfish antipode should be an $\mathcal{A}$-$\mathcal{A}^{\mathrm{op}}$ bimodule. A way to accomplish this is to use the star structure on $\mathcal{A}$ which comes from the groupoid inverse of G, $a^*(g) = \overline{a(g^{-1})}$, in order to convert the $\mathcal{A}$-$\mathcal{A}$ bimodule of functions on E_{inv} into the $\mathcal{A}$-$\mathcal{A}^{\mathrm{op}}$ bimodule $\boldsymbol{S}$ of the hopfish antipode. Just as we illustrated above in the case of Poisson groups with the diagram 4, it is not easy to formulate what is required of the antipode in a hopfish algebra. But a definition is given in [**17**], with a modification in [**5**] to cover the case of noncommutative torus algebras, and the bimodule described above does satisfy the definition.

Like an ordinary coproduct, the hopfish coproduct bimodule $\boldsymbol{\Delta}$ can be used to multiply two right $\mathcal{A}$-modules $T, T' \in \mathrm{Mod}_{\mathcal{A}}$ by

$$T \otimes_{\boldsymbol{\Delta}} T' := (T \otimes T') \otimes_{\mathcal{A}\otimes\mathcal{A}} \boldsymbol{\Delta}\,.$$

In [**5**], we have tried out this new multiplication on certain modules over the hopfish algebra associated to the stacky group $S^1//\mathbb{Z}$. For $\alpha \in \mathbb{R}$, $p, q \in \mathbb{Z}$ relatively prime, we have shown there that

$$T_{pq}^{\alpha} := \mathcal{A}/(e^{-i\alpha} a_{pq} - 1)\mathcal{A}\,,$$

is a simple module generated by $\xi := [1]$ with $a_{pq} \cdot \xi = e^{i\alpha}\xi$, and that adding any integer multiple of λ to α results in an isomorphic module. (For p, q not relatively prime the situation is slightly more complicated.) Some calculations lead to the following result.

THEOREM 1. *For $p_1 \neq 0$ or $p_2 \neq 0$ we have:*

$$T^{\alpha_1}_{p_1 q_1} \otimes_{\blacktriangle} T^{\alpha_2}_{p_2 q_2} \cong \gcd(p_1, p_2)\, T^{\alpha}_{pq}\,,$$

where

$$p := \operatorname{lcm}(p_1, p_2)\,,\ q := \frac{p_1 q_2 + p_2 q_1}{\gcd(p_1, p_2)}\,,\ \alpha := \frac{\alpha_1 p_2 + \alpha_2 p_1}{\gcd(p_1, p_2)}\,. \tag{9}$$

For $p_1 = 0$ and $p_2 = 0$ we have:

$$T^{\alpha_1}_{0,q_1} \otimes_{\blacktriangle} T^{\alpha_2}_{0,q_2} \cong \begin{cases} T^{\alpha}_{0,q}\,, & \text{for } \frac{\alpha_1 q_2 - \alpha_2 q_1}{\lambda \gcd(q_1, q_2)} \in \mathbb{Z} \bmod \operatorname{lcm}(q_1, q_2)\frac{2\pi}{\lambda} \\ 0\,, & \text{otherwise} \end{cases}\,,$$

where

$$q := \gcd(q_1, q_2)\,, \quad \alpha := s_1 \alpha_2 - s_2 \alpha_1\,, \quad s_1, s_2 \in \mathbb{Z} : \quad \frac{s_1 q_2 - s_2 q_1}{\gcd(q_1, q_2)} = 1\,. \tag{10}$$

Observe that $q/p = q_1/p_1 + q_2/p_2$ and $\alpha/p = \alpha_1/p_1 + \alpha_1/p_2$. Surprisingly, we did not notice these simple relations until we found the following geometric interpretation of them.

When we identify $\mathcal{A}$ with an algebra of functions on the torus $\mathbb{T}^2$, with coordinates (θ_1, θ_2), the algebra element $e^{-i\alpha}a_{pq} - 1$ becomes the function $e^{-i\alpha}e^{i(p\theta_1 + q\theta_2)} - 1$, and so the classical analogue of the quotient module $T^{\alpha}_{pq} = \mathcal{A}/(e^{-i\alpha}a_{pq} - 1)\mathcal{A}$ appears to be the functions on the embedded circle in $\mathbb{T}^2$ defined by the equation $p\theta_1 + q\theta_2 - \alpha = 0$, or $\theta_1 = -\frac{q}{p}\theta_2 + \frac{\alpha}{p}$. The classical analogue of the hopfish structure on $\mathcal{A}$ (and hence the symplectic model of the group structure on $S^1/\mathbb{Z}$) turns out to be the symplectic group*oid* structure $\mathbb{T}^2 \rightrightarrows \mathbb{T}^1$ for which the source and target maps are the projection in the θ_2 direction, and the composition law is addition in θ_1. The tensor product operation on modules corresponds to the application of the composition law on the groupoid to the embedded circles which represent them. This results in the addition of fractions mentioned above.

Note that, if we "unwrap" the θ_2 circle to a line, we obtain the cotangent bundle $\mathbb{T}^1 \times \mathbb{R} \cong T^*\mathbb{T}^1$, the symplectic groupoid $\Gamma(\mathbb{T}^1)$ of $\mathbb{T}^1$ with the zero Poisson structure. The groupoid structure described in the preceding paragraph is the second symplectic groupoid structure on $\Gamma(\mathbb{T}^1)$ which is obtained by lifting the (Poisson) Lie group structure on $\mathbb{T}^1$ given by addition in θ. This is an instance of the double symplectic groupoid structures attached to general Poisson Lie groups [**12**].

What has become of the parameter λ? In fact, it seems that the curve $p\theta_1 + q\theta_2 - \alpha$ represents only one generator of the module in question. The others are obtained by applying the unitary basis elements of the algebra itself. These correspond to constant "integer" bisections of $\Gamma(\mathbb{T}^2) \cong T^*\mathbb{T}^2$; the bisection $md\theta_1 + nd\theta_2$ acts via the source and target maps, determined by the Poisson structure (see [**21**]), as translation by $(-\lambda n, \lambda m)$. This yields the collection of all circles of the form $p\theta_1 + q\theta_2 - (\alpha + \lambda(np - mq)) = 0$. Under the assumption $\gcd(p, q) = 1$, all the integer multiples of λ occur, so that this collection of circles, like the isomorphism class of T^{α}_{pq}, depends only on the image of α in $S^1/\mathbb{Z}$.

References

[1] Androulidakis, I., and Skandalis, G., The holonomy groupoid of a singular foliation, preprint math.DG/0612370.

[2] Behrend, K., Cohomology of stacks. *Intersection theory and moduli*, (electronic), *ICTP Lect. Notes*, XIX, Abdus Salam Int. Cent. Theoret. Phys., Trieste, 2004, 249–294.

[3] Behrend, K., and Xu, P., Differentiable stacks and gerbes, preprint math.DG/0605694 (2006).

[4] Blohmann. C., Stacky Lie groups, preprint math.DG/0702399

[5] Blohmann, C., Tang, X., Weinstein, A., Hopfish structure and modules over irrational rotation algebras, preprint math.QA/0604405.

[6] Bursztyn, H., and Weinstein, A., Poisson geometry and Morita equivalence, preprint math.SG/0402347, *Poisson Geometry, Deformation Quantization, and Group Representations*, LMS Lecture Note Series, Cambridge University Press, Cambridge, 2005, pp. 3–78.

[7] Connes, A., *Noncommutative geometry*, Academic Press, San Diego, CA, 1994.

[8] Coste, A., Dazord, P. et Weinstein, A., Groupoïdes symplectiques, *Publications du Département de Mathématiques, Université Claude Bernard-Lyon I* **2A** (1987), 1–62 (Available at `http://math.berkeley.edu/~alanw/cdw.pdf`).

[9] Douady, A., and Lazard, M., Espaces fibrés en algèbres de Lie et en groupes, *Invent. Math.* **1** (1966), 133-151.

[10] Est, E.T. van, and Korthagen, T.J., Non-enlargible Lie algebras, *Indag. Math.* **26** (1964), 15-31.

[11] Karasev, M.V., Analogues of objects of Lie group theory for nonlinear Poisson brackets, *Math. USSR Izvestiya* **28**, (1987), 497-527.

[12] Lu, J.-H., and Weinstein, A., Groupoïdes symplectiques doubles des groupes de Lie-Poisson, *C. R. Acad, Sci. Paris* **309** (1989), 951–954.

[13] Moerdijk, I., Orbifolds as groupoids: an introduction. *Orbifolds in mathematics and physics*, Contemp. Math. **310**, Amer. Math. Soc., Providence, RI, 2002, 205–222.

[14] Mrčun, J. *Stability and invariants of Hilsum-Skandalis maps*, PhD thesis, Utrecht University, Utrecht, 1996, math.DG/0506484.

[15] Renault, J., A groupoid approach to C^*-algebras, *Lecture Notes in Mathematics* **793**, Springer, Berlin, 1980

[16] Rieffel, M., Deformation quantization for actions of R^d, *Mem. Amer. Math. Soc.* **106** (1993), no. 506.

[17] Tang, X., Weinstein, A., and Zhu, C., Hopfish algebras, to appear in *Pacific J. Math*, math.QA/0510421.

[18] Tseng, H., Zhu. C., Integrating Lie algebroids via stacks. *Compositio Mathematica* **142** (2006), 251–270.

[19] Ulbrich, K.-H., Tannakian categories for non-commutative Hopf algebras, *Israel J. Math.* **72** (1990), no. 1-2, 252–256.

[20] Weinstein, A., The local structure of Poisson manifolds, *J. Differential Geom.* **18** (1983), no. 3, 523–557.

[21] Weinstein, A., Symplectic groupoids, geometric quantization, and irrational rotation algebras, *Symplectic geometry, groupoids, and integrable systems, Séminaire sud-Rhodanien de géométrie à Berkeley (1989)*, P. Dazord and A. Weinstein, eds., Springer-MSRI Series (1991), 281–290.

[22] Zhu, C., Lie n-groupoids and stacky Lie groupoids, math.DG/0609420.

Department of Mathematics, University of California, Berkeley, CA 94720, USA.
Fakultät für Mathematik, Universität Regensburg, 93040 Regensburg, Germany.
E-mail address: `blohmann@math.berkeley.edu`

Department of Mathematics, University of California, Berkeley, CA 94720, USA.
E-mail address: `alanw@math.berkeley.edu`

Contemporary Mathematics
Volume **450**, 2008

Poisson Fibrations and Fibered Symplectic Groupoids

Olivier Brahic and Rui Loja Fernandes

Abstract. We show that Poisson fibrations integrate to a special kind of symplectic fibrations, called fibered symplectic groupoids.

1. Introduction

Our purpose in this paper is to explain that there is a geometric theory of Poisson fibrations that is analogous to the theory of symplectic fibrations. This paper makes no special claims to originality, and improves previous works (see, e.g., Theorem 1.1) of Vorobjev (Poisson case, [**15**]) and Vaisman (Dirac case, [**14**]). Our main contribution is two folded. On the one hand, we propose an approach based on gauge theory and Dirac geometry, which gives some natural explanations for some of the mysterious formulas that appear in those works. On the other hand, we look for the first time into the integration of such structures, recovering symplectic fibrations from Poisson fibrations.

A symplectic fibration is a locally trivial fibration with fiber type a symplectic manifold, admitting a collection of trivializations whose transition functions are symplectomorphisms. Symplectic fibrations have a long history going back to the early works of Weinstein *et al.* [**18, 10**] and Guillemin *et al.* ([**12, 11**]). A closed 2-form on the total space of a fibration which restricts to a symplectic form on the fibers, determines a symplectic fibration. Conversely, given a symplectic fibration, one may ask if there exists a closed 2-form which is compatible with the fibration. It is well-known that there are non-trivial obstructions for the existence of such coupling forms and that one can classify all such coupling forms. A very nice exposition of the the theory of symplectic fibrations, where these results are discussed in detail, is given in Chapter 6 of the monograph by McDuff and Salamon [**13**].

We are interested in the more general notion of a *Poisson fibration*, i.e., a locally trivial fibration with fiber type a Poisson manifold, admitting a collection of trivializations whose transition functions are Poisson diffeomorphisms. At first sight, the analogous questions for Poisson geometry are either hopeless or trivial. On the one hand, there are simple examples of fibrations with a Poisson structure

1991 *Mathematics Subject Classification.* Primary 53D17; Secondary 58H05.

Key words and phrases. Poisson and symplectic fibrations; symplectic groupoid.

Supported in part by FCT/POCTI/FEDER and by grants POCI/MAT/57888/2004 and POCI/MAT/55958/2004.

on the total space that restricts to each fiber and which is not a Poisson fibration. On the other hand, every Poisson fibration always admits trivially a compatible Poisson structure: one can just declare the fibers to be Poisson submanifolds.

A closer inspection, however, reveals a very different point of view. Note that for a symplectic fibration one looks for a *presymplectic structure* which intersects each fiber in a symplectic submanifold. Therefore, for a Poisson fibration we should look for a *Dirac structure* for which each presymplectic leaf intersects every fiber in a symplectic leaf of the Poisson structure on the fiber (recall that a Poisson manifold is a (singular) foliation by symplectic manifolds, while a Dirac manifold is a (singular) foliation by presymplectic manifolds). This yields the notion of a *Dirac coupling* for a Poisson fibration. The theory of Poisson fibrations can then be seen as a foliated version of the theory of symplectic fibrations (this is the point of view advocated by Vaisman [**14**]).

Therefore, the questions (and answers) in the theory of Poisson fibrations, are analogous to (and generalize) the theory of symplectic fibrations. For example, given a Poisson fibration one would like to know (i) if it admits coupling Dirac structures and (ii) classify all such couplings. In this direction we have the following generalization of a well-known result in symplectic fibrations (see [**13**, Theorem 6.13]):

THEOREM 1.1. *Let $p : M \to B$ be a Poisson fibration. Then the following statements are equivalent:*

(i) $p : M \to B$ admits a coupling a Dirac structure.
(ii) There exists a Poisson connection on $p : M \to B$ whose holonomy groups act on the fibers in a hamiltonian fashion.

Its is also possible to classify all such coupling forms. The proof of this result follows the same pattern as in the symplectic case: one builds a Poisson gauge theory where coupling forms are obtained on associated fiber bundles $M = P \times_G F$, starting from a connection on a principal G-bundle and a Hamiltonian G-action on a Poisson manifold (F, π). Note that G is not necessarily a finite dimensional Lie group.

Our second main purpose is the *integration* of Poisson fibrations. Recall that Poisson structures are infinitesimal objects which integrate to global objects called symplectic groupoids. There are obstructions to integrability which were recently understood [**4, 3**], but we will ignore them for the time being. Now, we will see that:

- The global object associated to a Poisson fibration is a *fibered symplectic groupoid.*

Let us explain what we mean by this. By a *fibered groupoid* we mean a groupoid $\mathcal{G} \rightrightarrows M$ which is fibered over B:

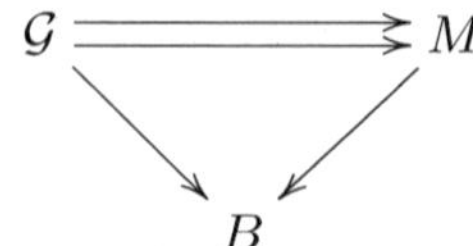

Here, $\mathcal{G}$ and B are fibered as well as all structure maps. So a fibered groupoid maybe thought of as a fiber bundle with fiber type a groupoid where the structure group acts by groupoid automorphism. Then by a *fibered symplectic groupoid* we mean a fibered groupoid $\mathcal{G}$ whose fiber type is a symplectic groupoid. For a fibered symplectic groupoid, $\mathcal{G} \to B$ is a symplectic fibration and the symplectic structure is compatible with the groupoid structure. The base $p : M \to B$ of a fibered symplectic groupoid has a natural structure of a Poisson fibration. Moreover, the fibers of $\mathcal{G} \to B$ are symplectic groupoids over the fibers of $p : M \to B$, inducing the Poisson structures on the fibers. In particular, the fibers $p : M \to B$ are integrable Poisson manifolds. Conversely, we will show the following:

THEOREM 1.2. *Let $p : M \to B$ be a Poisson fibration with fiber type (F, π) an integrable Poisson manifold. There exists a unique (up to isomorphism) source 1-connected fibered symplectic groupoid integrating $p : M \to B$. Moreover, this symplectic fibration always admits a coupling 2-form.*

It was proved in [**2**] that the global objects integrating Dirac structures are presymplectic groupoids. Therefore, if $p : M \to B$ is a Poisson fibration which admits a coupling Dirac structure, there are two natural groupoids associated with it:

(i) The source 1-connected fibered symplectic groupoid integrating $p : M \to B$.
(ii) The source 1-connected presymplectic groupoid integrating the coupling Dirac structure of $p : M \to B$.

The precise relationship between these two objects is more involved, and it would take us too far afield, so it will be discussed elsewhere.

2. Connections and Dirac structures

One can define a connection on a fibration by specifying on the total space an almost Dirac structure of a special type. In this section we make a preliminary study of connections on fibrations induced by Dirac structures.

2.1. Connections defined by almost Dirac structures. In what follows, by a *fibration* we always mean a locally trivial fiber bundle. Given such a fibration $p : M \to B$, we denote by F_b the fiber over $b \in B$, and we let $\mathrm{Vert} \subset TM$ be the *vertical sub-bundle* whose fibers are $\mathrm{Vert}_x \equiv \mathrm{Ker}\, \mathrm{d}_x p = T_x F_{p(x)}$. By a *connection* Γ on $p : M \to B$ we mean an Ehresmann connection, i.e., a distribution $\Gamma : x \mapsto \mathrm{Hor}_x$ in TM which splits the tangent bundle as

$$T_x M = \mathrm{Hor}_x \oplus \mathrm{Vert}_x, \tag{2.1}$$

and satisfies the following lifting property:

> **Lifting Property:** for each $x_0 \in M$ and each curve $\gamma : [0,1] \to B$ starting at $b_0 = p(x_0)$ there exists an integral curve $\widetilde{\gamma} : [0,1] \to M$ of Γ starting at x_0 and covering γ.

REMARK 2.1. Given a C^1-path $\gamma : [0,1] \to B$ and $x \in F_{\gamma(0)}$ the splitting (2.1) guarantees that there exists a unique *horizontal lift* $\widetilde{\gamma}_x : [0, \varepsilon] \to M$ starting at x. The Lifting Property says that we can take $\varepsilon = 1$ ([1]).

In the usual way, one obtains the notion of *parallel transport* of fibers: given a piecewise C^1-path on the base $\gamma : [0,1] \to B$ we have the diffeomorphism:

$$\phi_\gamma(t) : F_{\gamma(0)} \to F_{\gamma(t)}, \qquad x \mapsto \widetilde{\gamma}_x(t).$$

[1]Note that our curves are always parameterized in the interval $[0,1]$.

When γ is a loop, we call $\phi_\gamma(1)$ the *holonomy* of Γ along γ. For $b \in B$, the *holonomy group* with base point b is the subgroup $\Phi(b) \subset \mathrm{Diff}(F_b)$ formed by all holonomy transformations $\phi_\gamma(1)$, where $\gamma : [0,1] \to B$ is any loop based at b. Note that to define the concatenation of loops we need to reparameterize our curves, but this causes no problem since "horizontal lift commutes with concatenation" and, hence, two paths differing by a reparameterization determine the same holonomy transformation.

Now our basic observation is that connections can be defined by specifying on the total space of the fibration an almost Dirac structure of a special type. Namely:

DEFINITION 2.2. Let $p : M \to B$ be a fibration. An almost Dirac structure L on the total space of the fibration is called **fiber non-degenerate** if

$$(\mathrm{Vert} \oplus \mathrm{Vert}^0) \cap L = \{0\}. \tag{2.2}$$

In fact, we have:

PROPOSITION 2.3. *Let $p : M \to B$ be a fibration and L a fiber non-degenerate almost Dirac structure. Then L defines a connection Γ_L with horizontal space:*

$$\mathrm{Hor} = \left\{X \in TM : \exists \alpha \in (\mathrm{Vert})^0, (X, \alpha) \in L\right\}. \tag{2.3}$$

Moreover, any connection on $p : M \to B$ can be obtained in this way.

PROOF. Take a vector space W, with a maximal isotropic subspace $L \subset W \times W^*$, and let $V \subset W$ be a subspace such that:

$$(V \times V^0) \cap L = \{0\}.$$

We claim that

$$W = V \oplus H,$$

where $H = \left\{w \in W : \exists \xi \in V^0, (w, \xi) \in L\right\}$. If we take $W = T_xM$, $L = L_x$ and $V = \mathrm{Vert}_x$, this claim yields the first part of the proposition.

To prove the claim, we start by checking that $V \cap H = \{0\}$. In fact, if $v \in H$ there exists $\xi \in V^0$ such that $(v, \xi) \in L$. Hence, if $v \in V \cap H$ we obtain:

$$(v, \xi) \in (V \times V^0) \cap L = \{0\},$$

so we must have $v = 0$. It remains to check that $W = V + H$. First, we observe that since L and $V \times V^0$ intersect trivially and $\dim L = \dim(V \times V^0) = \dim W$, we must have:

$$W \times W^* = (V \times V^0) \oplus L.$$

Therefore, for any $w \in W$ we have a decomposition:

$$(w, 0) = (v, \xi) + (h, \eta),$$

where $v \in V$, $\xi \in V^0$ and $(h, \eta) \in L$. It follows that $\eta = -\xi \in V^0$, so that $h \in H$. We conclude that $w = v + h$, with $v \in V$ and $h \in H$, as claimed.

Conversely, if Γ is a connection with horizontal distribution Hor, then $L = \mathrm{Hor} \oplus \mathrm{Hor}^0$ defines a fiber non-degenerate almost Dirac structure whose associated connection is Γ. Hence, every connection arises in this way. ☐

2.2. The horizontal 2-form and the vertical bivector field. Note that two fiber non-degenerate almost Dirac structures L_1 and L_2 may lead to the same connection $\Gamma_{L_1} = \Gamma_{L_2}$. In fact, as we will see now, there is more structure associated with the specification of a fiber non-degenerate almost Dirac structure.

PROPOSITION 2.4. *Let $p : M \to B$ be a fibration and L a fiber non-degenerate almost Dirac structure, with associated connection Γ_L. Then the horizontal distribution* Hor *is contained in the characteristic distribution of L, and the pull-back of the natural 2-form yields a smooth 2-form $\omega_L \in \Omega^2(\mathrm{Hor})$.*

We will refer to ω_L as the **horizontal 2-form of** L.

PROOF. Fix a fiber non-degenerate almost Dirac structure L on the total space of a fibration $p : M \to B$. Relations (2.2) and (2.3) together show that, for each $X \in \mathrm{Hor}$, there exists a unique $\alpha \in \mathrm{Vert}^0$ such that $(X, \alpha) \in L$. One can then define a skew-symmetric bilinear form $\omega : \mathrm{Hor} \times \mathrm{Hor} \to \mathbb{R}$ by:

$$\omega_L(X_1, X_2) := \frac{1}{2}\left(\alpha_1(X_2) - \alpha_2(X_1)\right), \tag{2.4}$$

with $\alpha_1, \alpha_2 \in \mathrm{Vert}^0$ the unique elements such that $(X_1, \alpha_1), (X_2, \alpha_2) \in L$. Since L is maximal isotropic we have:

$$0 = 2\langle (X_1, \alpha_1), (X_2, \alpha_2)\rangle_+ = \alpha_1(X_2) - \alpha_2(X_1),$$

so this two form can also be written:

$$\omega_L(X_1, X_2) = \alpha_1(X_2) = -\alpha_2(X_1). \tag{2.5}$$

In this way we obtain a smooth 2-form $\omega_L \in \Omega^2(\mathrm{Hor})$.

From the definition (2.3) of Hor it is clear that the horizontal distribution Hor is contained in the characteristic distribution of L. From (2.5), it is clear that ω_L is the pull-back of the natural 2-form on the characteristic distribution of L. □

This construction of the horizontal 2-form can be dualized:

PROPOSITION 2.5. *Let $p : M \to B$ be a fibration and let L be a fiber non-degenerate almost Dirac structure, with associated connection Γ_L. For each fiber $i : F_x = p^{-1}(x) \hookrightarrow M$ the pull-back almost Dirac structure i^*L is well-defined and coincides with the graph of a bivector field $\pi_L \in \mathfrak{X}^2(\mathrm{Vert})$.*

We will refer to π_L as the **vertical bivector field of** L.

PROOF. First observe that the annihilator of the horizontal space is:

$$\mathrm{Hor}^0 = \{\alpha \in T^*M : \exists X \in \mathrm{Vert}, (X, \alpha) \in L\}. \tag{2.6}$$

Relations (2.2) and (2.6) together show that, for each $\alpha \in \mathrm{Hor}^0$, there exists a unique $X \in \mathrm{Vert}$ such that $(X, \alpha) \in L$. One can then define a skew-symmetric bilinear form $\pi_L : \mathrm{Hor}^0 \times \mathrm{Hor}^0 \to \mathbb{R}$ by:

$$\pi_L(\alpha_1, \alpha_2) := \frac{1}{2}\left(\alpha_1(X_2) - \alpha_2(X_1)\right), \tag{2.7}$$

with $X_1, X_2 \in \mathrm{Vert}$ the unique elements such that $(X_1, \alpha_1), (X_2, \alpha_2) \in L$. Since L is maximal isotropic we have:

$$0 = 2\langle (X_1, \alpha_1), (X_2, \alpha_2)\rangle = \alpha_1(X_2) - \alpha_2(X_1),$$

the form $\pi_L : \mathrm{Hor}^0 \times \mathrm{Hor}^0 \to \mathbb{R}$ can also be written:

$$\pi_L(\alpha_1, \alpha_2) = \alpha_1(X_2) = -\alpha_2(X_1). \tag{2.8}$$

Now we remark that the splitting $TM = \mathrm{Hor} \oplus \mathrm{Vert}$ allows us to identify $\mathrm{Hor}^0 = \mathrm{Vert}^*$, so π_L becomes a bivector field on the fibers of $p : M \to B$.

Let us fix a fiber $i : F_x = p^{-1}(x) \hookrightarrow M$. Then $TF_x = \mathrm{Vert}$ and we identify $T^*F_x = \mathrm{Vert}^* \simeq \mathrm{Hor}^0$. The pull-back Dirac structure i^*L is then given by:

$$\begin{aligned} i^*L &= \{(X, \alpha|_{\mathrm{Vert}}) \in \mathrm{Vert} \oplus \mathrm{Vert}^* : (X, \alpha) \in L\} \\ &= \big\{(X, \alpha) \in \mathrm{Vert} \oplus \mathrm{Hor}^0 : X = \pi_L(\alpha, \cdot)\big\} = \mathrm{graph}(\pi_L), \end{aligned}$$

where for the last inequality we have used (2.8). □

Putting together these results, we conclude that:

COROLLARY 2.6. *To a fiber non-degenerate almost Dirac structure L on a fibration $p : M \to B$ there is associated the following data:*

- *A connection Γ_L on $p : M \to B$.*
- *A horizontal 2-form $\omega_L \in \Omega^2(\mathrm{Hor})$.*
- *A vertical bivector field $\pi_L \in \mathfrak{X}^2(\mathrm{Vert})$.*

Conversely, every such triple (Γ, ω, π) on a fibration $p : M \to B$ determines a unique fiber non-degenerate almost Dirac structure L, which is given by:

$$L = \mathrm{graph}\,(\pi_L) \oplus \mathrm{graph}\,(\omega_L). \tag{2.9}$$

2.3. Fiber non-degenerate Dirac structures. The next natural question is: Given a fiber non-degenerate almost Dirac structure L on a fibration $p : M \to B$, what are the conditions on the associated triple $(\Gamma_L, \omega_L, \pi_L)$ that guarantee that L is integrable, i.e., is a Dirac structure?

Recall (see [**5**]) that the obstruction to integrability for an almost Dirac structure L is a 3-form $T_L \in \Omega^3(L)$, which is defined on sections $s_1, s_2, s_3 \in \Gamma(L)$ by:

$$T_L(s_1, s_2, s_3) := \langle [\![s_1, s_2]\!], s_3 \rangle_+ \tag{2.10}$$

where:

- $[\![\cdot, \cdot]\!]$ denotes the *Courant bracket*, on $\mathfrak{X}(M) \oplus \Omega^1(M)$, given by:

$$[\![(X, \alpha), (Y, \beta)]\!] := ([X, Y], \mathcal{L}_X\beta - \mathcal{L}_Y\alpha + \mathrm{d}\langle (X, \alpha), (Y, \beta)\rangle_-) \tag{2.11}$$

- $\langle \cdot, \cdot \rangle_+$ denotes the *natural pairing* on $\mathfrak{X}(M) \times \Omega^1(M)$, defined by:

$$\langle (X, \alpha), (Y, \beta)\rangle_+ := \frac{1}{2}\,(i_Y\alpha + i_X\beta)\,. \tag{2.12}$$

Our next result gives the 3-form T_L of a fiber non-degenerate almost Dirac structure L in terms of the geometric data $(\Gamma_L, \pi_L, \omega_L)$. We need to introduce some notation.

For a vector field $v \in \mathfrak{X}(B)$ we denote by $\widetilde{v} \in \mathfrak{X}(M)$ its horizontal lift, and we let $\pi_L^\# : \mathrm{Vert}^* \to \mathrm{Vert}$ denote the bundle map induced by the vertical bivector field π_L. Also, we have isomorphisms:

$$\mathrm{Hor} \simeq L \cap (TM \oplus \mathrm{Vert}^0), \quad \mathrm{Hor}^0 \simeq L \cap (\mathrm{Vert} \oplus T^*M).$$

These allow us to identify a horizontal vector field $X \in \Gamma(\mathrm{Hor})$ with a section $s_X = (X, \alpha) \in \Gamma(L)$, where $\alpha \in \Gamma(\mathrm{Vert}^0)$, and a vertical form $\beta \in \Gamma(\mathrm{Hor}^0)$ with a section $s_\beta = (Y, \beta) \in \Gamma(L)$, where $Y \in \Gamma(\mathrm{Vert})$. In this notation we have:

PROPOSITION 2.7. *Let $(\Gamma_L, \pi_L, \omega_L)$ be the geometric data determined by a fiber non-degenerate almost Dirac structure L on a fiber bundle $p: M \to B$. Then:*

(i) *If $\alpha, \beta, \gamma \in \Gamma(\mathrm{Hor}^0)$ then:*

$$T_L(s_\alpha, s_\beta, s_\gamma) = \frac{1}{2}[\pi_L, \pi_L](\alpha, \beta, \gamma).$$

(ii) *If $v \in \mathfrak{X}(B)$ and $\beta, \gamma \in \Gamma(\mathrm{Hor}^0)$, then:*

$$T_L(s_{\tilde{v}}, s_\beta, s_\gamma) = \frac{1}{2}\mathcal{L}_{\tilde{v}}\pi_L(\beta, \gamma).$$

(iii) *If $v_1, v_2 \in \mathfrak{X}(B)$ and $\gamma \in \Gamma(\mathrm{Hor}^0)$, then:*

$$T_L(s_{\tilde{v}_1}, s_{\tilde{v}_2}, s_\gamma) = \frac{1}{2}(\gamma(\Omega_{\Gamma_L}(\tilde{v}_1, \tilde{v}_2)) + \pi_L(\mathrm{d}i_{\tilde{v}_1} i_{\tilde{v}_2}\omega_L, \gamma))$$

where Ω_{Γ_L} is the curvature 2-form of Γ_L.

(iv) *If $v_1, v_2, v_3 \in \mathfrak{X}(B)$, then:*

$$T_L(s_{\tilde{v}_1}, s_{\tilde{v}_2}, s_{\tilde{v}_3}) = \frac{1}{2}\mathrm{d}_{\Gamma_L}\omega_L(\tilde{v}_1, \tilde{v}_2, \tilde{v}_3),$$

where $\mathrm{d}_{\Gamma_L}: \Omega^\bullet(\mathrm{Hor}) \to \Omega^{\bullet+1}(\mathrm{Hor})$ is the differential induced by Γ_L.

The proofs are routine calculations so we omit them. For a different approach to the geometric data we refer the reader to [**8**].

Now observe that sections of the form $s_{\tilde{v}}$ and s_α with $v \in \mathfrak{X}(B)$ and $\alpha \in \Gamma(\mathrm{Hor}^0)$ generate $\Gamma(L)$, as a $C^\infty(M)$-module. Therefore, as a corollary, we obtain the conditions that the triple $(\Gamma_L, \pi_L, \omega_L)$ must satisfy for the associated L to be a Dirac structure:

COROLLARY 2.8. *Let $(\Gamma_L, \pi_L, \omega_L)$ be the geometric data determined by a fiber non-degenerate almost Dirac structure L on a fiber bundle $p: M \to B$. Then L is Dirac iff the following conditions hold:*

(i) *π_L is a vertical Poisson structure: $[\pi_L, \pi_L] = 0$.*

(ii) *Parallel transport along Γ_L preserves the vertical Poisson structure π_L:*

$$\mathcal{L}_{\tilde{v}}\pi_L = 0, \quad \forall v \in \mathfrak{X}(B).$$

(iii) *The horizontal 2-form ω_L is closed: $\mathrm{d}_\Gamma \omega_L = 0$.*

(iv) *The following* **curvature identity** *is satisfied:*

$$\Omega_\Gamma(v_1, v_2) = \pi_L^\#(\mathrm{d}i_{\tilde{v}_1} i_{\tilde{v}_2}\omega_L), \quad \forall v_1, v_2 \in \mathfrak{X}(B). \tag{2.13}$$

The curvature identity (2.13) expresses the fact that the curvature 2-form of the connection associated with a fiber non-degenerate Dirac structure L takes values in the vertical Hamiltonian vector fields. We will explore this property later in our study of Poisson fibrations.

2.4. Presymplectic forms and Poisson structures. Let us illustrate the previous results with the two extreme cases of Dirac structures determined by Poisson and presymplectic structures.

If L is determined by a presymplectic form, one checks easily:

PROPOSITION 2.9. *Let Ω be a presymplectic form on the total space of a fibration $p: M \to B$. Then $L = \mathrm{graph}(\Omega)$ is a fiber non-degenerate Dirac structure iff the pull-back of Ω to each fiber is non-degenerate.*

In this case, the vertical Poisson structure π_L is non-degenerate on the fibers and coincides with the inverse of the restriction of Ω to the fibers.

The converse is also true: a fiber non-degenerate Dirac structure L for which the vertical Poisson structure π_L is non-degenerate on the fibers, is determined by a presymplectic form Ω. In fact, it follows from (2.9) that:

$$\Omega = \omega_L \oplus (\pi_L)^{-1}.$$

Hence, fiber non-degenerate presymplectic forms are presymplectic forms which restrict to symplectic forms on the fibers.

Dually, for Poisson structures, it is also immediate to check:

PROPOSITION 2.10. *Let Π denote a Poisson structure on the total space of a fibration $p : M \to B$. Then $L = \operatorname{graph}(\Pi)$ is a fiber non-degenerate Dirac structure iff Π is horizontal non-degenerate, i.e., $\Pi|_{\mathrm{Vert}^0} : \mathrm{Vert}^0 \times \mathrm{Vert}^0 \to \mathbb{R}$ is a non-degenerate bilinear form.*

In this case, the horizontal 2-form ω_L is non-degenerate: in fact, Π gives an isomorphism $\mathrm{Vert}^0 \to \mathrm{Hor}$, and under this isomorphism ω_L coincides with the restriction $\Pi|_{\mathrm{Vert}^0}$.

The converse is also true: a fiber non-degenerate Dirac structure L for which the horizontal 2-form ω_L is non-degenerate, is a Poisson structure Π. In fact, it follows from (2.9) that:

$$\Pi = (\omega_L)^{-1} \oplus \pi_L.$$

Hence, fiber non-degenerate Poisson structures are the same thing as the horizontal non-degenerate Poisson structures of Vorobjev ([**15**]).

3. Poisson fibrations

In this section we will study Poisson fibrations and their relationship to fiber non-degenerate Dirac structures.

3.1. Poisson and symplectic fibrations. Let (F, π) be a Poisson manifold. We denote by $\mathrm{Diff}_\pi(F)$ the group of Poisson diffeomorphisms of F. This is the subgroup of $\mathrm{Diff}(F)$ formed by all diffeomorphisms $\phi : F \to F$ such that:

$$(\phi)_*\pi = \pi.$$

We are interested in the following class of fibrations:

DEFINITION 3.1. A **Poisson fibration** $p : M \to B$ is a locally trivial fiber bundle, with fiber type a Poisson manifold (F, π) and with structure group a subgroup $G \subset \mathrm{Diff}_\pi(F)$. More precisely, p is a submersion such that there exists a covering $\{U_i\}_i$ of B, and trivializations $\phi_i : p^{-1}(U_i) \to U_i \times F$, with transition functions $\phi_j \circ \phi_i^{-1}|_{\{x\}\times F}$, $x \in \mathcal{U}_i \cap \mathcal{U}_j$ belonging to G. When π is symplectic the fibration is called a **symplectic fibration**.

If $p : M \to B$ is a Poisson fibration modeled on a Poisson manifold (F, π), each fiber F_b carries a natural Poisson structure π_b: if $\phi_i : p^{-1}(U_i) \to U_i \times F$ is a local trivialization, π_b is defined by:

$$\pi_b = (\phi_i(b)^{-1})_*\pi,$$

for $b \in U_i$. It follows from the definition that this 2-vector field is independent of the choice of trivialization. Note that the Poisson structures π_b on the fibers can be glued to a Poisson structure π_V on the total space of the fibration:

$$\pi_V(x) = \pi_{p(x)}(x), \quad (x \in M).$$

This 2-vector field is *vertical*: π_V takes values in $\wedge^2 \mathrm{Vert} \subset \wedge^2 TM$. In this way, the fibers (F_b, π_b) become Poisson submanifolds of (M, π_V).

EXAMPLE 3.2. An important class of Poisson fibrations is obtained as follows. Take any Poisson manifold (P, Π), fix a *closed* symplectic leaf B of P, and let $p : M \to B$ be a tubular neighborhood of B in P. Each fiber carries a natural Poisson structure, namely, the transverse Poisson structure ([**17**]). These transverse Poisson structures are all Poisson diffeomorphic and it follows from the Weinstein splitting theorem that this is a Poisson fibration.

More generally, one can take any Dirac manifold (P, L) and fix a closed presymplectic leaf B of P. If $p : M \to B$ is a tubular neighborhood of B in P, then each fiber carries a natural Poisson structure, called also the transverse Poisson structure ([**8**]). These transverse Poisson structures are all Poisson diffeomorphic and it follows from the generalization of the Weinstein splitting theorem in [**8**] that this is a Poisson fibration.

We saw in the previous section than any fiber non-degenerate Dirac structure L on the total space of a fibration $p : M \to B$ induces Poisson structures on the fibers. In fact, we have the following:

PROPOSITION 3.3. *Let $p : M \to B$ be a fibration with connected base and compact fibers. If L is a fiber non-degenerate Dirac structure, then $p : M \to B$ admits the structure of a Poisson fibration such that $\pi_V = \pi_L$.*

PROOF. Corollary 2.8 gives the vertical Poisson structure. While trivializations are obtained by parallel transport along paths in B, induced by the connection (F being any chosen fiber). Compactness of the fibers here just ensures completeness of horizontal lifts. □

Note that a Dirac structure on the total space of a fibration $p : M \to B$ for which the fibers are Poisson-Dirac submanifolds may fail to be a Poisson fibration. In other words, Proposition 3.3 becomes false if one omits the assumption of fiber non-degeneracy. This is illustrated by the following simple example.

EXAMPLE 3.4. Take the fibration $p : \mathbb{R}^3 \to \mathbb{R}$ obtained by projection on the x-axis and the Poisson bracket on $\mathbb{R}^3$ defined by:

$$\{x, y\} = \{x, z\} = 0, \quad \{y, z\} = x.$$

Then each fiber is a Poisson-Dirac submanifold: the fiber over $x = 0$ has the zero Poisson structure, while the fibers over $x \neq 0$ are symplectic. Since the fibers are not Poisson diffeomorphic, $p : \mathbb{R}^3 \to \mathbb{R}$ cannot be a Poisson fibration.

3.2. Coupling Dirac structures. Motivated by Proposition 3.3 we introduce the following definition:

DEFINITION 3.5. If $p : M \to B$ is a Poisson fibration, we will say that a fiber non-degenerate Dirac structure L is **compatible with the fibration** if $\pi_L = \pi_V$. In this case, we call L a **coupling Dirac structure**.

Note that in the special case of a symplectic fibration, by the results of Section 2.4, a coupling Dirac structure is necessarily given by a presymplectic form Ω. In this case, Proposition 3.3 is well-known (see [**13**], Lemma 6.2). Moreover, given a symplectic fibration $p : M \to B$, there are well-known non-trivial obstructions for the existence of a coupling 2-form $\omega \in \Omega^2(M)$. Our purpose now is to determine the corresponding obstructions for a general Poisson fibration and answer the following question:

- Given a Poisson fibration $p : M \to B$, is there a *coupling* Dirac structure L compatible with the fibration?

Let us recall the notion of a Poisson connection:

DEFINITION 3.6. A connection Γ on a Poisson fibration $p : M \to B$ is called a **Poisson connection** if, for every path γ, parallel transport

$$\phi_\gamma : (F_{\gamma(0)}, \pi_{\gamma(0)}) \to (F_{\gamma(1)}, \pi_{\gamma(1)})$$

is a Poisson diffeomorphism.

Clearly, a connection Γ on a Poisson fibration $p : M \to B$ is Poisson iff

$$\mathcal{L}_{\tilde{v}}\pi_V = 0, \forall v \in \mathfrak{X}(B).$$

Hence, by Corollary 2.8 (ii), an obvious necessary condition for the existence of a coupling Dirac structure is the existence of a Poisson connection. However, one can show that a Poisson fibration always admits such a connection. In fact, we have:

PROPOSITION 3.7. *Let $p : M \to B$ be a Poisson fibration. There exists a fiber non-degenerate almost Dirac structure L on $p : M \to B$ such that:*

(a) The vertical bivector field π_L coincides with π_V.
(b) The connection Γ_L is a Poisson connection.

PROOF. Let $p : M \to B$ be a Poisson fibration with fiber (F, π) and choose choose local trivializations $\phi_i : p^{-1}(U_i) \to U_i \times F$. Let L_i be the Dirac structure on $U_i \times F$ obtained by pull-back of $L_\pi = \text{graph}\,(\pi)$ under the projection $U_i \times F \to F$:

$$L_i := \{((v, w), (0, \eta)) \in T(U_i \times F) \oplus T^*(U_i \times F) : w = \pi(\eta, \cdot)\}.$$

Observe that to L_i it is associated the geometric data $(\Gamma_{L_i}, \pi_{L_i}, \omega_{L_i})$ where Γ_{L_i} is the canonical flat connection on $U_i \times F \to U_i$, $\pi_{L_i} = \pi$ and $\omega_{L_i} = 0$. Hence L_i is fiber non-degenerate, induces π on the fibers, and Γ_{L_i} is a Poisson connection.

Next, we choose a partition of unity $\rho_i : B \to \mathbb{R}$ subordinated to the cover $\{U_i\}$, and we define

$$L := \sum_i (\rho_i \circ p)\phi_i^* L_i.$$

By this we mean that the associated geometric data $(\Gamma_L, \pi_L, \omega_L)$ has connection $\Gamma_L = \sum_i (\rho_i \circ p)\phi_i^* \Gamma_{L_i}$, vertical bivector field $\pi_L = \sum_i (\rho_i \circ p)\phi_i^* \pi_{L_i}$ and horizontal 2-form $\omega_L = \sum_i (\rho_i \circ p)\phi_i^* \omega_{L_i} = 0$.

It is clear that L is a fiber non-degenerate almost Dirac structure. Moreover, π_L coincides with π_V. Finally, given $v \in \mathfrak{X}(B)$, if $\tilde{v}_i$ denotes the horizontal lift relative to $\phi_i^* \Gamma_{L_i}$, we have:

$$\mathcal{L}_{\tilde{v}_i}\pi_V = 0.$$

The horizontal lift of v relative to Γ_L is $\widetilde{v} = \sum_i (\rho_i \circ p)\widetilde{v}_i$. It follows that:

$$\begin{aligned}
\mathcal{L}_{\widetilde{v}}\pi_V &= [\widetilde{v}, \pi_V] \\
&= \sum_i [(\rho_i \circ p)\widetilde{v}_i, \pi_V] \\
&= \sum_i \Big((\rho_i \circ p)[\widetilde{v}_i, \pi_V] + \pi_V^{\#}(\mathrm{d}(\rho_i \circ p)) \Big) \\
&= \sum_i (\rho_i \circ p)\mathcal{L}_{\widetilde{v}_i}\pi_V = 0,
\end{aligned}$$

where we have used the fact that $\mathrm{d}(\rho_i \circ p) \in \mathrm{Vert}^0$. Hence L is the desired almost Dirac structure. □

There is also a procedure to construct coupling Dirac structures due to A. Wade ([**16**]), which is entirely analogous to a construction in symplectic geometry due to A. Weinstein (see [**13**, Theorem 6.17]):

THEOREM 3.8. *Let $G \times F \to F$ be a Hamiltonian action of a compact Lie group G on the Poisson manifold (F, π). Every connection on a principal G-bundle $P \to B$ determines a coupling Dirac structure L on the associated Poisson fibration $P \times_G F \to B$.*

PROOF. The connection Γ on P determines a projection $TP \to \mathrm{Vert}$, along the horizontal distribution Hor. Hence, there is an injection:

$$i_\Gamma : \mathrm{Vert}^* \hookrightarrow T^*P,$$

where Vert^* is the vertical cotangent bundle with typical fiber T^*P_b. Since Γ is G-invariant, the inclusion is G-equivariant. Therefore, the canonical symplectic form ω_{can} in T^*P induces a closed 2-form on Vert^*

$$\omega_\Gamma = i_\Gamma^*\omega_{\mathrm{can}},$$

which is G-invariant and restricts to the canonical symplectic form on the fibers T^*P_b. Also, the action of G on Vert^* has a moment map $\mu_P \circ i_\Gamma : \mathrm{Vert}^* \to \mathfrak{g}^*$, where μ_P is the moment map of the lifted cotangent action $G \times T^*P \to T^*P$.

Now let $G \times F \to F$ be a Hamiltonian action with moment map $\mu_F : F \to \mathfrak{g}^*$. This determines a Hamiltonian G-action on the Dirac manifold $M := \mathrm{Vert}^* \times F$, with Dirac structure

$$L = \mathrm{graph}\,(\omega_\Gamma) \oplus \mathrm{graph}\,(\pi).$$

and with moment map

$$\mu_M = (\mu_P \circ i_\Gamma) \oplus \mu_F.$$

This action is free and 0 is a regular value of μ_M. Therefore, the reduced space

$$M_{\mathrm{red}} := \mu_M^{-1}(0)/G \simeq P \times_G F,$$

carries a Dirac structure. It is easy to check that this Dirac structure is fiber non-degenerate and restricts to the canonical Poisson structures on the fibers. Hence, L is the desired coupling Dirac structure. □

REMARK 3.9. Note that the resulting coupling Dirac structure is presymplectic iff the fiber (F, π) is symplectic. Also, it is easy to check that the remaining geometric data associated with the coupling Dirac structure L in the theorem above is the following:

- The connection Γ_L is just the connection on the fiber bundle $P \times_G F$ induced from the connection Γ on P.
- The horizontal 2-form ω_L is given by the curvature of the connection composed with the moment map

$$\omega_L(\tilde{v}_1, \tilde{v}_2)([u,x]) = \langle \mu_F(x), F_\Gamma(\tilde{v}_1, \tilde{v}_2)_u \rangle, \quad ([u,x] \in P \times_G F).$$

Hence, the resulting coupling Dirac structure is Poisson iff the curvature 2-form of the connection is non-degenerate on the image of μ_F. Such a connection is sometimes called a *fat* connection. The proof of A. Wade in [**16**] consists in proving that this data satifies the conditions of Corollary 2.8 and so defines a coupling.

3.3. Obstruction to the existence of coupling. The construction of Theorem 3.8 can be extended for general Poisson fibrations, leading to a characterization of those fibrations which admit a coupling Dirac structure.

Let L be a fiber non-degenerate almost Dirac structure on a Poisson fibration $p : M \to B$. We will say that L is compatible with the fibration if $\pi_L = \pi_V$. By Proposition 3.7, any Poisson fibration admits a compatible L such that Γ_L is a Poisson connection. For L to be Dirac this connection must have a much more constrained holonomy:

THEOREM 3.10. *Let $p : M \to B$ be a Poisson fibration and let L be a compatible almost Dirac structure such that Γ_L is a Poisson connection. Then the following statements are equivalent:*

(i) L is a Dirac structure: $T_L = 0$.
(ii) For every base point $b \in B$, the action of the holonomy group $\Phi(b)$ of Γ_L on the fiber F_b is Hamiltonian.

A proof of this result will be given in the next paragraph, using a Poisson gauge theory. An immediate corollary is the following result (see, also, [**16**, Theorem 3.4]):

COROLLARY 3.11. *Let (F, π) be a compact Poisson manifold whose first Poisson cohomology group vanishes: $H^1_\pi(F) = 0$. Then any Poisson fibration $p : M \to B$ with fiber (F, π), and finite dimensional structure group admits a coupling Dirac structure.*

PROOF. Since $H^1_\pi(F) = 0$, the same holds for the fibers F_b, and it follows that any Poisson action on the fibers is Hamiltonian. Now apply Proposition 3.7 to construct a fiber non-degenerate almost Dirac structure L compatible with the fibration and such that Γ_L is a Poisson connection. Using the implication (ii) $\Rightarrow$ (i) in Theorem 3.10, we conclude that L is a coupling Dirac structure. Compactness of F ensures completeness of the connection. □

Notice that the condition that the holonomy group $\Phi(b)$ of Γ_L acts in a Hamiltonian fashion on the fiber F_b is a property of its connected component of the identity $\Phi(b)^0$. This connected component is known as the *restricted holonomy group* and is formed by the holonomy homomorphisms ϕ_γ, where γ is a contractible loop based at b.

In particular, each ϕ_γ with γ a contractible loop based at b, lies in the group of Hamiltonian diffeomorphisms $\mathrm{Ham}\,(F_b, \pi_b)$, which is known to be a normal subgroup of the group of Poisson diffeomorphisms $\mathrm{Diff}\,(F_b, \pi_b)$. The quotient group $\mathrm{Diff}\,(F_b, \pi_b)/\mathrm{Ham}\,(F_b, \pi_b)$ is known as the group of *outer Poisson diffeomorphisms.*

We conclude that a coupling Dirac structure L for a Poisson fibration $p: M \to B$ has an associated *coupling holonomy homomorphism*:

$$\phi: \pi_1(B,b) \to \mathrm{Diff}\,(F_b,\pi_b)/\mathrm{Ham}\,(F_b,\pi_b), \quad [\gamma] \mapsto [\phi_\gamma].$$

EXAMPLE 3.12. A tubular neighborhood $p: M \to B$ of a symplectic leaf B of a Poisson manifold (P,Π) (see Example 3.2) admits L_Π as a coupling Dirac structure. It follows that the connection Γ_Π is Poisson and has Hamiltonian holonomy around any contractible loop in B. This can also be proved directly using the Weinstein splitting theorem.

In general, the holonomy around a non-contractible loop will not be Hamiltonian and we will have a nontrivial homomorphism

$$\phi: \pi_1(B) \to \mathrm{Diff}\,(F,\pi)/\mathrm{Ham}\,(F,\pi).$$

This is precisely the (reduced) Poisson holonomy of the leaf B introduced in [**6**].

3.4. Poisson Gauge Theory. We now turn to the proof of Theorem 3.10. The idea will be to give an analogue of Theorem 3.8, but where the structure group is allowed to be infinite dimensional.

We consider a Poisson fibration $p: M \to B$ with fiber type a Poisson manifold (F,π). The structure group of this fibration is the group $G = \mathrm{Diff}\,(F,\pi)$ of Poisson diffeomorphisms. The corresponding principal G-bundle is the **Poisson frame bundle**:

$$P \to B$$

whose fiber over a point $b \in B$ is formed by all Poisson diffeomorphisms $u: F \to F_b$. The group G acts on (the right of) P by pre-composition:

$$P \times G \to P: \ (u,g) \mapsto u \circ g.$$

Then our original Poisson fiber bundle is canonically isomorphic to the the associated fiber bundle: $M = P \times_G F$.

Every Poisson connection Γ on the Poisson fiber bundle $p: M \to B$ is induced by a principal bundle connection on $P \to B$. To see this, observe that the tangent space $T_uP \subset C^\infty(u^*TM)$ at a point $u \in P$ is formed by the vector fields along u, $X(x) \in T_{u(x)}M$ such that:

$$d_{u(x)}p \cdot X(x) = \text{constant}, \quad \mathcal{L}_X\pi_V = 0.$$

The Lie algebra $\mathfrak{g}$ of G is the space of Poisson vector fields: $\mathfrak{g} = \mathfrak{X}(F,\pi)$. The infinitesimal action on P is given by:

$$\rho: \mathfrak{g} \to \mathfrak{X}(P), \quad \rho(X)_u = \mathrm{d}u \cdot X,$$

so the vertical space of P is:

$$\mathrm{Vert}_u = \{\mathrm{d}u \cdot X : X \in \mathfrak{X}(F,\pi)\}.$$

Now a Poisson connection Γ on $p: M \to B$ determines a connection in $P \to M$ whose horizontal space is:

$$\mathrm{Hor}_u = \{\widetilde{v} \circ u : v \in T_bB\},$$

where $u: F \to F_b$ and $\widetilde{v}: F_b \to T_{F_b}M$ denotes the horizontal lift of v. Clearly, this defines a principal bundle connection on $P \to B$, whose induced connection on the associated bundle $M = P \times_G F$ is the original Poisson connection Γ.

Fix a Poisson connection Γ on the Poisson fiber bundle $p: M \to B$. Recall that the holonomy group $\Phi(b)$ with base point $b \in B$ is the group of holonomy transformations $\phi_\gamma : F_b \to F_b$, where γ is a loop based at b. Clearly, we have $\Phi(b) \subset \operatorname{Diff}(F_b, \pi_b)$. On the other hand, for $u \in P$ we have the holonomy group $\Phi(u) \subset G = \operatorname{Diff}(F, \pi)$ of the corresponding connection in P which induces Γ: it consist of all elements $g \in G$ such that u and ug can be joined by a horizontal curve in P. Obviously, these two groups are isomorphic, for if $u : F \to F_b$ then:

$$\Phi(u) \to \Phi(b), \ g \mapsto u \circ g \circ u^{-1},$$

is an isomorphism.

The curvature of a principal bundle connection is a $\mathfrak{g}$-valued 2-form F_Γ on P which transforms as:

$$R_g^* F_\Gamma = \operatorname{Ad}(g^{-1}) \cdot F_\Gamma, \quad (g \in G).$$

Therefore, we can also think of the curvature as a 2-form Ω_L with values in the adjoint bundle $\mathfrak{g}_P := P \times_G \mathfrak{g}$. In the case of the Poisson frame bundle, the adjoint bundle has fiber over b the space $\mathfrak{X}(F_b, \pi_b)$ of Poisson vector fields on the fiber. Hence the curvature of our Poisson connection can be seen as a 2-form $\Omega_\Gamma : T_bB \times T_bB \to \mathfrak{X}(F_b, \pi_b)$. The two curvature connections are related by:

$$\Omega_\Gamma = \mathrm{d}u \circ F_\Gamma \circ u^{-1}. \tag{3.1}$$

Finally, it is easy to check that, in fact, we have:

$$\Omega_\Gamma(v_1, v_2) = [\widetilde{v}_1, \widetilde{v}_2] - \widetilde{[v_1, v_2]},$$

which is the expression we have used before for the curvature.

After these preliminarities, we can now proceed to the proof.

Proof of Theorem 3.10. We will prove the two implications separately.

(i) $\Rightarrow$ (ii). Let us start by observing that given any $u \in P$, a Poisson diffeomorphism $u : F \to F_b$, the curvature identity (2.13) together with (3.1) shows that, for any $v_1, v_2 \in T_bB$, the vector field $F_\Gamma(v_1, v_2)_u \in \mathfrak{g} = \mathfrak{X}(F, \pi)$ is Hamiltonian:

$$F_\Gamma(v_1, v_2)_u = \pi^\# \mathrm{d}(\omega_L(\widetilde{v}_1, \widetilde{v}_2) \circ u).$$

Now fix $u_0 \in P$. The Holonomy Theorem states that the Lie algebra of the holonomy group $\Phi(u_0)$ is generated by all values $F_\Gamma(v_1, v_2)_u$, with $u \in P$ any point that can be connected to u_0 by a horizontal curve. Hence, we can define a moment map $\mu_F : F \to (\operatorname{Lie}(\Phi(u_0))^*)$ for the action of $\Phi(u_0)$ on F by:

$$\langle \mu_F(x), F_\Gamma(v_1, v_2)_u \rangle = \omega_L(\widetilde{v}_1, \widetilde{v}_2)_{u(x)}. \tag{3.2}$$

This shows that the action of $\Phi(u_0)$ on (F, π) is Hamiltonian, and so (ii) holds (recall the comments above about the relationship between the holonomy groups $\Phi(b)$ and $\Phi(u)$).

(ii) $\Rightarrow$ (i). Again we fix $u_0 \in P$, and we assume now that the action of $\Phi(u_0)$ on (F, π) is Hamiltonian with moment map $\mu_F : F \to (\operatorname{Lie}(\Phi(u_0))^*)$. By the Reduction Theorem we can reduce the principal Poisson frame bundle to a principal $\Phi(u_0)$-bundle $P' \to B$. Now we can apply (the infinite dimensional version) of Theorem 3.8 to produce a coupling Dirac structure on the associated Poisson fiber bundle $p : M \to B$. Instead, if the reader does not like an infinite dimensional argument, he can check by himself that the geometric data formed by the connection Γ_L, the vertical Poisson vector field π_V and the 2-form ω_L defined from (3.2) (we are now given μ_F and we define ω_L) satisfy the conditions of Corollary 2.8. $\square$

REMARK 3.13. Note that our proof really shows that for *any* Poisson fibration the coupling Dirac structure arises as in the construction of Theorem 3.8. Relation (3.2) between the moment map μ_F, the curvature of the connection, and the horizontal 2-form ω_L, was already present there (see Remark 3.9).

4. Integration of Poisson fibrations

In this section, we study the integration of Poisson fibrations. Just as Poisson manifolds integrate to symplectic groupoids, we will see that Poisson fibrations integrate to fibered symplectic groupoids.

4.1. Fibered symplectic groupoids. Recall that for us a fibration always means a locally trivial fiber bundle. If we fix a base B, we have a category **Fib** of fibrations over B, where the objects are the fibrations $p : M \to B$ and the morphisms are the fiber preserving maps over the identity:

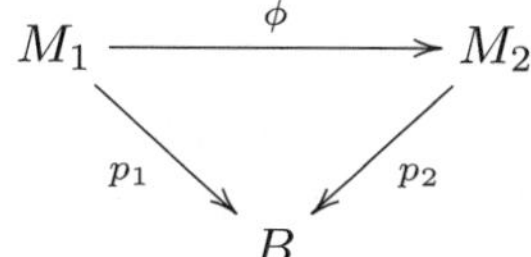

A *fibered groupoid* is an internal groupoid in **Fib**, i.e., an internal category where every morphism is an isomorphism. This means that both the total space $\mathcal{G}$ and the base M of a fibered groupoid are fibrations over B and all structure maps are fibered maps. For example, the source and target maps are fiber preserving maps over the identity:

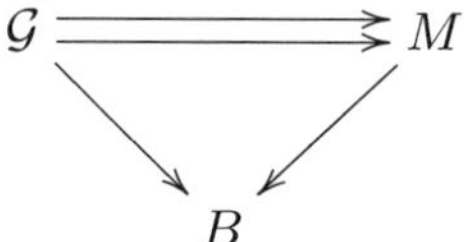

In particular, each fiber of $\mathcal{G} \to M$ is a groupoid over a fiber of $M \to B$. Moreover, the orbits of $\mathcal{G}$ lie inside the fibers of the base M.

A general procedure to construct fibered Lie groupoids is as follows. Let $P \to B$ be a principal G-bundle and assume that G acts on a groupoid $\mathcal{F} \rightrightarrows F$ by groupoid automorphisms. Then the associated fiber bundles $\mathcal{G} = P \times_G \mathcal{F}$ and $M = P \times_G F$ are the spaces of arrows and objects of a fibered Lie groupoid. Clearly, every fibered Lie groupoid is of this form provided we allow infinite dimensional structure groups. We will say that the fibered Lie groupoid $\mathcal{G}$ has *fiber type* the Lie groupoid $\mathcal{F}$.

DEFINITION 4.1. A **fibered symplectic groupoid** is a fibered Lie groupoid $\mathcal{G}$ whose fiber type is a symplectic groupoid $(\mathcal{F}, \omega)$.

Therefore, if $\mathcal{G}$ is a fibered symplectic groupoid over B, then $\mathcal{G} \to B$ is a symplectic fibration, and each symplectic fiber $\mathcal{F}_b$ is in fact a symplectic groupoid over the corresponding fiber F_b of $M \to B$.

PROPOSITION 4.2. *The base $M \to B$ of a fibered symplectic groupoid $\mathcal{G} \rightrightarrows M$ has a natural structure of a Poisson fibration.*

PROOF. Note that (i) the base of any symplectic groupoid has a natural Poisson structure for which the source (respectively, the target) is a Poisson (respectively, anti-Poisson) map, and (ii) any symplectic groupoid isomorphism between two symplectic groupoids covers a Poisson diffeomorphism of the base Poisson manifolds. Hence, each fiber of the base $M \to B$ of a fibered symplectic groupoid carries a natural Poisson structure, and a trivialization of the fibered symplectic groupoid covers a trivialization of $M \to B$ whose transition functions are Poisson diffeomorphisms of the fibers. Therefore the result follows. □

Note that the fibers $p : M \to B$ are integrable Poisson manifolds.

4.2. Integration of Poisson fibrations. We just saw that the base of a fibered symplectic groupoid is a Poisson fibration. We will say that a fibered symplectic groupoid $\mathcal{G} \to B$ *integrates* a Poisson fibration $p : M \to B$ whenever this fibration is (Poisson) isomorphic to the Poisson fibration determined by $\mathcal{G} \to B$. If such a fibered symplectic groupoid exists we say that the Poisson fibration is *integrable*. Note that the fiber type $\mathcal{F}$ of $\mathcal{G}$ is a symplectic groupoid integrating the fiber type (F, π) of $p : M \to B$.

THEOREM 4.3. *A Poisson fibration is integrable iff its fiber type is an integrable Poisson manifold. There exists a 1:1 correspondence between, source 1-connected, fibered symplectic groupoids and integrable Poisson fibrations.*

In one direction, the proof follows from Proposition 4.2. In the other direction, we will offer two proofs. The first proof is an heuristic proof that uses Poisson gauge theory. The second proof uses the approach to integrability through cotangent paths developed in [**3, 4**].

HEURISTIC PROOF VIA GAUGE THEORY. Given a Poisson fibration $p : M \to B$ with fiber type (F, π), we start by writing it as an associated fiber bundle:

$$M = P \times_G F$$

where P is the Poisson frame bundle and $G \subset \mathrm{Diff}(F, \pi)$ is the structure group of the fibration.

Since (F, π) is integrable, there exists a unique source 1-connected symplectic groupoid $\mathcal{F} \rightrightarrows F$ which integrates (F, π). The action of G on F lifts to an action of G on $\mathcal{F}$ by symplectic groupoid automorphisms (see [**7**]). Hence, we can form the associated bundle:

$$\mathcal{G} = P \times_G \mathcal{F}.$$

Since the action of G on $\mathcal{F}$ is by groupoid automorphisms, $\mathcal{G} \to B$ becomes a fibered groupoid over $M \to B$. Since this action is by symplectomorphisms, $\mathcal{G} \to B$ becomes a symplectic fibration. Since the groupoid structures and the symplectic structure on the fibers are compatible, $\mathcal{G}$ is a source 1-connected fibered symplectic groupoid with fiber type $\mathcal{F}$. It should be clear that the Poisson fibration determined by $\mathcal{G}$ is isomorphic to the original fibration. □

REMARK 4.4. Note that this heuristic proof becomes a real proof if the structure group G of the fibration is a finite dimensional Lie group. We will illustrate this below in Example 4.3.

PROOF OF THEOREM 4.3. Given a Poisson fibration $p : M \to B$ with fiber type (F, π), we denote by $\Sigma(M) \rightrightarrows M$ the symplectic groupoid that integrates the vertical Poisson structure π_V. Note that since we assume that (F, π) is integrable, we have that (M, π_V) is integrable, so that $\Sigma(M)$ is a Lie groupoid.

Let us recall (see [**3, 4**] for details and notations) that $\Sigma(M)$ is the space of equivalence classes of cotangent paths:

$$\Sigma(M) = \frac{\{a : [0,1] \to T^*M : \pi_V^\sharp(a(t)) = \frac{\mathrm{d}}{\mathrm{d}t}p(a(t))\}}{\{\text{cotangent homotopies}\}};$$

where $p : T^*M \to M$ is the cotangent bundle projection. Now we observe that $\mathrm{Vert}^0 \subset T^*M$ is a Lie subalgebroid, which is in fact, a bundle of Abelian Lie algebras. This is a direct consequence of π_V being vertical. Hence, the equivalence classes of cotangent paths with image in Vert^0 form a closed Lie subgroupoid $\mathcal{K} \subset \Sigma(M)$, which is in fact a bundle of Abelian Lie groups.

Let us consider the quotient Lie groupoid:

$$\mathcal{G} := \Sigma(M)/\mathcal{K}.$$

Notice that $\mathcal{G}$ is fibered over B, where the fiber over b is the symplectic groupoid $\Sigma(F_b)$ integrating the fiber (F_b, π_b). It is easy to check that $\mathcal{G}$ is, in fact, the desired source 1-connected, fibered symplectic groupoid integrating $p : M \to B$. We leave the details to the reader. □

As we have seen in Proposition 3.7, a Poisson fibration $p : M \to B$ always admits Poisson connections. What does the specification of a Poisson connection on $p : M \to B$ amounts to in the corresponding fibered symplectic groupoid $\mathcal{G}$?

PROPOSITION 4.5. *Let $M \to B$ be a Poisson fibration which integrates to a source 1-connected fibered symplectic groupoid $\mathcal{G} \to B$. The choice of a Poisson connection on the Poisson fibration $M \to B$ determines a coupling form Ω on the fibered symplectic groupoid $\mathcal{G} \to B$, and conversely.*

PROOF. We will give a "gauge theoretical" proof, which is valid at least in the case where the structure group is a finite dimensional Lie group. One can also give a longer proof using paths, which avoids this assumption.

Hence, assume that:

$$M = P \times_G F,$$

where P is a principal G-bundle and F is a Poisson G-space. As we have mentioned above, the Poisson action $G \times F \to F$ lifts to an action $G \times \mathcal{F} \to \mathcal{F}$ by automorphisms of the symplectic groupoid $\mathcal{F} = \Sigma(F)$, which is Hamiltonian with equivariant moment map $J : \mathcal{F} \to \mathfrak{g}^*$, which is a groupoid cocycle. We have that:

$$\mathcal{G} = P \times_G \mathcal{F}.$$

Now, Poisson connections Γ on $M \to B$ are in 1:1 correspondence with principal bundle connections on P.

To complete the proof we observe that, since the action $G \times \mathcal{F} \to \mathcal{F}$ is Hamiltonian, a choice of a principal bundle connection on P determines a coupling form on $\mathcal{G}$ and conversely (see Theorem 3.8 or [**13**, Chapter 6] for more details). □

The next natural question is: what does a coupling Dirac structure on the Poisson fibration $p : M \to B$ amounts to in the corresponding fibered symplectic groupoid $\mathcal{G}$? This question is more delicate, and it is intimately related with the pre-symplectic groupoids integrating Dirac structures described in [**2**]. This will be discussed elsewhere.

4.3. An Example. Let us denote by $\mathbb{S}^3 \to \mathbb{S}^2$ the Hopf fibration which we view as a principal $\mathbb{S}^1$-bundle $P \to \mathbb{S}^2$. We will consider as fiber types (F, π) the following two Poisson $\mathbb{S}^1$-manifolds:

1) The manifold $F = \mathbb{S}^2$, with the standard area form and the $\mathbb{S}^1$-action by rotations around the north-south poles axis;
2) The manifold $F = \mathfrak{su}(2)^* \simeq \mathbb{R}^3$, with its canonical linear Poisson structure and the $\mathbb{S}^1$-action by rotations around the z-axis;

The corresponding Poisson fibrations $M = P \times_{\mathbb{S}^1} F$ are:

1) the non-trivial $\mathbb{S}^2$-bundle $p : M \to \mathbb{S}^2$ (a symplectic fibration), and
2) the non-trivial rank 3 vector bundle $p : E \to \mathbb{S}^2$ (a Poisson fibration which is not symplectic).

The symplectic leaves of $\mathfrak{su}(2)^*$ are the concentric spheres around the origin and the origin itself. Hence the Poisson fibration $p : E \to \mathbb{S}^2$ is foliated by symplectic fibrations isomorphic to $p : M \to \mathbb{S}^2$ and the zero section. Since $\mathbb{S}^2$ is symplectic, we have:

$$H^1_\pi(\mathbb{S}^2) \simeq H^1(\mathbb{S}^2) = \{0\}.$$

Since $\mathfrak{su}(2)$ is semisimple of compact type, we also have:

$$H^1_\pi(\mathfrak{su}(2)^*) = \{0\}.$$

It follows from Corollary 3.11 that both $p : E \to \mathbb{S}^2$ and $p : M \to \mathbb{S}^2$ admit coupling Dirac structures. Of course, since $p : M \to \mathbb{S}^2$ is a symplectic fibration, its Dirac coupling is actually associated with a closed 2-form. We let the reader check that the presymplectic leaves of the Dirac coupling for $p : E \to \mathbb{S}^2$ are the symplectic fibrations isomorphic to $p : M \to \mathbb{S}^2$ (with their coupling forms) and the zero section (with the zero 2-form).

Let us now turn to the fibered symplectic groupoids integrating these fibrations. For that, we use the method in the heuristic proof of Theorem 4.3. Since the structure group is $\mathbb{S}^1$, a finite dimensional Lie group, this is allowed. We need the source 1-connected symplectic groupoid $\mathcal{F} = \Sigma(F)$ integrating the fiber type, and this is well-known in both examples:

1) Since $\mathbb{S}^2$ is symplectic and 1-connected, the associated source 1-connected symplectic groupoid is the pair groupoid $\Sigma(\mathbb{S}^2) = \mathbb{S}^2 \times \overline{\mathbb{S}^2}$, where the bar over the second factor means that we change the sign of symplectic form.
2) From general facts about linear Poisson structures, the symplectic groupoid of $\mathfrak{su}(2)^*$ is $\Sigma(\mathfrak{su}(2)^*) = T^*\mathrm{SU}(2)$, furnished with the canonical cotangent bundle symplectic structure. This groupoid is isomorphic to the action groupoid $\mathrm{SU}(2) \ltimes \mathfrak{su}(2)^*$ for the coadjoint action of $\mathrm{SU}(2)$ on $\mathfrak{su}(2)^*$.

Now we can describe, in both cases, the fibered symplectic groupoid $\mathcal{G} \to \mathbb{S}^2$ given by Theorem 4.3.

For the non-trivial $\mathbb{S}^2$-fibration $M \to \mathbb{S}^2$, the action of $\mathbb{S}^1$ on $\mathbb{S}^2$ lifts to the diagonal $\mathbb{S}^1$-action on $\mathbb{S}^2 \times \overline{\mathbb{S}^2}$, and we have:

$$\mathcal{G}(M) = P \times_{\mathbb{S}^1} (\mathbb{S}^2 \times \overline{\mathbb{S}^2}),$$

which is a non-trivial symplectic $(\mathbb{S}^2 \times \overline{\mathbb{S}^2})$-fibration over $\mathbb{S}^2$ and a groupoid over the non-trivial $\mathbb{S}^2$-fibration.

For the rank 3 vector bundle $E \to \mathbb{S}^2$, the action of $\mathbb{S}^1$ on $\mathfrak{su}(2)^*$ lifts to an action on $\mathrm{SU}(2) \times \mathfrak{su}(2)^*$, which is trivial on the first factor, and we have:

$$\mathcal{G}(E) = P \times_{\mathbb{S}^1} (\mathrm{SU}(2) \times \mathfrak{su}(2)^*) \simeq E \times \mathrm{SU}(2).$$

Note that, contrary to the case of the Poisson fibrations, the symplectic fibered groupoid $\mathcal{G}(M)$ does not sit naturally in $\mathcal{G}(E)$. This is because a Poisson submanifold does not always integrate to a symplectic subgroupoid, and this is exactly the case with the spheres in $\mathfrak{su}(2)^*$.

References

[1] H. Bursztyn, M. Crainic, Dirac structures, momentum maps, and quasi-Poisson manifolds, in *The breadth of symplectic and Poisson geometry* , 1–40, **Progr. Math.**, 232, Birkhser Boston, Boston, MA, 2005.

[2] H. Bursztyn, M. Crainic, A. Weinstein and C. Zhu, Integration of twisted Dirac brackets, *Duke Math. J.* **123** (2004), no.3, 549–607.

[3] M. Crainic and R. L. Fernandes, Integrability of Lie brackets, *Ann. of Math. (2)* **157** (2003), 575–620.

[4] M. Crainic and R. L. Fernandes, Integrability of Poisson brackets, *J. Differential Geometry* **66** (2004), 71–137.

[5] T. Courant, Dirac manifolds, *Trans. Amer. Math. Soc.* **319** (1990), no. 2, 631–661.

[6] R. L. Fernandes, Connections in Poisson Geometry I: Holonomy and Invariants, *J. Differential Geometry* **54** (2000), 303–166.

[7] R. L. Fernandes, J.P. Ortega and T. Ratiu, Momentum maps in Poisson geometry, *in preparation.*

[8] J.-P. Dufour and A. Wade, On the local structure of Dirac manifolds, *math.SG/0405257.*

[9] C. Ehresmann, Les connexions infinitésimales dans un espace fibré différentiable, *Séminaire Bourbaki*, Vol. 1, Exp. No. 24, 153–168, Soc. Math. France, Paris, 1995.

[10] M. Gotay, R. Lashof, J. Śniatycki and A. Weinstein, Closed forms on symplectic fiber bundles, *Comment. Math. Helv.* **58** (1983), 617–621.

[11] V. Guillemin, E. Lerman and S. Sternberg, *Symplectic fibrations and multiplicity diagrams*, Cambridge University Press, Cambridge, 1996.

[12] V. Guillemin and S. Sternberg, *Symplectic techniques in physics*, 2nd edition. Cambridge University Press, Cambridge, 1990.

[13] D. McDuff and D. Salamon, *Introduction to symplectic topology*, 2nd edition, Oxford Mathematical Monographs, Oxford University Press, New York, 1998.

[14] I. Vaisman, Foliation-coupling Dirac structures, *J. Geom. Phys.* **56** (2006), no. 6, 917–938..

[15] Y. Vorobjev, Coupling tensors and Poisson geometry near a single symplectic leaf,Banach Center Publ. **54** (2001), 249–274.

[16] A. Wade, Poisson fiber bundles and coupling Dirac structures, preprint *math.SG/0507594.*

[17] A. Weinstein, The local structure of Poisson manifolds, *J. Differential Geometry* **18** (1983), 523–557.

[18] A. Weinstein, Fat bundles and symplectic manifolds. *Adv. in Math.* **37** (1980), 239–250.

Depart. de Matemática, Instituto Superior Técnico, 1049-001 Lisboa, PORTUGAL
E-mail address: brahic@math.ist.utl.pt, rfern@math.ist.utl.pt

Contemporary Mathematics
Volume **450**, 2008

Generalized Kähler and hyper-Kähler quotients

Henrique Bursztyn, Gil R. Cavalcanti, and Marco Gualtieri

Abstract. We develop a theory of reduction for generalized Kähler and hyper-Kähler structures which uses the generalized Riemannian metric in an essential way, and which is not described with reference solely to a single generalized complex structure. We show that our construction specializes to the usual theory of Kähler and hyper-Kähler reduction, and it gives a way to view usual hyper-Kähler quotients in terms of generalized Kähler reduction.

Introduction

Generalized geometrical structures, such as Dirac structures [**4**] and generalized complex or Kähler structures [**5, 7**], are similar to their classical counterparts, i.e. integrable tangent distributions, complex or Kähler structures, except that they operate not on the tangent bundle but on the direct sum $TM \oplus T^*M$, or more generally, on an exact Courant algebroid E which is an extension of TM by T^*M.

In the presence of an action on the underlying manifold by a Lie group G, one may ask whether such generalized geometries *reduce* to a suitable quotient space. To address this question, we developed in [**1**] a theory of reduction for exact Courant algebroids and their associated generalized geometrical structures, based on the idea of *extending* the G-action on TM to an action on the Courant algebroid E. This is particularly interesting since E has symmetries beyond the usual diffeomorphisms; there are also the *B-field* transformations, well known to physicists. See [**10, 13, 18, 19**] for related work.

In this article, we provide a streamlined approach to the reduction procedures described in [**1**], focusing on the useful special case where the extended G-action is defined by a bracket-preserving map $\widetilde{\psi} : \mathfrak{g} \to \Gamma(E)$ and an equivariant map $\mu : M \to \mathfrak{h}^*$ for $\mathfrak{h}$ a $\mathfrak{g}$-module. The map μ is called a *moment map*, and is not associated with any geometrical structure but rather to the extended action itself. We phrase the constructions in this paper explicitly in terms of the data $(\widetilde{\psi}, \mathfrak{h}, \mu)$, showing that specific choices lead to well-known reductions.

The main construction in this paper is a reduction procedure for generalized complex structures which goes beyond the natural generalization of holomorphic or

1991 *Mathematics Subject Classification.* 53C15, 53C26, 53D20.
Key words and phrases. Courant algebroid, generalized complex, generalized Kähler, Hyperkähler quotient.

symplectic quotients developed in [**1**]. For this, we require a G-invariant generalized Riemannian metric $\mathcal{G}$, compatible with the generalized complex structure $\mathcal{J}$, and such that the Hermitian structure $(\mathcal{J}, \mathcal{G})$ is compatible with the G-action in a suitable sense. In particular, we obtain new reduction procedures for generalized Kähler and generalized hyper-Kähler structures.

Finally, we prove that this generalized Hermitian reduction does specialize to the usual Kähler and hyper-Kähler reductions, and that it commutes with the forgetful functors taking hyper-Kähler geometry to generalized Kähler geometry.

In a sequel to this work [**2**], we apply the theory developed here to certain infinite-dimensional quotients, including the generalized (hyper-) Kähler structure on the moduli space of instantons on a generalized (hyper-) Kähler 4-manifold first obtained by Hitchin [**8**] (see [**16**] for the hyper-Kähler case).

We would like to thank the organizers of the Poisson 2006 conference, in particular Giuseppe Dito and Yoshiaki Maeda, for their assistance and hospitality. H.B. thanks CNPq for financial support, and G.C. thanks EPSRC for financial support.

1. Generalized geometry and Courant algebroids

1.1. Geometry of $TM \oplus T^*M$. Given a manifold M of dimension m, the direct sum $TM \oplus T^*M$ is equipped with the following canonical structures: a fiberwise inner product of signature (m, m) given by

$$\langle X + \xi, Y + \eta \rangle := \eta(X) + \xi(Y), \quad X + \xi, Y + \eta \in \Gamma(TM \oplus T^*M), \tag{1.1}$$

and the *Courant bracket* [**4**] on smooth sections of $TM \oplus T^*M$, defined by

$$[\![X + \xi, Y + \eta]\!] := [X, Y] + \mathcal{L}_X \eta - i_Y d\xi. \tag{1.2}$$

A central idea in "generalized geometry" [**5, 7**] is that $TM \oplus T^*M$, together with the operations (1.1) and (1.2), should be thought of as a "generalized tangent bundle" to M. This leads to a unified view of various geometrical structures on M, as we now briefly recall.

A *Dirac structure* [**4**] on M is a Lagrangian subbundle $L \subset TM \oplus T^*M$ (i.e., $L = L^\perp$, where $L^\perp$ is the orthogonal complement of L with respect to the pairing (1.1)) whose space of sections is closed under the Courant bracket:

$$[\![\Gamma(L), \Gamma(L)]\!] \subseteq \Gamma(L). \tag{1.3}$$

Examples of Lagrangian subbundles of $TM \oplus T^*M$ include the graphs of 2-forms $\omega : TM \to T^*M$ and bivector fields $\pi : T^*M \to TM$. In these cases the integrability condition (1.3) amounts to $d\omega = 0$ and π being a Poisson bivector field; other examples can be found in [**4**]. All these definitions clearly carry over to the complexified bundle $(TM \oplus T^*M) \otimes \mathbb{C}$, leading to *complex Dirac structures.*

We now consider a special class of complex Dirac structures. A *generalized complex structure* [**5, 7**] on M is a complex structure $\mathcal{J}$ on the bundle $TM \oplus T^*M$ which is orthogonal with respect to (1.1), and such that the $+i$-eigenbundle $L \subset (TM \oplus T^*M) \otimes \mathbb{C}$ satisfies the integrability condition (1.3). The orthogonality of $\mathcal{J}$ implies that $L = L^\perp$, so L is a complex Dirac structure also satisfying

$$L \cap \overline{L} = \{0\}. \tag{1.4}$$

Conversely, any complex Dirac structure on M satisfying (1.4) uniquely determines a generalized complex structure on M. Examples of generalized complex structures include those determined by complex structures $I : TM \to TM$ and symplectic structures $\omega : TM \to T^*M$, namely

$$\mathcal{J}_I = \begin{pmatrix} -I & 0 \\ 0 & I^* \end{pmatrix}, \quad \mathcal{J}_\omega = \begin{pmatrix} 0 & -\omega^{-1} \\ \omega & 0 \end{pmatrix}.$$

One can similarly define generalized versions of Riemannian metrics, Hermitian structures, as well as Kähler and hyper-Kähler structures [**5**], see Sections 4.2 and 4.3.

Interpreting geometries on M as structures on $TM \oplus T^*M$ introduces two important features. First of all, one can easily adapt definitions in order to incorporate geometrical structures which are "twisted" by a closed 3-form on M: given $H \in \Omega^3_{cl}(M)$, one replaces the Courant bracket (1.2) by the *H-twisted Courant bracket* [**17**]

$$[\![X+\xi, Y+\eta]\!]_H = [X,Y] + \mathcal{L}_X\eta - i_Y d\xi + i_Y i_X H, \tag{1.5}$$

and changes the integrability condition (1.3) accordingly; this leads to *H-twisted Dirac structures*, *H-twisted generalized complex structures* and so on. Second, there is an action of the additive group of closed 2-forms $\Omega^2_{cl}(M)$ on $TM\oplus T^*M$ preserving the operations (1.1), (1.2), (1.5), given by

$$X + \xi \mapsto X + \xi + i_X B,$$

for $B \in \Omega^2_{cl}(M)$. As a result, structures on $TM\oplus T^*M$, such as generalized complex structures, inherit these extra symmetries, known as *B-field* transformations.

1.2. Exact Courant algebroids. An axiomatization of the properties of the Courant bracket on $TM \oplus T^*M$ leads to the general notion of a Courant algebroid, introduced in [**14**]. This more intrinsic approach to the Courant bracket will play an important role in the context of reduction.

A *Courant algebroid* over a manifold M is a vector bundle $E \to M$ equipped with a fibrewise nondegenerate symmetric bilinear form $\langle\cdot,\cdot\rangle$, a bilinear bracket $[\![\cdot,\cdot]\!]$ on the smooth sections $\Gamma(E)$, and a bundle map $\pi : E \to TM$ (called the *anchor*), such that, for all $e_1, e_2, e_3 \in \Gamma(E)$ and $f \in C^\infty(M)$, the following properties are satisfied:

C1) $[\![e_1,[\![e_2,e_3]\!]]\!] = [\![[\![e_1,e_2]\!],e_3]\!] + [\![e_2,[\![e_1,e_3]\!]]\!]$,
C2) $[\![e_1, fe_2]\!] = f[\![e_1,e_2]\!] + (\mathcal{L}_{\pi(e_1)}f)e_2$,
C3) $\mathcal{L}_{\pi(e_1)}\langle e_2,e_3\rangle = \langle[\![e_1,e_2]\!],e_3\rangle + \langle e_2,[\![e_1,e_3]\!]\rangle$,
C4) $\pi([\![e_1,e_2]\!]) = [\pi(e_1),\pi(e_2)]$,
C5) $[\![e_1,e_1]\!] = \frac{1}{2}\pi^* d\langle e_1,e_1\rangle$,

where in (C5) we identify $E \cong E^*$ via $\langle\cdot,\cdot\rangle$ in order to view π^* as taking values in E. The model example of a Courant algebroid is $TM \oplus T^*M$, with pairing (1.1), anchor given by the canonical projection $TM\oplus T^*M \to TM$, and bracket given by the H-twisted Courant bracket (1.5). It is straightforward to extend the concepts of Dirac structures, generalized complex structures etc. to general Courant algebroids.

In the definition of a Courant algebroid, properties C1)–C4) express natural compatibility conditions between the anchor π, the bracket $[\![\cdot,\cdot]\!]$ and the pairing $\langle\cdot,\cdot\rangle$ that will be further discussed in Section 2.1. Property C5), on the other hand, prevents the bracket $[\![\cdot,\cdot]\!]$ from being skew-symmetric, and it implies that $\pi\circ\pi^* = 0$, so we have a chain complex

$$0 \longrightarrow T^*M \xrightarrow{\pi^*} E \xrightarrow{\pi} TM \longrightarrow 0. \tag{1.6}$$

In this paper we will restrict our attention to *exact Courant algebroids*, i.e., those for which the sequence (1.6) is exact. In this case, we always identify T^*M with a subspace of E via π^*.

Given an exact Courant algebroid $E \to M$, we can always choose a right splitting $\nabla : TM \to E$ of (1.6) which is *isotropic*, i.e., whose image in E is isotropic with respect to $\langle\cdot,\cdot\rangle$. Each such ∇ defines a "curvature" 3-form $H \in \Omega^3_{cl}(M)$ by

$$H(X, Y, Z) := \langle [\![\nabla(X), \nabla(Y)]\!], \nabla(Z)\rangle, \quad \text{for } X, Y, Z \in \Gamma(TM). \tag{1.7}$$

Under the vector bundle isomorphism $\nabla + \pi^* : TM \oplus T^*M \to E$, the Courant algebroid structure on E is identified with the usual Courant algebroid structure on $TM \oplus T^*M$ defined by the H-twisted Courant bracket (1.5).

Remark: Exact Courant algebroids were first studied by P. Ševera, who classified them by noticing that the choice of a different isotropic splitting of (1.6) modifies H by an exact 3-form. As a result, the cohomology class $[H] \in H^3(M, \mathbb{R})$ is independent of the splitting and completely determines the exact Courant algebroid E up to isomorphism. We call $[H]$ the *Ševera class* of E.

2. Extended actions on Courant algebroids

In this section we briefly review the notion of *extended action*, introduced in [**1**] for the purpose of describing the reduction of Courant algebroids.

2.1. Infinitesimal actions. The action of a Lie group G on a manifold M may be described infinitesimally as a Lie algebra homomorphism $\mathfrak{g} \longrightarrow \Gamma(TM)$. The definition of an extended action on a Courant algebroid E is analogous, with E playing the role of TM.

Recall that an *infinitesimal automorphism* of a vector bundle E is a pair (F, X), where $X \in \Gamma(TM)$ and $F : \Gamma(E) \to \Gamma(E)$ satisfies

$$F(fe) = fF(e) + (\mathcal{L}_X f)e, \quad e \in \Gamma(E), f \in C^\infty(M). \tag{2.1}$$

Infinitesimal bundle automorphisms form a Lie algebra with respect to the bracket

$$[(F_1, X_1), (F_2, X_2)] := (F_1F_2 - F_2F_1, [X_1, X_2]).$$

If E is a Courant algebroid, then its Lie algebra of symmetries, denoted by $\mathbf{sym}(E)$, consists of infinitesimal bundle automorphism (F, X) which preserve the bracket $[\![\cdot,\cdot]\!]$, the pairing $\langle\cdot,\cdot\rangle$, and the anchor $\pi : E \to TM$:

$$\begin{aligned} F([\![e_1, e_2]\!]) &= [\![F(e_1), e_2]\!] + [\![e_1, F(e_2)]\!], \\ \mathcal{L}_X\langle e_1, e_2\rangle &= \langle F(e_1), e_2\rangle + \langle e_1, F(e_2)\rangle, \\ \pi \circ F &= \mathcal{L}_X \circ \pi, \end{aligned}$$

where $e_1, e_2 \in \Gamma(E)$. Given a section $e \in \Gamma(E)$, we observe that axioms C1)–C4) in the definition of a Courant algebroid imply that the pair $(F = [\![e, \cdot]\!], X = \pi(e))$ is in $\mathbf{sym}(E)$. As a result, we obtain a map:

$$\mathrm{ad} : \Gamma(E) \longrightarrow \mathbf{sym}(E), \quad e \mapsto ([\![e, \cdot]\!], \pi(e)). \tag{2.2}$$

The elements of $\mathbf{sym}(E)$ in the image of ad are called *inner symmetries* of E. Note that the map (2.2) extends the usual identification of vector fields $X \in \Gamma(TM)$ with infinitesimal symmetries of the Lie bracket on TM:

$$\Gamma(TM) \longrightarrow \mathbf{sym}(TM), \quad X \mapsto ([X, \cdot], X). \tag{2.3}$$

It is important to note, however, that although (2.3) is an isomorphism, the map (2.2) is neither injective nor surjective in general.

Given a Lie algebra $\mathfrak{g}$, an equivariant structure on E preserving its Courant algebroid structure is defined infinitesimally by a Lie algebra homomorphism $\mathfrak{g} \to \mathbf{sym}(E)$. The particular situation that will concern us in this paper is that of $\mathfrak{g}$-actions by inner symmetries, i.e. compositions

$$\mathfrak{g} \xrightarrow{\Psi} \Gamma(E) \xrightarrow{\mathrm{ad}} \mathbf{sym}(E). \tag{2.4}$$

Observe that, since the Courant bracket on $\Gamma(E)$ is *not* a Lie bracket, it is natural to replace the Lie algebra $\mathfrak{g}$ in (2.4) by a more general structure with "Courant-type" bracket. We call these *Courant algebras* [**1**] and describe them below.

2.2. Courant algebras. A *Courant algebra* over a Lie algebra $\mathfrak{g}$ is a Leibniz algebra $\mathfrak{a}$ [**15**] together with a bracket-preserving map $\pi : \mathfrak{a} \longrightarrow \mathfrak{g}$. In other words, $\mathfrak{a}$ is a vector space endowed with a bilinear bracket $[\![\cdot,\cdot]\!] : \mathfrak{a} \times \mathfrak{a} \longrightarrow \mathfrak{a}$ such that

$$[\![a_1, [\![a_2, a_3]\!]]\!] = [\![[\![a_1, a_2]\!], a_3]\!] + [\![a_2, [\![a_1, a_3]\!]]\!], \quad a_1, a_2, a_3 \in \mathfrak{a}, \tag{2.5}$$

and $\pi([\![a_1, a_2]\!]) = [\pi(a_1), \pi(a_2)]$ for all $a_1, a_2 \in \mathfrak{a}$. A Courant algebra is *exact* if π is surjective and $[\![w_1, w_2]\!] = 0$ for all $w_i \in \ker(\pi)$. In this paper we only consider exact Courant algebras. Morphisms of Courant algebras are defined in a natural way.

It is clear that if E is a Courant algebroid, then $\Gamma(E)$ is a Courant algebra over the Lie algebra of vector fields, and is exact if and only if E is.

If $\mathfrak{a} \longrightarrow \mathfrak{g}$ is an exact Courant algebra, then $\mathfrak{h} = \ker(\pi)$ automatically acquires a $\mathfrak{g}$-module structure: the action of $u \in \mathfrak{g}$ on $w \in \mathfrak{h}$ is

$$u \cdot w := [\![\tilde{u}, w]\!],$$

where $\tilde{u}$ is any element of $\mathfrak{a}$ such that $\pi(\tilde{u}) = u$. Since the bracket vanishes on $\mathfrak{h}$, it easily follows that this is a well defined map $\mathfrak{g} \times \mathfrak{h} \longrightarrow \mathfrak{h}$ defining a $\mathfrak{g}$-action.

The following example of an exact Courant algebra will be central in this paper.

EXAMPLE 2.1 (Hemisemidirect product [**1, 11**]). If $\mathfrak{g}$ is a Lie algebra and $\mathfrak{h}$ is a $\mathfrak{g}$-module, then we can endow $\mathfrak{a} := \mathfrak{g} \oplus \mathfrak{h}$ with the structure of an exact Courant algebra by taking $\pi : \mathfrak{g} \oplus \mathfrak{h} \longrightarrow \mathfrak{g}$ to be the natural projection and defining

$$[\![(u_1, w_1), (u_2, w_2)]\!] := ([u_1, u_2], u_1 \cdot w_2). \tag{2.6}$$

It is clear that $\pi : \mathfrak{a} \longrightarrow \mathfrak{g}$ is surjective and preserves brackets, and that $[\![\mathfrak{h}, \mathfrak{h}]\!] = 0$. Finally, condition (2.5) is a consequence of the Jacobi identity for $\mathfrak{g}$ and the fact that $\mathfrak{h}$ is a $\mathfrak{g}$-module. This Courant algebra first appeared in [**11**], where it was studied in the context of Leibniz algebras.

2.3. Extended actions. Let E be an exact Courant algebroid over M. We will now show how one can produce an infinitesimal action $\mathfrak{g} \to \mathbf{sym}(E)$ starting from an exact Courant algebra $\mathfrak{h} \to \mathfrak{a} \xrightarrow{\pi} \mathfrak{g}$, and a Courant algebra morphism

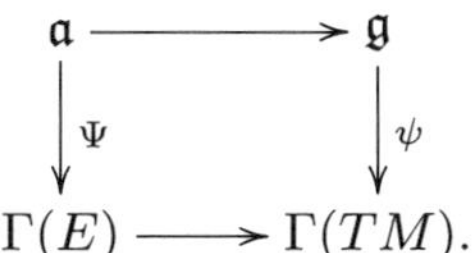

We will denote a Courant algebra morphism simply by $\Psi : \mathfrak{a} \to \Gamma(E)$, keeping in mind that it always projects to an action on M, denoted by $\psi : \mathfrak{g} \to \Gamma(TM)$. It also follows from the definitions that $\Psi(\mathfrak{h}) \subseteq \Omega^1(M)$.

Composing the map (2.2) with Ψ, we get a map

$$\mathrm{ad} \circ \Psi : \mathfrak{a} \longrightarrow \mathbf{sym}(E). \tag{2.7}$$

It is important to note that, unlike the map (2.3), the map $\mathrm{ad} : \Gamma(E) \to \mathbf{sym}(E)$ has a nontrivial kernel: by identifying E with $TM \oplus T^*M$ through the choice of an isotropic splitting, one can directly check that $\ker(\mathrm{ad}) = \Omega^1_{cl}(M)$. Hence if Ψ maps $\mathfrak{h}$ into closed 1-forms, $\Psi(\mathfrak{h}) \subseteq \Omega^1_{cl}(M)$, the map (2.7) factors through $\mathfrak{a} \xrightarrow{\pi} \mathfrak{g}$. As a result, there is an induced infinitesimal $\mathfrak{g}$-action

$$\mathfrak{g} \longrightarrow \mathbf{sym}(E), \tag{2.8}$$

as desired.

We are then led to the following definitions. Let G be a connected Lie group with Lie algebra $\mathfrak{g}$. An *extended* $\mathfrak{g}$*-action* on an exact Courant algebroid E over M is a Courant algebra morphism $\Psi : \mathfrak{a} \to \Gamma(E)$ from an exact Courant algebra $\mathfrak{h} \to \mathfrak{a} \xrightarrow{\pi} \mathfrak{g}$ into $\Gamma(E)$ so that $\Psi(\mathfrak{h}) \subseteq \Omega^1_{cl}(M)$. We call it an *extended* G*-action* if the induced $\mathfrak{g}$-action (2.8) on E integrates to a G-action. In particular, an extended G-action on E makes it into an equivariant G-bundle, with G acting by Courant algebroid automorphisms.

Remark: Suppose that $\Psi : \mathfrak{a} \to \Gamma(E)$ is an extended $\mathfrak{g}$-action for which the projected action $\psi : \mathfrak{g} \to \Gamma(TM)$ integrates to a global G-action on M. In this case, a sufficient condition ensuring that this data defines an extended G-action on E (and not only an extended action of a cover of G) is the existence of a $\mathfrak{g}$-invariant isotropic splitting for E (these always exist, e.g., if G is compact), see [**1**, Sec. 2]. Indeed, such a splitting gives an identification of E with the Courant algebroid $(TM \oplus T^*M, \langle\cdot,\cdot\rangle, [\![\cdot,\cdot]\!]_H)$ satisfying

$$i_{X_a} H = d\xi_a, \qquad \text{for all } a \in \mathfrak{a}, \tag{2.9}$$

where $\Psi(a) = X_a + \xi_a$. Since M is a G-manifold, $TM \oplus T^*M$ is naturally a G-equivariant bundle, and condition (2.9) exactly says that this canonical G-action on $TM \oplus T^*M$ coincides infinitesimally with the one induced by Ψ under the identification $E \cong TM \oplus T^*M$.

Any Lie algebra $\mathfrak{g}$ can be thought of as a Courant algebra over itself in a trivial way, with the projection $\pi : \mathfrak{g} \to \mathfrak{g}$ given by the identity. An extended action of this Courant algebra on an exact Courant algebroid E,

$$\begin{array}{ccc} \mathfrak{g} & \xrightarrow{\mathrm{Id}} & \mathfrak{g} \\ \downarrow{\scriptstyle \widetilde{\psi}} & & \downarrow{\scriptstyle \psi} \\ \Gamma(E) & \xrightarrow{\pi} & \Gamma(TM), \end{array}$$

is called a *lifted (or trivially extended) action* on E, and denoted by $\widetilde{\psi} : \mathfrak{g} \to \Gamma(E)$. A trivial example is of course when $E = TM \oplus T^*M$, with $H = 0$, and $\widetilde{\psi} = \psi$ is an ordinary action.

In this paper, we will be only concerned with lifted actions for which the image of $\widetilde{\psi} : \mathfrak{g} \to E$ is *isotropic*, i.e., the pairing $\langle\cdot,\cdot\rangle$ vanishes on elements $\widetilde{\psi}(u)$.

EXAMPLE 2.2 (Lifted actions). Take E to be the Courant algebroid $TM \oplus T^*M$, with $H = 0$, let $\psi : \mathfrak{g} \to \Gamma(TM)$ be an action on M, and $\nu : M \to \mathfrak{g}^*$ be an equivariant map. Then

$$\widetilde{\psi}(u) := \psi(u) + d\langle \nu, u\rangle, \quad u \in \mathfrak{g},$$

is a lifted action on E. This observation is a special case of the following general fact, proven in [**1**, Sec. 2]: if we start with an action ψ on M, and assume that E admits an invariant splitting, with associated 3-form curvature H, then the problem of finding an isotropic lifted action $\widetilde{\psi}$ extending ψ is equivalent to finding a closed equivariant extension of H in the Cartan model.

2.4. Moment maps for extended actions. Let $\mathfrak{h} \to \mathfrak{a} \to \mathfrak{g}$ be an exact Courant algebra, and let $\Psi : \mathfrak{a} \to \Gamma(E)$ be an extended $\mathfrak{g}$-action on an exact Courant algebroid E over M. Recall that this implies that $\Psi(\mathfrak{a}) \subset \Omega^1_{cl}(M)$, and that $\mathfrak{h}$ is a $\mathfrak{g}$-module. Let us equip $\mathfrak{h}^*$ with the dual $\mathfrak{g}$-action. A *moment map* for the extended action Ψ is a $\mathfrak{g}$-equivariant map $\mu : M \longrightarrow \mathfrak{h}^*$ such that, for each $w \in \mathfrak{h}$,

$$\Psi(w) = d\langle \mu, w\rangle.$$

We now describe how to use equivariant maps to further extend lifted actions:

PROPOSITION 2.1. *Let $\widetilde{\psi} : \mathfrak{g} \to \Gamma(E)$ be an isotropic lifted $\mathfrak{g}$-action on an exact Courant algebroid E, $\mathfrak{h}$ be a $\mathfrak{g}$-module, and $\mu : M \to \mathfrak{h}^*$ be an equivariant map. Then the map $\Psi : \mathfrak{g} \oplus \mathfrak{h} \to \Gamma(E)$,*

$$\Psi(u, w) = \widetilde{\psi}(u) + d\langle \mu, w\rangle, \tag{2.10}$$

defines an extended $\mathfrak{g}$-action of the hemisemidirect product $\mathfrak{a} = \mathfrak{g} \oplus \mathfrak{h}$ on E with moment map μ. Moreover, the image $\Psi(\mathfrak{a}) \subseteq E$ is isotropic over $\mu^{-1}(0)$.

PROOF. Let Ψ be defined as in (2.10). Then, using that $[\![\xi, \cdot]\!] = 0$ if $\xi \in \Omega^1_{cl}(M)$, we get

$$\begin{aligned}
[\![\Psi(u_1, w_1), \Psi(u_2, w_2)]\!] &= [\![\widetilde{\psi}(u_1), \widetilde{\psi}(u_2)]\!] + [\![\widetilde{\psi}(u_1), d\langle \mu, w_2\rangle]\!] \\
&= \widetilde{\psi}([u_1, u_2]) + \mathcal{L}_{\psi(u_1)} d\langle \mu, w_2\rangle \\
&= \Psi([u_1, u_2], u_1 \cdot w_2),
\end{aligned}$$

where for the last equality we used the equivariance of μ. Comparing with (2.6), we conclude that Ψ preserves brackets. So it defines a Courant algebra morphism. It is also clear that $\Psi(\mathfrak{h}) \subseteq \Omega^1_{cl}(M)$, hence Ψ is an extended $\mathfrak{g}$-action.

Let us now consider the pairing $\langle \Psi(u_1, w_1), \Psi(u_2, w_2)\rangle$ over $\mu^{-1}(0)$. Using that $\widetilde{\psi}$ is isotropic, i.e., $\langle \widetilde{\psi}(u_1), \widetilde{\psi}(u_2)\rangle = 0$, we get

$$\begin{aligned}
\langle \Psi(u_1, w_1), \Psi(u_2, w_2)\rangle &= \langle \widetilde{\psi}(u_1), d\langle \mu, w_2\rangle\rangle + \langle \widetilde{\psi}(u_2), d\langle \mu, w_1\rangle\rangle \\
&= \mathcal{L}_{\psi(u_1)}\langle \mu, w_2\rangle + \mathcal{L}_{\psi(u_2)}\langle \mu, w_1\rangle \\
&= \langle \mu, u_1 \cdot w_2\rangle + \langle \mu, u_2 \cdot w_1\rangle,
\end{aligned}$$

which clearly vanishes on points $x \in M$ where $\mu(x) = 0$. □

Remark: Note that if $\widetilde{\psi}$ is a lifted G-action for a Lie group G, then Ψ is an extended G-action.

3. Reduction of exact Courant algebroids

In this section, we review the reduction procedure for exact Courant algebroids introduced in [**1**], giving special attention to the case described in Prop. 2.1, i.e. a lifted action extended by an equivariant map.

3.1. The general procedure. Let G be a connected Lie group, $\mathfrak{h} \to \mathfrak{a} \to \mathfrak{g}$ be an exact Courant algebra and $\Psi : \mathfrak{a} \to \Gamma(E)$ be an extended G-action on an exact Courant algebroid E. We consider the distribution

$$K = \Psi(\mathfrak{a}) \subset E$$

given by the image of the bundle map $\mathfrak{a} \times M \to E$ associated with Ψ.

Our goal is to use the extended action Ψ to produce a "smaller" Courant algebroid. To this end, let us suppose that $P \hookrightarrow M$ is a G-invariant submanifold of M such that

(1) $K|_P$ is isotropic;
(2) $TP = \pi(K^{\perp})$.

One can directly check that $\pi(K^{\perp}) = \mathrm{Ann}(\Psi(\mathfrak{h}))$, so the last condition can be re-written as

$$TP = \mathrm{Ann}(\Psi(\mathfrak{h}))|_P. \tag{3.1}$$

Let us assume the following regularity conditions: $K|_P$ is a vector bundle over P and the G-action on P is free and proper. As observed in [**1**, Sec. 3.1], it follows from the fact that Ψ is a Courant algebra morphism that the vector bundles $K|_P$ and $K^{\perp}|_P$ are G-equivariant subbundles of $E|_P$. It is then proven in [**1**, Thm. 3.3] that the vector bundle

$$E^{red} := \frac{K^{\perp}|_P}{K|_P} \Big/ G \tag{3.2}$$

defines an exact Courant algebroid over the manifold P/G, called the *reduced Courant algebroid.*

Remarks:

a) As shown in [**1**], the assumption that $K|_P$ is isotropic is not necessary, and this is relevant for some examples. Note also that $K|_P$ is a vector bundle if and only if the distribution $\Psi(\mathfrak{h}) \subset E$ has constant rank over P.
b) The reduced Courant bracket on E^{red} is obtained canonically, by noticing that the restriction of the bracket on E to the space $\Gamma(K^{\perp})^G$ of G-invariant sections of $K^{\perp}$ is well-defined modulo $\Gamma(K)^G$, and $\Gamma(K)^G$ is an ideal in $\Gamma(K^{\perp})^G$.
c) Although the reduced Courant algebroid E^{red} is exact, it may not have a canonical splitting. Nevertheless, one can still describe the Ševera class of E^{red}, see [**1**, Sec. 3.2].

3.2. A special case of moment map reduction. Let E be an exact Courant algebroid over M. We will now specialize the reduction procedure for actions arising as in Prop. 2.1.

DEFINITION 3.1. Let $\widetilde{\psi} : \mathfrak{g} \to \Gamma(E)$ be an isotropic lifted G-action on E and let $\mu : M \to \mathfrak{h}^*$ be an equivariant map, for $\mathfrak{h}$ some $\mathfrak{g}$-module. Assume that 0 is a regular value of μ, and that the G-action on $\mu^{-1}(0)$ is free and proper. We refer to the triple $(\widetilde{\psi}, \mathfrak{h}, \mu)$ as *reduction data.*

It follows from Prop. 2.1 that reduction data define an extended G-action Ψ of the hemisemidirect product $\mathfrak{g} \oplus \mathfrak{h}$ on E, with image

$$K = \{\widetilde{\psi}(u) + d\langle\mu, w\rangle,\ u \in \mathfrak{g}, w \in \mathfrak{h}\} \subseteq E. \tag{3.3}$$

PROPOSITION 3.1. *Let E be an exact Courant algebroid over M, and let $(\widetilde{\psi}, \mathfrak{h}, \mu)$ be reduction data. Then $K|_{\mu^{-1}(0)}$ is an equivariant G-bundle over $\mu^{-1}(0)$, and the quotient vector bundle*

$$E^{red} := \frac{K^{\perp}|_{\mu^{-1}(0)}}{K|_{\mu^{-1}(0)}} \Big/ G$$

defines an exact Courant algebroid over $\mu^{-1}(0)/G$.

PROOF. Note that $\mathrm{Ann}(\Psi(\mathfrak{h}))|_{\mu^{-1}(0)} = \mathrm{Ann}(d\mu)|_{\mu^{-1}(0)} = T(\mu^{-1}(0))$, so (3.1) holds for $P = \mu^{-1}(0)$. The fact that $K|_{\mu^{-1}(0)}$ is isotropic follows from Prop. 2.1. Using that the G-action on $\mu^{-1}(0)$ is free and that $\Psi(\mathfrak{h})|_{\mu^{-1}(0)}$ is a bundle, one concludes that $K|_{\mu^{-1}(0)}$ is a vector bundle. The result now follows from the construction outlined in Section 3.1. □

EXAMPLE 3.2. Let $(TM \oplus T^*M, \langle\cdot,\cdot\rangle, [\![\cdot,\cdot]\!]_H)$ be the Courant algebroid E. If the closed 3-form H is basic with respect to a G-action on M, then we may regard this G-action as a lifted action on E, i.e., $\widetilde{\psi} = \psi : \mathfrak{g} \to \Gamma(TM) \subset \Gamma(E)$. In this case, for any choice of $\mathfrak{g}$-module $\mathfrak{h}$ and equivariant map $\mu : M \to \mathfrak{h}^*$, the reduced Courant algebroid is naturally split, as follows. Since $K = \psi(\mathfrak{g}) \oplus d\langle\mu, \mathfrak{h}\rangle$ and $K^{\perp} = T\mu^{-1}(0) \oplus \mathrm{Ann}(\psi(\mathfrak{g}))$, we have

$$E^{red} = \frac{T\mu^{-1}(0)}{\psi(\mathfrak{g})} \oplus \frac{\mathrm{Ann}(\psi(\mathfrak{g}))}{d\langle\mu, \mathfrak{h}\rangle} \Big/ G = TM^{red} \oplus T^*M^{red},$$

where $M^{red} = \mu^{-1}(0)/G$. The curvature 3-form of E^{red} with respect to this natural splitting is just the pushdown of the (basic) 3-form ι^*H, where $\iota : \mu^{-1}(0) \hookrightarrow M$ is the inclusion.

This example has two simple extreme cases: if we take $\mathfrak{h} = \{0\}$, then $E^{red} = T(M/G) \oplus T^*(M/G)$, whereas if we pick $\mathfrak{g} = \{0\}$, then $E^{red} = T\mu^{-1}(0) \oplus T^*\mu^{-1}(0)$.

Remarks:

a) In general, the reduced Courant algebroid E^{red} will not be canonically split, but one can construct splittings by choosing connections for the G-bundle $\mu^{-1}(0) \longrightarrow \mu^{-1}(0)/G$, see [**1**, Sec. 3] for a detailed discussion.
b) A description of the Ševera class of a reduced Courant algebroid is presented in [**1**, Sec. 3.2], including explicit examples where a trivial Ševera class is reduced to a non-trivial one.

4. Reduction of geometrical structures

In this section we explain how various geometrical structures can be transported to the reduced Courant algebroid in the presence of an extended action. For simplicity, we restrict our attention to the special case of extended actions determined by the reduction data $(\widetilde{\psi}, \mathfrak{h}, \mu)$, as in section 3.2.

4.1. Reduction of Dirac and generalized complex structures. As in Section 3.2, E is an exact Courant algebroid over M, $(\widetilde{\psi}, \mathfrak{h}, \mu)$ defines reduction data, K is given by (3.3), and the reduced Courant algebroid E^{red} over $M^{red} = \mu^{-1}(0)/G$ is given in Prop. 3.1.

Suppose that $L \subset E$ is a G-invariant Dirac structure, defining a G-equivariant subbundle of E. (Infinitesimally, this means that $[\![\widetilde{\psi}(u), \Gamma(L)]\!] \subseteq \Gamma(L)$.) Consider the following distribution of E^{red}:

$$L^{red} := \left. \frac{(L \cap K^{\perp} + K)|_{\mu^{-1}(0)}}{K|_{\mu^{-1}(0)}} \right/ G. \tag{4.1}$$

One can directly verify that $L^{red\perp} = L^{red}$. So, at each point, (4.1) defines a Lagrangian subspace of E^{red}. However, the distribution L^{red} may not form a smooth vector bundle over M^{red}. (A sufficient, but not necessary, condition is that $L \cap K|_{\mu^{-1}(0)}$ has constant rank.) If it does, then it is shown in [**1**, Sec. 4.1] that L^{red} is automatically integrable with respect to the reduced Courant bracket, and hence defines a Dirac structure in E^{red}. The same construction holds for complex Dirac structures if we replace K by its complexification $K \otimes \mathbb{C} \subseteq E \otimes \mathbb{C}$ in (4.1), yielding a reduced complex Dirac structure in $E^{red} \otimes \mathbb{C}$.

If now $\mathcal{J}$ is a G-invariant generalized complex structure on E, then its $+i$-eigenbundle L is a G-invariant complex Dirac structure in $E \otimes \mathbb{C}$. So one may attempt to transport $\mathcal{J}$ to the reduced Courant algebroid E^{red} by reducing this Dirac structure as in (4.1). Assuming that L^{red} is a smooth bundle over M^{red}, then it defines a reduced generalized complex structure $\mathcal{J}^{red}$ in E^{red} if and only if it satisfies the extra condition

$$L^{red} \cap \overline{L^{red}} = \{0\}. \tag{4.2}$$

As observed in [**1**, Sec. 5.1], this last condition can be equivalently expressed in terms of the operator $\mathcal{J}$ as follows: over each point in $\mu^{-1}(0)$, we must have

$$\mathcal{J}K \cap K^{\perp} \subset K. \tag{4.3}$$

A particularly simple condition implying both the smoothness of L^{red} and condition (4.3) is

$$\mathcal{J}K = K \quad \text{over } \mu^{-1}(0), \tag{4.4}$$

as discussed in [**1**, Sec. 5]. We summarize the discussion by citing the following result:

THEOREM 4.1 ([**1**], Thm. 5.2). *Let $\mathcal{J}$ be a generalized complex structure on the exact Courant algebroid E and let $(\widetilde{\psi}, \mathfrak{h}, \mu)$ be reduction data. If $\mathcal{J}$ is G-invariant and satisfies (4.4), then L^{red} defines a reduced generalized complex structure $\mathcal{J}^{red}$ on E^{red}.*

We now illustrate this reduction procedure with two simple examples, in which the Courant algebroid E is taken to be $TM \oplus T^*M$, with $H = 0$.

EXAMPLE 4.1 (Hamiltonian reduction). Consider a hamiltonian G-manifold (M, ω), with action $\psi : \mathfrak{g} \to \Gamma(TM)$ and moment map $\mu : M \to \mathfrak{g}^*$. We can describe Hamiltonian reduction as generalized reduction as follows: the triple $(\psi, \mathfrak{g}, \mu)$

defines reduction data, and, according to Example 3.2, the reduced Courant algebroid is $E^{red} = TM^{red} \oplus T^*M^{red}$, where $M^{red} = \mu^{-1}(0)/G$. Viewing ω as a generalized complex structure $\mathcal{J}_\omega$, it follows from the *moment map condition*

$$i_{\psi(u)}\omega = d\langle \mu, u\rangle, \quad u \in \mathfrak{g} \tag{4.5}$$

that the reduction data and the generalized complex structure are related by

$$\mathcal{J}_\omega d\langle \mu, u\rangle = \psi(u), \tag{4.6}$$

and this immediately implies condition (4.4). So we can carry out generalized reduction to obtain $\mathcal{J}_\omega^{red} = \mathcal{J}_{\omega_{red}}$, where ω_{red} is the reduced symplectic form on M^{red} obtained by Marsden-Weinstein reduction.

Remark: The compatibility (4.6) of the previous example can be generalized as follows: instead of ordinary actions ψ, one can consider more general lifted actions, for example those of the form $\widetilde{\psi}(u) = \psi(u) + d\langle \nu, u\rangle$, where $\nu : M \to \mathfrak{g}^*$ is equivariant (see Example 2.2); one can also consider more general $\mathfrak{g}$-modules $\mathfrak{h}$, and then impose the condition

$$\mathcal{J} d\langle \mu, u\rangle = \widetilde{\psi}(u) = \psi(u) + d\langle \nu, u\rangle,$$

which directly implies (4.4). If $\mathfrak{h} = \mathfrak{g}$, these are the actions studied in [**10, 13**] (where the map $\mu + i\nu$ is called the moment map). For the examples of interest in this paper, we will need $\mathfrak{g} \neq \mathfrak{h}$ and a weaker version of (4.4) to hold (c.f. Example 4.4).

EXAMPLE 4.2 (Holomorphic quotients). Suppose that a complex group (G, I_G) acts holomorphically on a complex manifold (M, I). We now consider the reduction data $(\psi, \mathfrak{h} = \{0\}, \mu = 0)$. In this case, the reduced Courant algebroid is $T(M/G) \oplus T^*(M/G)$. On the other hand, condition (4.4) is a direct consequence of the fact that the action is holomorphic: $\psi(I_G u) = I\psi(u)$, $u \in \mathfrak{g}$. Carrying out generalized reduction for $\mathcal{J}_I$, we obtain $\mathcal{J}_I^{red} = \mathcal{J}_{I^{red}}$, where I^{red} is the quotient complex structure on M/G.

The framework of reduction described in this section is general enough to include more exotic examples (see e.g. [**1**, Sec. 5.2]), such as the case of a lifted action preserving a symplectic structure whose reduction is a complex structure.

4.2. Generalized Hermitian reduction. We now describe a situation in which a generalized complex structure may have a natural reduction even when condition (4.4) fails to hold. While we state our results for extended actions defined by reduction data $(\widetilde{\psi}, \mathfrak{h}, \mu)$, this is not essential; more general extended actions may be treated in the same way.

A *generalized (Riemannian) metric* on a Courant algebroid E is an orthogonal, self-adjoint bundle automorphism $\mathcal{G} : E \longrightarrow E$ satisfying

$$\langle \mathcal{G}e, e\rangle > 0, \quad \forall e \neq 0.$$

A generalized metric is compatible with a generalized complex structure $\mathcal{J}$ if they commute. The pair $(\mathcal{J}, \mathcal{G})$ is then called a *generalized Hermitian structure.*

If $K \subset E$ and $\mathcal{G}$ is a generalized metric on E, then we define

$$K^{\mathcal{G}} := \mathcal{G}K^\perp \cap K^\perp, \tag{4.7}$$

which is the $\mathcal{G}$-orthogonal complement of K in $K^\perp$.

THEOREM 4.2 (Generalized Hermitian reduction). *Let E be an exact Courant algebroid over M, with reduction data $(\widetilde{\psi}, \mathfrak{h}, \mu)$. Suppose that E is equipped with a G-invariant generalized Hermitian structure $(\mathcal{J}, \mathcal{G})$. If*

$$\mathcal{J}K^{\mathcal{G}} = K^{\mathcal{G}} \quad over\ \mu^{-1}(0), \tag{4.8}$$

then $\mathcal{J}$ can be reduced to E^{red}, and $\mathcal{G}$ induces a compatible generalized metric on E^{red}.

PROOF. The proof follows closely the ideas in [**1**, Sec. 6.1]. Let us first notice that condition (4.8) implies (4.3): Since $\mathcal{J}$ is orthogonal, we see that

$$\langle \mathcal{J}K, K^{\mathcal{G}} \rangle = \langle K, K^{\mathcal{G}} \rangle = 0,$$

hence $\mathcal{J}K \subset K^{\mathcal{G}\perp}$ and $\mathcal{J}K \cap K^{\perp} \subset K^{\mathcal{G}\perp} \cap K^{\perp} = K$. This shows that the Dirac reduction L^{red} of the i-eigenbundle of $\mathcal{J}$ satisfies (4.2), hence it defines a reduced generalized complex structure as long as it is a smooth bundle, a fact which we now verify.

Using the $\mathcal{G}$-orthogonal decomposition $K^{\perp} = K \oplus K^{\mathcal{G}}$ over $\mu^{-1}(0)$, we obtain an identification of vector bundles

$$K^{\mathcal{G}}|_{\mu^{-1}(0)}/G \cong E^{red}.$$

Since $K^{\mathcal{G}}$ is $\mathcal{J}$ invariant, this identification induces a generalized almost complex structure $\mathcal{J}^{red}$ on E^{red}, whose $+i$-eigenbundle agrees with the reduced Dirac structure L^{red}. This implies that L^{red} is smooth, and hence integrable, and that $\mathcal{J}^{red}$ is the generalized complex structure associated to it.

Since $\mathcal{G}(K^{\mathcal{G}}) = K^{\mathcal{G}}$, we can also transport the generalized metric $\mathcal{G}$ to a generalized metric $\mathcal{G}^{red}$ on E^{red}, and since $\mathcal{G}$ and $\mathcal{J}$ commute pointwise, the same holds for their restrictions to $K^{\mathcal{G}}$. Thus the reduced metric and generalized complex structure are compatible. □

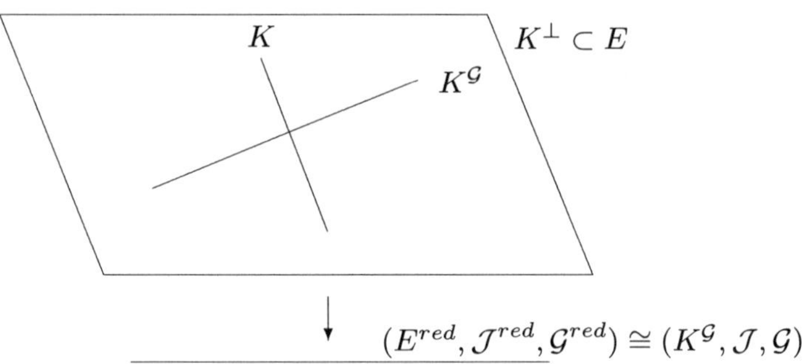

FIGURE 1. The generalized metric and complex structure on the reduced Courant algebroid are modeled on the $\mathcal{G}$-orthogonal complement of K inside $K^{\perp}$.

Intuitively, thinking of E as a generalized tangent bundle to M, the distribution K plays the role of the tangent distribution to the orbits of the extended action. The construction above can be interpreted as saying that the orthogonal complement of the generalized "G-orbit" is a model for the quotient; see Figure 1. Despite the clarity of condition (4.8), it is often easier to verify its orthogonal complement:

$$\mathcal{J}(K + \mathcal{G}K) = K + \mathcal{G}K. \tag{4.9}$$

For further details concerning the reduction of generalized metrics, see [**3**].

4.3. Generalized Kähler and hyper-Kähler reductions. A *generalized Kähler structure* is a generalized complex structure $\mathcal{J}$ together with a compatible generalized Riemannian metric $\mathcal{G}$ such that $\mathcal{J}' := \mathcal{J}\mathcal{G}$ is also a generalized complex structure.

A *generalized hyper-Kähler* structure is a triple of generalized complex structures, $\mathcal{J}_1, \mathcal{J}_2, \mathcal{J}_3$, each of which forms a generalized Kähler structure with the same generalized Riemannian metric $\mathcal{G}$, and such that

$$\mathcal{J}_1\mathcal{J}_2 = -\mathcal{J}_2\mathcal{J}_1 = \mathcal{J}_3.$$

Remark: Observe that given a generalized complex structure $\mathcal{J}$ with compatible metric $\mathcal{G}$, the product $\mathcal{J}\mathcal{G}$ is always a generalized almost complex structure, i.e., $(\mathcal{J}\mathcal{G})^2 = -\mathrm{Id}$. The nontrivial requirement is the integrability of this structure. Note also that $\mathcal{J}\mathcal{G}$ is compatible with $\mathcal{G}$.

We may now apply Theorem 4.2 to obtain generalized Kähler and generalized hyper-Kähler reductions. As before, let $(\widetilde{\psi}, \mathfrak{h}, \mu)$ be reduction data for the exact Courant algebroid E.

THEOREM 4.3 (Generalized Kähler reduction). *Suppose that $(\mathcal{J}, \mathcal{G})$ is a G-invariant generalized Kähler structure on E. If $\mathcal{J}K^{\mathcal{G}} = K^{\mathcal{G}}$ over $\mu^{-1}(0)$, then the generalized Kähler structure $(\mathcal{J}, \mathcal{G})$ reduces to the Courant algebroid E^{red} over $\mu^{-1}(0)/G$.*

PROOF. Letting $\mathcal{J}' = \mathcal{J}\mathcal{G}$, it is clear that $\mathcal{J}'$ is also G-invariant and that both $(\mathcal{J}, \mathcal{G})$ and $(\mathcal{J}', \mathcal{G})$ satisfy the conditions of Theorem 4.2. Hence they can be reduced to E^{red}, and the reduced metric $\mathcal{G}^{red}$ is compatible with both reduced generalized complex structures. Since the reduced structures are identified with the restrictions of $\mathcal{J}, \mathcal{J}'$ and $\mathcal{G}$ to $K^{\mathcal{G}}$, the fact that $\mathcal{J}' = \mathcal{J}\mathcal{G}$ on $K^{\mathcal{G}}$ implies that the same holds in E^{red}. Hence $(\mathcal{J}^{red}, \mathcal{G}^{red})$ is a generalized Kähler structure. □

We have the following analogous result for generalized hyper-Kähler structures:

THEOREM 4.4 (Generalized hyper-Kähler reduction). *Suppose that $(\mathcal{J}_1, \mathcal{J}_2, \mathcal{J}_3, \mathcal{G})$ is a G-invariant generalized hyper-Kähler structure on E. If $\mathcal{J}_1 K^{\mathcal{G}} = \mathcal{J}_2 K^{\mathcal{G}} = \mathcal{J}_3 K^{\mathcal{G}} = K^{\mathcal{G}}$ over $\mu^{-1}(0)$, then the generalized hyper-Kähler structure $(\mathcal{J}_1, \mathcal{J}_2, \mathcal{J}_3, \mathcal{G})$ reduces to a generalized hyper-Kähler structure in E^{red} over $\mu^{-1}(0)/G$.*

4.4. Examples. We now describe how one recovers the classical Kähler and hyper-Kähler quotients [**6, 9, 12**] using our methods. As a final example we show that the classical hyper-Kähler quotient may be viewed as a non-trivial example of a generalized Kähler quotient.

For the examples in this section, the Courant algebroid in question is $E = TM \oplus T^*M$, with $H = 0$, and with reduction data $(\psi, \mathfrak{h}, \mu)$, where the lifted action ψ is an ordinary action.

Remark: In a separate paper [**2**], we describe examples of generalized Kähler and hyper-Kähler quotients involving non-vanishing twists H and non-trivial lifted actions, such as the construction of generalized Kähler and hyper-Kähler structures on certain moduli spaces of instantons and Lie groups.

EXAMPLE 4.3 (Kähler reduction). Let (I,ω) be a Kähler structure on M, preserved by a Hamiltonian G-action $\psi : \mathfrak{g} \to \Gamma(TM)$, with moment map $\mu : M \to \mathfrak{g}^*$. Let $g = \omega I$ be its associated Kähler metric.

We then view the Kähler structure as a generalized Kähler structure $(\mathcal{J}_\omega, \mathcal{G})$, where

$$\mathcal{G} = \begin{pmatrix} 0 & g^{-1} \\ g & 0 \end{pmatrix}, \tag{4.10}$$

and carry out its reduction using the reduction data $(\psi, \mathfrak{g}, \mu)$. The reduced Courant algebroid, as described in Example 3.2, is simply $TM^{red} \oplus T^*M^{red}$, where $M^{red} = \mu^{-1}(0)/G$.

As in Example 4.1, the moment map condition (4.5) for ω immediately implies that $\mathcal{J}_\omega K = K$. Hence $K^{\mathcal{G}} = \mathcal{J}_I \mathcal{J}_\omega K^\perp \cap K^\perp = \mathcal{J}_I K^\perp \cap K^\perp$ and therefore

$$\mathcal{J}_\omega K^{\mathcal{G}} = \mathcal{J}_\omega(\mathcal{J}_I K^\perp \cap K^\perp) = \mathcal{J}_I K^\perp \cap K^\perp = K^{\mathcal{G}}.$$

So we can apply Theorem 4.3 and reduce the Kähler structure as a generalized Kähler structure. We now check that the reduced structure $\mathcal{J}_I^{red}$ and $\mathcal{J}_\omega^{red}$ agree with the generalized structures associated with the usual reduced Kähler structure on M^{red}. First, as observed in Example 4.1, the generalized reduction of the symplectic structure is just the usual symplectic reduction. Let us now discuss the reduction of $\mathcal{J}_I$. We know that, at each point of M^{red}, $\mathcal{J}_I^{red}$ is described by its restriction to

$$K^{\mathcal{G}} = \{X + \xi \in TP \oplus T^*P : X \perp \psi(\mathfrak{g}) \text{ and } \xi(\psi(\mathfrak{g})) = 0\},$$

where $P = \mu^{-1}(0)$. We have the identifications

$$\{X \in TP : X \perp \psi(\mathfrak{g})\} = TM^{red} \quad \text{and} \quad \{\xi \in T^*P : \xi(\psi(\mathfrak{g})) = 0\} = T^*M^{red},$$

and the space on the left is invariant under $\mathcal{J}_I$. It follows that $\mathcal{J}_I^{red}$ preserves TM^{red}, and hence it is of complex type. One sees from this description that $\mathcal{J}_I^{red}$ agrees with the generalized complex structure associated with the complex structure obtained by the usual Kähler reduction procedure [**6, 12**].

EXAMPLE 4.4 (hyper-Kähler reduction). Let (I_1, I_2, I_3, g) be a hyper-Kähler structure on M preserved by a Hamiltonian G-action $\psi : \mathfrak{g} \to \Gamma(TM)$, in the sense that there exist moment maps $\mu_1, \mu_2, \mu_3 \in C^\infty(M, \mathfrak{g}^*)^G$ satisfying

$$i_{\psi(u)}\omega_j = d\langle \mu_j, u\rangle, \quad \forall u \in \mathfrak{g}, \text{ and } j = 1,2,3, \tag{4.11}$$

where $\omega_j = gI_j$ are the Kähler forms.

We consider the generalized complex structures $\mathcal{J}_j$ associated with the complex structures I_j, $j = 1,2,3$, and the generalized metric $\mathcal{G}$ as in (4.10). Then $(\mathcal{J}_1, \mathcal{J}_2, \mathcal{J}_3, \mathcal{G})$ defines a generalized hyper-Kähler structure, and we consider the reduction data $(\psi, \mathfrak{h}, \mu)$, where now

$$\mathfrak{h} = \mathfrak{g} \oplus \mathfrak{g} \oplus \mathfrak{g}, \quad \text{and} \quad \mu = (\mu_1, \mu_2, \mu_3) : M \to \mathfrak{h}^*.$$

It follows that

$$K = \{\psi(u) + d\langle \mu_1, w_1\rangle + d\langle \mu_2, w_2\rangle + d\langle \mu_3, w_3\rangle,\ u \in \mathfrak{g},\ \ w_1, w_2, w_3 \in \mathfrak{h}\}.$$

In order to apply Theorem 4.4, we must check that

$$\mathcal{J}_j(K+\mathcal{G}K) = K+\mathcal{G}K \tag{4.12}$$

over $\mu^{-1}(0)$, $j=1,2,3$. Using (4.11), we see that

$$K+\mathcal{G}K = \psi(\mathfrak{g})+\omega_1(\psi(\mathfrak{g}))+\omega_2(\psi(\mathfrak{g}))+\omega_3(\psi(\mathfrak{g}))+g(\psi(\mathfrak{g}))+I_1\psi(\mathfrak{g})+I_2\psi(\mathfrak{g})+I_3\psi(\mathfrak{g}).$$

It now follows from the relations $I_1I_2 = I_3$, $I_1\omega_2 = -\omega_3$ and $I_1\omega_1 = -g$ that (4.12) holds. Hence we can reduce the hyper Kähler-structure to $M^{red} = \mu^{-1}(0)/G$ as a generalized hyper-Kähler structure.

As in Example 4.3, we have the identification

(4.13)
$$K^{\mathcal{G}} = \{X \in TP : X \perp \psi(\mathfrak{g})\} \oplus \{\xi \in T^*P : \xi(\psi(\mathfrak{g})) = 0\} = TM^{red} \oplus T^*M^{red},$$

for $P = \mu^{-1}(0)$. Since $K^{\mathcal{G}}$ is invariant under $\mathcal{J}_1$, $\mathcal{J}_2$ and $\mathcal{J}_3$ and these structures are of complex type, it follows that the space

$$\{X \in TP : X \perp \psi(\mathfrak{g})\} \cong TM^{red},$$

is also invariant by these structures. Hence $\mathcal{J}_1^{red}, \mathcal{J}_2^{red}$ and $\mathcal{J}_3^{red}$ are complex structures and the generalized hyper-Kähler structure obtained in the reduced manifold is precisely the usual hyper-Kähler reduction of M from [**9**].

In [**5**], it was shown that a generalized Kähler structure $(\mathcal{J},\mathcal{G})$ on $E = TM \oplus T^*M$ with $H=0$ determines and is uniquely determined by a quadruple (I_+, I_-, g, b), where g is a Riemannian metric on M, I_+ and I_- are Hermitian complex structures (hence defining a *bihermitian* structure), and b is a 2-form such that

$$d^c_-\omega_- = -d^c_+\omega_+ = db, \tag{4.14}$$

where $\omega_\pm = gI_\pm$ and $d^c = i(\overline{\partial}-\partial)$ is defined by the appropriate complex structure.

The bihermitian structure is obtained as follows (see [**5**] for details). Since $\mathcal{G}^2 = \mathrm{Id}$, we can write $E = C_+ \oplus C_-$, where $C_\pm$ is the ±1-eigenbundle of $\mathcal{G}$. The spaces C_+ and C_- intersect T^*M trivially, so the projection $\pi : E \to TM$ induces identifications $C_\pm \cong TM$. The metric g on M is induced by the restriction of $\langle\cdot,\cdot\rangle$ to C_+, whereas $J_\pm$ come from the restrictions of $\mathcal{J}$ to $C_\pm$. Conversely, the generalized Kähler structure may be written in terms of (I_+, I_-, g, b) as follows.

$$\mathcal{J} = \frac{1}{2}\begin{pmatrix} 1 & \\ b & 1\end{pmatrix}\begin{pmatrix} I_+ + I_- & -(\omega_+^{-1} - \omega_-^{-1}) \\ \omega_+ - \omega_- & -(I_+^* + I_-^*)\end{pmatrix}\begin{pmatrix} 1 & \\ -b & 1\end{pmatrix},$$
$$G = \begin{pmatrix} 1 & \\ b & 1\end{pmatrix}\begin{pmatrix} & g^{-1} \\ g & \end{pmatrix}\begin{pmatrix} 1 & \\ -b & 1\end{pmatrix}.$$

It follows from this bihermitian interpretation of generalized Kähler geometry that any hyper-Kähler structure (g, I_1, I_2, I_3) determines a generalized Kähler structure, by simply choosing $I_+ = I_1$ and $I_- = I_2$, for example. We now show that this "forgetful functor" commutes with reduction, i.e. intertwines the notion of generalized Kähler reduction from Theorem 4.3 with the usual hyper-Kähler quotient procedure.

EXAMPLE 4.5 (Hyper-Kähler reduction versus generalized Kähler reduction). A hyper-Kähler structure (g, I_1, I_2, I_3) defines a bihermitian structure (g, I_1, I_2) satisfying (4.14) for $b=0$, hence it defines a generalized Kähler structure $(\mathcal{G},\mathcal{J})$

as above. The generalized complex structure $\mathcal{J}$ may be described as a bundle automorphism of $E = C_+ \oplus C_-$ as follows: on C_+,

$$\mathcal{J}(X + g(X)) = I_1 X + g(I_1 X),$$

and on C_-,

$$\mathcal{J}(X - g(X)) = I_2 X - g(I_2 X).$$

Given a Hamiltonian G-action $\psi : \mathfrak{g} \longrightarrow \Gamma(TM)$ preserving the hyper-Kähler structure, and with moment maps $\mu_j : M \longrightarrow \mathfrak{g}^*$, $j = 1, 2, 3$, we consider the same reduction data as in Example 4.4,

$$(\psi, \mathfrak{h} = \mathfrak{g} \oplus \mathfrak{g} \oplus \mathfrak{g}, \mu = (\mu_1, \mu_2, \mu_3)),$$

but we will now use it to reduce the generalized Kähler structure $(\mathcal{G}, \mathcal{J})$, rather than the original hyper-Kähler structure.

Following Example 4.4, we know that $K^{\mathcal{G}}$ is given by (4.13), and each summand in that expression is invariant under I_j, $j = 1, 2, 3$. Note that we can write

$$K^{\mathcal{G}} = (K^{\mathcal{G}} \cap C_+) \oplus (K^{\mathcal{G}} \cap C_-).$$

If $X + g(X) \in K^{\mathcal{G}} \cap C_+$, then $I_1 X - I_1^* g(X) = I_1 X + g(I_1 X) \in K^{\mathcal{G}}$, and, similarly, $X - g(X) \in K^{\mathcal{G}} \cap C_-$ implies that $I_2 X + I_2^* g(X) = I_2 X - g(I_2 X) \in K^{\mathcal{G}}$. Hence $\mathcal{J} K^{\mathcal{G}} = K^{\mathcal{G}}$ and, by Theorem 4.3, we may reduce the generalized complex structure $(\mathcal{G}, \mathcal{J})$ to $M^{red} = \mu^{-1}(0)/G$.

The bihermitian structure on M_{red} associated with $(\mathcal{G}^{red}, \mathcal{J}^{red})$ can be described as follows: $\mathcal{G}$ restricted to $K^{\mathcal{G}}$ is just the reduced metric g^{red} obtained by hyper-Kähler reduction as in Example 4.4, whereas the restriction of $\mathcal{J}$ to $C_\pm \cap K^{\mathcal{G}}$ defines complex structures via the projection to TM^{red}. It is easily verified that the restriction of $\mathcal{J}$ to $C_+ \cap K^{\mathcal{G}}$ defines I_1^{red}, and the restriction to $C_- \cap K^{\mathcal{G}}$ gives I_2^{red}, where $(g^{red}, I_1^{red}, I_2^{red}, I_3^{red})$ is the reduced hyper-Kähler structure, as required.

References

[1] H. Bursztyn, G. R. Cavalcanti, and M. Gualtieri. Reduction of Courant algebroids and generalized complex structures. *Advances in Math.*, 211:726–765, 2007.

[2] H. Bursztyn, G. R. Cavalcanti, and M. Gualtieri. In preparation.

[3] G. R. Cavalcanti. Reduction of metric structures on Courant algebroids. In preparation.

[4] T. Courant. Dirac manifolds. *Trans. Amer. Math. Soc.*, 319:631–661, 1990.

[5] M. Gualtieri. *Generalized Complex Geometry.* D.Phil. thesis, Oxford University, 2003. math.DG/0401221.

[6] V. Guillemin and S. Sternberg. Geometric quantization and multiplicities of group representations. *Invent. Math.*, 67(3):515–538, 1982.

[7] N. J. Hitchin. Generalized Calabi-Yau manifolds. *Q. J. Math.*, 54:281–308, 2003.

[8] N. J. Hitchin. Instantons, Poisson structures and generalized Kaehler geometry. *Comm. Math. Phys.*, 265: 131-164, 2006. math.DG/0503432.

[9] N. J. Hitchin, A. Karlhede, U. Lindström, and M. Roček. Hyper-Kähler metrics and supersymmetry. *Comm. Math. Phys.*, 108(4):535–589, 1987.

[10] S. Hu. Hamiltonian symmetries and reduction in generalized geometry. math.DG/0509060.

[11] M. K. Kinyon and A. Weinstein. Leibniz algebras, Courant algebroids, and multiplications on reductive homogeneous spaces. *Amer. J. Math.*, 123(3):525–550, 2001.

[12] F. C. Kirwan. *Cohomology of quotients in symplectic and algebraic geometry*, volume 31 of *Mathematical Notes*. Princeton University Press, Princeton, NJ, 1984.

[13] Y. Lin, S. Tolman. Symmetries in generalized Kähler geometry. math.DG/0509069.

[14] Z.-J. Liu, A. Weinstein, P. Xu: Manin triples for Lie algebroids. *J. Differential Geom.*, 45: 547–574, 1997.

[15] J.-L. Loday. Une version non commutative des algèbres de Lie: les algèbres de Leibniz. In *R.C.P. 25, Vol. 44 (French) (Strasbourg, 1992)*, volume 1993/41 of *Prépubl. Inst. Rech. Math. Av.*, pages 127–151. Univ. Louis Pasteur, Strasbourg, 1993.
[16] R. Moraru and M. Verbitsky. Stable bundles on hypercomplex surfaces. math.DG/0611714.
[17] P. Ševera and A. Weinstein. Poisson geometry with a 3-form background. *Progr. Theoret. Phys. Suppl.*, 144:145–154, 2001. Noncommutative geometry and string theory (Yokohama, 2001).
[18] M. Stienon, P. Xu. Reduction of generalized complex structures. math.DG/0509393.
[19] I. Vaisman. Reduction and submanifolds of generalized complex manifolds. math.DG/0511013.

Instituto de Matemática Pura e Aplicada, Estrada Dona Castorina 110, Rio de Janeiro, 22460-320, Brasil
E-mail address: henrique@impa.br

Mathematical Institute, 24-29 St Giles', Oxford, OX1 3LB, UK.
E-mail address: gilrc@maths.ox.ac.uk

Department of Mathematics, Massachusetts Institute of Technology, Cambridge, MA 02139, USA.
E-mail address: mgualt@mit.edu

Contemporary Mathematics
Volume **450**, 2008

Deformation Quantization and Reduction

Alberto S. Cattaneo

ABSTRACT. This note is an overview of the Poisson sigma model (PSM) and its applications in deformation quantization. Reduction of coisotropic and pre-Poisson submanifolds, their appearance as branes of the PSM, quantization in terms of L_∞- and A_∞-algebras, and bimodule structures are recalled. As an application, an "almost" functorial quantization of Poisson maps is presented if no anomalies occur. This leads in principle to a novel approach for the quantization of Poisson–Lie groups.

1. Introduction

The Poisson sigma model (PSM) [**I, SS**] is a two-dimensional topological field theory with target a Poisson manifold (M,π). It is defined by the action functional

$$S=\int_\Sigma \eta\,\mathrm{d}X+\frac{1}{2}\pi(X)\eta\,\eta, \tag{1.1}$$

where the pair (X,η) is a bundle map $T\Sigma\to T^*M$, and Σ is a two-manifold.

The Hamiltonian study of the PSM on $\Sigma=[0,1]\times\mathbb{R}$ leads [**CF01**] to the construction of the symplectic groupoid of M whenever T^*M is an integrable Lie algebroid. The functional-integral perturbative quantization yields [**CF00**] Kontsevich's deformation quantization [**K**] if Σ is the disk and one chooses vanishing boundary conditions for η.

General boundary conditions ("branes") compatible with symmetries and the perturbative expansion were studied in [**CF04**] and turned out to correspond to coisotropic submanifolds. In [**Ca, CFa05**] (see also references therein) general boundary conditions compatible just with symmetries were classified; they correspond to what we now call pre-Poisson submanifolds (a generalization of the familiar notion of pre-symplectic submanifolds in the symplectic case).

The perturbative functional-integral quantization with branes gives rise to deformation quantization of the reduced spaces [**CF04, CFa05**]. The many-brane

2000 *Mathematics Subject Classification.* Primary 53D55; Secondary 20G42, 51P05, 53D17, 57R56, 58A50, 81T70.

Key words and phrases. Symplectic geometry, Poisson geometry, deformation quantization, coisotropic submanifolds, reduction, Poisson sigma model, cohomological methods in quantization, graded manifolds, L_∞-algebras, A_∞-algebras, quantum groups.

The author acknowledges partial support of SNF Grant #20-113439. This work has been partially supported by the European Union through the FP6 Marie Curie RTN ENIGMA (Contract number MRTN-CT-2004-5652) and by the European Science Foundation through the MISGAM program.

case leads to the construction of bimodules and morphisms between them [**CF04**] and potentially to a method for quantizing Poisson maps in an almost functorial way (see Section 8). This problem is also discussed in [**B04**] where results (and obstructions) for the symplectic case are given. The linear case (i.e., the case when the Poisson manifold is the dual of a Lie algebra and the coisotropic submanifold is an affine subspace) leads to many interesting results in Lie theory [**CF04, CT**].

A more general approach leads to a cohomological description of coisotropic submanifolds by L_∞-algebras naturally associated to them [**OP**] and to a deformation quantization in terms of (possibly nonflat) A_∞-algebras. A natural way to reinterpret these results is by associating a graded manifold to the coisotropic submanifold and use a graded version of Kontsevich's L_∞-quasi-isomorphism [**CF07, LS**]. This may also be regarded as a duality for the PSM with target a graded manifold [**CF07, C06**].

In the absence of the so-called anomaly—i.e., when the A_∞-algebra can be made flat—one can go down to cohomology.

This paper is an overview of all these themes. It contains a more extended discussion, Section 7, of deformation quantization for graded manifolds and an introduction, Section 8, to methods for an almost functorial quantization of morphisms whenever the anomaly is not present.

Plan of the paper. In Section 2 we recall properties and reduction of coisotropic, presymplectic and pre-Poisson submanifolds collecting results in [**Ca, CFa05, CZ, CZbis**]. In Section 3 we give a simple derivation of the L_∞-structure associated [**OP**] to a coisotropic submanifold; we also recall the BFV (Batalin–Fradkin–Vilkovisky) formalism [**BF, BV**] and its relation [**S**] to the L_∞-structure. In Section 4 we recall the Hamiltonian description of the PSM and how coisotropic [**CF04**] and pre-Poisson [**Ca, CFa05**] submanifolds show up as boundary conditions (branes) for the PSM. Perturbative functional-integral quantization of the PSM [**CF00**] is recalled in Section 5 with results for coisotropic [**CF04**] and pre-Poisson [**CFa05**] branes. Section 6 deals with the reinterpretation of these results in terms of a duality for the PSM with target a graded manifold [**CF07, C06**]. In Section 7 we recall the Formality Theorem for graded manifolds [**K, CF07**] and its applications in deformation quantization [**CF07, LS**]; the potential anomaly[**CF04, CF07**] is discussed; subsection 7.3 contains a description of the induced properties in cohomology and is original; subsection 7.5 contains a new description of methods for quantizing the inclusion map of a Poisson submanifold; in particular, it contains a proof of a conjecture by [**CR**] that the deformation quantization of a Poisson submanifold determined by central constraints may be obtained by quotienting Kontsevich's deformation quantization by the ideal generated by the image of the constraints under the Duflo–Kirillov–Kontsevich map. Finally, Section 8 recalls the many-brane case [**CF04**] and, in the absence of anomalies, presents a novel method for the quantization of Poisson maps compatible with compositions; the application to the quantization of Poisson–Lie groups is briefly described.

ACKNOWLEDGMENT. I thank G. Felder, F. Schätz, J. Stasheff and M. Zambon for useful discussions and comments. I thank Hans-Christian Herbig for pointing out reference [**B00**] (which also led to an improvement of [**S**]).

2. Reduction

We recall reduction in the context of symplectic and Poisson manifolds. We will essentially follow [**CZ**] (see also [**CZbis**]). Let us first recall the linear case.

A symplectic space is a (possibly infinite dimensional) linear space V equipped with a nondegenerate skew-symmetric bilinear form ω, called a symplectic form. By nondegenerate we mean here that $\omega(v,w)=0\ \forall v\in V \Rightarrow w=0$ (this is sometimes called a weak symplectic form). We denote by $\omega^\sharp$ the induced linear map $V\to V^*$: $v\mapsto \omega(v,\ )$. With this notation, ω is nondegenerate iff $\omega^\sharp$ is injective.

The restriction of ω to a subspace W is in general degenerate, the kernel of $(\omega|_W)^\sharp$ being $W^\perp\cap W$, where

$$W^\perp := \{v\in V : \omega(v,w)=0\ \forall w\in W\}$$

is the symplectic orthogonal to W. Thus, the reduced space $\underline{W}:=W/(W\cap W^\perp)$ is automatically endowed with a symplectic form.

An extreme case is when $W\cap W^\perp=0$; i.e., when $(\omega|_W)^\sharp$ is nondegenerate. In this case, W is called a symplectic subspace and $\underline{W}=W$. In finite dimensions, one has the additional properties that $W^\perp$ is also symplectic and that

$$W\oplus W^\perp = V. \tag{2.1}$$

That is, in finite dimensions a symplectic subspace may be equivalently described as a subspace which admits a symplectic complement. For this reason, it may as well be called a cosymplectic subspace.

The other nontrivial extreme case is when $W^\perp\subset W$. In this case W is called a coisotropic subspace and $\underline{W}=W/W^\perp$. The coisotropic case is in some sense the only case one has to consider, as we have the following simple

Lemma 2.1. *Let W be a subspace of a finite dimensional symplectic space V. Then it is always possible to find a symplectic subspace W' of V which contains W as a coisotropic subspace.*

The Lemma is proved by adding W what is missing to make it symplectic. Any complement to $W+W^\perp$ will do (see Lemma 3.1 in [**CZ**] for details).

A symplectic manifold is a smooth manifold endowed with a closed, nondegenerate 2-form. Here nondegenerate means that the bundle map $\omega^\sharp\colon TM\to T^*M$ induced by the 2-form ω on M is injective. The restriction of ω to a submanifold C is in general degenerate. The kernel of $(\omega|_C)^\sharp$ is the characteristic distribution $T^\perp C\cap TC$ which in general is not smooth. Here $T^\perp C$ is the bundle of symplectic orthogonal spaces to TC in the restriction T_CM of TM to C. The submanifold N is called presymplectic if its characteristic distribution is smooth (i.e., $T^\perp C\cap TC$ is a subbundle of TC); by the closedness of ω it is automatically integrable. The corresponding leaf space $\underline{C}$ is called the reduced space. If it is a manifold, it is endowed with a unique symplectic form whose pullback to C is equal to the restriction of ω. The extreme nontrivial examples of presymplectic submanifolds are the symplectic and coisotropic submanifolds. A submanifold C is called symplectic (coisotropic) if T_xC is symplectic (coisotropic) in T_xM for every $x\in C$. The reduction of a symplectic submanifold is the manifold itself. If C is a submanifold of a finite dimensional manifold M, by Lemma 2.1 we may find a symplectic subspace of T_xM containing T_xC as a coisotropic subspace for every $x\in C$. These subspaces may be chosen to be glued together smoothly if the submanifold is presymplectic. Namely, we have the following

PROPOSITION 2.2 ([**CZ**]). *Let C be a presymplectic submanifold of a symplectic manifold M. Then it is always possible to find a symplectic submanifold which contains C as a coisotropic submanifold.*

Moreover, one can show that these symplectic extensions are neighborhood equivalent. Namely, given two such extensions C' and C'', there exists a tubular neighborhood U of C such that $U \cap C'$ and $U \cap C''$ are related by a symplectomorphisms (i.e., a diffeomorphism compatible with the symplectic form) of U which fixes C.

We now move to the Poisson case. We will only work in finite dimensions. A Poisson space is a linear space V endowed with a bivector π (i.e., an element of $\Lambda^2 V$). We denote by $\pi^\sharp$ the induced linear map $V^* \to V$, $\alpha \mapsto \pi(\alpha, \ )$. A finite dimensional symplectic space (V, ω) is automatically Poisson with $\pi^\sharp = (\omega^\sharp)^{-1}$. The Poisson generalization of the notion of symplectic orthogonal of a subspace W is the image under $\pi^\sharp$ of its annihilator $W^0 := \{\alpha \in V^* : \alpha(w) = 0 \ \forall w \in W\}$. In the symplectic case, $\pi^\sharp(W^0) = W^\perp$. Condition (2.1) may be replaced by $W \oplus \pi^\sharp(W^0) = V$; a subspace W satisfying it is called cosymplectic; it may equivalently be characterized by the condition that the projection of π to $\Lambda^2(V/W)$ is symplectic. A coisotropic subspace is analogously defined as a subspace W with $\pi^\sharp(W^0) \subset W$.

A Poisson manifold is a manifold M endowed with a bivector field π (i.e., a section of $\Lambda^2 TM$) such that the bracket $\{f, g\} := \pi(\mathrm{d}f, \mathrm{d}g)$, $f, g \in C^\infty(M)$ satisfies the Jacobi identity. Equivalently, $[\pi, \pi] = 0$ with the Schouten–Nijenhuis bracket. This also amounts to saying that $(C^\infty(M), \cdot, \{\ ,\ \})$ is a Poisson algebra. We denote by $\pi^\sharp \colon T^*M \to TM$ the corresponding bundle map. A finite dimensional symplectic manifold (M, ω) is Poisson with $\pi^\sharp = {\omega^\sharp}^{-1}$. A submanifold C is called cosymplectic (coisotropic) if T_xC is so in T_xM $\forall x \in C$. We will denote by N^*C—the conormal bundle—the vector bundle over C with fiber at x given by $N^*_xC := (T_xC)^0 \subset T^*_xM$. Observe that $\pi^\sharp(N^*C)$ is a (singular) distribution on C if C is coisotropic. Invariant functions form naturally a Poisson algebra. If the leaf space $\underline{C}$ is a manifold, it is naturally a Poisson manifold. A cosymplectic manifold C is also automatically a Poisson manifold (not a Poisson submanifold though). One way to see this is to think of a Poisson manifold as a manifold foliated by symplectic leaves. A cosymplectic submanifold intersects symplectic leaves cleanly, and each intersection is a symplectic submanifold of the symplectic leaf; thus, a cosymplectic submanifold is foliated in symplectic leaves, which makes it into a Poisson manifold. The induced Poisson bracket may also be obtained by Dirac's procedure.

We now wish to describe a Poisson generalization of the notion of presymplectic submanifold leading to a generalization of Proposition 2.2. Namely, we call a submanifold C pre-Poisson if $\pi^\sharp(N^*C) + TC$ has constant rank along C. (In [**CFa05**] and references therein, this was called a "submanifold with strong regularity constraints".) In the symplectic case, this is equivalent to C being presymplectic.

An equivalent definition of a pre-Poisson submanifold C amounts to asking that the bundle map $\phi \colon N^*C \to NC$ obtained by composing the injection $N^*C \to T^*_CM$ with $\pi^\sharp$ and finally with the projection $T_CM \to NC$ should have constant rank. As special cases we recognize coisotropic submanifolds ($\phi = 0$) and cosymplectic submanifolds (ϕ surjective). We have the following

PROPOSITION 2.3 ([**CFa05, CZ**]). *Let C be a pre-Poisson submanifold of a Poisson manifold M. Then it is always possible to find a cosymplectic submanifold*

M' which contains C as a coisotropic submanifold. Moreover, this extension M' is unique up to neighborhood equivalence.

In the following we will also need a description in terms of Lie algebroids. Recall that a Lie algebroid is a vector bundle E over a smooth manifold M endowed with a bundle map $\rho\colon E \to TM$ (the anchor map) and a Lie algebra structure on $\Gamma(E)$ with the property $[a, fb] = f[a, b] + \rho(a)f\, b$ for every $a, b \in \Gamma(E)$, $f \in C^\infty(M)$. The tangent bundle itself is a Lie algebroid. The cotangent bundle of Poisson manifold is also a Lie algebroid with $\rho = \pi^\sharp$ and $[\mathrm{d}f, \mathrm{d}g] := \mathrm{d}\{f, g\}$. If C is a pre-Poisson submanifold, then $AC := N^*C \cap {\pi^\sharp}^{-1}TC$ is a Lie subalgebroid [**CFa05**]. Its anchor is defined just by restriction; as for the bracket, one has to extend sections of AC to sections of T^*M in a neighborhood of C, use the bracket on T^*M and finally restrict; it is not difficult to check that the result is independent of the extension. Observe that for C coisotropic, one simply has $AC = N^*C$.

3. Cohomological descriptions

The complex $\Gamma(\Lambda TM)$ of multivector fields on a smooth manifold M is naturally endowed with a Lie bracket of degree -1, the Schouten–Nijenhuis bracket. Let $\mathcal{V}(M) := \Gamma(\Lambda TM)[1]$ be the corresponding graded Lie algebra (GLA). A Poisson bivector field is now the same as an element π of $\mathcal{V}(M)$ of degree one which satisfies $[\pi, \pi] = 0$. In general, a self-commuting element of degree 1 in a GLA is called a Maurer–Cartan (MC) element. Observe that $[\pi, \ ]$ is a differential; actually, it is the Lie algebroid differential corresponding to the Lie algebroid structure on T^*M.

Let C be a submanifold and NC its normal bundle, canonically defined as the quotient by TC of the restriction T_CM of TM to C. It is also the dual bundle of the conormal bundle N^*C introduced above. Let A be the graded commutative algebra $\Gamma(\Lambda NC)$. Denote by $P\colon \Gamma(\Lambda TM) \to A$ the composition of the restriction to C with the projection $T_CM \to NC$. It is not difficult to check that $\operatorname{Ker} P$ is a Lie subalgebra. Moreover, C is coisotropic iff $\pi \in \operatorname{Ker} P$. In this case, there is a well-defined differential δ on A acting on $X \in A$ by

$$\delta X := P\left[\pi, \tilde{X}\right], \quad \tilde{X} \in P^{-1}(X).$$

This is the differential corresponding to the Lie algebroid structure on N^*C. Observe that H^0_δ is the algebra of invariant functions on C.

The projection P admits a section if M is a vector bundle over C. In particular, this is true if M is the normal bundle of C. The section i for $P\colon \Gamma(\Lambda T(NC)) \to A$ is given as follows: If f is a function on C, one defines if as the pullback of f by the projection $NC \to C$. If X is a section of NC, one defines iX as the unique vertical vector field $\tilde{X}$ on NC which is constant along the fibers and such that $P\tilde{X} = X$. Finally, i is extended to sections of ΛNC so that it defines an algebra morphism. It turns out that $i(\Gamma(NC))$ is an abelian subalgebra.

Let us now choose an embedding of NC into our Poisson manifold (M, π). By restriction π defines a Poisson structure on NC, so a MC element in $\mathcal{V}(NC)$. Following Voronov [**V**], we define the derived brackets

$$\begin{aligned} &\lambda_k\colon A^{\otimes k} \to A, \\ &\lambda_k(a_1, \ldots, a_k) := P\left([[\ldots[[\pi, ia_1], ia_2], \ldots], ia_k]\right), \end{aligned} \tag{3.1}$$

and $\lambda_0 := P(\pi) \in A$. Observe that $\lambda_0 = 0$ iff C is coisotropic; in this case, $\delta = \lambda_1$. In general Voronov proved the following

THEOREM 3.1. *Let $\mathfrak{g}$ be a GLA, $\mathfrak{h}$ an abelian subalgebra, i the inclusion map. Let P be a projection to $\mathfrak{h}$ such that $P \circ i = id$ and that* Ker P *is a Lie subalgebra. Then the derived brackets of every MC element of $\mathfrak{g}$ define an L_∞-structure on $\mathfrak{h}$.*

This means that the operations λ_k satisfy certain quadratic relations which in particular for $\lambda_0 = 0$ imply that λ_1 is a differential for λ_2 and that λ_2 satisfies the Jacobi identity up to λ_1-homotopy. So the λ_1-cohomology inherits the structure of a GLA.

In our case A is also a graded commutative algebra and the multibrackets are multiderivations. In this case we say that we have a P_∞-algebra (P for Poisson). In particular, when C is coisotropic, $H_{\lambda_1} = H_\delta$ is a graded Poisson algebra. The Poisson structure in degree zero—i.e., on invariants functions—is the same as the one described in the previous Section.

Observe that the P_∞-structure depends on a choice of embedding $NC \hookrightarrow M$. It is possible to show that different choices lead to L_∞-isomorphic algebras. We will return on this in a forthcoming paper [**CS**].

The P_∞-structure appeared first in [**OP**] as a tool to describe deformations of a coisotropic submanifold. In [**CF07**] it was rediscovered as the semiclassical limit of the quantization of coisotropic submanifolds. We will return on quantization in Section 7.

3.1. The BFV method. The Batalin–Fradkin–Vilkovisky (BFV) method is an older method to describe coisotropic submanifold in terms of a differential graded Poisson algebra (DGPA) [**BF, BV**]. Its advantage is that DGPAs are more manageable than P_∞-algebras. The disadvantage is that it works only under certain assumptions. We will essentially follow Stasheff's presentation [**S97**].

Suppose first that the coisotropic submanifold C is given by global constraints, or, equivalently, that the normal bundle of C is trivial. Let $\{y^\mu\}_{\mu=1,\dots,n}$, $n = \operatorname{codim} C$, be a basis of constraints (equivalently, a basis of sections of NC or a basis of transverse coordinates). Let $\mathcal{B}$ be the free graded commutative algebra generated by a set $\{b^\mu\}$ of odd variables of degree 1 (the "ghost momenta"). On $C^\infty(NC) \otimes \mathcal{B}$ one defines the Koszul differential $\delta_0 := y^\mu \frac{\partial}{\partial b^\mu}$. The δ_0 cohomology is then concentrated in degree 0 and is $C^\infty(C)$. Let us also introduce a set $\{c_\mu\}$ of odd variables of degree -1 (the "ghosts") and the free graded commutative algebra $\mathcal{C}$ they generate. On $C^\infty(NC) \otimes \mathcal{B} \otimes \mathcal{C}$ one has a unique Poisson structure with $\{\, b^\mu \,,\, c_\nu \,\} = \{\, c_\nu \,,\, b^\mu \,\} = \delta^\mu_\nu$. Then δ_0 is a Hamiltonian vector field with Hamiltonian function $\Omega_0 = y^\mu c_\mu$. On this enlarged algebra the δ_0-cohomology is $C^\infty(NC) \otimes \mathcal{C} \simeq \Gamma(\Lambda NC)$.

By choosing an embedding $NC \hookrightarrow M$, we get a Poisson structure on $C^\infty(NC)$ which we can extend to $C^\infty(NC) \otimes \mathcal{B} \otimes \mathcal{C}$. We now consider the sum of the two Poisson structure. Let $F_0 := \frac{1}{2}\{\, \Omega_0 \,,\, \Omega_0 \,\} = \frac{1}{2}\{\, y^\mu \,,\, y^\nu \,\}\, c_\mu c_\nu$. Observe that $\{F_0, \delta_0, \{\ ,\ \}\}$ defines a nonflat L_∞-structure[1] on $C^\infty(NC) \otimes \mathcal{B} \otimes \mathcal{C}$. Moreover, F_0

[1]Originally L_∞-algebras were always assumed to miss the 0th term. When L_∞-algebras with the 0th term turned out to be interesting, various terminologies were introduced. An L_∞-algebra without the 0th term is called nowadays flat (resp. strict), while one where the 0th term is there is called curved (resp. weak). If no assumption on the 0th term is made (i.e., it may or may not vanish), we simply speak of an L_∞-algebra.

is δ_0-exact iff C is coisotropic. In this case, one can use cohomological perturbation theory to kill F_0 as follows.

Let $h := b^\mu \frac{\partial}{\partial y^\mu}$. Then $[\,\delta_0\,,\,h\,] = E := b^\mu \frac{\partial}{\partial b^\mu} + y^\mu \frac{\partial}{\partial y^\mu}$. One then has a homotopy s for δ_0 (i.e. $s\delta_0 + \delta_0 s = id - pr$, with pr the projection to a subspace isomorphic to the δ_0-cohomology) defined by $s\alpha = h\alpha/|\alpha|$ if $E\alpha = |\alpha|\alpha$ and $s\alpha = 0$ if $E\alpha = 0$. Let us then define $\Omega := \sum_{i=0}^{\infty} \Omega_i$ by induction as follows: Let $R_k := \sum_{i=0}^{k} \Omega_i$. Then define $\Omega_{k+1} = -\frac{1}{2}\, s\,(\{\,R_k\,,\,R_k\,\})$. If C is coisotropic, the following hold:

(1) $\{\,\Omega\,,\,\Omega\,\} = 0$, so $\{D, \{\ ,\ \}\}$ defines a DGPA structure on $C^\infty(NC)\otimes\mathcal{B}\otimes\mathcal{C}$, where $D := \{\,\Omega\,,\ \ \} = \delta_0 + \cdots$.
(2) The D-cohomology H_D is isomorphic to the Lie algebroid cohomology H_δ of N^*C.
(3) This way H_δ gets the structure of a GPA.

It is natural to ask whether the GPA structure on H_δ is the same as the one induced before. The answer is affirmative. Actually, there is a stronger result:

THEOREM 3.2 (Schätz [**S**]). *$(C^\infty(NC)\otimes\mathcal{B}\otimes\mathcal{C}, D, \{\ ,\ \})$ is L_∞-quasi-isomorphic to $(A, \{\lambda_k\})$.*

Observe that the L_∞-structure $(A, \{\lambda_k\})$ can be defined for every submanifold, while the BFV formalism as presented above requires that the normal bundle should be trivial, or at least flat. A more general version of the BFV formalism allows for linearly dependent global constraints and uses the Koszul–Tate resolution. There is however a generalization by Bordemann and Herbig [**B00**] of the construction presented above which needs no assumption on the normal bundle and just requires the choice of a connection. The above Theorem holds also in the general case [**S**].

4. The reduced space of the Poisson sigma model

Given a finite dimensional manifold M, we denote by PM the Banach manifold of C^1-paths in M; viz., $PM := C^1(I, M)$, $I = [0,1]$. We denote by T^*PM the vector bundle over PM with fiber at $X \in PM$ the space of C^0 sections of $T^*I \otimes X^*T^*M$. Using integration over I and the canonical pairing between TM and T^*M, one can endow T^*PM with a symplectic structure. Let now π be Poisson bivector field on M. We define

$$\mathcal{C}(M,\pi) := \{(X,\zeta) \in T^*PM : \mathrm{d}X + \pi^\sharp(X)\zeta = 0\}.$$

These infinite-dimensional manifolds naturally appear when considering the PSM (1.1) on $\Sigma = I \times \mathbb{R}$. Namely, the map $X\colon \Sigma \to M$ may be regarded as a path in PM. On the other hand, introducing the coordinate t on $\mathbb{R}$, we may write $\eta = \zeta + \lambda dt$, and reinterpret the pair (X, ζ) as a path in T^*PM. The bilinear term in X and ζ in the action functional (1.1) yields the canonical symplectic structure on T^*PM, while λ appears linearly as a Lagrange multiplier leading to the constraints that define $\mathcal{C}(M,\pi)$.

THEOREM 4.1 ([**CF01**]). *$\mathcal{C}(M,\pi)$ is a coisotropic submanifold of T^*PM. Its reduced space $\underline{\mathcal{C}}(M,\pi)$ is the source simply connected symplectic groupoid of M, whenever M is integrable (i.e., T^*M is an integrable Lie algebroid).*

We now want to discuss boundary conditions. Given two submanifolds C_0 and C_1 of M, let

$$\mathcal{C}(M,\pi; C_0, C_1) := \{(X,\zeta) \in \mathcal{C}(M,\pi) : X(0) \in C_0,\ X(1) \in C_1\}.$$

Observe that $\mathcal{C}(M,\pi;C_0,C_1)$ has natural maps to C_0 and C_1 (evaluation of X at 0 and 1). We call points in the image in $C_0 \times C_1$ connectable.

Its symplectic orthogonal bundle may be explicitly computed. For simplicity we choose local coordinates on M. (The correct, but more cumbersome, description involves choosing a torsion-free connection for TM.) One gets [**Ca**]

$$T^{\perp}_{(X,\zeta)}\mathcal{C}(M,\pi;C_0,C_1) = \{(-\pi^\sharp(X)\beta, \mathrm{d}\beta_i + \partial_i\pi^{jk}(X)\zeta_j\beta_k) : \beta \in \Gamma(X^*T^*M),\ \beta(0) \in N^*_{X(i)}C_i,\ i=0,1\}. \tag{4.1}$$

Intersection with $T\mathcal{C}(M,\pi;C_0,C_1)$ just forces the boundary conditions for β to be such that the X-variations are tangent to the submanifolds: viz., $\pi^\sharp(X(i))\beta(i) \in T_{X(i)}C_i$, $i=0,1$. Thus, $\mathcal{C}(M,\pi;C_0,C_1)$ is presymplectic iff the pre-Poisson condition is satisfied at all connectable points [**CFa05, Ca**], and it is coisotropic iff the coisotropicity condition is satisfied at all connectable points [**CF04**]. The sections of $\delta X \oplus \delta\zeta$ of the characteristic distribution are parametrized by a section β of X^*T^*M with $\beta(i) \in A_{X(i)}C_i$, $i=0,1$:[2]

$$\delta X = -\pi^\sharp(X)\beta, \tag{4.2}$$

$$\delta\zeta_i = \mathrm{d}\beta_i + \partial_i\pi^{jk}(X)\zeta_j\beta_k, \tag{4.3}$$

where for simplicity we have written the second equation using local coordinates on M. (A more invariant, but more cumbersome, possibility is to write it upon choosing a torsion-free connection for TM and observing that the distribution does not depend on this choice.)

To remove the condition on connectable points, one can e.g. take the second submanifold to be the whole manifold M. So we have

THEOREM 4.2.

(1) *$\mathcal{C}(M,\pi;C,M)$ is coisotropic in T^*PM iff C is coisotropic in M* [**CF04**].
(2) *$\mathcal{C}(M,\pi;C,M)$ is presymplectic in T^*PM iff C is pre-Poisson in M* [**CFa05, Ca**].

Point 2) says in particular that pre-Poisson submanifolds are the most general boundary conditions for the Poisson sigma model compatible with symmetries. A submanifold chosen as a boundary condition is usually called a brane. Point 1) is important as it describes boundary conditions compatible with symmetries *and* perturbation theory around $\pi = 0$. Namely, consider the family of Poisson structures $\pi_\epsilon := \epsilon\pi$, $\epsilon \geq 0$. Let C be a pre-Poisson submanifold for π. Then it is pre-Poisson for π_ϵ, $\forall\epsilon \geq 0$. In particular, it is coisotropic for $\epsilon = 0$. One may indeed check that the codimension of the characteristic distribution of $\mathcal{C}(M,\pi_\epsilon;C,M)$ is the same for all $\epsilon > 0$ but it may change for $\epsilon = 0$. It stays the same iff C is coisotropic for π (and hence coisotropic for π_ϵ for all ϵ).

5. Expectation values in perturbation theory

We now consider the perturbative functional-integral quantization of the Poisson sigma model. Namely, we consider integrals of the form

$$\langle\, \mathcal{O}\, \rangle := \int \mathrm{e}^{\frac{\mathrm{i}}{\hbar}S}\,\mathcal{O}, \tag{5.1}$$

[2] Actually, this is the image of the anchor of an infinite-dimensional Lie algebroid over $\mathcal{C}(M,\pi;C,M)$. Explicitly this is described in [**BC**] and in [**BCZ**].

where $\mathcal{O}$ is a function of the fields and the integration is over all fields (X, η), $X \in \mathrm{Map}(\Sigma, M)$, $\eta \in \Gamma(T\Sigma \otimes X^*T^*M)$. If Σ has a boundary, the discussion in Section 4 shows that we have to choose a pre-Poisson submanifold C of M (a brane) and impose the conditions $X(\partial\Sigma) \subset C$, $\iota_\partial^*\eta \in \Gamma(T^\Sigma \otimes X^*AC)$, where ι_∂ is the inclusion map $\partial\Sigma \hookrightarrow \Sigma$.

Also observe that the Poisson sigma model has symmetries (see (4.2)). This makes the integrand $\mathrm{e}^{\frac{\mathrm{i}}{\hbar}S}$ invariant under the Lagrangian extension of the symmetries which has the form:

$$\delta X = -\pi^\sharp(X)\beta, \tag{5.2}$$

$$\delta\eta_i = \mathrm{d}\beta_i + \partial_i\pi^{jk}(X)\eta_j\beta_k, \tag{5.3}$$

with $\beta \in \Gamma(X^*T^*M)$, $\iota_\partial^*\beta \in \Gamma(X^*AC)$. As a consequence the integral (5.1) cannot possibly converge (it would not even if we were in finite dimensions). The trick is to select representatives of the fields modulo symmetries in a consistent way. This is done using the BV formalism [**BV**] which in particular guarantees that the result is independent of the choice involved ("gauge fixing"). Its application to the PSM is described in [**CF00**]. An important issue is that an expectation value as in (5.1) is also gauge-fixing independent if $\mathcal{O}$ is invariant under symmetries. We will consider only boundary observables. Namely, let f be a function on C. To it we associate the observable $\mathcal{O}_{f,u}(X, \eta) := f(X(u))$, where u is some point of $\partial\Sigma$. The observable is invariant iff f is invariant constant along the characteristic distribution of C. It will be interesting for us to take two such functions f and g and consider $\mathcal{O} = \mathcal{O}_{f,u}\mathcal{O}_{g,v}$.

We compute integrals as in (5.1) in perturbation theory, i.e., by expanding S around its critical points and then computing the integral formally in terms of the momenta of the Gaussian distribution associated to the Hessian of S (observe that after gauge fixing this is no longer degenerate). We will pick only the trivial critical point; viz., the class of solutions $\{X \text{ constant}, \eta = 0\}$. We fix the constant by imposing X to be equal to some $x \in C$ at some boundary point which we denote by ∞. Finally, we select Σ to be a disk[3] and compute

$$f \star g(x; u, v) := \int_{X(\infty)=x} \mathrm{e}^{\frac{\mathrm{i}}{\hbar}S}\, \mathcal{O}_{f,u}\mathcal{O}_{g,v}, \tag{5.4}$$

where the points u, v, ∞ are cyclically ordered. It turns out that such an integral may be explicitly computed in perturbation theory. The results are as follows.

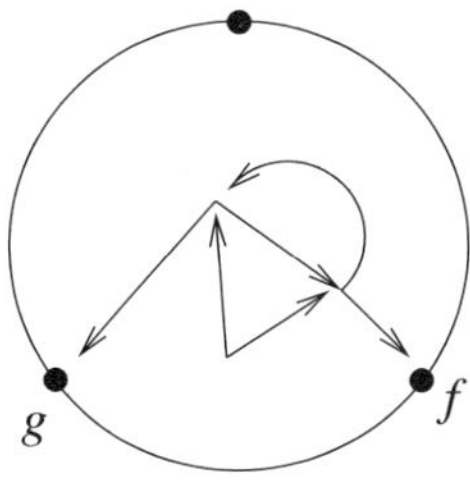

FIGURE 1. An example of a Kontsevich graph

[3]Higher genera would be interesting to consider, but this has not been done so far, the main difficulty being the appearance of nontrivial cohomological classes in the space of solutions.

The case $C = M$ has been studied in [**CF00**]. The result turns out to be independent of u and v and to define a star product [**BFFLS**] on M: viz., an associative product $\star$ on $C^\infty(M)[[\epsilon]]$ (in the above computation $\epsilon = \mathrm{i}\hbar/2$) deforming the pointwise product in the direction of the Poisson bracket:

$$f \star g = fg + \epsilon\{f, g\} + \sum_{n=2}^{\infty} B_n(f,g), \tag{5.5}$$

where the B_ns are bidifferential operator with the property $B_n(1,\bullet) = B_n(\bullet,1) = 0$. Moreover, this star product is exactly the one defined by Kontsevich in [**K**].

The case C pre-Poisson has been considered in [**CFa05**]. It turns out that the PSM on M with boundary conditions on C is equivalent to the PSM on any M' as in Prop. 2.3 with the same boundary conditions. As a result, it is enough to consider coisotropic submanifolds.

The case C coisotropic has been considered in [**CF04**]. A few interesting, but unpleasant phenomena appear. First, it might happen that the result of (5.4) depends on u and v. Second, the result regarded as a function on $C^\infty(C)$ may not be invariant. Third, associativity may not be guaranteed. Under suitable conditions, one may ensure that everything works. We return on this in subsection 7.4, where we show that the vanishing of the first and second Lie algebroid cohomologies is a sufficient condition.

6. Super PSM and duality

To deal with symmetries in the PSM consistently, one has to resort to the BV formalism. In the case at hand the recipe (following from the AKSZ formalism [**AKSZ**] adapted to the PSM with boundary [**CF01b**]) consists just of formally replacing the fields (X,η) in the action by "superfields" $(\mathsf{X},\boldsymbol{\xi})$. To simplify the discussion, let us choose a coordinate chart on M. Then we may regard X as an element of $\Omega^0(\Sigma)\otimes\mathbb{R}^m$, $m = \dim M$, and η as an element of $\Omega^1(\Sigma)\otimes(\mathbb{R}^m)^*$. The action in local coordinate reads

$$S = \int_\Sigma \eta_i \mathrm{d}X^i + \frac{1}{2}\pi^{ij}(X)\eta_i\eta_j,$$

where we use Einstein's convention on repeated indices.

The superfields are then $\mathsf{X} \in V := \Omega(\Sigma)\otimes\mathbb{R}^m$, $\boldsymbol{\eta} \in W := \Omega(\Sigma)\otimes(\mathbb{R}^m)^*$ and one just puts them in the action instead of the classical fields. The integral selects by definition the two-form component.[4] The superaction is a function on $V \oplus W$ regarded as a graded vector space. Observe that for consistency, we have to consider elements of V as even and those of W as odd. It is also useful to introduce a $\mathbb{Z}$-grading and assign V degree 0 and W degree 1, while integration over Σ is given degree -2.

If we only consider the "kinetic term" $S_0 = \int_\Sigma \boldsymbol{\eta}_i \mathrm{d}\mathsf{X}^i$, it is quite natural to observe that there is a duality obtained by exchanging $\boldsymbol{\eta}$ and X. The only problem is that we want to think of the zero form in X as a map. However, the zero form in $\boldsymbol{\eta}$ has to be considered as an odd element. The problem is solved if we allow the target M to be a graded manifold itself. The duality then exchanges M with a

[4] An invariant description requires introducing graded manifolds and regarding the space of superfields as the graded infinite-dimensional manifold $\mathrm{Map}(T[1]\Sigma, T^*[1]M)$.

dual manifold $\tilde{M}$. If we now take into account the term involving π, we see that it corresponds to a similar term for a multivector field on $\tilde{M}$.

If Σ has a boundary, the superfields have to be assigned boundary conditions. We choose a coisotropic submanifold C of M and require X on the boundary to take values in C, while $\boldsymbol{\eta}$ has to take values in the pullback of the conormal bundle of C. As we work in perturbation theory, we may actually consider just a formal neighborhood of C. Namely, we choose an embedding of the normal bundle NC of C into M and regard NC as the Poisson manifold. The dual manifold turns out to be the graded manifold $N^*[1]C$. To π there corresponds a multivector field $\tilde{\pi}$ of total degree 2. The dual theory can be mathematically understood in terms of a formality theorem for graded manifolds.

7. Formality Theorem

Kontsevich's star product is an application of his formality theorem stating that the DGLA of multidifferential operators on a smooth manifold is formal [**K**]. His explicit local formula for the L_∞-quasi-isomorphism may also be obtained in terms of expectation values of boundary and bulk observables in the PSM with zero Poisson structure [**CF00**]. This construction may be generalized to smooth graded manifolds [**CF07**] (see also [**C06**] for the PSM version). An application [**CF07, LS**] is the deformation quantization of coisotropic submanifolds [**CF04**].

Let $\mathcal{M}$ be a smooth graded manifold $A = \bigoplus_{k\in\mathbb{Z}} A^k$ its graded algebra of functions, $\mathcal{V}(\mathcal{M})$ its complex of graded multivector fields and $\mathcal{D}(\mathcal{M})$ its complex of graded multidifferential operators (regarded as a subcomplex of the Hochschild complex of A). We consider $\mathcal{V}(\mathcal{M})$ as DGLAs by taking the total degree (shifted by one); viz.:

$$\mathcal{V}^k(\mathcal{M}) := \{r\text{-multivector fields of homogeneous degree } s \text{ with } r+s=k+1\},$$
$$\mathcal{D}^k(\mathcal{M}) := \{r\text{-multidiff. operators of homogeneous degree } s \text{ with } r+s=k+1\}.$$

THEOREM 7.1 (Formality Theorem). *There exists an L_∞-quasi-isomorphism $U\colon \mathcal{V}(\mathcal{M}) \rightsquigarrow \mathcal{D}(\mathcal{M})$, hence the DGLA $\mathcal{D}(\mathcal{M})$ is formal. Moreover, U may be chosen such that the degree-one component $U_1\colon \mathcal{V}(\mathcal{M}) \mapsto \mathcal{D}(\mathcal{M})$ is the Hochschild–Kostant–Rosenberg (HKR) map.*

See [**K**] for the proof in the ordinary case and [**CF07**] for the case of graded manifolds.

One may also extend U to formal power series in a parameter ϵ and get an L_∞-quasi-isomorphism $\mathcal{V}_\epsilon(\mathcal{M}) \rightsquigarrow \mathcal{D}_\epsilon(\mathcal{M})$, where

$$\mathcal{V}_\epsilon(\mathcal{M}) := \epsilon\mathcal{V}(\mathcal{M})[[\epsilon]], \quad \mathcal{D}_\epsilon(\mathcal{M}) := \epsilon\mathcal{D}(\mathcal{M})[[\epsilon]].$$

In particular, since MC elements are mapped to MC elements by an L_∞-morphism (with no convergence problems in the setting of formal power series), the Formality Theorem also shows that every MC element of $\mathcal{V}(\mathcal{M})$ gives rise to a MC element of $\mathcal{D}_\epsilon(\mathcal{M})$. In the case of ordinary manifolds, the former MC element is the same as a Poisson structure, while the latter defines a deformation quantization of the former. (A classification theorem also follows by the fact that an L_∞-quasi-isomorphism actually yields an isomorphism of the sets of MC elements modulo gauge transformations).

The case of graded manifold is important for coisotropic submanifolds. As we recalled in Section 3, the algebra $A := \Gamma(NC)$ naturally appears in the cohomological description of a coisotropic submanifold C of a Poisson manifold (M, π). Now one can reinterpret A as the algebra of functions on the graded manifold $N^*[1]C$ (where [1] denotes a shift by 1 in the fiber coordinates). Moreover, the restriction of π to NC (obtained upon choosing an embedding of NC into M) may be regarded as a MC element on NC but also on $N^*[1]C$.

One way to see this is via Roytenberg's Legendre mapping theorem [**R**] which states the existence of a canonical antisymplectomorphism between $T^*[n]E$ and $T^*[n](E^*[n])$ for every integer n and for every graded vector bundle E. In particular, for $n = 1$ and observing that multivector fields on a graded manifold may be reinterpreted as functions on its tangent bundle shifted by 1, one gets an anti-isomorphism between the GLAs $\mathcal{V}(E)$ and $\mathcal{V}(E[1])$; in particular, this yields an isomorphism of the sets of MC elements. Specializing to $E = NC$ (as a graded vector bundle concentrated in degree zero) yields the sought for result.

A more direct way is just to observe that each derived bracket λ_k in (3.1) is a multiderivation on the algebra A and so a k-vector field on $N^*[1]C$. Their linear combination is the desired MC element. We now consider a more general situation.

7.1. Maurer–Cartan elements for graded manifolds. In general, given a graded manifold $\mathcal{M}$ and a MC element π in $\mathcal{V}(\mathcal{M})$, by Theorem 3.1 (with $\mathfrak{g} = \mathcal{V}(\mathcal{M})$, $\mathfrak{h} = C^\infty(\mathcal{M})$, P and i the natural projection and inclusion), we may construct a P_∞-algebra on $A := C^\infty(\mathcal{M})$ with multibrackets as in (3.1). Taking care of degrees, one sees that λ_k is a multiderivation of degree $2 - i$, i.e.:

$$\begin{aligned}
&\lambda_0 = P_0 \in A^2,\\
&\lambda_1 \colon A^j \to A^{j+1},\\
&\lambda_2 \colon A^{j_1} \otimes A^{j_2} \to A^{j_1+j_2},\\
&\dots\\
&\lambda_i \colon A^{j_1} \otimes \cdots \otimes A^{j_i} \to A^{j_1+\dots j_i+2-i},\\
&\dots
\end{aligned}$$

If $P_0 = 0$, then $\lambda_0 = 0$ and we have a flat L_∞-algebra. In this case, λ_1 is a differential and its cohomology

$$\mathsf{A}^\bullet := H^\bullet_{\lambda_1}(A)$$

is a graded Poisson algebra. In the following we will denote by $\boldsymbol{\{}\ ,\ \boldsymbol{\}}$ the induced Poisson bracket on A. In particular, A^0 is a Poisson algebra (but in general it is not the algebra of functions on a smooth manifold) whose Poisson bracket we will simply denote by $\{\ ,\ \}$.

Definition 7.2. A graded manifold $\mathcal{M}$ with a (flat) P_∞-structure on $A = C^\infty(\mathcal{M})$ is called a (flat) P_∞-manifold.

We now turn to multidifferential operators. An element m in $\mathcal{D}^1_\epsilon(\mathcal{M})$ consists of a sequence m_i, where m_i is an i-multidifferential operators of degree $2-i$, i.e.,

$$\begin{aligned}
&m_0 \in \epsilon A^2[[\epsilon]],\\
&m_1\colon A^j[[\epsilon]] \to \epsilon A^{j+1}[[\epsilon]],\\
&m_2\colon A^{j_1}[[\epsilon]] \otimes A^{j_2}[[\epsilon]] \to \epsilon A^{j_1+j_2}[[\epsilon]],\\
&\dots\\
&m_i\colon A^{j_1}[[\epsilon]] \otimes \dots \otimes A^{j_i}[[\epsilon]] \to \epsilon A^{j_1+\dots j_i+2-i}[[\epsilon]],\\
&\dots
\end{aligned}$$

Let us denote by χ the (extension to $A[[\epsilon]]$) of the multiplication on A and define $\mu = \chi + m$; viz.:

$$\mu_i = \begin{cases} m_i & i \neq 2,\\ \chi + m_2 & i = 2.\end{cases} \tag{7.1}$$

The element m is MC, i.e.,

$$\delta m + \frac{1}{2}[m, m] = 0, \qquad (\delta = [\chi, \ ]),$$

iff the operations μ_i define an A_∞-structure on $A[[\epsilon]]$. If in addition μ_0 vanishes, one says that the A_∞-algebra is flat. In this case, $H^\bullet_{\mu_1}$ is an associative algebra.

Let $\tilde{\mu}$ denote the skew-symmetrization of μ. Then $(A[[\epsilon]], \tilde{\mu})$ is an L_∞-algebra. Since the multiplication χ is graded commutative, the $\tilde{\mu}$ will take values in $\epsilon A[[\epsilon]]$. So, dividing by ϵ and working modulo ϵ, they define operations $\underline{\tilde{\mu}_i}$ on A which make it into an L_∞-algebra. It may be shown, see [**CF07**], that the $\underline{\tilde{\mu}_i}$s are actually multiderivations, so we have a P_∞-algebra structure on A.

7.2. Deformation quantization of P_∞-manifolds. Let A be the algebra of functions on a P_∞-manifold $\mathcal{M}$. We denote by χ the graded commutative product and by λ the multibrackets.

DEFINITION 7.3. A deformation quantization of $\mathcal{M}$ consists of an A_∞-structure on $(A[[\epsilon]], \mu)$ with $\mu = \chi + m$, such that:

(1) m is of order ϵ;
(2) m consists of multidifferential operators;
(3) m vanishes when one of its arguments is the unit in A; (i.e., $m_i(a_1, \dots, a_i) = 0$ if $a_j = 1$ for some $j \in \{1, \dots, i\}$, $i > 0$);
(4) the induced P_∞-structure $\tilde{\mu}$ is equal to λ.

If $\mathcal{M}$ is flat, by deformation quantization in the flat sense we mean that in addition the condition $\mu_0 = 0$ is fulfilled.

The formality theorem for graded manifolds then implies the following

THEOREM 7.4. *Every P_∞-manifold $\mathcal{M}$ admits a deformation quantization.*

PROOF. Let $P \in \mathcal{V}_\epsilon(\mathcal{M})$ be the MC element with derived brackets λ. Define m by applying the L_∞-quasi-isomorphism U to ϵP:

$$m = \sum_{n=0}^{\infty} \frac{1}{n!} U_n(P, \dots, P).$$

This is a MC element in $\epsilon\mathcal{D}_\epsilon(\mathcal{M})$ with the desired properties. □

Observe however that the problem of quantizing flat P_∞-manifolds in the flat sense is on the other hand not solved. In fact, we are not able to conclude that $P_0 = 0$ implies $m_0 = 0$ but only that $m_0 = O(\epsilon^2)$. This is the deformation quantization version of a (potential) anomaly (see [**CF07**]).

There are some cases when m_0 actually happens to vanish (see [**CF04**]). There are also cases [**CF07**] where m_0 can be killed by shifting the A_∞-structure without changing its classical limit. Namely, for $\gamma \in \epsilon A^1[[\epsilon]]$ define

$$\hat{m}_0 = m_0 + m_1(\gamma) + \frac{1}{2} m_2(\gamma,\gamma) + \cdots,$$
$$\hat{m}_1(a) = m_1(a) + m_2(a,\gamma) \pm m_2(\gamma, a) + \cdots.$$

Then, $\forall\gamma$, $\mu + \hat{m}$ is again an A_∞-structure on $A[[\epsilon]]$ that induces the same P_∞-structure on A. One may then look for a γ such that $\hat{m}_0 = 0$. It is not difficult to prove by induction, see [**CF07**], that a sufficient condition for this to happen is that $\mathsf{A}^2 = \{0\}$. So we have the following

THEOREM 7.5. *Every flat P_∞-manifold $\mathcal{M}$ with $\mathsf{A}^2 = \{0\}$ admits a deformation quantization in the flat sense.*

7.3. Quantization of cohomology. Assume now that the flat P_∞-manifold $\mathcal{M}$ has been quantized in the flat A_∞ sense. We denote by χ the graded commutative multiplication on the algebra of functions A, by λ_i the multibrackets defining the flat P_∞-structure on A and by μ_i the multibrackets defining the flat A_∞-structure on $A[[\epsilon]]$. We will also denote by d the differential λ_1 and by A the d-cohomology of A.

The μ_1-cohomology is an associative algebra. However, in view of quantizing $\mathsf{A}^\bullet$, or at least A^0, this is not the algebra we are in general interested in since the projection modulo ϵ is not a chain map. We have instead to proceed as follows. Observe that by appropriately rescaling the multibrackets, we get a new A_∞-structure; viz., for given integers s_i, define $\tau_i = \epsilon^{s_i}\mu_i$, where we assume μ_i to be divisible by ϵ^{-s_j} whenever $s_j < 0$. If moreover $s_i + s_j$ is a function of $i+j$, the multibrackets τ_i also define an A_∞-algebra.

In particular, we may take $s_i = i-2$ since $\mu_1 = O(\epsilon)$. Namely, we define the new A_∞-algebra structure

$$\tau_i := \epsilon^{i-2}\mu_i.$$

Observe that

$$\tau_1 = \mathrm{d} + O(\epsilon), \qquad \tau_2 = \chi + O(\epsilon), \qquad \tau_i = O(\epsilon^{i-1}),\ i > 2.$$

By construction, the graded skew-symmetrization $\tilde{\tau}_2$ of τ_2 is divisible by ϵ; so there is a unique operation ψ with $\tilde{\tau}_2 = \epsilon\psi$. The bracket λ_2 on A is then ψ modulo ϵ.

We will also denote by $\delta = \sum_{n=0}^\infty \epsilon \mathrm{d}_n$, $\mathrm{d}_0 = \mathrm{d}$, the differential τ_1 and by B the δ-cohomology of $A[[\epsilon]]$. Observe that the differential δ is a deformation of the differential d.

Observe that by construction τ_2, $\tilde{\tau}_2$ and ψ are chain maps and we will denote by $[\tau_2]$, $[\tilde{\tau}_2]$ and $[\psi]$ the induced operations in cohomology. The operation $[\tau_2]$ is an associative product which we will also denote by $\bigstar$ (and by $\star$ its restriction to B^0).

Whenever needed we denote δ-cohomology classes by $[\ ]_\delta$ and d-cohomology classes by $[\ ]_\mathrm{d}$. With these notations we have

$$[a]_\delta \bigstar [b]_\delta = [\tau_2]([a]_\delta, [b]_\delta) = [\tau_2(a,b)]_\delta,$$

where $[\tau_2]$ is the map induced by τ_2 in cohomology, while a and b are representatives of the classes $[a]_\delta$ and $[b]_\delta$. By construction $[\tilde{\tau}_2]$ defines the graded $\star$-commutator $[\ ,\ ]$ on B:

$$[\,[a]_\delta\,,\,[b]_\delta\,] = [\tilde{\tau}_2]([a]_\delta, [b]_\delta) = [\tilde{\tau}_2(a,b)]_\delta.$$

We will denote simply by $[\ ,\ ]$ its restriction to B^0. Observe that multiplication by ϵ commutes with δ. So B is also an $\mathbb{R}[[\epsilon]]$-module with $\epsilon[a]_\delta = [\epsilon a]_\delta$. Moreover, we have

$$[\,[a]_\delta\,,\,[b]_\delta\,] = \epsilon[\psi]([a]_\delta, [b]_\delta).$$

If B (B^0) is ϵ-torsion free—i.e., multiplication by ϵ is injective—, then $[\psi]$ is the unique operation with the above property. Consider now the $\mathbb{R}$-linear projection

$$\begin{array}{rccc} \varpi\colon & A[[\epsilon]] & \to & A \\ & \sum_{n=0}^\infty \epsilon^n a_n & \mapsto & a_0. \end{array}$$

Observe that $\varpi\colon (A[[\epsilon]], \delta) \to (A, \mathrm{d})$ is a chain map. We will denote by $[\varpi]\colon \mathsf{B} \to \mathsf{A}$ the induced map in cohomology. Since ϖ is an algebra homomorphism, so is $[\varpi]$. The relation with the Poisson structure on A is clarified by the now obvious formula

$$[\varpi]([\psi]([a]_\delta, [b]_\delta)) = \{\,[\varpi]([a]_\delta)\,,\,[\varpi]([b]_\delta)\,\}, \qquad \forall [a]_\delta, [b]_\delta \in \mathsf{B}$$

which immediately implies the

PROPOSITION 7.6. *The algebra homomorphism $[\varpi]$ has the following additional properties:*

(1) *The image C (C^0) of $[\varpi]$ is a Poisson subalgebra of A (A^0).*
(2) *If B is ϵ-torsion free, then*

$$[\varpi]\left(\frac{[\,[a]_\delta\,,\,[b]_\delta\,]}{\epsilon}\right) = \{\,[\varpi]([a]_\delta)\,,\,[\varpi]([b]_\delta)\,\}, \qquad \forall [a]_\delta, [b]_\delta \in \mathsf{B}.$$

(3) *If B^0 is ϵ-torsion free, then the above formula holds for all $[a]_\delta, [b]_\delta \in \mathsf{B}^0$.*

Since obviously $\operatorname{Ker}\varpi = \epsilon A[[\epsilon]]$, we have the short exact sequence

$$0 \to A[[\epsilon]] \xrightarrow{\epsilon} A[[\epsilon]] \xrightarrow{\varpi} A \to 0$$

which induces the long exact sequence

$$\cdots \to \mathsf{A}^{i-1} \xrightarrow{\partial} \mathsf{B}^i \xrightarrow{\epsilon} \mathsf{B}^i \xrightarrow{[\varpi]} \mathsf{A}^i \xrightarrow{\partial} \mathsf{B}^{i+1} \to \cdots$$

in cohomology. Immediately we then get the

PROPOSITION 7.7. *The following statements are equivalent:*

(1) *$[\varpi]$ is surjective;*
(2) *B is ϵ-torsion free;*
(3) *∂ is trivial.*

To continue our study of the problem, we now need Lemma A.1 of [**CFT**], which we state in a slightly modified version:

LEMMA 7.8. *Let $\mathbb{k}$ be a field and M a $\mathbb{k}[[\epsilon]]$-module endowed with the $\mathbb{k}[[\epsilon]]$-adic topology. Then $\mathsf{M} \simeq_{\mathbb{k}} \mathsf{M}_0[[\epsilon]]$ for some $\mathbb{k}$-vector space M_0 iff M is Hausdorff, complete and ϵ-torsion free. Moreover, $\mathsf{M}_0 \simeq_{\mathbb{k}} \mathsf{M}/\epsilon\mathsf{M}$.*

We finally have the

THEOREM 7.9. *If* B *is Hausdorff and complete in the* ϵ*-adic topology, then* B *is a deformation quantization of* A *iff any (and so all) of the statements in Proposition 7.7 holds.*

PROOF. If B is a deformation quantization of A, then in particular $\mathsf{B} \simeq_{\mathbb{R}} \mathsf{A}[[\epsilon]]$. So by Lemma 7.8 B is ϵ-torsion free. On the other hand, the statements in Proposition 7.7 imply that $\mathsf{A} \simeq_{\mathbb{R}} \mathsf{B}/\operatorname{Ker}[\varpi] = \mathsf{B}/\epsilon\mathsf{B}$; so $\mathsf{B} \simeq_{\mathbb{R}} \mathsf{A}[[\epsilon]]$ by Lemma 7.8. Statement (2) in Proposition 7.6 completes the proof. □

We are not able to show that $\operatorname{Im}\delta$ is closed in the ϵ-adic topology, so we must put the extra condition in the Theorem. We wish however to make the following[5]

CONJECTURE 7.10. *The* $\mathbb{R}[[\epsilon]]$*-module* B *is Hausdorff and complete in the* ϵ*-adic topology.*

In general we do not expect B to be a deformation quantization of A. However, it is often enough to have B^0 as a deformation quantization of A^0. We end this Section by exploring some sufficient conditions for this to happen.

LEMMA 7.11. *If* B^i *is Hausdorff and complete, then it is isomorphic to* $\mathsf{C}^i[[\epsilon]]$ *as an* $\mathbb{R}$*-vector space iff it is* ϵ*-torsion free.*

PROOF. The "only if" implication is obvious. As for the "if" part, by Lemma 7.8 we conclude that B^i is isomorphic to $(\mathsf{B}^i/\epsilon\mathsf{B}^i)[[\epsilon]]$. On the other hand the long exact sequence gives $0 \to \mathsf{B}^i \xrightarrow{\epsilon} \mathsf{B}^i \xrightarrow{[\varpi]} \mathsf{C}^i \to 0$ which completes the proof. □

By statement (3) in Proposition 7.6, we then get

COROLLARY 7.12. *If* B^0 *is Hausdorff, complete and* ϵ*-torsion free, then it is a deformation quantization of* C^0*.*

The long exact sequence yields the

LEMMA 7.13. *If* $\mathsf{A}^{i-1} = \{0\}$*, then* B^i *is* ϵ*-torsion free.*

Thus, $A^{-1} = \{0\}$ is a sufficient condition for B^0 to be a deformation quantization of C^0 provided it is Hausdorff and complete. Finally, we have the

LEMMA 7.14. *If* $\mathsf{A}^{i+1} = \{0\}$*, then* $[\varpi]^i \colon \mathsf{B}^i \to \mathsf{A}^i$ *is surjective.*

PROOF. This is a standard proof in cohomological perturbation theory. Let $[a_0]_{\mathrm{d}} \in \mathsf{A}^i$. Choose one of its representatives $a_0 \in A^i$. We look for a δ-closed $a = \sum_{n=0}^{\infty} \epsilon^n a_n \in A^i[[\epsilon]]$. The equations we have to solve have the form

$$\mathrm{d}a_n = -\sum_{\substack{r+s=n+1\\ r>0}} \mathrm{d}_r a_s, \tag{7.2}$$

and we may solve them by induction. Namely:

(1) For $n = 0$, the equation is just $\mathrm{d}a_0 = 0$ which is satisfied by assumption.
(2) Assume now that all equations for a_k, $k < n$, have been solved. This implies that that the r.h.s. of (7.2) is d-closed. Since we assume that $\mathsf{A}^{i+1} = \{0\}$, we may then find a_n satisfying the equation.

□

[5]Observe that the Conjecture is general false if we drop the conditions that A is the algebra of functions on a graded manifold and that the components d_n of δ are differential operators.

REMARK 7.15. Observe that, if we knew that B^{i+1} were Hausdorff (e.g., if Conjecture 7.10 were true), then we could derive Lemma 7.14 directly from the long exact sequence (this is just a variant of Nakayama's Lemma). In fact, $A^{i-1} = \{0\}$ implies that multiplication by ϵ on B^{i+1} is surjective. So every element $a \in \mathsf{B}^{i+1}$ may be written as $a = \epsilon a_1$. Continuing this process, we get a sequence a_n with $a = \epsilon^n a_n$. The sequence $\epsilon^n a_n$ converges to zero, so $a = 0$ since B^{i+1} is Hausdorff. Thus, $\mathsf{B}^{i+1} = \{0\}$ and by the long exact sequence again we get the result.

Putting together Theorem 7.5, Corollary 7.12, Lemmata 7.13 and 7.14, we get the following

THEOREM 7.16. *If $\mathsf{A}^2 = \mathsf{A}^1 = \mathsf{A}^{-1} = \{0\}$, then B^0 is a deformation quantization of A^0 provided it is Hausdorff and complete.*

If Conjecture 7.10 holds, we get the sufficient conditions in a much nicer way which makes reference only to the d-cohomology:

"THEOREM". *A sufficient condition for a deformation quantization of A^0 to exist is $\mathsf{A}^2 = \mathsf{A}^1 = \mathsf{A}^{-1} = \{0\}$.*

Observe that the sufficient condition is by no means necessary. For example, in the case $\lambda_i = 0\ \forall i$, we have $\mathsf{A}^\bullet = A^\bullet$. On the other hand, $\delta = 0$ and $\mathsf{B}^\bullet = A^\bullet[[\epsilon]]$ is a deformation quantization of $\mathsf{A}^\bullet$.

7.4. Deformation quantization of coisotropic submanifolds. We now return to the case when C is a coisotropic submanifold of M (which is Poisson at least in a neighborhood of C). We take our graded manifold $\mathcal{M}$ to be $N^*[1]C$ with MC element the restriction of π to $\mathcal{V}(NC) = \mathcal{V}(N^*[1]C)$. Now $\mathsf{A}^{-1} = \{0\}$ since we do not have negative degrees, while A^0 is the Poisson algebra of functions on C that are invariant under the canonical distribution on C. If we denote by $\underline{C}$ the leaf space, we also write $C^\infty(\underline{C}) = \mathsf{A}^0$. Observe that in general $\underline{C}$ is not a manifold, so the above is just a definition. Using the notations of Section 7.3, we have $\mathsf{B}^0 = \operatorname{Ker}\delta$. So B^0 is Hausdorff and complete, and by Theorem 7.16 we have the

THEOREM 7.17 ([**CF07**]). *If the first and second Lie algebroid cohomology of N^*C vanish, the zeroth cohomology $C^\infty(\underline{C})$ has a deformation quantization.*

The second Lie algebroid cohomology is also the space where the usual BRST/BFV anomaly lives; see [**B04**] for this in the context of deformation quantization. The advantage of the present approach is that one has formulae for m_0, so one can in principle check whether it may be canceled.

Observe that the conditions of Theorem 7.17 are very strong and by no means necessary. At the moment we actually do not know a single example where the anomaly shows up, at least locally. There are examples where the Lie algebroid cohomology class of m_0 is nontrivial, see [**W**], but in these examples C is a point, so there is no problem in quantizing its reduced space (the problem arises however in the many-brane setting of Section 8).

As $C^\infty(\underline{C})$ might be rather poor (e.g., just constant functions), it might be interesting to quantize the whole Lie algebroid cohomology. Assuming that the A_∞-structure is flat, by Theorem 7.9 the deformed cohomology B yields a deformation quantization of A iff it is Hausdorff, complete and ϵ-torsion free.

7.5. Poisson submanifolds. A particular case of a coisotropic submanifold is a Poisson submanifold, i.e., a submanifold C of (M, π) such that the inclusion map is a Poisson map. Equivalently, a Poisson submanifold is a coisotropic submanifold with trivial characteristic distribution. In this case, $\underline{C} = C$, so the quantization of the reduction is not a problem. The interesting question is whether one can deform the pullback of the inclusion map to a morphism of associative algebras. To approach this problem, we associate to C a different graded manifold as in the coisotropic case.

If the submanifold C is determined by constraints, i.e., $C = \Phi^{-1}(0)$ with $\Phi\colon M \to V$ a given map to a vector space V, we set $\mathcal{M} := M \times V[-1]$ and reinterpret Φ as an element of $C^\infty(M) \otimes V$ and so as a vector field X of degree 1 on $\mathcal{M}$. If we introduce coordinates $\{\mu^\alpha\}_{\alpha=1,\dots,k:=\dim V}$ of degree -1, we have $X = \Phi^\alpha(x)\frac{\partial}{\partial \mu^\alpha}$, where the Φ^αs are the components of Φ w.r.t. the same basis. Observe that $[\,X\,,\,X\,] = 0$ and that $[\,\pi\,,\,X\,]$ vanishes on C. Moreover, the cohomology of $C^\infty(\mathcal{M})$ w.r.t the differential $\delta := [\,X\,,\ \,]$ is $C^\infty(C)$. If C is not determined by constraints, we take $\mathcal{M} := N[-1]C$ and X the vector field of degree 1 corresponding to the zero section. The crucial observation now is that $A := C^\infty(\mathcal{M}) = \bigoplus_{j\le 0} A^j$ is a nonpositively graded commutative algebra. As a consequence, $\mathsf{A}^i = 0$, $i > 0$, with the differential $\lambda_1 = \delta$. Observe moreover that A^0 is just $C^\infty(M)$ and that $\mathsf{A}^0 = A^0/I = C^\infty(C)$, with $I = \delta A^{-1}$ the vanishing ideal of C.

7.5.1. *Casimir functions.* A particularly simple case is when the components of Φ are Casimir functions: viz., $\{\,\Phi^\alpha\,,\,f\,\} = 0\ \forall\alpha\ \forall f \in C^\infty(M)$. Equivalently $[\,\pi\,,\,\Phi^\alpha\,] = 0\ \forall\alpha$, i.e., $[\,\pi\,,\,X\,] = 0$. Thus $\hat\pi := \pi + X$ is a MC element. The induced P_∞-structure has $\lambda_1 = \delta$. Quantization has the following easy to check remarkable properties: *i)* $(A^0[[\epsilon]], \tau_2)$ is Kontsevich's deformation quantization A_M of $C^\infty(M)$; *ii)* $\tau_2(\mu^\alpha, f) = \tau_2(f, \mu^\alpha) = \mu^\alpha f$, $\forall\alpha$. As a consequence $\mathsf{I} := \tau_1(A^{-1}[[\epsilon]])$ is the two-sided ideal generated by $\{\tau_1(\mu^\alpha)\}$. A further easy computation, using Stokes' theorem in the explicit expression of the coefficients by Kontsevich's graphs, shows that $\tau_1(\mu^\alpha) = D(\Phi^\alpha)$, where D is the Duflo–Kirillov–Kontsevich map

$$\begin{array}{rccc} D\colon & C^\infty(M) & \to & C^\infty(M)[[\epsilon]], \\ & f & \mapsto & \sum_{n=0}^\infty \frac{\epsilon^n}{n!} U_{n+1}(f, \pi, \dots, \pi). \end{array}$$

So we get a deformation quantization of C as $\mathsf{B}^0 = A_M/\mathsf{I}$, and the projection $A_M \to \mathsf{B}^0$ is a quantization of the inclusion map. This proves a conjecture in [**CR**] (see also [**C06**, Sect. 5] for a previous sketch of this proof).

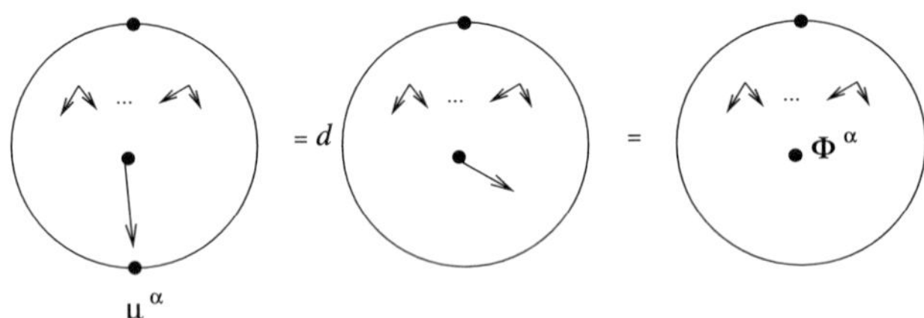

FIGURE 2. $\tau_1(\mu^\alpha) = D(\Phi^\alpha)$

7.5.2. *The general case.* In general $\pi + X$ is not a MC element. However, the fact that $[\,\pi\,,\,X\,]$ vanishes on C, makes $\pi + X$ a MC element up to δ-exact terms. As a consequence, one may use cohomological perturbation theory as in [**LS**] and find a MC element $\hat{\pi}$ of the form $\pi + X$+corrections in the ideal of multivector fields of total degree 2 generated by $\{\mu^\alpha\}$. Observe that the only vector field in $\hat{\pi}$ is X, so $\lambda_1 = \delta$. The vanishing of the cohomologies in positive degrees implies that we have an A_∞-structure on $A[[\epsilon]]$ and a surjective map $\mathsf{B}^0 \to A^0/I$. Moreover, the restriction of τ_2 in degree 0 makes $A^0[[\epsilon]]$ into an algebra with an algebra morphism to $\mathsf{B}^0 = A^0[[\epsilon]]/\mathsf{I}$, $\mathsf{I} := \tau_1(A^{-1}[[\epsilon]])$. The two problems of this constructions are the following: *i)* $(A^0[[\epsilon]], \tau_2)$ might not be associative; *ii)* B^0 might not be isomorphic to $A^0/I[[\epsilon]]$.

It may be shown that $(A^0[[\epsilon]], \tau_2)$ is associative iff $\hat{\pi}$ has the form $\pi + X + \pi'$ with π' a *bivector* field. If in addition one chooses the constraints to be linear (e.g., if one works with $\mathcal{M} = N[-1]C$), then one can also easily see that B^0 turns out to be a deformation quantization of $C^\infty(C)$. However, finding a π' as above is a highly nontrivial problem, and it is not clear under which conditions a solution may exist.

A very simple case is when C consists of a point x (a zero of the Poisson structure). A quantization of the inclusion map can then be reinterpreted as a character (i.e., an algebra morphism to the ground ring) of the deformed algebra (deforming the evaluation at x). Even in such a simple situation, the existence of a π' is guaranteed only for $\dim M = 2$ [**S2**]. In higher dimensions, it is an open problem.

8. Many branes

We now turn to the case when more than one coisotropic submanifold is chosen as a boundary condition. Namely, as in Section 5, we take Σ to be a disk. However, we now subdivide the boundary into closed, cyclically ordered intervals $I_1, \dots, I_n$ with exactly one intersection point between subsequent intervals. To the interval I_i we associate boundary conditions corresponding to a coisotropic submanifold C_i.

Assuming clean intersections, in [**CF04**] it is shown that the case $i = 2$ leads to the construction of a bimodule for the deformation quantizations of $\underline{C}_1$ and $\underline{C}_2$ if no anomaly appears.

In this Section we assume that no anomalies show up. If the submanifolds are pairwise transverse this amounts to asking that for each of them the anomaly discussed in subsection 7.4 vanishes. The vanishing of the second Lie algebroid cohomology for each submanifold is a sufficient condition by Theorem 7.17. Notice however that this is only a sufficient condition by no means necessary. A very simple example is when M is symplectic and C_1 is Lagrangian. In this case $\underline{C}_1$ is a point (assume C_1 to be connected), so there is no problem in quantizing it. The bimodule structure associated to $C_2 = M$ can always be found (upon choosing the star product appropriately). On the other hand the relevant Lie algebroid cohomology is the de Rham cohomology of C_1. Another example is the linear case $C_1 = \mathfrak{h}^0 \subset \mathfrak{g}^*$, where $\mathfrak{h}$ is a Lie subalgebra of $\mathfrak{g}$. It is shown in [**CF04**] that there is no anomaly in this case even if the cohomology may be very complicated. Observe however that an example of nonvanishing anomaly has recently been found in [**W**].

We also use the notation $A_{\underline{C}}$ for the PSM deformation quantization of the Poisson algebra $C^\infty(\underline{C})$, where C is a coisotropic submanifold of M. Whenever a distribution is trivial, we omit underlining the submanifold. Observe that A_M is Kontsevich's deformation quantization. With $B_{\underline{C_1 \cap C_2}}$ we then denote the $A_{\underline{C}_1}$-module-$A_{\underline{C}_2}$ corresponding to the picture on the left in fig. 3 (where the big black dot denotes the point at infinity which has to be sent to the point in the intersection where evaluation takes place).

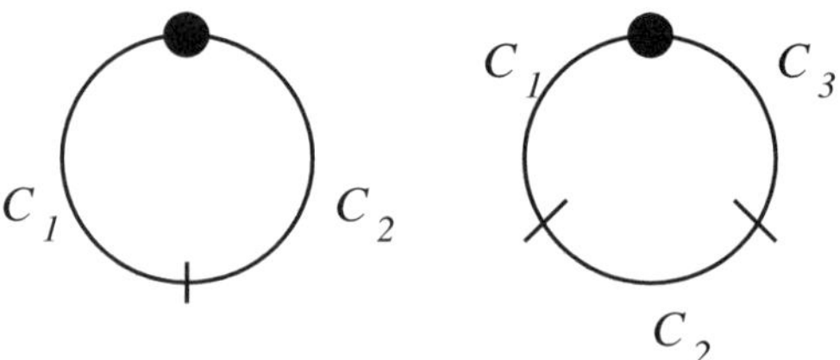

FIGURE 3. Two and three branes

The situation with three coisotropic branes C_1, C_2 and C_3—see the picture on the right in fig. 3—leads to a morphism of bimodules $B_{\underline{C_1\cap C_2}} \otimes_{A_{\underline{C_2}}} B_{\underline{C_2\cap C_3}} \to B_{\underline{C_1\cap C_3}}$ in a neighborhood of a triple intersection. The construction of the corresponding Kontsevich graphs has been analyzed in [**CT**]. In subsection 8.2 we discuss a special case.

One may also consider many branes and, instead of just invariant functions, sections of the corresponding complexes of exterior algebras of normal bundles. One may hope that this would lead to A_∞-bimodules and morphisms thereof, but problems seem to arise with four and more branes.

In the rest of this Section we will concentrate on the special case when one of the branes is the whole manifold; we will consider only functions and no more than three branes.

8.1. Quantization of morphisms. Let ϕ be a Poisson map $M \to N$. Then its graph $\operatorname{Graph}\phi$ is a coisotropic submanifold of $\overline{M} \times N$, where $\overline{M}$ denotes M with opposite Poisson structure. We select the two branes $C_1 = \operatorname{Graph}\phi$ and $C_2 = M \times N$ in $\overline{M} \times N$. Consider the right $A_{\overline{M}\times N}$-module structure on $B_{\operatorname{Graph}\phi}$, forgetting about its left $A_{\underline{\operatorname{Graph}\phi}}$-module structure. Since $A_{\overline{M}} \otimes A_N$ is a subalgebra of $A_{\overline{M}\times N}$ and Kontsevich's star product has the property $A_{\overline{M}} = A_M^{\mathrm{opp}}$, we may regard $B_{\operatorname{Graph}\phi}$ as an A_M-bimodule-A_N. This bimodule has a distinguished element, the constant function 1, and the map $A_M \to B_{\operatorname{Graph}\phi}$, $f \mapsto f \cdot 1$ is an isomorphism of $\mathbb{R}[[\epsilon]]$-modules (since it is a deformation of the pullback of the diffeomorphism $p_1\colon \operatorname{Graph}\Phi \to M$ defined by projection). Thus, to every $f \in A_N$ we associate a unique element $\hat{\phi}(f)$ of A_M by the equation

$$1 \cdot f = \hat{\phi}(f) \cdot 1.$$

Observe that $\hat{\phi}$ is a deformation of the pullback ϕ^*. It is not difficult to see that it is a morphism of associative algebras. This way, with the assumption that there is no anomaly, we have found a quantization procedure for Poisson maps.

Observe that one can exchange the role of C_1 and C_2 using the symmetry corresponding to reflecting the disk through the line joining the two intersections of the intervals I_1 and I_2 ($\bullet$ and $|$ in fig. 3). In terms of Kontsevich graphs one gets the same formulae upon changing the sign of the Poisson structure; viz., we get the same result as before if we take $M \times \overline{N}$ as the ambient Poisson manifold.

8.2. Compositions. Now suppose we have two Poisson maps $\phi\colon M \to N$ and $\psi\colon N \to K$. With no anomalies, we may quantize ϕ, ψ and $\psi\circ\phi$. The next problem we wish to address concerns the relation between $\widehat{\psi\circ\phi}$ and $\hat{\phi}\circ\hat{\psi}$.

To deal with it we consider the three branes $C_1 = \operatorname{Graph}\phi\times K$, $C_2 = M\times N\times K$ and $C_3 = M\times\operatorname{Graph}\psi$ in $\overline{M}\times N\times\overline{K}$ as in fig.3. Let $p_1\colon M\times N \to M$ and $\tilde{p}_2\colon N\times K \to K$ be the canonical projections. To $f\in C^\infty(M)[[\epsilon]]$ and $g\in C^\infty(K)[[\epsilon]]$, we associate the element $D(f,g)\in C^\infty(\operatorname{Graph}\psi\circ\phi)[[\epsilon]]$ obtained by putting $p_1^* f\otimes 1$ and $1\otimes\tilde{p}_2^* g$ at the two intersection points (here 1 is the constant function on $C^\infty(N)$). It is possible to check that D defines a morphism of A_M-bimodules-A_K $B_{\operatorname{Graph}\phi}\otimes_{A_N}\otimes B_{\operatorname{Graph}\psi}\to B_{\operatorname{Graph}\psi\circ\phi}$. So it is enough to compute $\sigma_{\phi,\psi} := D(1,1) = 1+O(\epsilon^2)\in B_{\operatorname{Graph}\psi\circ\phi}$. Observe again that there is a unique element $\hat{\sigma}_{\phi,\psi}\in A_M$ such that $\hat{\sigma}_{\phi,\psi}\cdot 1 = \sigma_{\phi,\psi}$. Moreover, since $\hat{\sigma}_{\phi,\psi}$ is of the form $1+O(\epsilon^2)$, it is invertible. Now for $h\in A_K$ we have the identities

$$\begin{aligned}\sigma_{\phi,\psi}\cdot h = D(1,1\cdot h) = D(1,\hat{\psi}(h)\cdot 1) =&\\ = D(1\cdot\hat{\psi}(h),1) = D(\hat{\phi}(\hat{\psi}(h))\cdot 1,1) &= \hat{\phi}\circ\hat{\psi}\cdot\sigma_{\phi,\psi}.\end{aligned}$$

Using $1\cdot h = \widehat{\psi\circ\phi}(h)\cdot 1$ and the definition of $\hat{\sigma}_{\phi,\psi}$, we get

$$\begin{aligned}\sigma_{\phi,\psi}\cdot h = \hat{\sigma}_{\phi,\psi}\cdot 1\cdot h = \hat{\sigma}_{\phi,\psi}\cdot(\widehat{\psi\circ\phi}(h)\cdot 1) = (\hat{\sigma}_{\phi,\psi}\star_M\widehat{\psi\circ\phi}(h))\cdot 1,\\ \hat{\phi}\circ\hat{\psi}(h)\cdot\sigma_{\phi,\psi} = \hat{\phi}\circ\hat{\psi}(h)\cdot(\hat{\sigma}_{\phi,\psi}\cdot 1) = (\hat{\phi}\circ\hat{\psi}(h)\star_M\hat{\sigma}_{\phi,\psi})\cdot 1,\end{aligned}$$

where $\star_M$ denotes the star product on M. Finally,

$$\widehat{\psi\circ\phi}(h) = \hat{\sigma}_{\phi,\psi}^{-1}\star_M(\hat{\phi}\circ\hat{\psi}(h))\star_M\hat{\sigma}_{\phi,\psi},$$

where $\hat{\sigma}_{\phi,\psi}^{-1}$ is the inverse of $\hat{\sigma}_{\phi,\psi}$ w.r.t. $\star_M$.

Thus, upon conjugation (by an element which depends on the given Poisson maps), we see that the composition of quantizations is the quantization of compositions.

8.3. Quantum groups. We now want to apply the results of the last subsection to the case of a Poisson–Lie group G; i.e., a Poisson manifold and a Lie group such that the product $m\colon G\times G\to G$ and the inverse ${}^{-1}\colon\overline{G}\to G$ are Poisson maps. By Kontsevich we have an associative algebra A_G. On the other hand, if no anomaly is present, we get a morphism of algebras $\Delta := \hat{m}\colon A_G\to A_G\hat{\otimes}A_G := A_{G\times G}$. The associativity equation $m\circ(m\otimes id) = m\circ(id\otimes m)$ then yields

$$(\Delta\otimes id)\circ\Delta = \Phi^{-1}\star_{G^3}((id\otimes\Delta)\circ\Delta)\star_{G^3}\Phi,$$

with $\Phi = \hat{\sigma}_{id\otimes m,m}^{-1}\star_{G^3}\hat{\sigma}_{m\otimes id,m}\in A_{G^3}$. In some lucky cases, Φ might turn out to be 1 (or at least central). In this case we would have quantized the Poisson–Lie group as a bialgebra. Otherwise, one may hope to get the right properties for the "associator" Φ in order to get a bialgebra out of it.

Finally, observe that we may also quantize the inverse and get a candidate for the antipode. If relations are preserved, we should get a Hopf algebra structure on A_G, i.e., the corresponding quantum group.

Just assuming that anomalies are absent is not enough to get a Hopf algebra as relations are preserved only up to conjugation by certain elements. If one cannot get rid of them, the resulting structure is probably that of a hopfish algebra as defined in [**BW**].

References

[AKSZ] M. Alexandrov, M. Kontsevich, A. Schwarz and O. Zaboronsky, *The geometry of the master equation and topological quantum field theory*, Int. J. Mod. Phys. **A12**, 1405–1430 (1997).

[BF] I. A. Batalin and E. S. Fradkin, *A generalized canonical formalism and quantization of reducible gauge theories*, Phys. Lett. **B 122**, 157–164 (1983).

[BV] I. A. Batalin and G. A. Vilkovisky, *Relativistic S-matrix of dynamical systems with boson and fermion constraints*, Phys. Lett. **B 69** (1977), 309–312.

[BFFLS] F. Bayen, M. Flato, C. Frønsdal, A. Lichnerowicz, D. Sternheimer, *Deformation theory and quantization.* I, II, Ann. Phys. **111** (1978), 61–110, 111–151.

[BW] C. Blohmann and A. Weinstein, *Group-like objects in Poisson geometry and algebra*, math.SG/0701499.

[BCZ] F. Bonechi, A. S. Cattaneo and M. Zabzine, *Geometric quantization and non-perturbative Poisson sigma model*, Adv. Theor. Math. Phys. **10**, 683–712 (2006).

[B00] M. Bordemann, *The deformation quantization of certain super-Poisson brackets and BRST cohomology*, Conférence Moshé Flato 1999, Vol. II (Dijon), 45–68, Math. Phys. Stud. **22**, Kluwer Acad. Publ. (Dordrecht, 2000).

[B04] ———, *(Bi)modules, morphismes et réduction des star-produits : le cas symplectique, feuilletages et obstructions*, math.QA/0403334; *(Bi)Modules, morphisms, and reduction of star-products: the symplectic case, foliations, and obstructions*, Travaux mathématiques **16** (2005), 9–40, http://www.cu.lu/mathlab/travaux/Last/1BORDE.PDF

[BC] P. Bressler and A. Chervov, *Courant algebroids*, J. Math. Sci. **128**, 3030–3053 (2005).

[Ca] I. Calvo, *Poisson Sigma Models on Surfaces with Boundary: Classical and Quantum Aspects*, Ph. D. thesis, Zaragoza, 2006.

[CFa05] I. Calvo and F. Falceto, *Star products and branes in Poisson-Sigma models*, Commun. Math. Phys. **268**, 607–620 (2006).

[C06] A. S. Cattaneo, *From topological field theory to deformation quantization and reduction*, in *Proceedings of the International Congress of Mathematicians, Madrid, Spain, 2006*, (ed. M. Sanz-Solé, J. Soria, J. L. Varona, J. Verdera), **Vol. III** (European Mathematical Society, 2006), 338–365; Zurich University Preprint No. 06-2006 http://www.math.unizh.ch/fileadmin/math/preprints/icm.pdf.

[CF00] A. S. Cattaneo and G. Felder, *A path integral approach to the Kontsevich quantization formula*, Commun. Math. Phys. **212** (2000), 591–611.

[CF01] ———, *Poisson sigma models and symplectic groupoids*, in *Quantization of Singular Symplectic Quotients*, (ed. N. P. Landsman, M. Pflaum, M. Schlichenmeier), Progress in Mathematics **198** (Birkhäuser, 2001), pp. 61–93.

[CF01b] ———, *On the AKSZ formulation of the Poisson sigma model*, Lett. Math. Phys. **56**, 163–179 (2001).

[CF04] ———, *Coisotropic submanifolds in Poisson geometry and branes in the Poisson sigma model*, Lett. Math. Phys. **69** (2004), 157–175.

[CF07] ———, *Relative formality theorem and quantisation of coisotropic submanifolds*, Adv. Math. **208** (2007), 521–548.

[CFT] A. S. Cattaneo, G. Felder and L. Tomassini, *From local to global deformation quantization of Poisson manifolds*, Duke Math. J. **115**, 329–352 (2002).

[CS] A. S. Cattaneo and F. Schätz, *Equivalences of higher derived brackets*, math.QA/0704.1403.

[CT] A. S. Cattaneo and C. Torossian, *Quantification pour les paires symetriques et diagrammes de Kontsevich*, math.RT/0609693.

[CZ] A. S. Cattaneo and M. Zambon, *Coisotropic embeddings in Poisson manifolds*, math.SG/0611480.

[CZbis] ———, *Pre-Poisson submanifolds*, Zurich University Preprint No. 26-2006: http://www.math.unizh.ch/fileadmin/math/preprints/26-06.pdf, to appear in Travaux mathématiques

[CR] A. Chervov and L. Rybnikov, *Deformation quantization of submanifolds and reductions via Duflo–Kirillov–Kontsevich map*, hep-th/0409005.

[I] N.Ikeda , *Two-dimensional gravity and nonlinear gauge theory*, Ann. Phys. **235**, 435–464 (1994).

[K] M. Kontsevich, *Deformation quantization of Poisson manifolds*, Lett. Math. Phys. **66** (2003), 157–216.

[LS] S. L. Lyakhovich and Sharapov, *BRST theory without Hamiltonian and Lagrangian*, JHEP 0503 (2005) 011.

[OP] Y.-G. Oh and J.-S. Park, *Deformations of coisotropic submanifolds and strong homotopy Lie algebroids*, Invent. Math. **161** (2005), 287–360.

[R] D. Roytenberg, *Courant Algebroids, Derived Brackets and Even Symplectic Supermanifolds*, Ph.D. Thesis, Berkeley, 1999; `math.DG/9910078`.

[SS] P. Schaller and T. Strobl, *Poisson structure induced (topological) field theories*, Mod. Phys. Lett. **A9**, 3129–3136 (1994).

[S] F. Schätz, *BFV-complex and higher homotopy structures*, `math.QA/0611912`.

[S2] ———, private communication.

[S97] J. Stasheff, *Homological reduction of constrained Poisson algebras*, J. Diff. Geom. **45**, 221–240 (1997).

[V] T. Voronov, *Higher derived brackets and homotopy algebras*, J. Pure Appl. Algebra **202** (2005), 133–153.

[W] T. Willwacher, *A counterexample to the quantizability of modules*, `math.SG/0706.0970`.

Institut für Mathematik, Universität Zürich–Irchel, Winterthurerstrasse 190, CH-8057 Zürich, Switzerland

E-mail address: `alberto.cattaneo@math.unizh.ch`

Contemporary Mathematics
Volume **450**, 2008

Examples of higher order stable singularities of Poisson structures

Jean-Paul Dufour

ABSTRACT. In a recent paper of A. Wade and the author we exhibited a theorem which characterizes stable singularities of any order for Poisson structures and Lie algebroids. We gave also many examples: essentially examples of stable singularities of order 2 in any dimension and others of order greater than 3 in dimension 3. The aim of this paper is twofold: first to give a more precise theorem, secondly to give examples of stable singularity of any order greater than 3 but in any dimension greater than 2.

1. Introduction

Let M be a Poisson manifold with Poisson tensor Π. A singular point x of Π is said to be a *stable* singular point if every Poisson tensor on M sufficiently near Π has a singular point near x.

In this paper we consider Poisson tensors as maps $F : M \longrightarrow \bigwedge^2 TM$ and we endow the set of such maps with the $\mathcal{C}^s$ compact open topology (s will be made more precise later); this signifies that two such Poisson tensors are near if their values and the values of their derivatives up to order s are uniformly near on a fixed compact set of M.

The results of this paper were inspired by the following theorem of M. Crainic and R. L. Fernandes which was announced at the conference Poisson 2004 in Luxemburg and which is a particular case of the one they give in Poisson 2006 in Tokyo (this volume).

THEOREM 1.1. (Crainic-Fernandes) *Let x be a singular point of a Poisson tensor Π. Denote by $\mathfrak{g}$ the Lie algebra corresponding to the linear part $\Pi_x^{(1)}$ of Π at x. If $H^2(\mathfrak{g}, \mathbb{R}) = 0$ then x is stable for the $\mathcal{C}^2$ compact open topology.*

There is also a sort of converse result: let $\mathfrak{g}$ be any finite dimensional real Lie algebra; if every singular point of a Poisson tensor having $\mathfrak{g}$ as linear part is stable then we have $H^2(\mathfrak{g}, \mathbb{R}) = 0$.

So we may think that the stability of a singular point is equivalent to $H^2(\mathfrak{g}, \mathbb{R}) = 0$.

1991 *Mathematics Subject Classification.* 53D17.

In fact it appears (see [**DW06**]) that there are stable singular points with a null linear part; in these cases $\mathfrak{g}$ is commutative and $H^2(\mathfrak{g}, \mathbb{R})$ doesn't vanish. We call them "higher order stable singularities." There is no contradiction here with the above "converse" result; this simply signifies that, when $H^2(\mathfrak{g}, \mathbb{R})$ doesn't vanish, we may have sometimes stable singularities and sometimes unstable ones.

In the following I will recall first results of [**DW06**] (Theorem 1.2) giving a more precise statement which, for example, may be useful for the study of the stability of compact symplectic leaves not reduced to a point. In the second part of this paper I will show the stability of the singular point at the origin of Poisson tensors $I \wedge X^{(k-1)}$ with $I = \sum_{i=1}^n x_i \partial/\partial x_i$ and $X^{(k-1)} = \sum_{i=1}^n x_i^{k-1} \partial/\partial x_i$ and $k > 3$, $n > 2$. This shows that, even when n is big, there are stable singularities of any order $k \neq 3$ for Poisson tensors of maximal rank 2. This, a priori surprising, result shows that the set of Poisson structures is intricate.

2. An improvement of a result of [DW06]

Let Π be a Poisson tensor on M. A *singularity of order* k of Π is a point m of M where the $(k-1)$-jet of Π vanishes but not its k-jet ($k > 0$).

Recall ([**DZ05**]) that it makes sense to speak of the *k-part* $\Pi_m^{(k)}$ of a Poisson structure Π at a singularity m of order k: it is the k-homogeneous Poisson structure on $T_m M$ which is, in any coordinate system centered at m, given by the k-order terms of the Taylor expansion of Π at the origin. We attach to $\Pi_m^{(k)}$ the *homogeneous Lichnerowicz-Poisson cohomology complexes*: by definition, they are, for every s,

$$\mathcal{V}_1^{(s-k+1)}(T_m M) \quad \overset{\partial_1^{s-k+1}(\Pi_m^{(k)})}{\longrightarrow} \quad \mathcal{V}_2^{(s)}(T_m M) \quad \overset{\partial_2^{s}(\Pi_m^{(k)})}{\longrightarrow} \quad \mathcal{V}_3^{(s+k-1)}(T_m M) \cdots$$

where $\mathcal{V}_r^{(s)}(T_m M)$ is the space of s-homogeneous r-vector fields on $T_m M$ (chosen to be $\{0\}$ for $s < 0$), and the operators $\partial_r^\ell(\Pi_m^{(k)})$ are defined by

$$\partial_r^\ell(\Pi_m^{(k)})(A) = [\Pi_m^{(k)}, A],$$

for all homogeneous multi-vector field A. The associated second cohomology space is

$$H^{2,s}(\Pi_m^{(k)}) = \frac{\operatorname{Ker}\left(\partial_2^s(\Pi_m^{(k)})\right)}{\operatorname{Im}\left(\partial_1^{s-k+1}(\Pi_m^{(k)})\right)}.$$

With these notations we have the following theorem.

THEOREM 2.1. ([**DW06**]) *Let m be an order k singularity of the Poisson structure Π and $\Pi^{(k)}$ its k-homogeneous part at m. If $H^{2,s}(\Pi^{(k)}) = \{0\}$, for any $s = 0, \ldots, k-1$, then m is a stable singularity for the C^{2k} compact open topology .*

REMARK 2.2. The hypothesis of Theorem 2.1 says that $\partial_2^s(\Pi_m^{(k)})$ are one-to-one for $s = 0, \ldots, k-2$, and

$$\operatorname{Ker}\partial_2^{k-1}(\Pi_m^{(k)}) = \operatorname{Im}\partial_1^0(\Pi_m^{(k)}).$$

We denote by ℓ the dimension of $\operatorname{Ker}\partial_2^{k-1}(\Pi_m^{(k)})$ $(= \operatorname{Im}\partial_1^0(\Pi_m^{(k)}))$. Let $\Sigma_{k,\Pi}^\ell$ be the set of points x of M where the $(k-1)$-jet of Π vanishes but where its k-order term $\Pi_x^{(k)}$ is also such that $H^{2,s}(\Pi_x^{(k)}) = \{0\}$, for any $s = 0, \ldots, k-1$ and $\dim\operatorname{Ker}\partial_2^{k-1}(\Pi_x^{(k)}) = \ell$.

With these notations we have the following improvement of 2.1.

THEOREM 2.3. *Under the hypothesis of Theorem 2.1, $\Sigma^{\ell}_{k,\Pi}$ is, near m, a submanifold of codimension ℓ and, for every Poisson structure Π' near enough Π in the $\mathcal{C}^{2k}$ compact open topology, $\Sigma^{\ell}_{k,\Pi'}$ is also a submanifold of codimension ℓ which is near $\Sigma^{\ell}_{k,\Pi}$.*

Exactly as we did in [**DW06**] for Theorem 2.1 we can write a version of this theorem adapted to Lie algebroids.

Also we probably could deduce from this result some criteria of stability for a compact singular symplectic leaf L not reduced to a point (see also the Crainic-Fernandes lecture in this volume) using the following scheme. Choose a transversal T to L and denote by Π_T the transversal Poisson structure induced on T; L crosses T at a singular point m of Π_T. Suppose now that we have the hypothesis of Theorem 2.3 for Π_T at m. Then Σ^{ℓ}_{k,Π_T} is, near m, a submanifold (of codimension ℓ) of T which is made of singular points of Π_T. Each symplectic leaf of the ambient manifold M which passes through a point of Σ^{ℓ}_{k,Π_T} near m has the same dimension as L. Moreover, when we stay near L, the union U of these leaves is a submanifold of M: U is, by definition, regularly foliated by singular symplectic leaves. The stability of the singular point m of Π_T and that of Σ^{ℓ}_{k,Π_T} (Theorem 2.3) induce that for any sufficiently near Poisson structure we have an analogous submanifold U', near U, regularly foliated by singular symplectic leaves; these foliations can be chosen to be near. So we could apply classical results on stability of leaves of a regular foliation to get stability of L. The simplest examples are when Σ^{ℓ}_{k,Π_T} reduces to a point or when L is 1-connected.

3. Proof of Theorem 2.3

The proof is based on the following scheme which may be classical (see [**CN82**] for another use of this scheme).

Let M be a manifold, A and B real vector spaces and $\mathcal{T}$ a topological space. Let $(F_\Pi)_{\Pi\in\mathcal{T}}$ and $(G_{x,\Pi})_{x\in M,\Pi\in\mathcal{T}}$ be two families of smooth maps

$$F_\Pi : M \longrightarrow A,$$

$$G_{x,\Pi} : A \longrightarrow B,$$

such that $\Pi \mapsto F_\Pi$ and $(x,\Pi) \mapsto G_{x,\Pi}$ are continuous for the $\mathcal{C}^1$ compact open topology on the set of smooth maps from M to A and A to B. We impose also the relations:

$$G_{x,\Pi} \circ F_\Pi(x) = 0 \quad , \quad G_{x,\Pi}(0) = 0 \;, \tag{3.1}$$

for every x and Π; there are points x_0 in M and Π_0 in $\mathcal{T}$ such that

$$F_{\Pi_0}(x_0) = 0 \quad , \quad \mathrm{Im} d_{x_0} F_{\Pi_0} = \mathrm{Ker} d_0 G_{x_0,\Pi_0}. \tag{3.2}$$

LEMMA 3.1. *Under the preceding hypothesis, for any neighborhood $U(x_0)$ of x_0 in M, there is a smaller neighborhood $U'(x_0)$ of x_0 and a neighborhood $\Omega(\Pi_0)$ of Π_0 in $\mathcal{T}$ with the following properties: For every Π in $\Omega(\Pi_0)$, the set $F_\Pi^{-1}(0)$, restricted to $U'(x_0)$, is a submanifold of codimension $\dim(\mathrm{Im} d_{x_0} F_{\Pi_0})$ which depends continuously on Π.*

In Lemma 3.1, by the continuous dependence of the manifold $F_\Pi^{-1}(0)$ with respect to Π, we mean that, if we fix a tranversal T to $F_{\Pi_0}^{-1}(0)$ at x_0, $F_\Pi^{-1}(0)$ intersects T at a point which converges to x_0 when Π converges to Π_0.

Proof of Lemma 3.1. We fix a subspace N_0 of A such that

$$A = N_0 \oplus \mathrm{Im} d_{x_0} F_{\Pi_0}. \tag{3.3}$$

1) When restricted to N_0, $d_0 G_{x_0,\Pi_0}$ is 1-1, so the implicit function Theorem says that there is an open neighborhood N_0' of the origin in N_0 such that G_{x_0,Π_0}, restricted to N_0', is also 1-1. Moreover a classical singularity technic (Lemma A, p. 61, [**GG73**]) gives that $G_{x,\Pi}$, restricted to N_0' is also 1-1 when (x,Π) is sufficiently near (x_0,Π_0) (even after a new restriction of N_0').

2) The preceding fact and Equations (3.1) imply

$$F_\Pi^{-1}(0) = F_\Pi^{-1}(N_0'), \tag{3.4}$$

when (x,Π) is sufficiently near (x_0,Π_0).

3) Hypothesis (3.3) implies that F_{Π_0} is transversal to N_0' at the origin. So, up to a new shrinking of N_0', we get that $F_{\Pi_0}^{-1}(N_0')$ is a submanifold of M (with the wanted codimension). By transversality we have the same result for $F_\Pi^{-1}(N_0')$, for every Π sufficiently near Π_0 and there follows the lemma. □

Proof of Theorem 2.3. First we remark that we can reduce M to a neighborhood of m and so we can suppose $M = \mathbb{R}^n$ with $m = 0$.

We will apply Lemma 3.1 with the following choices:

We take for x_0 the origin; $\mathcal{T}$ is the set of Poisson structures on $\mathbb{R}^n$ endowed with the $\mathcal{C}^{2k}$ compact open topology; $A = \mathcal{V}_2^{(0)} \times \cdots \times \mathcal{V}_2^{(k-1)}$ (here $\mathcal{V}_r^{(s)}$ is the vector space of s-homogeneous r-vectors on $\mathbb{R}^n$); $B = \mathcal{V}_3^{(k-1)} \times \cdots \times \mathcal{V}_3^{(2k-2)}$; $F_\Pi(x)$ is the $(k-1)$-jet of Π at the point x: it is identified with the k-uple $(\Pi_x^{(0)}, \ldots, \Pi_x^{(k-1)})$ where $\Pi_x^{(s)}$ consists in the s-homogeneous terms in the Taylor expansion of Π near x; finally we take for $G_{x,\Pi} : A \longrightarrow B$ the map which have the components $G_{x,\Pi}^{k-1}, \cdots, G_{x,\Pi}^{2k-2}$ with

$$G_{x,\Pi}^{\ell-1}(v_0, \cdots, v_{(k-1)}) = \sum_{i \le \ell-k} [v_i, \Lambda_p^{(\ell-i)}] + \frac{1}{2} \sum_{\ell-k<i,j\le\ell,\ i+j=\ell} [v_i, v_j], \tag{3.5}$$

for all $k \le \ell \le 2k-1$, and where $[\ ,\ ]$ is the Schouten bracket.

The first part of Relation (3.1) is evident and the second is a direct consequence of the Jacobi identity $[\Pi,\Pi] = 0$. Relation (3.2) is a direct translation of the hypothesis of the theorem. Now we remark that $\Sigma_{k,\Pi}^\ell$ is exactly $F_\Pi^{-1}(0)$ and then Theorem 2.3 is a direct consequence of Lemma 3.1. □

4. A new example of higher order stable singularity

The Crainic-Fernandes Theorem recalled in the introduction is the case $k = 1$ of Theorem 2.3.

In [**DW06**] we give examples of stable higher order singularities essentially of two types: in dimension 3 using results of [**CN82**] and in all dimensions, but with $k = 2$, using results of [**M01**]. This section is devoted to the study of a new example of higher order stable singularity which exists in any dimension $n > 2$ and order $k > 3$.

In the following we will work in $M = \mathbb{R}^n$, with coordinates $x_1, \ldots, x_n$, $n > 2$. We will use the notations $I = \sum_{i=1}^n x_i \partial/\partial x_i$, $X^{(k-1)} = \sum_{i=1}^n x_i^{k-1} \partial/\partial x_i$ with $k > 2$.

THEOREM 4.1. *For $n > 2$ and $k > 3$ the tensor $\Pi = I \wedge X^{(k-1)}$ is a Poisson 2-vector which has the origin as a stable singular point of order k.*

The end of this section gives a proof of this result, as a corollary of Theorem 2.1.

First remark that we have Relation

$$[I, A] = (s - r)A \tag{4.1}$$

for any s-homogeneous r-vector A on $\mathbb{R}^n$. This, in particular, implies that Π is a Poisson structure.

For any r-vector A on $\mathbb{R}^n$ we denote by DA its "curl" relative to the volume $\Omega = dx_1 \wedge \cdots \wedge dx_n$: we recall that $D : \mathcal{V}_r^{(s)} \longrightarrow \mathcal{V}_{r-1}^{(s-1)}$ is the operator $(\Omega^\flat)^{-1} \circ d \circ \Omega^\flat$, where $\Omega^\flat$ is the contraction with Ω. We will use notations and sign conventions of [**DZ05**].

We have

$$\Pi = I \wedge X^{(k-1)} = I \wedge Z \tag{4.2}$$

with

$$Z = X^{(k-1)} - \frac{k-1}{n+k-2} (\sum_{i=1}^n x_i^{k-2}) I. \tag{4.3}$$

The advantage of this new form for Π is that we have

$$DZ = 0. \tag{4.4}$$

DEFINITION 4.2. We say that an analytic vector field V on $\mathbb{R}^n$ has the analytic division property if for any analytic p-vector A on $\mathbb{R}^n$, with $p < n$, Relation

$$V \wedge A = 0$$

implies

$$A = B \wedge V$$

for some analytic $(p-1)$-vector B.

LEMMA 4.3. *The vector field Z has the analytic division property.*

PROOF. Using classical results, see section 2 of [**M76**], it suffices to prove that Z has an isolated zero at the origin. This can be done as follows. First we rewrite the components of Z on the form $x_i g_i$ with

$$g_i = x_i^{k-2} - \frac{k-1}{n+k-2} (\sum_{j=1}^n x_j^{k-2}), \tag{4.5}$$

for $i = 1 \ldots n$.

Suppose that $x = (x_1, \ldots, x_n)$ is a zero of Z; so we have $x_i g_i = 0$ for every i. Up to a permutation of the indexes, we can suppose that there is an integer p, $0 \leq p \leq n$, with

$$x_1 \cdots x_p \neq 0 \quad , \quad x_{p+1} = \cdots = x_n = 0. \tag{4.6}$$

Then we have

$$g_1 = \cdots = g_p = 0 \quad , \quad x_{p+1} = \cdots = x_n = 0. \tag{4.7}$$

With the notation $X = (x_1^{k-2}, \dots, x_p^{k-2})$, the p first equations can be rewritten in the matrix form

$$MX = 0 \tag{4.8}$$

where M is an invertible matrix. So we get $X = 0$ which is in contradiction with the first relations of (4.6), except if $p = 0$, i.e. $x = 0$. $\square$

To prove Theorem 4.1 it suffices to prove $H^{2,s}(\Pi) = \{0\}$, for any $s = 0, \dots, k-1$, and apply Theorem 2.1. Let A be an s-homogeneous 2-vector; it is a cocycle if we have Relation

$$0 = [A, \Pi] = [A, I \wedge Z] = [A, I] \wedge Z - I \wedge [A, Z] \ , \tag{4.9}$$

according to signs conventions of [**DZ05**]. So, using Relation (4.1), this is equivalent to

$$(2 - s) A \wedge Z = I \wedge [A, Z] \ . \tag{4.10}$$

LEMMA 4.4. *For $s \neq 2$, $s < k$, every s-homogeneous 2-cocycle A is a coboundary.*

PROOF. We apply the curl operator D to the two members of cocycle Equation (4.10) to get

$$(2 - s) D(A \wedge Z) = D(I \wedge [A, Z]) \ . \tag{4.11}$$

Using Relations (2.91) of [**DZ05**] this becomes

$$[A, Z] = \frac{2 - s}{2 - k - n} DA \wedge Z + \frac{1}{2 - k - n} I \wedge D[A, Z] \ . \tag{4.12}$$

When we replace $[A, Z]$ in (4.10) according to the above formula we get

$$A \wedge Z = \frac{1}{2 - k - n} I \wedge DA \wedge Z \ ; \tag{4.13}$$

which can be written in the form

$$\{A + \frac{1}{k + n - 2} I \wedge DA\} \wedge Z \ = 0 \ . \tag{4.14}$$

Using division property of Z and checking homogeneity degrees this gives

$$A + \frac{1}{k + n - 2} I \wedge DA = \begin{cases} 0 & \text{if} \quad s < k - 1 \\ U \wedge Z & \text{if} \quad s = k - 1 \end{cases} \tag{4.15}$$

where U is a constant vector field.

Case $s < k - 1$ $(s \neq 2)$. We apply the operator D to the two members of (4.15) to get

$$DA + \frac{1}{k + n - 2} D(I \wedge DA) = 0 \ . \tag{4.16}$$

But we have the Formula ((2.91) of [**DZ05**])

$$[I, DA] = -D(I \wedge DA) - DI \wedge DA \ , \tag{4.17}$$

which leads to

$$(n + s - 2) DA = -D(I \wedge DA) \ . \tag{4.18}$$

When we put this in Equation (4.17) we get

$$[(n + k - 2) - (n + s - 2)] DA = 0 \ , \tag{4.19}$$

so

$$DA = 0 \ , \tag{4.20}$$

which, with Formula (4.15), leads to

$$A = 0 \ . \tag{4.21}$$

So we have proved the lemma in the case $s < k$ ($s \neq 2$).
Case $s = k - 1$ ($s \neq 2$). We apply the operator D to the two members of (4.15) to get

$$DA + \frac{1}{k+n-2} D(I \wedge DA) = D(U \wedge Z) \ . \tag{4.22}$$

The same calculations as in the case $s < k - 1$ lead to

$$DA = (k+n-2)[Z, U] \ , \tag{4.23}$$

which, with Formula (4.15), leads to

$$A = U \wedge Z + I \wedge [U, Z] \ , \tag{4.24}$$

so to

$$A = [U, I \wedge Z] = [U, \Pi] \ , \tag{4.25}$$

proving that A is also a coboundary in that case. □

Now to finish the proof of Theorem 4.1 it suffices to prove the following lemma.

LEMMA 4.5. *Every* 2*-homogeneous* 2*-cocycle* A *is a coboundary.*

PROOF. Unlike cases of s-homogeneous 2-cocycles for $s \neq 2$, we cannot use division property and we will use a direct elementary computation. For this we will use notations

$$A = \sum_{i<j} a_{ij}^{(2)} \partial x_i \wedge \partial x_j \tag{4.26}$$

with

$$a_{ij}^{(2)} = \sum_s a_{ij}^s x_s^2 + 2 \sum_{r<s} a_{ij}^{rs} x_r x_s. \tag{4.27}$$

Now Equation (4.10), replacing Z by $X^{(k-1)}$, implies that a 2-homogeneous 2-cocycle A is characterized by Equation

$$I \wedge [A, X^{(k-1)}] = 0 \ ; \tag{4.28}$$

which becomes

$$\oint_{iju} [X^{(k-1)}(a_{ij}^{(2)}) - (k-1) a_{ij}^{(2)} (x_i^{k-2} + x_j^{k-2})] x_u = 0, \tag{4.29}$$

for any triple (i, j, u) of two by two different indexes; $\oint_{iju}$ signifies the sum of the three terms obtained from the first one after a circular permutation on i, j, u. This gives

$$\begin{aligned} &\oint_{iju} [2 \textstyle\sum_s a_{ij}^s x_s^k + 2 \sum_{r<s} a_{ij}^{rs} x_r x_s (x_r^{k-2} + x_s^{k-2}) \\ &-(k-1)(\textstyle\sum_s a_{ij}^s x_s^2 + 2 \sum_{r<s} a_{ij}^{rs} x_r x_s)(x_i^{k-2} + x_j^{k-2})] x_u = 0 \end{aligned} \ . \tag{4.30}$$

Terms in $x_u x_s^k$ in this expression, with s different from i, j and u, lead to

$$a_{ij}^s = 0. \tag{4.31}$$

Terms in x_u^{k+1}, lead to

$$a_{ij}^u = 0. \tag{4.32}$$

So we get

$$a_{ij}^s = 0 \tag{4.33}$$

for every $i < j$ and s different from i and j. For r and s different from i, j and u $(r < s)$, terms in $x_u x_r^{k-1} x_s$ lead to

$$a_{ij}^{rs} = 0. \tag{4.34}$$

In the cases $n > 3$, we can replace (i, j, u) by any (i, j, v) for $v \neq u$, we obtain

$$a_{ij}^{rs} = 0 \tag{4.35}$$

for every $i < j$, r and s not in $\{i, j\}$. Still in cases $n > 3$, terms in $x_i^{k-1} x_s x_u$ give

$$a_{ij}^{is} = 0 \tag{4.36}$$

for s not in $\{i, j, u\}$; so also under the only condition $s \neq j$. For $n = 3$ we can get the same result when we study terms in $x_i^2 x_j^{k-1}$.

Finally, the only terms which could be non-zero are only a_{ij}^i, a_{ij}^j and a_{ij}^{ij} for every i and j. So we can rewrite $a_{ij}^{(2)}$ as

$$a_{ij}^{(2)} = a_{ij}^i x_i^2 + 2b_{ij} x_i x_j + a_{ij}^j x_j^2. \tag{4.37}$$

With this notation Equation (4.30) reduces to

$$\begin{array}{c} \oint_{iju} [2(a_{ij}^i x_i^k + a_{ij}^j x_j^k + b_{ij} x_i x_j (x_i^{k-2} + x_j^{k-2})) \\ -(k-1)(a_{ij}^i x_i^2 + 2b_{ij} x_i x_j + a_{ij}^j x_j^2)(x_i^{k-2} + x_j^{k-2})] x_u = 0 \end{array} . \tag{4.38}$$

Terms in $x_u x_i^k$ in this expression lead to

$$a_{ij}^i = 0 \ , \tag{4.39}$$

for every i and j. Terms in $x_u x_j^{k-1} x_i$ lead to

$$b_{ij} + b_{ju} = 0 \ , \tag{4.40}$$

for every i, j and u, two by two different. So the antisymmetric matrix with coefficients b_{ij} is such that every column consists of same terms, except the diagonal one. This is only possible if all the b_{ij} vanish. Thus we get that A vanishes which proves the lemma. □

Case $k = 3$.

Lemma 4.5 doesn't extend to the case $k = 3$. For example $A = x_1^2 \partial x_1 \wedge \sum_i \partial x_i$ is a 2-homogeneous cocycle of $I \wedge X^{(2)}$ which is not a coboundary.

References

[CN82] C. Camacho and A. Lins Neto, *The topology of integrable differential forms near a singularity* Inst. Hautes Études Sci. Publ. Math. **55** (1982), 5-35.

[DW06] J.-P. Dufour and A. Wade, *Stability of higher order singular points of Poisson manifolds and Lie algebroids*, Ann. Inst. Fourier, 56 (2006), fasc.3, 545-559. Arxiv:math.DG/0501168

[DZ05] J.-P. Dufour and T. Zung, *Poison structures and their normal forms.* Progress in Math., Springer Verlag (2005).

[GG73] M. Golubitsky and V. Guillemin, *Stable mappings and their singularities.* Graduate Texts in Mathematics, Vol. 14. Springer-Verlag, New York-Heidelberg, 1973.

[M01] P. Monnier, *Formal Poisson cohomology of quadratic Poisson structures*, Lett. Math. Phys., 59 (2002), 253-267.

[M76] R. Moussu, *Sur l'existence d'intégrales premières pour un germe de forme de Pfaff*, Ann. Inst. Fourier, 26 (1976), fasc.2, 171-220.

Department of Mathematics, Université Montpellier II, 34095 Montpellier cedex 05, France

E-mail address: dufourj@math.univ-montp2.fr

Contemporary Mathematics
Volume **450**, 2008

Poisson reduction and the Hamiltonian structure of the Euler-Yang-Mills equations

François Gay-Balmaz and Tudor S. Ratiu

Abstract. The problem treated here is to find the Hamiltonian structure for an ideal gauge-charged fluid. Using a Kaluza-Klein point of view, we obtain the non-canonical Poisson bracket and the motion equations by a Poisson reduction involving the automorphism group of a principal bundle.

1. Introduction

The motion equations of an **ideal incompressible fluid** in a Riemannian manifold (M, g) are given by the Euler equations

$$\frac{\partial v}{\partial t} + \nabla_v v = -\operatorname{grad} p, \tag{1.1}$$

where the Eulerian velocity v is a divergence free vector field, p is the pressure and ∇ is the Levi-Civita covariant derivative associated to g. From [**A**] it is known that equations (1.1) are formally the spatial representation of the geodesic motion on the volume-preserving diffeomorphism group $\mathcal{D}_\mu(M)$ of M with respect to the L^2 Riemannian metric. See also [**AM**], §5.5.8, for a quick exposition of this fact and [**EM**] for the analytic formulation and many rigorous results concerning the Euler and Navier-Stokes equations derived from this geometric point of view. From the Hamiltonian perspective, equations (1.1) are the Lie-Poisson equations on the Lie algebra $\mathfrak{X}_{div}(M)$ of $\mathcal{D}_\mu(M)$, consisting of divergence free vector fields. Here, $\mathfrak{X}_{div}(M)$ is identified with its dual by the weak L^2-pairing.

1991 *Mathematics Subject Classification.* Primary 37K65, 53C80, 70S15; Secondary 53D17, 76W05.

Key words and phrases. Lie-Poisson equations, Euler-Yang-Mills equations, automorphism group, gauge group, reduction, Kaluza-Klein metric, Poisson bracket.

The authors were supported in part by the Swiss National Science Foundation Grant #200021-109111.

In [**MRW, MWRSS**], this approach is generalized to the case of the motion of an **ideal compressible adiabatic fluid**

$$
\begin{cases}
\dfrac{\partial v}{\partial t} + \nabla_v v = \dfrac{1}{\rho}\operatorname{grad} p, \\
\dfrac{\partial \rho}{\partial t} + \operatorname{div}(\rho v) = 0, \\
\dfrac{\partial \sigma}{\partial t} + \mathbf{d}\sigma(v) = 0,
\end{cases}
\tag{1.2}
$$

where ρ is the mass density, σ is the specific entropy and p is the pressure. In this case, the configuration space is the full diffeomorphism group $\mathcal{D}(M)$ and equations (1.2) are obtained via Lie-Poisson reduction for semidirect products.

In this paper we generalize the previous approach to the case of a **classical charged ideal fluid**. More precisely, using a Kaluza-Klein point of view, we obtain the motion equations by Poisson reduction. Therefore we need to consider on M a G-principal bundle $P \to M$ and to enlarge the configuration space from the group of diffeomorphisms to the product of the group of automorphisms of P with the field variables. This point of view is directly inspired by the paper [**M**] on the motion of a particle in a Yang-Mills field.

If $G = S^1$ we get the **Euler-Maxwell equations** describing the motion of an electrically charged fluid. The Euler-Maxwell equations are given by the system

$$
\begin{cases}
\dfrac{\partial v}{\partial t} + \nabla_v v = \dfrac{q}{m}(\mathbf{E} + v \times \mathbf{B}) - \dfrac{1}{\rho}\operatorname{grad} p, \\
\dfrac{\partial \rho}{\partial t} + \operatorname{div}(\rho v) = 0, \\
\dfrac{\partial \sigma}{\partial t} + \mathbf{d}\sigma(v) = 0, \\
\dfrac{\partial \mathbf{E}}{\partial t} = \operatorname{curl}\mathbf{B} - \dfrac{q}{m}\rho v, \quad \dfrac{\partial \mathbf{B}}{\partial t} = -\operatorname{curl}\mathbf{E}, \\
\operatorname{div}\mathbf{E} = \dfrac{q}{m}\rho, \quad \operatorname{div}\mathbf{B} = 0,
\end{cases}
\tag{1.3}
$$

where v is the Eulerian velocity, $\mathbf{E}$ is the electric field, $\mathbf{B}$ is the magnetic field, and the constant q is the charge per unit mass. The Hamiltonian structure of the Euler-Maxwell equation was derived in the paper [**MWRSS**] and is closely related to the Hamiltonian structure of the Maxwell-Vlasov equations (see [**MW**]).

In the general case of a Lie group G, we will show that the Poisson reduction leads to the equation for an ideal compressible adiabatic fluid carrying a gauge-charge, as given in [**GHK**]. We call these equations the **Euler-Yang-Mills equations**. The standard examples are obtained for $G = \mathrm{SU}(2)$ or $G = \mathrm{SU}(3)$.

The physical interpretation of the point of view given in this paper is the following. The evolution of the fluid particles as well as of the gauge-charge density of the fluid is given by a curve ψ_t in the automorphism group. In fact ψ_t is the flow of a time-dependent vector field U_t on the principal bundle P. This vector field induces a time-dependent vector field v_t on M, which represents the **Eulerian velocity of the fluid**. Given the evolution of the magnetic potential $\mathcal{A}_t$ and of the mass density ρ_t, the vector field U_t induces also a Lie algebra valued and time-dependent function $\rho_t\mathcal{A}_t(U_t)$, which represents the **gauge-charge density of**

the fluid. Note the analogy with the Kaluza-Klein point of view for the charged particle in a Yang-Mills field (see [**M**]).

The plan of the paper is as follows. Section 2 recalls some needed facts about principal bundles, connections, automorphisms, and gauge groups. Section 3 recalls the Hamiltonian formulation of the Maxwell equations and generalizes this to the case of the Yang-Mills fields equations. Section 4 shows that the compressible and incompressible Euler-Yang-Mills equations are Lie-Poisson equations for a semidirect product associated to the automorphism group of a principal bundle. One of the equations is obtained by conservation of the momentum map associated to the invariance under gauge transformations. We naturally obtain the associated non-canonical Poisson bracket. By applying the general process of reduction by stages, we get some already known results about the Euler-Maxwell equations. We also show that the two different Poisson brackets derived in [**GHK**]and in [**MWRSS**] are in fact obtained by Poisson reduction, at different stages, of the same canonical Poisson structure.

2. Connections, automorphisms and gauge transformations

In this section we fix our conventions on principal bundles and connections. Consider a smooth free and proper right action

$$\Phi : G \times P \to P, \quad (g, p) \mapsto \Phi_g(p)$$

of a Lie group G on a manifold P. Thus we get the principal bundle

$$\pi : P \to M := P/G,$$

where M is endowed with the unique manifold structure for which π is a submersion.

Recall that the adjoint vector bundle is

$$\operatorname{Ad} P := P \times_G \mathfrak{g} \to M,$$

where the quotient is taken relative to the right action $(g, (p, \xi)) \mapsto (\Phi_g(p), \operatorname{Ad}_{g^{-1}}(\xi))$. The elements of $\operatorname{Ad} P$ are denoted by $[p, \xi]_G$, for $(p, \xi) \in P \times \mathfrak{g}$.

Consider the space $\Omega^k(P, \mathfrak{g})$ of $\mathfrak{g}$-valued k-forms on P and let $\overline{\Omega^k}(P, \mathfrak{g})$ be the subspace of $\Omega^k(P, \mathfrak{g})$ consisting of $\mathfrak{g}$-valued k-forms ω such that

(1) $\Phi_g^* \omega = \operatorname{Ad}_{g^{-1}} \circ \omega$,

(2) if one of $u_1, ..., u_k \in T_pP$ is vertical then $\omega(u_1, ..., u_k) = 0$.

$\overline{\Omega^k}(P, \mathfrak{g})$ is naturally isomorphic to $\Omega^k(M, \operatorname{Ad} P)$, the space of $\operatorname{Ad} P$-valued k-forms on M. Indeed, to each $\omega \in \overline{\Omega^k}(P, \mathfrak{g})$ corresponds a k-form $\widetilde{\omega} \in \Omega^k(M, \operatorname{Ad} P)$ given on $v_1, .., v_k \in T_xM$ by

$$\widetilde{\omega}(x)(v_1, ..., v_k) := [p, \omega(p)(u_1, ..., u_k)]_G, \tag{2.1}$$

where $p \in P$ is such that $\pi(p) = x$ and $u_i \in T_pP$ are such that $T_p\pi(u_i) = v_i$. We shall use the notation $\overline{\Omega^0}(P, \mathfrak{g}) = \mathcal{F}_G(P, \mathfrak{g})$ and $\Omega^0(M, \operatorname{Ad} P) = \Gamma(\operatorname{Ad} P)$.

A **principal connection** on P is a $\mathfrak{g}$-valued 1-form $\mathcal{A} \in \Omega^1(P, \mathfrak{g})$ such that

$$\Phi_g^* \mathcal{A} = \operatorname{Ad}_{g^{-1}} \circ \mathcal{A} \text{ and } \mathcal{A}(\xi_P) = \xi,$$

where ξ_P is the infinitesimal vector field associated to ξ. The set of all connections will be denoted by $\mathcal{C}onn(P)$. It is an affine space with underlying vector space $\overline{\Omega^1}(P, \mathfrak{g})$. Recall that a connection induces a splitting $T_pP = V_pP \oplus H_pP$ of the tangent space into the vertical and horizontal subspace defined by

$$H_pP := \ker(\mathcal{A}(p)).$$

The **covariant exterior differential** associated to $\mathcal{A}$ is the map $\mathbf{d}^{\mathcal{A}} : \Omega^k(P, \mathfrak{g}) \to \Omega^{k+1}(P, \mathfrak{g})$ defined by

$$\mathbf{d}^{\mathcal{A}}\omega(p)(u_1, ..., u_k) := \mathbf{d}\,\omega(p)\left(\operatorname{hor}_p(u_1), ..., \operatorname{hor}_p(u_k)\right),$$

where $\operatorname{hor}_p(u_i)$ is the horizontal part of $u_i \in T_pP$, $i = 1, \ldots, k$. Note that for $\omega \in \overline{\Omega^k}(P, \mathfrak{g})$ we have $\mathbf{d}^{\mathcal{A}}\omega \in \overline{\Omega^{k+1}}(P, \mathfrak{g})$.

The **curvature of the connection** $\mathcal{A}$ is, by definition, the 2-form

$$\mathcal{B} := \mathbf{d}^{\mathcal{A}}\mathcal{A} \in \overline{\Omega^2}(P, \mathfrak{g}).$$

Recall that a principal connection $\mathcal{A}$ on P induces an affine connection and a covariant derivative, denoted respectively by $\nabla^{\mathcal{A}}$ and $\frac{D^{\mathcal{A}}}{dt}$, on the vector bundles $\operatorname{Ad} P \to M$ and $\operatorname{Ad} P^* \to M$ (see [**KN**]).

Given a Riemannian metric g on M and a connection $\mathcal{A}$ on P, we can define the **covariant codifferential**

$$\delta^{\mathcal{A}} : \overline{\Omega^k}(P, \mathfrak{g}) \to \overline{\Omega^{k-1}}(P, \mathfrak{g}),$$

see definition 4.2.8 in [**B**].

Given a Riemannian metric g on M and an Ad-invariant inner product γ on $\mathfrak{g}$ we can define a Riemannian metric $g\gamma$ on the vector bundles $\Lambda^k(M, \operatorname{Ad} P) \to M$.

Let M be a compact oriented boundaryless manifold. For $\alpha \in \Omega^k(M, \operatorname{Ad} P)$ and $\beta \in \Omega^{k+1}(M, \operatorname{Ad} P)$ we have (see Theorem 4.2.9 in [**B**]) :

$$\int_M (g\gamma)(\mathbf{d}^{\mathcal{A}}\alpha, \beta)\mu = \int_M (g\gamma)(\alpha, \delta^{\mathcal{A}}\beta)\mu, \tag{2.2}$$

where μ denotes the volume form associated to the Riemannian metric g.

Given a connection $\mathcal{A}$, a Riemannian metric g on M and an Ad-invariant inner product γ on $\mathfrak{g}$, we can define the **Kaluza-Klein metric** $K_{\mathcal{A}}$ on TP by

$$K_{\mathcal{A}}(u_p, v_p) := g(T_p\pi(u_p), T_p\pi(v_p)) + \gamma(\mathcal{A}(u_p), \mathcal{A}(v_p)). \tag{2.3}$$

We say that a diffeomorphism φ of P is an **automorphism** if $\Phi_g \circ \varphi = \varphi \circ \Phi_g$, for all $g \in G$. The Fréchet Lie group of all automorphisms is denoted by $\mathcal{A}ut(P)$. An automorphism φ of P induces a unique diffeomorphism $\overline{\varphi}$ of M defined by the condition $\pi \circ \varphi = \overline{\varphi} \circ \pi$. The Lie algebra $\mathfrak{aut}(P)$ consists of G-invariant vector fields on P. Its (left) Lie bracket is denoted by $[U, V]_L$ and is the negative of the usual Jacobi-Lie bracket $[U, V]_{JL}$. For $U \in \mathfrak{aut}(P)$ we denote by $[U] \in \mathfrak{X}(M)$ the unique vector field on M defined by the condition

$$T\pi \circ U = [U] \circ \pi.$$

The normal subgroup $\mathcal{G}au(P)$ of gauge transformations contains, by definition, the automorphisms φ with $\overline{\varphi} = id$.

The Lie algebra $\mathfrak{gau}(P)$ consists of G-invariant vertical vector fields on P. Therefore when $U \in \mathfrak{gau}(P)$ we have $[U] = 0$. Note the identifications

$$\mathfrak{gau}(P) \cong \mathcal{F}_G(P, \mathfrak{g}) \cong \Gamma(\operatorname{Ad} P).$$

Indeed, to $f \in \mathcal{F}_G(P, \mathfrak{g})$ we can associate the G-invariant vertical vector field $\sigma(f)$ given by

$$\sigma(f)(p) := f(p)_P(p). \tag{2.4}$$

The second equivalence is given by the map (2.1). A direct computation shows that σ is a Lie algebra isomorphism, that is,

$$\sigma([f,g]) = [\sigma(f),\sigma(g)]_L.$$

If the principal bundle $P \to M$ is trivial the automorphism group is a semidirect product of two groups :

$$\mathcal{A}ut(P) \simeq \mathcal{D}(M) \circledS \mathcal{G}au(P).$$

Duality. Throughout this paper we will identify the cotangent space $T^*_\varphi\mathcal{A}ut(P)$ with the space of G-invariant 1-forms on P along $\varphi \in \mathcal{A}ut(P)$. The duality is given by

$$\langle \mathbf{M}_\varphi, U_\varphi\rangle := \int_M \mathbf{M}_\varphi(U_\varphi)\mu,$$

where $\mathbf{M}_\varphi \in T^*_\varphi\mathcal{A}ut(P)$ and $U_\varphi \in T_\varphi\mathcal{A}ut(P)$. Note that in this formula we used the fact that $\mathbf{M}_\varphi(U_\varphi)$ is a smooth function on P that does not depend on the fiber variables and hence induces a unique smooth function on M. In particular we have $\mathfrak{aut}(P)^* = \Omega^1_G(P)$, the space of right-invariant 1-forms on P.

We identify the cotangent space $T^*_\varphi\mathcal{G}au(P)$ with the tangent space $T_\varphi\mathcal{G}au(P)$ via the duality

$$\langle U_\varphi, V_\varphi\rangle := \int_M \gamma\left(\widetilde{\mathcal{A}(U_\varphi)}, \widetilde{\mathcal{A}(V_\varphi)}\right)\mu, \tag{2.5}$$

for any principal connection $\mathcal{A}$ on P. Note that, since U_φ and V_φ are vertical, the pairing (2.5) does not depend on $\mathcal{A}$, indeed, we have $\mathcal{A}(U_\varphi) = \sigma^{-1}(U_\varphi\circ\varphi^{-1})\circ\varphi$.

3. Equations for the fields

In this section we give the Hamiltonian formulation for the Yang-Mills fields in the vacuum. We first treat the electromagnetic case.

3.1. Hamiltonian formulation of the Maxwell equations. The Hamiltonian is defined on the cotangent bundle $T^*\Omega^1(M,\mathbb{R})$ and is given by

$$H(A,Y) = \frac{1}{2}\int_M \|E\|^2\mu + \frac{1}{2}\int_M \|B\|^2\mu,$$

where $E := -Y$ is the electric field and $B := \mathbf{d}A$ is the magnetic field. Hamilton's equations are

$$\frac{\partial B}{\partial t} = -\mathbf{d}E \quad\text{and}\quad \frac{\partial E}{\partial t} = \delta B, \tag{3.1}$$

and the relation $B = \mathbf{d}A$ gives

$$\mathbf{d}B = 0. \tag{3.2}$$

To obtain the last equation $\delta E = 0$ we use the invariance of the Hamiltonian under gauge transformations. The action of the gauge group $\mathcal{F}(M)$ on $\Omega^1(M,\mathbb{R})$ is given by

$$\mathcal{F}(M)\times\Omega^1(M,\mathbb{R}) \to \Omega^1(M,\mathbb{R}),\ \ (\varphi, A)\mapsto A + \mathbf{d}\varphi, \tag{3.3}$$

and is Hamiltonian. The associated momentum map is

$$\mathbf{J} : T^*\Omega^1(M,\mathbb{R}) \to \mathcal{F}(M)^*,\ \ \mathbf{J}(A,Y) = \delta Y,$$

so the condition $\mathbf{J}(A,Y)=0$ gives the Gauss law

$$\delta E = 0. \tag{3.4}$$

Remark that we identify the tangent and cotangent spaces of $\Omega^1(M,\mathbb{R})$ and $\mathcal{F}(M)$ using the natural L^2-pairing.

When M is three dimensional, we can define the vector fields $\mathbf{E} := E^\sharp$ and $\mathbf{B} := (\star B)^\sharp$. In this case, equations (3.1), (3.2) and (3.4) are equivalent to the Maxwell equations in the vacuum

$$\begin{cases} \dfrac{\partial \mathbf{E}}{\partial t} = \operatorname{curl}\mathbf{B}, & \operatorname{div}\mathbf{E}=0, \\ \dfrac{\partial \mathbf{B}}{\partial t} = -\operatorname{curl}\mathbf{E}, & \operatorname{div}\mathbf{B}=0. \end{cases} \tag{3.5}$$

3.2. Generalization to any principal bundle. We now generalize the previous formulation to the case of a G-principal bundle $P \to M$ over an arbitrary compact boundaryless manifold M. When P is a trivial S^1 bundle over a three dimensional manifold, we recover the Maxwell equations.

As in the electromagnetic case, the configuration space variable is the magnetic potential $\mathcal{A} \in \mathcal{C}onn(P)$ and the Hamiltonian is defined on the cotangent bundle $T^*\mathcal{C}onn(P)$ by

$$H(\mathcal{A},\mathcal{Y}) = \frac{1}{2}\int_M \|E\|^2\mu + \frac{1}{2}\int_M \|B\|^2\mu,$$

where:
(1) $E := \widetilde{\mathcal{E}} \in \Omega^1(M,\operatorname{Ad}P)$ is the $\operatorname{Ad}P$-valued 1-form associated, through the map (2.1), to the "electric part" $\mathcal{E} \in \overline{\Omega^1}(P,\mathfrak{g})$ of the Yang-Mills field, given by

$$\mathcal{E} := -\mathcal{Y} \in \overline{\Omega^1}(P,\mathfrak{g}),$$

(2) $B := \widetilde{\mathcal{B}} \in \Omega^2(M,\operatorname{Ad}P)$ is the $\operatorname{Ad}P$-valued 2-form associated to the "magnetic part" $\mathcal{B} \in \overline{\Omega^1}(P,\mathfrak{g})$ of the Yang-Mills field, given by the curvature

$$\mathcal{B} := \mathbf{d}^{\mathcal{A}}\mathcal{A} \in \overline{\Omega^2}(P,\mathfrak{g}).$$

As before, we identify the cotangent bundle of $\mathcal{C}onn(P)$ with the tangent bundle, using the L^2 pairing

$$\langle \alpha,\beta\rangle = \int_M (g\gamma)(\alpha,\beta)\mu, \quad \alpha,\beta \in \Omega^k(M,\operatorname{Ad}P). \tag{3.6}$$

The Hamiltonian equations associated to H are

$$\frac{\partial \mathcal{B}}{\partial t} = -\mathbf{d}^{\mathcal{A}}\mathcal{E} \quad \text{and} \quad \frac{\partial \mathcal{E}}{\partial t} = \delta^{\mathcal{A}}\mathcal{B},$$

and the Bianchi identity gives

$$\mathbf{d}^{\mathcal{A}}\mathcal{B} = 0.$$

To obtain the last equation we use the invariance of the Hamiltonian under the gauge transformations. The action of $\varphi \in \mathcal{G}au(P)$ on $\mathcal{A} \in \mathcal{C}onn(P)$ is given by $\varphi^*\mathcal{A}$ and the cotangent lift of this action is given by $(\varphi^*\mathcal{A},\varphi^*\mathcal{Y})$. Under this action, $\mathcal{E}$ and $\mathcal{B}$ are transformed into $\varphi^*\mathcal{E}$ and $\varphi^*\mathcal{B}$, so H is gauge-invariant. The momentum mapping associated to this Hamiltonian action is given by

$$\mathbf{J} : T^*\mathcal{C}onn(P) \to \mathfrak{gau}(P)^*, \quad \mathbf{J}(\mathcal{A},\mathcal{Y}) = \sigma(\delta^{\mathcal{A}}\mathcal{Y}),$$

so the conservation law $\mathbf{J}(\mathcal{A}, \mathcal{Y}) = 0$ gives the last equation

$$\delta^{\mathcal{A}} \mathcal{E} = 0.$$

Note that we identify $\mathfrak{gau}(P)^*$ with $\mathfrak{gau}(P)$ via the L^2 pairing (3.6).

4. Hamiltonian formulation of Euler-Yang-Mills

We begin by quickly recalling some facts about the Hamiltonian semidirect product reduction theory.

Hamiltonian semidirect product reduction with parameter. Let $\rho : G \to \mathrm{Aut}(V)$ denote a *right* Lie group representation of G in the vector space V. As sets, the semidirect product $S = G \circledS V$ is the Cartesian product $S = G \times V$ whose group multiplication is given by

$$(g_1, v_1)(g_2, v_2) = (g_1 g_2, v_2 + \rho_{g_2}(v_1)).$$

Below we shall use the operation $\diamond : V \times V^* \to \mathfrak{g}^*$ defined by

$$\langle v \diamond a, \xi \rangle_{\mathfrak{g}} := -\langle a\xi, v \rangle_V,$$

where $v \in V$, $a\xi \in V^*$ denotes the induced $\mathfrak{g}$-action of $\xi \in \mathfrak{g}$ on $a \in V^*$, and $\langle \cdot, \cdot \rangle_{\mathfrak{g}} : \mathfrak{g}^* \times \mathfrak{g} \to \mathbb{R}$ and $\langle \cdot, \cdot \rangle_V : V^* \times V \to \mathbb{R}$ are the duality parings.

The lift of right translation of S on T^*S induces a right action on $T^*G \times V^*$. Let Q be another manifold (without any G or V-action). Consider a Hamiltonian function $H : T^*G \times T^*Q \times V^* \to \mathbb{R}$ right invariant under the S-action on $T^*G \times T^*Q \times V^*$; recall that the S-action on T^*Q is trivial. In particular, the function $H_{a_0} := H|_{T^*G \times T^*Q \times \{a_0\}} : T^*G \times T^*Q \to \mathbb{R}$ is invariant under the induced action of the isotropy subgroup $G_{a_0} := \{g \in G \mid \rho_g^* a_0 = a_0\}$ for any $a_0 \in V^*$. The following theorem is an easy consequence of the semidirect product reduction theorem (see [**MRW**]) and the reduction by stages method (see [**MMOPR**]).

THEOREM 4.1. *For $\alpha(t) \in T^*_{g(t)}G$ and $\mu(t) := T^*R_{g(t)}(\alpha(t)) \in \mathfrak{g}^*$, the following are equivalent:*

- **i** *$(\alpha(t), q(t), p(t))$ satisfies Hamilton's equations for H_{a_0} on $T^*(G \times Q)$.*
- **ii** *The following system of Lie-Poisson equations with parameter coupled with Hamilton's equations holds on $\mathfrak{s}^* \times T^*Q$:*

$$\frac{\partial}{\partial t}(\mu, a) = -\operatorname{ad}^*_{\left(\frac{\delta h}{\delta \mu}, \frac{\delta h}{\delta a}\right)}(\mu, a) = -\left(\operatorname{ad}^*_{\frac{\delta h}{\delta \mu}} \mu + \frac{\delta h}{\delta a} \diamond a, a\frac{\delta h}{\delta \mu}\right), \quad a(0) = a_0$$

and

$$\frac{dq^i}{dt} = \frac{\partial h}{\partial p_i}, \quad \frac{dp_i}{dt} = -\frac{\partial h}{\partial q^i},$$

where $\mathfrak{s}$ is the semidirect product Lie algebra $\mathfrak{s} = \mathfrak{g} \circledS V$. The associated Poisson bracket is the sum of the Lie-Poisson bracket on the Lie algebra $\mathfrak{s}^*$ and the canonical bracket on the cotangent bundle T^*Q

$$\{f, g\}(\mu, a, q, p) = \left\langle \mu, \left[\frac{\delta f}{\delta \mu}, \frac{\delta g}{\delta \mu}\right]\right\rangle + \left\langle a, \frac{\delta f}{\delta a}\frac{\delta g}{\delta \mu} - \frac{\delta g}{\delta a}\frac{\delta f}{\delta \mu}\right\rangle + \frac{\partial f}{\partial q^i}\frac{\partial g}{\partial p_i} - \frac{\partial g}{\partial q^i}\frac{\partial f}{\partial p_i}.$$

The Hamiltonian. The Hamiltonian for the Euler-Yang-Mills equations is defined on the cotangent bundle $T^*(\mathcal{A}ut(P) \times \mathcal{C}onn(P))$ and is given, for $(\rho,\sigma) \in \mathcal{F}(M)^{*2}$, by

$$\begin{aligned} H_{(\rho,\sigma)}(\mathbf{M}_\psi, \mathcal{A}, \mathcal{Y}) = & \frac{1}{2}\int_M \frac{1}{\rho} K^*_{\mathcal{A}}(\mathbf{M}_\psi, \mathbf{M}_\psi)\mu + \int_M \rho e(\rho(J\overline{\psi})^{-1}, \sigma)\mu \\ & + \frac{1}{2}\int_M \|E\|^2\mu + \frac{1}{2}\int_M \|B\|^2\mu, \end{aligned} \tag{4.1}$$

where e is the fluid's specific internal energy and $K^*_{\mathcal{A}}$ is the metric induced on T^*P by $K_{\mathcal{A}}$. This Hamiltonian is obtained by integrating the Kaluza-Klein energy of the particles and by summing it with the internal energy of the fluid and the Hamiltonian for the Yang-Mills field. By Theorem 4.1, Hamilton's equations for $H_{(\rho,\sigma)}$ are equivalent to the Lie-Poisson equations on the dual of the semidirect product Lie algebra $\mathfrak{aut}(P) \circledS \mathcal{F}(M)^2$, together with the Hamilton equations on $T^*\mathcal{C}onn(P)$, relative to the reduced Hamiltonian h given on $(\mathfrak{aut}(P) \circledS \mathcal{F}(M)^2)^* \times T^*\mathcal{C}onn(P)$ by

$$h(\mathbf{M}, \rho, \sigma, \mathcal{A}, \mathcal{Y}) = \frac{1}{2}\int_M \frac{1}{\rho} K^*_{\mathcal{A}}(\mathbf{M}, \mathbf{M})\mu + \int_M \rho e(\rho, \sigma)\mu + \frac{1}{2}\int_M \|E\|^2\mu + \frac{1}{2}\int_M \|B\|^2\mu.$$

The momentum map of the gauge group. As it was the case in section 3, the nonabelian Gauss equation

$$\delta^{\mathcal{A}}\mathcal{E} = -\mathcal{Q}$$

is obtained by invariance of the Hamiltonian under the gauge transformations. Indeed, consider the action of the gauge group given for $\eta \in \mathcal{G}au(P)$, by

$$(\psi, \mathcal{A}) \mapsto (\eta^{-1} \circ \psi, \eta^*\mathcal{A}). \tag{4.2}$$

The cotangent-lift of this action leaves the Hamiltonian invariant. So the associated momentum map, which is computed in the following lemma, is a conserved quantity.

LEMMA 4.2. *The momentum mapping associated to the cotangent lift of the gauge group action is given by*

$$\mathbf{J}(\mathbf{M}_\psi, \mathcal{A}, \mathcal{Y}) = \sigma\left(\delta^{\mathcal{A}}\mathcal{Y} - \mathcal{A}\left(\left(J\overline{\psi}^{-1}\right) V_\psi \circ \psi^{-1}\right)\right) \in \mathfrak{gau}(P)^*,$$

where $V_\psi \in T_\psi \mathcal{A}ut(P)$ *is such that* $\mathbf{M}_\psi = K_{\mathcal{A}}(V_\psi, \cdot)$ *and* $\sigma : \mathcal{F}_G(P, \mathfrak{g}) \to \mathfrak{gau}(P)$ *is defined in* (2.4).

When $\mathbf{M}_\psi = K_{\mathcal{A}}(\rho_0 U_\psi, \cdot)$, is a solution of Hamilton's equations associated to $H_{(\rho_0,\sigma_0)}$, the conservation law $\mathbf{J}(\mathbf{M}_\psi, \mathcal{A}, \mathcal{Y}) = 0$ gives

$$\mathcal{A}\left(\left(J\overline{\psi}^{-1}\right)(\rho_0 \circ \overline{\psi}^{-1}) U_\psi \circ \psi^{-1}\right) = \delta^{\mathcal{A}}\mathcal{Y}.$$

Using the definition of the charge density $\mathcal{Q} := \rho\mathcal{A}(U)$, the identities $U_\psi \circ \psi^{-1} = U$, $\left(J\overline{\psi}^{-1}\right)(\rho_0 \circ \overline{\psi}^{-1}) = \rho$, and the notation $\mathcal{E} = -\mathcal{Y}$, we get

$$\mathcal{Q} = -\delta^{\mathcal{A}}\mathcal{E}.$$

We can now state our main result.

THEOREM 4.3. *Let* $(\mathbf{M}_\psi, \mathcal{A}, \mathcal{Y})$ *be a curve in the cotangent bundle* $T^*(\mathcal{A}ut(P) \times \mathcal{C}onn(P))$, *and consider the induced curve* $(\mathbf{M}, \mathcal{A}, \mathcal{Y}) \in \mathfrak{aut}(P)^* \times T^*\mathcal{C}onn(P)$ *given by* $\mathbf{M} := (J\overline{\psi})\, \mathbf{M}_\psi \circ \psi^{-1}$. *Then* $(\mathbf{M}_\psi, \mathcal{A}, \mathcal{Y})$ *is a solution of the Hamilton's equations associated to the Hamiltonian* $H_{(\rho_0,\sigma_0)}$ *given in* (4.1) *if and only if* $(\mathbf{M}, \mathcal{A}, \mathcal{Y})$ *is a solution of the system*

$$(4.3)\qquad \begin{cases} \dfrac{\partial v}{\partial t} + \nabla_v v = \dfrac{1}{\rho}\gamma\left(Q, E(\cdot) + B(\cdot, v)\right)^\sharp - \dfrac{1}{\rho}\operatorname{grad} p, \\ \dfrac{\partial \rho}{\partial t} + \operatorname{div}(\rho v) = 0,\ \rho(0) = \rho_0, \quad \dfrac{\partial \sigma}{\partial t} + \mathbf{d}\sigma(v) = 0,\ \sigma(0) = \sigma_0, \\ \dfrac{\partial Q}{\partial t} + \nabla_v^{\mathcal{A}} Q + Q \operatorname{div} v = 0, \\ \dfrac{\partial \mathcal{E}}{\partial t} = \delta^{\mathcal{A}}\mathcal{B} - \mathcal{Q}v^\flat, \\ \dfrac{\partial \mathcal{B}}{\partial t} = -\mathbf{d}^{\mathcal{A}}\mathcal{E}, \end{cases}$$

where

$$\begin{aligned} &p := \rho^2 \frac{\partial e}{\partial \rho}(\rho, \sigma), \\ &\mathcal{E} := -\mathcal{Y} \in \overline{\Omega^1}(P, \mathfrak{g}) \quad \textit{and} \quad E := \widetilde{\mathcal{E}} \in \Omega^1(M, \operatorname{Ad} P), \\ &\mathcal{B} := \mathbf{d}^{\mathcal{A}}\mathcal{A} \quad \textit{and} \quad B := \widetilde{\mathcal{B}}, \\ &Q := \widetilde{\mathcal{Q}} \in \Gamma(\operatorname{Ad} P). \end{aligned}$$

The Eulerian velocity v *and the gauge-charge density* $\mathcal{Q}$ *are given in terms of the momentum* $\mathbf{M}$ *by*

$$(4.4)\qquad v = [U] \quad \textit{and} \quad \mathcal{Q} = \mathcal{A}(\rho U) \quad \textit{where} \quad U = K_{\mathcal{A}}^*\left(\frac{\mathbf{M}}{\rho}, \cdot\right).$$

Conservation of the momentum map associated to the gauge transformations gives the equation

$$\delta^{\mathcal{A}}\mathcal{E} = -\mathcal{Q}.$$

Thus we have recovered the Euler-Yang-Mills equations as given in [**GHK**]. One can adapt this theorem to the incompressible and homogeneous case.

COROLLARY 4.4. *In the Abelian case of the trivial bundle* $P = M \times S^1$ *and assuming that the fluid is composed of particles of mass* m *and charge* q, *we obtain the Euler-Maxwell equations*

$$(4.5)\qquad \begin{cases} \dfrac{\partial v}{\partial t} + \nabla_v v = \dfrac{q}{m}(\mathbf{E} + v \times \mathbf{B}) - \dfrac{1}{\rho}\operatorname{grad} p, \\ \dfrac{\partial \rho}{\partial t} + \operatorname{div}(\rho v) = 0,\ \rho(0) = \rho_0, \quad \dfrac{\partial \sigma}{\partial t} + \mathbf{d}\sigma(v) = 0,\ \sigma(0) = \sigma_0, \\ \dfrac{\partial \mathbf{E}}{\partial t} = \operatorname{curl}\mathbf{B} - \dfrac{q}{m}\rho v, \quad \dfrac{\partial \mathbf{B}}{\partial t} = -\operatorname{curl}\mathbf{E}, \\ \operatorname{div}\mathbf{E} = \dfrac{q}{m}\rho, \quad \operatorname{div}\mathbf{B} = 0, \end{cases}$$

where

$$\mathbf{E} := E^\sharp \quad \textit{and} \quad \mathbf{B} := (\star B)^\sharp.$$

PROOF. If we define $Q_t = \rho_t \frac{q_t}{m}$, the equation for Q in (4.3) becomes

$$0 = \frac{\partial q_t}{\partial t} + \mathbf{d}q_t(v) = \frac{d}{dt} q_t(x(t)),$$

where $x(t)$ is the trajectory of the particle starting at $x(0)$. Since all particles have the same charge $q \in \mathbb{R}$ by hypothesis, we conclude that $q(t,x)$ is a constant. Therefore, the equation for Q in (4.3) disappears. It is easily seen that the other equations become the ones in (4.5). □

Poisson bracket. From Theorem 4.1 we know that the Euler-Yang-Mills equations can be written

$$\dot{f} = \{f, h\}$$

with respect to the Poisson bracket

$$\begin{aligned} \{f,g\}(\mathbf{M},\rho,\sigma,\mathcal{A},\mathcal{Y}) &= \int_M \mathbf{M}\left(\left[\frac{\delta f}{\delta \mathbf{M}}, \frac{\delta g}{\delta \mathbf{M}}\right]_L\right)\mu \\ &+ \int_M \rho\left(\mathbf{d}\left(\frac{\delta f}{\delta \rho}\right)\left[\frac{\delta g}{\delta \mathbf{M}}\right] - \mathbf{d}\left(\frac{\delta g}{\delta \rho}\right)\left[\frac{\delta f}{\delta \mathbf{M}}\right]\right)\mu \\ &+ \int_M \sigma\left(\operatorname{div}\left(\frac{\delta f}{\delta \sigma}\left[\frac{\delta g}{\delta \mathbf{M}}\right]\right) - \operatorname{div}\left(\frac{\delta g}{\delta \sigma}\left[\frac{\delta f}{\delta \mathbf{M}}\right]\right)\right)\mu \\ &+ \int_M (g\gamma)\left(\frac{\delta f}{\delta \mathcal{A}}, \frac{\delta g}{\delta \mathcal{Y}}\right)\mu - \int_M (g\gamma)\left(\frac{\delta g}{\delta \mathcal{A}}, \frac{\delta f}{\delta \mathcal{Y}}\right)\mu. \end{aligned} \tag{4.6}$$

We can obtain this bracket and the associated Hamilton equations (4.3) alternatively by a reduction by stages process (see [**MMOPR**]). The symplectic reduced spaces are of the form $\mathcal{O} \times T^*\mathcal{C}onn(P)$, where $\mathcal{O}$ is a coadjoint orbit of the semidirect product $S := \mathcal{A}ut(P) \circledS (\mathcal{F}(M) \times \mathcal{F}(M))$.

If the principal bundle is trivial, the automorphism group is a semidirect product of groups. In this case the first term can be written more explicitly by taking advantage of the internal structure of $\mathcal{A}ut(P)$, and we recover (up to sign conventions) the Poisson bracket given in equation (38) in [**GHK**].

The second reduction. Note that right translation of $\mathcal{A}ut(P) \circledS (\mathcal{F}(M) \times \mathcal{F}(M))$ on itself and the action of $\mathcal{G}au(P)$ on $\mathcal{A}ut(P) \times \mathcal{C}onn(P)$ given by (4.2) commute if one views them as actions on $S \times \mathcal{C}onn(P)$. Therefore, by the general theory of commuting reduction by stages (see [**MMOPR**]), the momentum map associated to the gauge group action is $\mathcal{A}ut(P)$-invariant and induces a momentum map $\mathbf{J}_{\mathfrak{s}^*}$ on $\mathfrak{s}^* \times T^*\mathcal{C}onn(P)$ which restricts to a momentum map $\mathbf{J}_{\mathcal{O}}$ on the reduced space $\mathcal{O} \times T^*\mathcal{C}onn(P)$. A direct computation shows that we have

$$\mathbf{J}_{\mathfrak{s}^*} : \mathfrak{s}^* \times T^*\mathcal{C}onn(P) \to \mathfrak{gau}(P)^*, \quad \mathbf{J}_{\mathfrak{s}}^*(\mathbf{M},\rho,\sigma,\mathcal{A},\mathcal{Y}) = \sigma\left(\delta^{\mathcal{A}}\mathcal{Y} - \mathcal{A}(V)\right), \tag{4.7}$$

where $V \in \mathfrak{aut}(P)$ is such that $\mathbf{M} = K_{\mathcal{A}}(V,\cdot)$, and σ on the right hand side denotes the map defined in (2.4). The gauge group action induced on $\mathfrak{s}^* \times T^*\mathcal{C}onn(P)$ and $\mathcal{O} \times T^*\mathcal{C}onn(P)$ is given by

$$(\mathbf{M},\rho,\sigma,\mathcal{A},\mathcal{Y}) \mapsto (\mathrm{Ad}_\eta^* \mathbf{M},\rho,\sigma,\eta^*\mathcal{A},\eta^*\mathcal{Y}). \tag{4.8}$$

Using the notation $\mathbf{S} := (\mathbf{M},\rho,\sigma) \in \mathcal{O}$, it can be written as

$$(\mathbf{S},\mathcal{A},\mathcal{Y}) \mapsto (\mathrm{Ad}_{(\eta,0,0)}^* \mathbf{S},\eta^*\mathcal{A},\eta^*\mathcal{Y}). \tag{4.9}$$

This action is simply the diagonal action given on the first factor by the coadjoint action of the subgroup $\mathcal{G}au(P)$ of $S = \mathcal{A}ut(P) \circledS (\mathcal{F}(M) \times \mathcal{F}(M))$, and on the second factor by the gauge transformations. Note that when the center $Z(G)$ of the structure group G of the bundle $\pi : P \to M$ is trivial, then the transformation $\mathcal{A} \mapsto \eta^*\mathcal{A}$ is free. In this case, the reduced action (4.9) is also free and the second reduced symplectic spaces

$$\mathbf{J}_{\mathcal{O}}^{-1}(\mathbf{N})/\mathcal{G}au(P)_{\mathbf{N}}, \quad \mathbf{N} \in \mathfrak{gau}(P)^*,$$

have no singularities.

By the reduction by stages process (see [**MMOPR**]), the reduced spaces $\mathbf{J}_{\mathcal{O}}^{-1}(\mathbf{N})/\mathcal{G}au(P)_{\mathbf{N}}$ are symplectically diffeomorphic to the reduced spaces obtained by a one step reduction from the cotangent bundle

$$T^*\big(S \times \mathcal{C}onn(P)\big)$$

with respect to the product of the two cotangent-lifted actions.

Note that the Euler-Yang-Mills equation (4.3) projects to the reduced space at zero momentum

$$\mathbf{J}_{\mathcal{O}}^{-1}(0)/\mathcal{G}au(P). \tag{4.10}$$

The general case corresponds to the Yang-Mills fluid with an external charge $\mathbf{N}$.

In order to obtain the reduced Poisson structure concretely, we will identify the space $\mathfrak{s}^* \times T^*\mathcal{C}onn(P)$ with a space on which the gauge action is simpler. This identification is given in the following proposition.

PROPOSITION 4.5. *Consider the group $K = \mathcal{D}(M) \circledS (\mathcal{F}(M) \times \mathcal{F}(M))$ and denote by $\mathfrak{k}^*$ the dual of its Lie algebra. There is a gauge-equivariant diffeomorphism*

$$i : \mathfrak{s}^* \times T^*\mathcal{C}onn(P) \to \mathfrak{k}^* \times \mathcal{F}_G(P, \mathfrak{g}^*) \times T^*\mathcal{C}onn(P), \tag{4.11}$$

given by

$$i(\mathbf{M}, \rho, \sigma, \mathcal{A}, \mathcal{Y}) := ((\mathrm{Hor}_{\mathcal{A}})^* \circ \mathbf{M}, \rho, \sigma, \mathbb{J} \circ \mathbf{M}, \mathcal{A}, -\mathcal{Y}) =: (\mathbf{n}, \rho, \sigma, \nu, \mathcal{A}, \mathcal{E}),$$

where the gauge group acts on $\mathfrak{s}^* \times T^*\mathcal{C}onn(P)$ by the action (4.9) and on $\mathfrak{k}^* \times \mathcal{F}_G(P, \mathfrak{g}^*) \times T^*\mathcal{C}onn(P)$ only on the factor $\mathcal{F}_G(P, \mathfrak{g}^*) \times T^*\mathcal{C}onn(P)$ by the right action

$$(\nu, \mathcal{A}, \mathcal{E}) \mapsto (\nu \circ \eta, \eta^*\mathcal{A}, \eta^*\mathcal{E}). \tag{4.12}$$

Moreover, the image of the level set $\mathbf{J}_{\mathfrak{s}^*}^{-1}(\mathbf{N})$ by the diffeomorphism i is given by

$$\{(\mathbf{n}, \rho, \sigma, \nu, \mathcal{A}, \mathcal{E}) \mid \nu + \gamma(\delta^{\mathcal{A}}\mathcal{E} + f, \cdot) = 0\},$$

where $\mathbf{N} \in \mathfrak{gau}(P)$ and $f \in \mathcal{F}_G(P, \mathfrak{g})$ is such that $\sigma(f) = \mathbf{N}$. Thus $\mathbf{J}_{\mathfrak{s}^*}^{-1}(\mathbf{N})$ is diffeomorphic to $\mathfrak{k}^* \times T^*\mathcal{C}onn(P)$.

The map $\mathbb{J} : T^*P \to \mathfrak{g}^*$ denotes the momentum map $\mathbb{J}(\alpha_p)(\xi) := \langle \alpha_p, \xi_P(p) \rangle$, and $(\mathrm{Hor}_{\mathcal{A}})^*$ denotes the dual map of the horizontal-lift $\mathrm{Hor}_{\mathcal{A}} : TM \to TP$ with respect to $\mathcal{A}$.

From this Proposition we obtain that the reduced spaces $\mathbf{J}_{\mathfrak{s}^*}^{-1}(\mathbf{N})/\mathcal{G}au(P)_{\mathbf{N}}$ can be identified with the quotient space $\mathfrak{k}^* \times [(\mathcal{F}_G(P, \mathfrak{g}^*) \times T^*\mathcal{C}onn(P)) / \mathcal{G}au(P)_{\mathbf{N}}]$, via the diffeomorphism induced by i and given by

$$[(\mathbf{M}, \rho, \sigma, \nu, \mathcal{A}, \mathcal{Y})] \mapsto ((\mathrm{Hor}_{\mathcal{A}})^* \circ \mathbf{M}, \rho, \sigma, [\nu, \mathcal{A}, \mathcal{E}]), \tag{4.13}$$

where $[\cdot]$ denote the corresponding equivalence classes.

We now compute the Poisson structure $\{\,,\}'$ induced by i on $\mathfrak{k}^* \times \mathcal{F}_G(P,\mathfrak{g}^*) \times T^*\mathcal{C}onn(P)$. For $f,g \in \mathcal{F}(\mathfrak{k}^* \times \mathcal{F}_G(P,\mathfrak{g}^*) \times T^*\mathcal{C}onn(P))$ we have the formulas

$$\frac{\delta(f\circ i)}{\delta\mathbf{M}} = \mathrm{Hor}_{\mathcal{A}} \circ \frac{\delta f}{\delta\mathbf{n}} + \mathbb{J}^* \circ \frac{\delta f}{\delta\nu}, \qquad \left[\frac{\delta(f\circ i)}{\delta\mathbf{M}}\right] = \frac{\delta f}{\delta\mathbf{n}}, \qquad \mathbf{n} := (\mathrm{Hor}_{\mathcal{A}})^* \circ \mathbf{M},$$

$$(g\gamma)\left(\frac{\delta(f\circ i)}{\delta\mathcal{A}}, \mathcal{C}\right) = (g\gamma)\left(\frac{\delta f}{\delta\mathcal{A}}, \mathcal{C}\right) - \nu\left(\mathcal{C}\left(\mathrm{Hor}_{\mathcal{A}}\left(\frac{\delta f}{\delta\mathbf{n}}\right)\right)\right), \qquad \nu := \mathbf{J}\circ\mathbf{M}.$$

Therefore, we obtain

$$\begin{aligned}
&\{f,g\}'(\mathbf{n},\rho,\sigma,\nu,\mathcal{A},\mathcal{E}) := \{f\circ i, g\circ i\}(\mathbf{M},\rho,\sigma,\mathcal{A},\mathcal{Y})\\
&\quad = \int_M \mathbf{n}\left(\left[\frac{\delta f}{\delta\mathbf{n}}, \frac{\delta g}{\delta\mathbf{n}}\right]_L\right)\mu + \int_M \rho\left(\mathbf{d}\left(\frac{\delta f}{\delta\rho}\right)\frac{\delta g}{\delta\mathbf{n}} - \mathbf{d}\left(\frac{\delta g}{\delta\rho}\right)\frac{\delta f}{\delta\mathbf{n}}\right)\mu\\
&\quad + \int_M \sigma\left(\mathrm{div}\left(\frac{\delta f}{\delta\sigma}\frac{\delta g}{\delta\mathbf{n}}\right) - \mathrm{div}\left(\frac{\delta g}{\delta\sigma}\frac{\delta f}{\delta\mathbf{n}}\right)\right)\mu\\
&\quad + \int_M (g\gamma)\left(\frac{\delta g}{\delta\mathcal{A}}, \frac{\delta f}{\delta\mathcal{E}}\right)\mu - \int_M (g\gamma)\left(\frac{\delta f}{\delta\mathcal{A}}, \frac{\delta g}{\delta\mathcal{E}}\right)\mu\\
&\quad + \int_M \nu\left(\left[\frac{\delta f}{\delta\nu}, \frac{\delta g}{\delta\nu}\right]\right)\mu\\
&\quad + \int_M \nu\left(\frac{\delta g}{\delta\mathcal{E}}\left(\mathrm{Hor}_{\mathcal{A}}\circ\frac{\delta f}{\delta\mathbf{n}}\right) - \frac{\delta f}{\delta\mathcal{E}}\left(\mathrm{Hor}_{\mathcal{A}}\circ\frac{\delta g}{\delta\mathbf{n}}\right)\right.\\
&\qquad\qquad + \mathbf{d}^{\mathcal{A}}\left(\frac{\delta f}{\delta\nu}\right)\left(\mathrm{Hor}_{\mathcal{A}}\circ\frac{\delta g}{\delta\mathbf{n}}\right) - \mathbf{d}^{\mathcal{A}}\left(\frac{\delta g}{\delta\nu}\right)\left(\mathrm{Hor}_{\mathcal{A}}\circ\frac{\delta f}{\delta\mathbf{n}}\right)\\
&\qquad\qquad \left. + \mathcal{B}\left(\mathrm{Hor}_{\mathcal{A}}\circ\frac{\delta f}{\delta\mathbf{n}}, \mathrm{Hor}_{\mathcal{A}}\circ\frac{\delta g}{\delta\mathbf{n}}\right)\right)\mu.
\end{aligned} \tag{4.14}$$

Note that the first three terms in (4.14) represent the Lie-Poisson bracket on $\mathfrak{k}^*$, the fourth and fifth terms represent the canonical bracket on $T^*\mathcal{C}onn(P)$, the sixth term is the Lie-Poisson bracket on $\mathcal{F}_G(P,\mathfrak{g}^*)$, and the last term provides the coupling of the fluid variables to the Yang-Mills fields.

By the general process of Poisson (point) reduction, the reduced spaces

$$\mathbf{J}_{\mathfrak{s}^*}^{-1}(\mathbf{N})/\mathcal{G}au(P)_{\mathbf{N}} \simeq \mathfrak{k}^* \times \left[\left(\mathcal{F}_G(P,\mathfrak{g}^*) \times T^*\mathcal{C}onn(P)\right)/\mathcal{G}au(P)_{\mathbf{N}}\right]$$

inherit a Poisson bracket $\{\,,\}_{\mathbf{N}}$ given by

$$\{f_{\mathbf{N}}, g_{\mathbf{N}}\}_{\mathbf{N}}(\mathbf{n},\rho,\sigma,[\nu,\mathcal{A},\mathcal{E}]) := \{f,g\}'(\mathbf{n},\rho,\sigma,\nu,\mathcal{A},\mathcal{Y}), \tag{4.15}$$

where f,g are any $\mathcal{G}au(P)$-invariant extensions of the functions $f_{\mathbf{N}}\circ\pi_{\mathbf{N}}, g_{\mathbf{N}}\circ\pi_{\mathbf{N}} : \mathbf{J}_{\mathfrak{s}^*}^{-1}(\mathbf{N}) \to \mathbb{R}$, relative to the projection $\pi_{\mathbf{N}} : \mathbf{J}_{\mathfrak{s}^*}^{-1}(\mathbf{N}) \to \mathbf{J}_{\mathfrak{s}^*}^{-1}(\mathbf{N})/\mathcal{G}au(P)_{\mathbf{N}}$.

The equations of motion on the Poisson point reduced space $\mathbf{J}_{\mathcal{O}}^{-1}(0)/\mathcal{G}au(P)$ cannot be written explicitly, because there is no concrete realization of this quotient, to our knowledge. However, in the particular case of a trivial principal bundle, this is possible.

We end this section by examining the case of the trivial bundle $P = S^1 \times M$, that is, the case of the Euler-Maxwell equations. Since $G = S^1$, the gauge transformation (4.8) is simply

$$(\mathbf{m},\nu,\rho,\sigma,A,Y) \mapsto (\mathbf{m} + \nu\mathbf{d}\eta, \nu, \rho, \sigma, A + \mathbf{d}\eta, Y),$$

where $\nu = Q = \rho\frac{q}{m}$, by relation (4.4) and Corollary 4.4. This gauge transformation coincides with the one given in equation (36) in [**MWRSS**], where the notation $a := \frac{q}{m}$ is used. The zero level set of the momentum map is given by

$$\mathbf{J}_{\mathfrak{s}^*}^{-1}(0) = \left\{ \left(\mathbf{m}, \rho\frac{q}{m}, \rho, \sigma, A, Y \right) \mid \operatorname{div} \mathbf{E} = \rho\frac{q}{m},\ \mathbf{E} := -Y^\sharp \right\}.$$

The bijection i reads

$$i\left(\mathbf{m}, \rho\frac{q}{m}, \rho, \sigma, A, Y \right) = \left(\mathbf{m} - A\rho\frac{q}{m}, \rho, \sigma, \rho\frac{q}{m}, A, -Y \right),$$

and the image of $\mathbf{J}_{\mathfrak{s}^*}^{-1}(0)$ is given by

$$\{(\mathbf{n}, \rho, \sigma, \nu, A, E) \mid \operatorname{div} \mathbf{E} = \nu\},$$

where the notation $\mathbf{E} := E^\sharp \in \mathfrak{X}(M)$ is used. The gauge transformation (4.12) is simply given by

$$(\mathbf{n}, \rho, \sigma, \nu, A, E) \mapsto (\mathbf{n}, \rho, \sigma, \nu, A + \mathbf{d}\eta, E).$$

Through the diffeomorphism i, the projection $\pi_0 : \mathbf{J}_{\mathfrak{s}^*}^{-1}(0) \to \mathbf{J}_{\mathfrak{s}^*}^{-1}(0)/\mathcal{G}au(P)$ is given by

$$(\mathbf{n}, \rho, \sigma, \nu, A, E) \mapsto (\mathbf{n}, \rho, \sigma, [A], E),$$

where $[A] \in \Omega^1(M)/\mathbf{d}\mathcal{F}(M)$. Assuming that the first and second cohomology groups of M are trivial, $H^1(M) = H^2(M) = \{0\}$, we get the isomorphism

$$[A] \mapsto B := \mathbf{d}A \in \Omega^2_{cl}(M), \tag{4.16}$$

where $\Omega^2_{cl}(M)$ denotes the space closed two-forms. Thus i induces a diffeomorphism between $\mathbf{J}_{\mathfrak{s}^*}^{-1}(0)/\mathcal{G}au(P)$ and the space $\mathfrak{k}^* \times \Omega^2_{cl}(M) \times \Omega^1(M)$ given by

$$[\mathbf{m}, \nu, \rho, \sigma, A, Y] \mapsto (\mathbf{m} - A\nu, \rho, \sigma, \mathbf{d}A, -Y) =: (\mathbf{n}, \rho, \sigma, B, E).$$

where $\Omega^2_{cl}(M)$ denotes the closed 2-forms on M. This identification coincides with the one given in Proposition 10.1 in [**MWRSS**].

Using the definition (4.15) and the bracket (4.14), the reduced Poisson bracket on $\mathfrak{k}^* \times \Omega^2_{cl}(M) \times \Omega^1(M)$ is given by

$$\begin{aligned}
\{f, g\}_0(\mathbf{n}, \rho, \sigma, B, E) &= \int_M \mathbf{n}\left(\left[\frac{\delta f}{\delta \mathbf{n}}, \frac{\delta g}{\delta \mathbf{n}} \right]_L \right) \mu \\
&+ \int_M \rho \left(\mathbf{d}\left(\frac{\delta f}{\delta \rho} \right) \frac{\delta g}{\delta \mathbf{n}} - \mathbf{d}\left(\frac{\delta g}{\delta \rho} \right) \frac{\delta f}{\delta \mathbf{n}} \right) \mu \\
&+ \int_M \sigma \left(\operatorname{div}\left(\frac{\delta f}{\delta \sigma} \frac{\delta g}{\delta \mathbf{n}} \right) - \operatorname{div}\left(\frac{\delta g}{\delta \sigma} \frac{\delta f}{\delta \mathbf{n}} \right) \right) \mu \\
&+ \int_M g\left(\delta \frac{\delta g}{\delta B}, \frac{\delta f}{\delta E} \right) \mu - \int_M g\left(\delta \frac{\delta f}{\delta B}, \frac{\delta g}{\delta E} \right) \mu \\
&+ \int_M \rho \frac{q}{m} \left(\frac{\delta g}{\delta E}\left(\frac{\delta f}{\delta \mathbf{n}} \right) - \frac{\delta f}{\delta E}\left(\frac{\delta g}{\delta \mathbf{n}} \right) + B\left(\frac{\delta f}{\delta \mathbf{n}}, \frac{\delta g}{\delta \mathbf{n}} \right) \right) \mu,
\end{aligned} \tag{4.17}$$

and the Euler-Maxwell equations can be written as

$$\dot{f} = \{f, h\}_0,$$

relative to the induced Hamiltonian h given by

$$h(\mathbf{n}, \rho, \sigma, B, E) = \frac{1}{2}\int_M \left(\frac{1}{\rho} g(\mathbf{n}, \mathbf{n}) + \frac{1}{\rho}(\delta E)^2 \right) \mu + \int_M \rho e(\rho, \sigma)\mu + \frac{1}{2}\int_M \left(\|E\|^2 + \|B\|^2 \right) \mu.$$

Note that the function

$$C(\mathbf{n}, \rho, \sigma, B, E) = \frac{1}{2}\int_M \frac{1}{\rho}(\delta E)^2$$

is a Casimir function, so an equivalent Hamiltonian is given by

$$\bar{h}(\mathbf{n}, \rho, \sigma, B, E) = \frac{1}{2}\int_M \frac{1}{\rho} g(\mathbf{n}, \mathbf{n})\mu + \int_M \rho e(\rho, \sigma)\mu + \frac{1}{2}\int_M \left(\|E\|^2 + \|B\|^2\right)\mu.$$

When M is three dimensional, we can use the notations $\mathbf{B} := (\star B)^\sharp$ and $\mathbf{E} := E^\sharp$. Therefore the two last terms can be written as

$$\int_M g\left(\operatorname{curl}\frac{\delta g}{\delta \mathbf{B}}, \frac{\delta f}{\delta \mathbf{E}}\right)\mu - \int_M g\left(\operatorname{curl}\frac{\delta f}{\delta \mathbf{B}}, \frac{\delta g}{\delta \mathbf{E}}\right)\mu$$
$$+ \int_M \rho\frac{q}{m}\left(g\left(\frac{\delta g}{\delta \mathbf{E}}, \frac{\delta f}{\delta \mathbf{n}}\right) - g\left(\frac{\delta f}{\delta \mathbf{E}}, \frac{\delta g}{\delta \mathbf{n}}\right) + g\left(\mathbf{B}, \frac{\delta f}{\delta \mathbf{n}} \times \frac{\delta g}{\delta \mathbf{n}}\right)\right)\mu.$$

This bracket coincides with the one derived in [**MWRSS**] by a direct computation. Note that the first line in the formula above is the Pauli-Born-Infeld Poisson bracket for the Maxwell equations (see, e.g. [**MR**], §1.6). The Hamiltonian $\bar{h}$ is very simple: it is the sum of the total energy of the fluid plus the energy of the electromagnetic field.

References

[AM] R. Abraham and J. E. Marsden, *Foundations of mechanics*, Benjamin-Cummings Publ. Co, Updated 1985 version, reprinted by Perseus Publishing, second edition, 1978.

[A] V. I. Arnold, *Sur la géométrie différentielle des groupes de Lie de dimenson infinie et ses applications à l'hydrodynamique des fluides parfaits*, Ann. Inst. Fourier, Grenoble **16** (1966), 319–361.

[B] D. Bleecker, *Gauge theory and variational principles*, Addison-Wesley Publ. Co., 1981.

[EM] D. G. Ebin and J. E. Marsden, *Groups of diffeomorphisms and the motion of an incompressible fluid*, Ann. of Math. **92** (1970),102–163.

[GHK] Gibbons, J., D. D Holm and B. Kupershmidt, *The Hamiltonian structure of classical chromohydrodynamics*, Physica D **6** (1983), 179–194.

[HMR] Holm D. D, J. E. Marsden and T. S. Ratiu, *The Euler-Poincaré equations and semidirect products with applications to continuum theories*, Adv. in Math. **137** (1998), 1–81.

[KN] S. Kobayashi and K. Nomizu, *Foundations of differential geometry*, Wiley, 1963.

[MMOPR] J. E. Marsden, G. Misiołek, J.-P. Ortega, M. Perlmutter, and T. S. Ratiu *Hamiltonian reduction by stages*, Springer Lecture Notes in Mathematics, Springer-Verlag, 2007.

[MR] J. E. Marsden and T. S. Ratiu, *Introduction to mechanics and symmetry*, Texts in Applied Mathematics, **17**, Springer-Verlag, 1994; Second Edition, 1999, second printing 2003.

[MRW] J. E. Marsden, T. S. Ratiu, and A. Weinstein, *Semidirect product and reduction in mechanics*, Trans. Amer. Math. Soc. **281** (1984), 147–177.

[MW] J. E. Marsden and A. Weinstein, *The Hamiltonian structure of the Maxwell-Vlasov equations*, Physica D **4** (1982), 394–406.

[MWRSS] J. E. Marsden, A. Weinstein, T. S. Ratiu, R. Schmid, and R. G. Spencer, *Hamiltonian system with symmetry, coadjoint orbits and Plasma physics*, Atti della Acad. della Sc. di Torino, Supplemento al Vol. **117**, Proc. IUTAM-ISIMM Symposium on Modern Developments in Analytical Mechanics (Academy of Sciences Turin, Turin, June 7–11, 1982), 1983 pp. 289–340.

[M] R. Montgomery, *Canonical formulations of a classical particle in a Yang-Mills field and Wong's equations*, Lett. Math. Phys., **8** (1984), 59-67.

Section de Mathématiques and Bernoulli Center, École Polytechnique Fédérale de Lausanne, CH–1015 Lausanne, Switzerland

E-mail address: francois.gay-balmaz@epfl.ch, tudor.ratiu@epfl.ch

Contemporary Mathematics
Volume **450**, 2008

Orbifold cohomology of abelian symplectic reductions and the case of weighted projective spaces

Tara S. Holm

Abstract. These notes accompany a lecture about the topology of symplectic (and other) quotients. The aim is two-fold: first to advertise the ease of computation in the symplectic category; and second to give an account of some new computations for weighted projective spaces. We start with a brief exposition of how orbifolds arise in the symplectic category, and discuss the techniques used to understand their topology. We then show how these results can be used to compute the Chen-Ruan orbifold cohomology ring of abelian symplectic reductions. We conclude by comparing the several rings associated to a weighted projective space. We make these computations directly, avoiding any mention of a stacky fan or of a labeled moment polytope.

Contents

The notion of an **orbifold** has been present in topology since the 1950's [**S1, S2**]. More recently, orbifolds have played an important role in differential and algebraic geometry, and in mathematical physics. A fundamental theme is to compute topological invariants associated to an orbifold, with one ostensible goal to understand Gromov-Witten invariants for these spaces. The aim of the present article is modest: to expound how techniques from symplectic geometry may be used to understand the degree-zero genus-zero Gromov-Witten invariants with three marked points, the so-called **Chen-Ruan orbifold cohomology ring**; and to make explicit the details of these techniques in the case of weighted projective spaces.

1991 *Mathematics Subject Classification.* Primary 53D20; Secondary 14N35, 53D45, 57R91.
Key words and phrases. Symplectic quotient, orbifold, cohomology.
TSH is grateful for the support of the NSF through the grant DMS-0604807.

In the symplectic category, orbifolds arise as symplectic quotients. We recount the techniques from symplectic geometry that may be used to compute topological invariants of a symplectic quotient. This is based on Kirwan's seminal work [**Ki**]; and for orbifold invariants, the author's joint work with Goldin and Knutson [**GHK**]. The quotients we consider are by a compact connected abelian group. We employ techniques coming from algebraic topology, most notably using equivariant cohomology. For those used to working with finite groups, it is important to note that, whereas for finite groups the invariant part of a cohomology ring is identical to the equivariant cohomology, this is not the case for connected groups.

The main example in this article is a **weighted projective space** $\mathbb{C}P^n_{(b)}$. Its definition depends on a sequence $(b) = (b_0, \ldots, b_n)$ of positive integers. Kawasaki showed that the ordinary cohomology groups, with integer coefficients, of the underlying topological space of a weighted projective space are identical to the cohomology groups of a smooth projective space [**Ka**], but there is a twisted ring structure. We review the details of his work. Then in Theorem 4.2, we compute the cohomology of the orbifold $[\mathbb{C}P^n_{(b)}]$, proving that

$$H^*([\mathbb{C}P^n_{(b)}]; \mathbb{Z}) = H^*_{S^1_{(b)}}(S^{2n+1}; \mathbb{Z}) \cong \frac{\mathbb{Z}[u]}{\langle b_0 \cdots b_n u^{n+1} \rangle}. \tag{0.1}$$

Whereas Kawasaki finds a twist in the the ring structure, we find torsion in high degrees of the ring (0.1). There is a natural map from Kawasaki's ring to this one, and we describe the map explicitly. Finally in Theorem 4.3, we compute the Chen-Ruan cohomology ring of this orbifold. We make this computation using integer coefficients, generalizing results in [**J, Ma1, Ma2**]. Moreover, we give explicit generators and relations, and avoid mentioning a stacky fan [**BCS**] or a labeled polytope [**LT, GHK**].

The definitions in this article make sense for arbitrary coefficient rings. Indeed, all computations in the final section use integer coefficients. Moore and Witten have suggested that the torsion in K-theory has more physical significance than torsion in cohomology [**MoWi**]. The author together with Goldin, Harada and Kimura, is investigating a K-theoretic version of [**GHK**] and of the computations herein, building on the work of Harada and Landweber [**HL**].

The remainder of the paper is organized as follows. In Section 1 we give a quick exposition of how orbifolds arise in the symplectic category. We then introduce several cohomology rings associated to an orbifold in Section 2. We advertise the ease of computation for these rings in Section 3. The novel results in this article are the computations in Section 4. We include detailed proofs that avoid much of the symplectic machinery used in [**GHK**].

Acknowledgments. Many thanks are due to Tony Bahri, Matthias Franz, Rebecca Goldin, Megumi Harada, Ralph Kaufmann, Takashi Kimura, Allen Knutson, Eugene Lerman, Reyer Sjamaar, and Alan Weinstein for many helpful conversations; and to Yoshiaki Maeda and the organizers and sponsors of Poisson 2006 in Tokyo, Japan, where this work was presented.

1. Symplectic manifolds and quotients

We begin with a very brief introduction to the symplectic category; a more detailed account of the subject can be found in [**CdS**]. A **symplectic form** on a manifold M is a closed non-degenerate two-form $\omega \in \Omega^2(M)$. Thus, for any tangent vectors $\mathcal{X}, \mathcal{Y} \in T_pM$, $\omega_p(\mathcal{X}, \mathcal{Y}) \in \mathbb{R}$. The key examples include the following.

EXAMPLE 1.1: $M = S^2 = \mathbb{C}P^1$ with $\omega_p(\mathcal{X}, \mathcal{Y})$ equal to the signed area of the parallelogram spanned by $\mathcal{X}$ and $\mathcal{Y}$. This is the Fubini-Study form on $\mathbb{C}P^1$.

EXAMPLE 1.2: M any orientable Riemann surface with ω as in Example 1.1. Note that orientability is a necessary condition on a symplectic manifold M, because the top exterior power of the symplectic form is a volume form.

EXAMPLE 1.3: $M = \mathbb{R}^{2d}$ with $\omega = \sum dx_i \wedge dy_i$.

EXAMPLE 1.4: $M = \mathcal{O}_\lambda$ a coadjoint orbit of a compact connected semisimple Lie group, equipped with ω the Kostant-Kirillov-Soriau form.

Example 1.3 gains particular importance because of

DARBOUX'S THEOREM 1.5. *Let M be a symplectic $2d$-manifold with symplectic form ω. Then for every point $p \in M$, there exists a coordinate chart U about p with coordinates $x_1, \ldots, x_d, y_1, \ldots, y_d$ so that on this chart,*

$$\omega = \sum_{i=1}^{d} dx_i \wedge dy_i. \tag{1.1}$$

Thus, whereas Riemannian geometry uses local invariants such as **curvature** to distinguish metrics, symplectic forms are locally indistinguishable.

The symmetries of a symplectic manifold may be encoded as a group action. Here we restrict ourselves to a compact connected abelian group $T = (S^1)^n$. An action of T on M is **symplectic** if it preserves ω; that is, $\rho_g^*\omega = \omega$, for each $g \in T$, where ρ_g is the diffeomorphism corresponding to the group element g. The action is **Hamiltonian** if in addition, for every $\xi \in \mathfrak{t}$, the vector field

$$\mathcal{X}_\xi = \frac{d}{dt}[\exp(t\xi)]|_{t=0} \tag{1.2}$$

is a Hamiltonian vector field. That is, we require that $\omega(\mathcal{X}_\xi, \cdot) = d\phi^\xi$ is an exact one-form. Each ϕ^ξ is a smooth function on M, determined up to a constant. Taking them together, we may define a **moment map**

$$\begin{array}{rcl} \Phi : M & \longrightarrow & \mathfrak{t}^* \\ p & \longmapsto & \left(\begin{array}{rcl} \Phi(p) : \mathfrak{t} & \to & \mathbb{R} \\ \xi & \mapsto & \phi^\xi(p) \end{array} \right). \end{array} \tag{1.3}$$

Returning to our examples, we have Hamiltonian actions in all but the second example.

EXAMPLE 1.6: The circle S^1 acts on $M = S^2 = \mathbb{C}P^1$ by rotations. If we use angle and height coordinates on S^2, then the vector field this action generates is tangent to the latitude lines, so in coordinates, $\mathcal{X}^\xi = \frac{\partial}{\partial\theta}$, and since $\omega = d\theta \wedge dh$, $\omega(\mathcal{X}_\xi, \cdot) = dh$, so a moment map is the height function on S^2, as shown in Figure 1.1 below.

EXAMPLE 1.7: If M is a two-torus $M = T^2 = S^1 \times S^1$, then $S^1 \times S^1$ acts on itself by multiplication. This action is symplectic, but is not Hamiltonian. In fact, no Riemann surface with non-zero genus has a nontrivial Hamiltonian torus action.

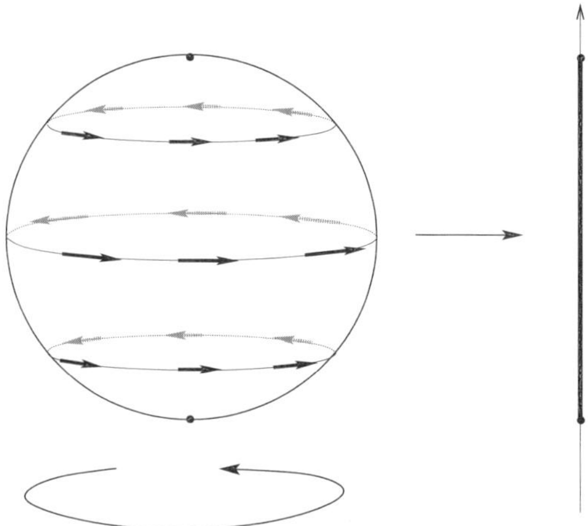

FIGURE 1.1. The vector field and moment map for S^1 acting by rotations on S^2.

EXAMPLE 1.8: The torus $T^d = (S^1)^d \subset \mathbb{C}^d$ acts by coordinate-wise multiplication on $M = \mathbb{R}^{2d} = \mathbb{C}^d$. This action rotates each copy of $\mathbb{C} = \mathbb{R}^2$ (at unit speed), and is Hamiltonian. Identifying $\mathfrak{t}^* \cong \mathbb{R}^d$, a moment map is

$$\Phi(z_1, \ldots, z_n) = (|z_1|^2, \ldots, |z_d|^2), \tag{1.4}$$

up to a constant multiple.

EXAMPLE 1.9: Each coadjoint orbit $M = \mathcal{O}_\lambda \subseteq \mathfrak{g}^*$ may be identified as a homogenous space G/L, where L is a Levi subgroup of the Lie group G. Thus G and its maximal torus T act on M by left multiplication. A G-moment map is inclusion

$$\Phi_G : \mathcal{O}_\lambda \hookrightarrow \mathfrak{g}^*, \tag{1.5}$$

and a T-moment map is the G-moment map composed with the natural projection $\mathfrak{g}^* \to \mathfrak{t}^*$ that is dual to the inclusion $\mathfrak{t} \hookrightarrow \mathfrak{g}$.

In each of these examples, the image of the (torus) moment map is a convex subset of $\mathbb{R}^n$. This is true more generally.

CONVEXITY THEOREM 1.10 ([**A**],[**GuSt**]). *If M is a compact Hamiltonian T-space, then $\Phi(M)$ is a convex polytope. It is the convex hull of $\Phi(M^T)$, the images of the T-fixed points.*

The convexity theorem is an example of a **localization phenomenon**: a global feature (the image of the moment map) that is determined by local features of the fixed points (their images under the moment map). The convexity property is a recurring theme in symplectic geometry; its many guises are illustrated in [**GuSj**].

The moment map is a T-invariant map: it maps entire T-orbits to the same point in $\mathfrak{t}^*$. Thus when α is a regular value, the level set $\Phi^{-1}(\alpha)$ is a T-invariant submanifold of M. Moreover, the action of T on a regular level set is **locally free**: it has only finite stabilizers. This follows directly from the moment map condition: at a regular value, $d\phi^\xi$ is never zero, implying that $\mathcal{X}_\xi$ is not zero, so there is no 1-parameter subgroup fixing points in the level set. Thus, at a regular value the

symplectic reduction $M/\!\!/T(\alpha) = \Phi^{-1}(\alpha)$ is an orbifold. In fact, Marsden and Weinstein proved

THEOREM 1.11 ([**MaWe**]). *If M is a Hamiltonian T-space and α is a regular value of the moment map Φ, then the symplectic reduction $M/\!\!/T(\alpha)$ is a* **symplectic** *orbifold.*

More generally, the symplectic reduction $M/\!\!/T(\alpha)$ at a critical value is a symplectic stratified space [**SL**].

Symplectic reduction is an important technique for constructing new symplectic manifolds from old. From our examples, we may construct several classes of symplectic manifolds.

EXAMPLE 1.12: For the action of S^1 on $M = S^2 = \mathbb{C}P^1$ by rotation, the level set of a regular value is a latitude line, which the circle rotates. The quotient is a point. Note that if S^1 acts by rotation S^2 at twice the usual speed, then the quotient, as an orbifold, is $[\mathrm{pt}/\mathbb{Z}_2]$. Thus, every orbifold $[\mathrm{pt}/\mathbb{Z}_k]$ is a quotient of S^2 by a rotation action.

EXAMPLE 1.13: T^d acts on $M = \mathbb{R}^{2d} = \mathbb{C}^n$ by rotation of each copy of $\mathbb{C} = \mathbb{R}^2$. The level set of a regular value is a copy of T^d, and so again the quotient is a point (or potentially an orbipoint, for different actions of T^d). However, we may also restrict our attention to subtori $K \subseteq T^n$. The action of K is still Hamiltonian, and for certain choices of K, $\mathbb{C}^n/\!\!/K(\alpha)$ is a **symplectic toric orbifold**. Lerman and Tolman show that every **effective** symplectic toric orbifold may be constructed in this way [**LT**].

EXAMPLE 1.14: For the T-action on a coadjoint orbit $M = \mathcal{O}_\lambda$, the symplectic reduction $M/\!\!/T(\alpha)$ is known as a **weight variety**. One may determine the possible orbifold singularities by analyzing the combinatorics of G and its Weyl group. See, for instance, [**Kn**] or [**GHK**]. This reduced space plays an important role in representation theory.

2. Orbifolds and their cohomology

We now turn to orbifolds in the topological category. In terms of local models, an **orbifold** is a topological space where each point has a neighborhood homeomorphic (or diffeomorphic) to the quotient of a (fixed dimensional) vector space by a finite group. Satake introduced this notion in the 1950's [**S1, S2**], originally calling the spaces V-manifolds. Thurston coined the term orbifold when he rediscovered them in the 1970's (see [**Th**]) in his study of 3-manifolds. This local model, however, makes it difficult to define very basic pieces in the theory of orbifolds: overlap conditions on orbifold charts, suborbifolds, and maps between orbifolds. As is evident already in the work of Haefliger [**Hæ**], the proper way to think of an orbifold is as a Morita equivalence class of **groupoids**, one of which is a proper étale groupoid (see [**Moe**]); or equivalently as a smooth **Deligne-Mumford stack** (see [**DM**]). For example, using this structure, a map of orbifolds should simply be a morphism of the appropriate objects.

While groupoids or stacks provide the correct mathematical framework, the technology is a bit beyond the scope of this article. Indeed, for us it is sufficient to work with the local models, largely because we restrict our attention to orbifolds that arise as global quotients. Nevertheless, we will need to distinguish between an orbifold $\mathfrak{X}$ or $[X]$ and its underlying topological space (or **coarse moduli space**)

X. In particular, when $\mathfrak{X}$ is presented as a global quotient of a manifold M by a group G, we will use square brackets $[M/G]$ to denote the orbifold, and M/G to denote the underlying coarse moduli.

For an orbifold $\mathfrak{X}$, at each point $x \in \mathfrak{X}$, we have a local isotropy group Γ_x at x. We will be interested in **almost complex orbifolds**, that is orbifolds that have local models isomorphic to $\mathbb{C}^d/\Gamma_x$, with $\Gamma_x \subseteq U(d)$ a finite group acting unitarily on $\mathbb{C}^d$. Our main example in this section is the orbisphere shown in the figure below. This is a symplectic toric orbifold in the sense of Tolman and Weitsman

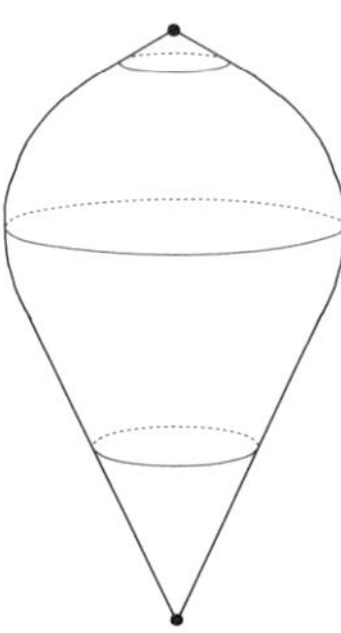

FIGURE 2.1. An orbisphere with two orbifold points, a $\mathbb{Z}_p$ singularity at the north pole and a $\mathbb{Z}_q$ singularity at the south pole. All other points in the space have local isotropy group the one-element group. When p and q are relatively prime, this is a weighted projective space $\mathbb{C}P^1_{p,q}$.

[**LT**]. In that context, it corresponds to the labeled polytope that is an edge, with the vertices labeled p and q. This orbifold cannot (always) be presented as a global quotient by a **finite group**, although it can be presented as a symplectic reduction. When p and q are relatively prime, it is the reduction of $\mathbb{C}^2$ by the Hamiltonian S^1-action

$$e^{2\pi i\theta} \cdot (z_1, z_2) = (e^{p\cdot 2\pi i\theta} \cdot z_1, e^{q\cdot 2\pi i\theta} \cdot z_2). \tag{2.1}$$

Thus, it is a global quotient of the level set $\Phi^{-1}(\alpha) \approx S^3$ by a locally free S^1 action. In this case, it is also the global quotient of $\mathbb{C}P^1$ by a $\mathbb{Z}_p \times \mathbb{Z}_q$ action[1]. When p and q are not relatively prime, the orbisphere is not a global quotient by a finite group, but it is still a symplectic quotient of $\mathbb{C}^2$ by a Hamiltonian $S^1 \times \mathbb{Z}_g$ action, where $g = \gcd(p, q)$ (following [**LT**]). Note also that an orbisphere is isomorphic to a weighted projective space $\mathbb{C}P^1_{p,q}$ exactly when p and q are relatively prime. If p and q are not relatively prime, the weighted projective space is not **reduced**: it has a global stabilizer. On the other hand, the orbisphere described above is always reduced.

Next we turn to algebraic invariants that we may attach to an orbifold $\mathfrak{X}$. The hope is that these invariants are computable and at the same time retain some information of the orbifold structure of $\mathfrak{X}$. All the invariants are isomorphic to singular cohomology if $\mathfrak{X} = X$ is in fact a manifold.

[1]With an apology to number theorists, we take the topologist's notation: $\mathbb{Z}_p$ denotes the integers modulo p.

DEFINITION 2.1. *The* **ordinary cohomology ring** *of an orbifold $\mathfrak{X}$ is the singular cohomology of the underlying topological space X,*

$$H^*(X;R), \tag{2.2}$$

with coefficients in a commutative ring R.

This ring is computable using standard techniques from algebraic topology, but it does not distinguish between the orbisphere in Figure 2.1 and a smooth sphere.

For the second invariant, we restrict our attention to orbifolds $\mathfrak{X}$ presented as the quotient of a manifold M by the locally free action of a Lie group G. It is conjectured that every orbifold can be expressed as such a global quotient, and it is known to be true for **effective** or **reduced orbifolds**, those that do not have a global finite stabilizer (see, for example, [**EHKV**, Theorem 2.18]). A presentation as a global quotient is desirable because then the topology or geometry of the quotient $\mathfrak{X}$ is simply the G-**equivariant** topology or geometry of the manifold M. This principle motivates the following definition.

DEFINITION 2.2. *Given a presentation of an orbifold $\mathfrak{X} = [M/G]$ as a global quotient, the* **cohomology ring of the orbifold $\mathfrak{X}$** *is the equivariant cohomology ring*

$$H^*(\mathfrak{X};R) := H^*_G(M;R), \tag{2.3}$$

with coefficients in a commutative ring R.

Recall that **equivariant cohomology** is a generalized cohomology theory in the equivariant category. Using the **Borel model**, we define

$$H^*_G(M;R) := H^*((M \times EG)/G;R), \tag{2.4}$$

where EG is a contractible (though infinite dimensional) space with a free G action, and G acts diagonally on $M \times EG$. There are well-developed methods for computing equivariant cohomology, hence this invariant is still computable.

Whenever $\mathfrak{X} = [M/G]$ is a global quotient, the associated quotient map

$$q : M \to X \tag{2.5}$$

is a G-invariant map. This induces a continuous map

$$q : (M \times EG)/G \to X. \tag{2.6}$$

When $\mathfrak{X} = X$ is a manifold (i.e. G acts on M locally freely), this map is a fibration with fiber BG. The map q induces a map in cohomology,

$$q^* : H^*(X;R) \longrightarrow H^*((M \times EG)/G;R) = H^*_G(M;R). \tag{2.7}$$

This induced map is an isomorphism when

1. G acts freely on M;
2. G is a finite group and R is a ring in which $|G|$ is invertible; and
3. G acts locally freely on M and R is a field of characteristic 0.

This last item implies that the cohomology of the orbifold differs from the cohomology of its coarse moduli only in its torsion. Notably, when $R = \mathbb{Z}$, this ring does in fact distinguish between the orbisphere in Figure 2.1 and a smooth sphere.

The third invariant was introduced by Chen and Ruan [**CR**] to explain mathematically the **stringy Betti numbers** and **stringy Hodge numbers** that physicists have attached to orbifolds. To define this third invariant, we need to introduce the **first inertia orbifold**

$$I^1(\mathfrak{X}) := \left\{ (x, (g)_{\Gamma_x}) \;\middle|\; x \in \mathfrak{X} \text{ and } (g)_{\Gamma_x} \text{ is a conjugacy class in } \Gamma_x \right\}. \tag{2.8}$$

This is again an orbifold, and $\mathfrak{X}$ is the suborbifold called the **identity sector** whose pairs consist of a point in $\mathfrak{X}$ together with the identity element coset. The other connected components of $I^1(\mathfrak{X})$ are called the **twisted sectors**. For a global quotient $\mathfrak{X} = [M/G]$ with G abelian, we may identify $I^1(\mathfrak{X}) = \coprod_{g \in G} [M^g/G]$. On the other hand, when $\mathfrak{X} = [M/G]$ is global quotient with G finite, we may identify $I^1(\mathfrak{X}) = \coprod_{g \in T} [M^g/C(g)]$, where the union is over T a set of representatives of conjugacy classes in G. For the orbisphere example, the inertia orbifold is shown in Figure 2.2.

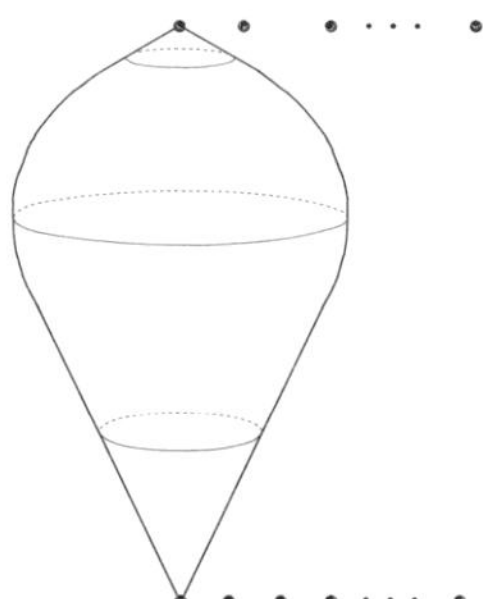

FIGURE 2.2. The inertia orbifold for the the orbisphere with two orbifold points as in Figure 2.1. Each of the $p-1$ points to the right of the north pole represents a $[\mathrm{pt}/\mathbb{Z}_p]$ and each of the $q-1$ to the right of the south pole a $[\mathrm{pt}/\mathbb{Z}_q]$.

DEFINITION 2.3 ([**CR**]). *Given a presentation of an orbifold $\mathfrak{X} = [M/G]$ as a global quotient, the* **Chen-Ruan orbifold cohomology** *of $\mathfrak{X}$, as a vector space, is defined to be the cohomology of the first inertia orbifold,*

$$H_{CR}(\mathfrak{X}; R) := H(I^1(\mathfrak{X}); R), \tag{2.9}$$

with coefficients in a commutative ring R.

The Chen-Ruan ring is endowed with a $\mathbb{Q}$ grading, different from the grading coming from singular cohomology. For a connected component $\mathcal{Z}$ of $I^1(\mathfrak{X})$ that lies in the (g) piece of the first inertia, the grading of $H(\mathcal{Z}; R)$ is shifted by a rational number which is twice the **age** of $\mathcal{Z}$. The age is determined by the weights of the action of the group element g on the normal bundle to $\mathcal{Z}$ inside of $\mathfrak{X}$. This is precisely where we need $\mathfrak{X}$ to be (stably) almost complex. These rational shifts ensure that the ranks of the Chen-Ruan cohomology groups agree with the stringy invariants of an orbifold. The dependence of the rational shifts on the normal bundles $\nu(\mathcal{Z} \subset \mathfrak{X})$ means that the Chen-Ruan ring is not in general functorial for arbitrary morphisms of orbifolds.

To define a product on the Chen-Ruan cohomology, we must define higher inertia. The $\mathbf{n}^{\mathrm{th}}$ **inertia orbifold** consists of tuples, a point in the orbifold and an n-tuple of conjugacy classes. Restricting to the 2nd inertia, there are natural maps

$$e_1, e_2, \overline{e}_3 : I^2(X) \longrightarrow I^1(X) \tag{2.10}$$

defined by

$$e_1(p,(g)_{\Gamma_p},(h)_{\Gamma_p}) = (p,(g)_{\Gamma_p}), \tag{2.11}$$

$$e_2(p,(g)_{\Gamma_p},(h)_{\Gamma_p}) = (p,(h)_{\Gamma_p}), \text{ and} \tag{2.12}$$

$$\overline{e}_3(p,(g)_{\Gamma_p},(h)_{\Gamma_p}) = (p,((gh)^{-1})_{\Gamma_p}) \tag{2.13}$$

Chen and Ruan define the product of two classes $\alpha, \beta \in H_{CR}(\mathfrak{X};R)$ to be be

$$\alpha \smile \beta := (\overline{e}_3)_*(e_1^*\alpha \cup e_2^*\beta \cup \varepsilon), \tag{2.14}$$

where e_1^* and e_2^* are pull-back maps, $(\overline{e}_3)_*$ is the push-forward, $\cup$ is the usual cup product, and ε is the Euler class of the **obstruction bundle**. This Euler class should be viewed as a quantum correction term. It ensures that the product respects the $\mathbb{Q}$-grading and that the product is associative. Neither of these properties is immediately obvious, and proving the latter requires a rather substantial argument. Since we can avoid mention of the obstruction bundle in our computations, we suppress further details here, and refer the curious reader to [**ℵGV, CR, GHK**] for additional information.

The Chen-Ruan cohomology ring is the degree 0 part of the (small) quantum cohomology ring. Hence, it is generally the most difficult of these three invariants to compute. It has been computed for orbifolds that are global quotients of a manifold by a finite group in [**FG**]. The definition was extended to the algebraic category in [**ℵGV**], and in [**BCS**], the ring is computed for toric Deligne-Mumford stacks with $\mathbb{Q}$, $\mathbb{R}$ and $\mathbb{C}$ coefficients. In the next section, we review how to compute this ring for abelian symplectic quotients, as demonstrated in [**GHK**]. In the last section, we will compute each of these invariants explicitly for weighted projective spaces.

We conclude this section with a table of these rings for an orbisphere that has a $\mathbb{Z}_2$ singularity at the north pole and is otherwise smooth. This is an example of a **weighted projective space**, and is denoted $\mathbb{C}P^1_{1,2}$.

(2.15)

$\mathfrak{X} = [\mathbb{C}P^1_{1,2}]$	Ring	Grading
$H^*(X;\mathbb{Z})$	$\mathbb{Z}[x]/\langle x^2\rangle$	$\deg(x)=2$
$H^*(\mathfrak{X};\mathbb{Z})$	$\mathbb{Z}[x]/\langle 2x^2\rangle$	$\deg(x)=2$
$H^*_{CR}(\mathfrak{X};\mathbb{Z})$	$\mathbb{Z}[x,u]/\langle 2x^2, 2xu, u^2-x\rangle$	$\deg(x)=2,\ \deg(u)=1$

3. Why the symplectic category is convenient

In the past thirty years, tremendous progress has been made in understanding the equivariant topology of Hamiltonian T-spaces and its relationship to the ordinary topology of their quotients. For a compact torus $T=(S^1)^n$, the classifying bundle ET is an n-fold product of infinite dimensional spheres, and the classifying space BT is an n-fold product of copies of $\mathbb{C}P^\infty$. Thus,

$$H_T^*(pt;\mathbb{Z}) = H^*(BT;\mathbb{Z}) = \mathbb{Z}[x_1,\dots,x_n], \tag{3.1}$$

where $\deg(x_i)=2$. The key ingredient to understanding topology of Hamiltonian T-spaces is the moment map. Frankel [**F**] proved that for a Hamiltonian T-action on a Kähler manifold, each component ϕ^ξ of the moment map is a Morse-Bott function on M, and generically the critical set is the fixed point set M^T. In his paper [**A**] on the Convexity Theorem 1.10, Atiyah generalized this work to the purely symplectic setting.

Building on the work of Frankel and Atiyah, Kirwan developed techniques to prove two fundamental theorems that allow us to understand the cohomology of Hamiltonian T-spaces and their quotients. The first is a version of **localization**: it allows us to make global computations by understanding fixed point data. While this theorem is not explicitly stated in her book [**Ki**], it does follow immediately from her work in Chapter 5.

INJECTIVITY THEOREM 3.1 ([**Ki**]). *Let M be a compact Hamiltonian T-space. The inclusion map $M^T \hookrightarrow M$ induces*

$$i^* : H_T^*(M;\mathbb{Q}) \longrightarrow H_T^*(M^T;\mathbb{Q}) \tag{3.2}$$

an **injection** *in equivariant cohomology.*

The compactness hypothesis is stronger than strictly necessary. We may replace it with a properness condition on the moment map. The proof relies on the fact that a generic component of the moment map is an **equivariantly perfect** Morse-Bott function on M. The image of this injection has been computed in many examples, including toric varieties, coadjoint orbits of compact connected semisimple Lie groups, and coadjoint orbits of Kac-Moody groups. These computations initially appeared in [**CS, GKM**] and further generalizations are described in [**GoH, GuH, GuZ, HHH**].

The second theorem relates the equivariant topology of a Hamiltonian T-space to the ordinary topology of its reduction.

SURJECTIVITY THEOREM 3.2 ([**Ki**]). *Let M be a compact Hamiltonian T-space, and α a regular value of the moment map. The inclusion $\Phi^{-1}(\alpha) \hookrightarrow M$ induces*

$$\kappa : H_T^*(M;\mathbb{Q}) \longrightarrow H_T^*(\Phi^{-1}(\alpha);\mathbb{Q}) \tag{3.3}$$

a **surjection** *in equivariant cohomology.*

Again for surjectivity, compactness is more than is necessary. Most importantly, this result does apply to linear actions of a torus on $\mathbb{C}^d$ with a proper moment map. The key idea in the proof is to use the function $||\Phi - \alpha||^2$ as a *Morse-like* function, now known as a **Morse-Kirwan function**. The critical sets are not non-degenerate, but one may still explicitly understand them via a local normal form. It is then possible to prove that $||\Phi - \alpha||^2$ is an equivariantly perfect function on M.

The kernel of the map κ can be computed using methods in [**Go, JK, TW**]. Using the fact that at a regular value, $H_T^*(\Phi^{-1}(\alpha);\mathbb{Q}) \cong H^*(M/\!\!/T(\alpha);\mathbb{Q})$, we have a diagram

$$\begin{array}{ccccccccc} 0 & \longrightarrow & \ker(\kappa) & \hookrightarrow & H_T^*(M;\mathbb{Q})) & \overset{\kappa}{\twoheadrightarrow} & H^*(M/\!\!/T(\alpha);\mathbb{Q}) & \longrightarrow & 0 \\ & & & & \Big\downarrow i^* & & & & \\ & & & & H_T^*(M^T;\mathbb{Q}) & & & & \end{array} \tag{3.4}$$

Thus, by computing $\mathrm{im}(i^*)$ and $\ker(\kappa)$, we may derive an explicit presentation of the cohomology $H^*([M/\!\!/T(\alpha)];\mathbb{Q})$ of an orbifold arising as a symplectic quotient.

We now turn to a generalization of Theorem 3.2 in the context of orbifolds and the Chen-Ruan ring.

CR SURJECTIVITY THEOREM 3.3 ([**GHK**]). *Let M be a compact Hamiltonian T-space, and α a regular value of the moment map. The inclusion $\Phi^{-1}(\alpha) \hookrightarrow M$ induces*

$$\mathcal{K} : \bigoplus_{g\in T} H_T^*(M^g;\mathbb{Q}) \longrightarrow \bigoplus_{g\in T} H_T^*(\Phi^{-1}(\alpha)^g;\mathbb{Q}) \tag{3.5}$$

a **surjection**. *Moreover, these are $\mathbb{R}\times T$-graded rings, $\mathcal{K}$ is a map of graded rings, and there is an isomorphism of graded rings*

$$\bigoplus_{g\in T} H_T^*(\Phi^{-1}(\alpha)^g;\mathbb{Q}) \cong H_{CR}^*(M/\!\!/T(\alpha);\mathbb{Q}). \tag{3.6}$$

The surjectivity (3.5) is a direct consequence of the Surjectivity Theorem 3.2 applied to each space M^g (and again the compactness is not strictly necessary). The hard work is defining the grading and ring structure, and proving that $\mathcal{K}$ is a map of graded rings. Goldin, Knutson, and the author define a ring structure $\smile$ that generalizes a definition (for G finite) of Fantechi and Göttsche [**FG**]. Using this definition, we may deduce (3.6); on the other hand, associativity of this product is not at all obvious. Making use of the injection i^* on each piece, there is an alternative product $\star$ on the ring $\bigoplus_{g\in T} H_T^*(M^g;\mathbb{Q})$ that is much simpler to compute, and clearly associative. This alternative product has the advantage that it avoids all mention of the obstruction bundle; instead it relies only on fixed point data (i.e. the topology of the fixed point set and isotropy data for the action of the torus on the normal bundles to the fixed point components).

Another key point is that although the ring on the left of (3.5) is quite large, there is a finite subgroup Γ of T, generated by all finite-order elements that stabilize some regular point in M, so that the **Γ-subring**

$$\bigoplus_{g\in\Gamma} H_T^*(M^g;\mathbb{Q}) \tag{3.7}$$

still surjects onto the Chen-Ruan cohomology of the reduction. Thus, while it appears that we have made the computation much more complicated, it turns that there is still an effective algorithm to complete it. For full details, please refer to [**GHK**].

Chen and Hu have also taken the symplectic viewpoint on the Chen-Ruan orbifold cohomology of abelian symplectic quotients in [**CH**]. They propose a deRham model for the ring, and investigate the changes in the ring as the value α varies. We now return to our examples.

EXAMPLE 3.4: For a **symplectic toric orbifold** $\mathbb{C}^n/\!\!/K(\alpha)$, we may use the combinatorics of its labeled polytope to establish an explicit presentation of the Chen-Ruan cohomology of these orbifolds [**GHK**, § 9]. In the cases where the symplectic picture is identical to the algebraic, the [**GHK**] results replicate those of [**BCS**].

EXAMPLE 3.5: For the T-action on a coadjoint orbit $M = \mathcal{O}_\lambda$, the symplectic reduction $M/\!\!/T(\alpha)$ is a **weight variety**. The equivariant cohomology of M may be read directly from its moment polytope, as may the orbifold singularities of the reduction. In this case, the Theorem 3.3 yields an explicit combinatorial description of the Chen-Ruan cohomology of the weight variety.

We conclude this section with some brief remarks on coefficients. In both Theorems 3.1 and 3.2, the rational coefficients are necessary. For the Injectivity Theorem 3.1, we may prove the result over $\mathbb{Z}$ with the additional hypothesis that

$H^*(M^T;\mathbb{Z})$ contains no torsion. This is also the hypothesis necessary to define the alternative product $\star$ on $\bigoplus_{g\in T} H_T^*(M^g;\mathbb{Z})$. The Surjectivity Theorem 3.2 over $\mathbb{Z}$ requires much stronger hypotheses. We will see in the next section that we may compute integrally for weighted projective spaces, but that a simple product of two weighted projective spaces yields a counter-example to the general theorem. Tolman and Weitsman verify surjectivity over $\mathbb{Z}$ for a rather restrictive class of torus actions [**TW**]. This topic is being more closely examined for a larger collection of actions by Susan Tolman and the author [**HT**].

4. The case of weighted projective spaces

Let $b = (b_0, \ldots, b_n)$ be an $(n+1)$-tuple of positive integers. Consider the circle action on $\mathbb{C}^{n+1}$ given by

$$t \cdot (z_0, \ldots, z_n) = (t^{b_0} z_0, \ldots, t^{b_n} z_n) \tag{4.1}$$

for each $t \in S^1$. This action preserves the unit sphere S^{2n+1}, and the **weighted projective space** $\mathbb{C}P^n_{(b)}$ is the quotient of S^{2n+1} by this locally free circle action. This is a symplectic reduction because S^{2n+1} is (up to equivariant homeomorphism) a regular level set for a moment map Φ for the weighted S^1 action on $\mathbb{C}^{n+1}$. Thus,

$$\mathbb{C}P^n_{(b)} \cong \mathbb{C}^{n+1}/\!\!/S^1. \tag{4.2}$$

Nonetheless, we may continue our analysis without invoking the full symplectic machinery: the arguments simplify greatly in this special case.

When the b_i are relatively prime (that is, $\gcd(b_0, \ldots, b_n) = 1$), this fits into the framework of symplectic toric orbifolds discussed in Example 3.4. When the b_i are not relatively prime, we let $g = \gcd(b_0, \ldots, b_n)$, and note that there is a global $\mathbb{Z}_g$ stabilizer. In this case, the orbifold $[\mathbb{C}P^n_{(b)}]$ is not reduced. Its coarse moduli space is the same as the coarse moduli space of $\mathbb{C}P^n_{(b/g)}$, where (b/g) denotes the sequence of integers $(b_0/g, \ldots, b_n/g)$; and as a non-reduced orbifold, it corresponds to a **gerbe**

$$[\mathbb{C}P^n_{(b)}] \to [\mathbb{C}P^n_{(b/g)}].$$

It is important to include the case when the b_i are not relatively prime, because such non-reduced weighted projective spaces may well show up as suborbifolds of a reduced weighted projective space.

The cohomology of the topological space $\mathbb{C}P^n_{(b)}$. Kawasaki studied the singular cohomology ring of (the coarse moduli space of) weighted projective spaces [**Ka**]. To present the product structure, we will need the integers

$$\ell_k = \ell_k^{(b)} := \operatorname{lcm}\left\{ \frac{b_{i_0} \cdots b_{i_k}}{\gcd(b_{i_0}, \ldots, b_{i_k})} \;\middle|\; 0 \le i_0 < \cdots < i_k \le n \right\}, \tag{4.3}$$

for each $1 \le k \le n$.

THEOREM 4.1 ([**Ka**]). *The integral cohomology of $\mathbb{C}P^n_{(b)}$ is*

$$H^i(\mathbb{C}P^n_{(b)};\mathbb{Z}) \cong \begin{cases} \mathbb{Z} & \textit{if } i = 2k,\ 0 \le k \le n, \\ 0 & \textit{otherwise.} \end{cases} \tag{4.4}$$

Moreover, letting γ_i denote the generator of $H^{2i}(\mathbb{C}P^n_{(b)};\mathbb{Z})$, we have

$$\gamma_k \cup \gamma_m = \frac{\ell_k \cdot \ell_m}{\ell_{m+k}} \gamma_{m+k}. \tag{4.5}$$

OUTLINE OF THE PROOF. Let G_k denote the group of kth roots of unity. Then as a topological space, $\mathbb{C}P^n_{(b)}$ is homeomorphic to a quotient of ordinary projective space $\mathbb{C}P^n$ by the finite group

$$G_{(b)} = G_{b_0} \times \cdots \times G_{b_n}. \tag{4.6}$$

Explicitly, using standard homogeneous coordinates, the map

$$p_b : \mathbb{C}P^n \longrightarrow \mathbb{C}P^n_{(b)} \tag{4.7}$$

$$[z_0 : \cdots : z_n] \longmapsto [z_0^{b_0} : \cdots : z_n^{b_n}] \tag{4.8}$$

induces the homeomorphism $\mathbb{C}P^n/G_{(b)} \cong \mathbb{C}P^n_{(b)}$. This then induces an isomorphism in singular cohomology with rational coefficients, since over $\mathbb{Q}$ we have the isomorphisms

$$H^*(\mathbb{C}P^n;\mathbb{Q}) \cong H^*(\mathbb{C}P^n;\mathbb{Q})^{G_{(b)}} \cong H^*(\mathbb{C}P^n/G_{(b)};\mathbb{Q}). \tag{4.9}$$

Over the integers, the computation is a bit more subtle. Using twisted lens spaces, Kawasaki verifies that just as for $\mathbb{C}P^n$, the cohomology ring $H^*(\mathbb{C}P^n_{(b)};\mathbb{Z})$ is torsion-free with a copy of $\mathbb{Z}$ in each even degree between 0 and $2n$; however the product structure is twisted by the weights b_i. Moreover, there are cases when we need all n generators $\gamma_1, \ldots, \gamma_n$ to present this ring.

Kawasaki showed that the map

$$p_b^* : H^{2k}(\mathbb{C}P^n_{(b)};\mathbb{Z}) \longrightarrow H^{2k}(\mathbb{C}P^n;\mathbb{Z}) \tag{4.10}$$

is multiplication by ℓ_k for all $1 \le k \le n$. From this, we may deduce that

$$\gamma_1 \cup \gamma_k = \frac{\ell_1 \cdot \ell_{k+1}}{\ell_k} \gamma_{k+1}, \text{ and hence } \gamma_k \cup \gamma_m = \frac{\ell_k \cdot \ell_m}{\ell_{m+k}} \gamma_{m+k}. \tag{4.11}$$

The result now follows. □

It is important to note that p_b is **not** an isomorphism of orbifolds. Indeed, the isotropy group at any point in $\mathbb{C}P^n_{(b)}$ is the stabilizer group of any lift of the point in S^{2n+1}. Thus, all isotropy groups for $\mathbb{C}P^n_{(b)}$ are cyclic. On the other hand, the orbifold $\mathbb{C}P^n/G_{(b)}$ has points with isotropy group $G_{(b)}$, which may not be cyclic. Nevertheless we will make use of the map p_b to understand the structure of the cohomology of the orbifold $[\mathbb{C}P^n_{(b)}] = [S^{2n+1}/S^1_{(b)}]$. Here, the subscript (b) on S^1 indicates that the circle action is weighted by the integers $(b) = (b_0, \ldots, b_n)$.

The cohomology of the orbifold $[\mathbb{C}P^n_{(b)}]$. Since the weighted projective space $[\mathbb{C}P^n_{(b)}]$ is a symplectic reduction, it is possible invoke the results from Section 3 to determine the cohomology of the orbifold $H^*([\mathbb{C}P^n_{(b)}];\mathbb{Q})$, and to apply results from [**GHK**, §9] to obtain a presentation over $\mathbb{Z}$. We give a direct argument here that is similar in spirit, but that avoids much of this big machinery; we then compare this to Kawasaki's Theorem 4.1.

THEOREM 4.2. *The cohomology of the orbifold* $[\mathbb{C}P^n_{(b)}]$ *is*

$$H^*([\mathbb{C}P^n_{(b)}];\mathbb{Z}) = H^*_{S^1_{(b)}}(S^{2n+1};\mathbb{Z}) \cong \frac{\mathbb{Z}[u]}{\langle b_0 \cdots b_n u^{n+1} \rangle}. \tag{4.12}$$

Moreover, the natural map

$$q_{(b)}^* : H^*(S^{2n+1}/S^1_{(b)};\mathbb{Z}) \longrightarrow H^*_{S^1_{(b)}}(S^{2n+1};\mathbb{Z}) \tag{4.13}$$

is completely determined by $q^*_{(b)}(\gamma_1) = \ell_1 \cdot u = \mathrm{lcm}(b_0, \ldots, b_n) \cdot u$.

PROOF. Consider the (weighted) circle action of S^1 on $\mathbb{C}^{n+1}$ given by

$$t \cdot (z_0, \cdots, z_n) = (t^{b_0} \cdot z_0, \cdots, t^{b_n} \cdot z_n). \tag{4.14}$$

The unit sphere S^{2n+1} is invariant under this action, so we get a long exact sequence in S^1-equivariant cohomology for the pair $(\mathbb{C}^{n+1}, S^{2n+1})$,

$$\cdots \to H^i_{S^1_{(b)}}(\mathbb{C}^{n+1}, S^{2n+1};\mathbb{Z}) \overset{\alpha}{\to} H^i_{S^1_{(b)}}(\mathbb{C}^{n+1};\mathbb{Z}) \overset{\beta}{\to} H^i_{S^1_{(b)}}(S^{2n+1};\mathbb{Z}) \to \cdots. \tag{4.15}$$

Thinking of $(\mathbb{C}^{n+1}, S^{2n+1})$ as a disk and sphere bundle over a point, we may use the Thom isomorphism to identify $H^i_{S^1_{(b)}}(\mathbb{C}^{n+1}, S^{2n+1};\mathbb{Z}) \cong H^{i-2(n+1)}_{S^1_{(b)}}(\mathbb{C}^{n+1};\mathbb{Z})$. Under this identification, the map α is the cup product with the equivariant Euler class

$$e_{S^1_{(b)}}(\mathbb{C}^{n+1}) = b_o \cdots b_n u^{n+1}. \tag{4.16}$$

Thus, the map α is injective, so the long exact sequence splits into short exact sequences

$$0 \to H^i_{S^1_{(b)}}(\mathbb{C}^{n+1}, S^{2n+1};\mathbb{Z}) \overset{\alpha}{\to} H^i_{S^1_{(b)}}(\mathbb{C}^{n+1};\mathbb{Z}) \overset{\beta}{\to} H^i_{S^1_{(b)}}(S^{2n+1};\mathbb{Z}) \to 0. \tag{4.17}$$

Thus, we have a surjection

$$\beta : \mathbb{Z}[u] \cong H^*_{S^1_{(b)}}(\mathbb{C}^{n+1};\mathbb{Z}) \longrightarrow H^*_{S^1_{(b)}}(S^{2n+1};\mathbb{Z}). \tag{4.18}$$

Moreover, the exactness of (4.17) means that the kernel of β is equal to the image of α, namely all multiples of the equivariant Euler class. This establishes (4.12).

Turning to (4.13), the map

$$q_{(b)}^* : H^*(S^{2n+1}/S^1_{(b)};\mathbb{Z}) \longrightarrow H^*_{S^1_{(b)}}(S^{2n+1};\mathbb{Z}), \tag{4.19}$$

is exactly the one defined in (2.7). We know that this is an isomorphism over $\mathbb{Q}$, so $q^*_{(b)}$ must map γ_1 to a multiple of u. Moreover, because $b_0 \cdots b_n u^{n+1}$ is zero in $H^*_{S^1_{(b)}}(S^{2n+1};\mathbb{Z})$, we must have that

$$(q^*_{(b)}(\gamma_1))^{n+1} \in \langle b_0 \cdots b_n u^{n+1} \rangle. \tag{4.20}$$

To determine the image of the class γ_1, we return to the map $p_b : \mathbb{C}P^n \to \mathbb{C}P^n_{(b)}$. This map lifts to maps on S^{2n+1} and $\mathbb{C}^{n+1}$ given by

$$\begin{array}{ccc} \mathbb{C}^{n+1} & \xrightarrow{\Pi_b} & \mathbb{C}^{n+1} \\ \uparrow i & & \uparrow i \\ S^{2n+1} & \xrightarrow{\pi_b} & S^{2n+1} \end{array} . \tag{4.21}$$

$$0 \neq (z_0, \cdots, z_n) \longmapsto \frac{\sum |z_i|^2}{\sum |z_i^{b_i}|^2} \cdot \left(z_0^{b_0}, \cdots, z_n^{b_n}\right) \tag{4.22}$$

$$(0, \ldots, 0) \longmapsto (0, \ldots, 0) \tag{4.23}$$

The maps in this diagram are all equivariant with respect to the standard circle action on the left-hand spaces and the (b)-weighted circle action on the right-hand spaces. Thus, we have a diagram of maps

$$(4.24)\qquad \begin{array}{ccc} (\mathbb{C}^{n+1}\times S^1)/S^1 & \xrightarrow{\Pi_b} & (\mathbb{C}^{n+1}\times S^1_{(b)})/S^1_{(b)} \\ \uparrow{\scriptstyle i} & & \uparrow{\scriptstyle i} \\ (S^{2n+1}\times S^1)/S^1 & \xrightarrow{\pi_b} & (S^{2n+1}\times S^1_{(b)})/S^1_{(b)} \\ \downarrow{\scriptstyle q} & & \downarrow{\scriptstyle q_{(b)}} \\ \mathbb{C}P^n & \xrightarrow[p_b]{} & \mathbb{C}P^n_{(b)} \end{array}.$$

Applying singular cohomology $H^*(\ ;\mathbb{Z})$ and identifying equivariant cohomology, we have a commutative diagram

(4.25)

$$\begin{array}{ccccccc} \mathbb{Z}[x] & = & H^*_{S^1}(\mathbb{C}^{n+1};\mathbb{Z}) & \xleftarrow{\Pi_b^*} & H^*_{S^1_{(b)}}(\mathbb{C}^{n+1};\mathbb{Z}) & = & \mathbb{Z}[u] \\ & & \downarrow{\scriptstyle\kappa} & & \downarrow{\scriptstyle\kappa_{(b)}} & & \\ \frac{\mathbb{Z}[x]}{\langle x^{n+1}\rangle} & = & H^*_{S^1}(S^{2n+1};\mathbb{Z}) & \xleftarrow{\pi_b^*} & H^*_{S^1_{(b)}}(S^{2n+1};\mathbb{Z}) & = & \frac{\mathbb{Z}[u]}{\langle b_0\cdots b_n u^{n+1}\rangle} \\ \| & & \uparrow{\scriptstyle q^*} & & \uparrow{\scriptstyle q^*_{(b)}} & & \\ \frac{\mathbb{Z}[x]}{\langle x^{n+1}\rangle} & = & H^*(\mathbb{C}P^n;\mathbb{Z}) & \xleftarrow[p_b^*]{} & H^*(\mathbb{C}P^n_{(b)};\mathbb{Z}) & = & \mathbb{Z}\oplus\mathbb{Z}\gamma_1\oplus\cdots\oplus\mathbb{Z}\gamma_n \end{array}.$$

Because $\mathbb{C}^{n+1}$ equivariantly deformation retracts to a point, the map Π_b^* maps the generator u to x. The commutativity of the top square then implies that $\pi_b^*(u)=x$. Thus, we know that

$$\begin{aligned} &(4.26) & \pi_b^*(q^*_{(b)}(\gamma_1) &= q^*(p_b^*(\gamma_1)) & \\ &(4.27) & &= q^*(\ell_1\cdot x), & \text{by Kawasaki's result,} \\ &(4.28) & &= \ell_1\cdot x, & \text{since } q^* \text{ is an equality,} \\ &(4.29) & &= \pi_b^*(\ell_1\cdot u). & \end{aligned}$$

In low degree, π_b^* is injective, so we may conclude that $q^*_{(b)}(\gamma_1)=\ell_1\cdot u$. Noting that $\ell_1=\operatorname{lcm}(b_0,\dots,b_n)$ completes the proof. □

Over the integers, this invariant does distinguish a weighted projective space from the standard one; however, it may not differentiate between two weighted projective spaces. For example, the cohomology rings of the orbifolds $[\mathbb{C}P^1_{2,2}]$ and $[\mathbb{C}P^1_{4,1}]$ are identical. They are both

$$(4.30)\qquad \frac{\mathbb{Z}[u]}{\langle 4u^2\rangle}.$$

We note that these surjectivity techniques do not generally work over the integers. To see this, we note that for any abelian reduction of affine space, the domain of the Kirwan map $H^*_T(\mathbb{C}^N;\mathbb{Z})$ has terms only in even degrees. If we consider the simple product $[\mathbb{C}P^1_{1,2}\times\mathbb{C}P^1_{1,2}]$, we may compute the cohomology of this orbifold using the above result and the Künneth formula. Since $[\mathbb{C}P^1_{1,2}]$ has 2-torsion in high degrees, the Tor term from the Künneth formula plays a role, yielding 2-torsion in high odd degrees in the cohomology of the orbifold $[\mathbb{C}P^1_{1,2}\times\mathbb{C}P^2_{1,2}]$. Thus, surjectivity must fail over the integers in this example. We note that any failure over $\mathbb{Z}$ must be due to problems with torsion, because surjectivity does hold over $\mathbb{Q}$.

The Chen-Ruan orbifold cohomology of $[\mathbb{C}P^n_{(b)}]$. When computing the Chen-Ruan ring, it is important to recall that a weighted projective space is a circle reduction. Thus, the finite group Γ for which the Γ-piece surjects onto $H^*_{CR}([\mathbb{C}P^n_{(b)}];\mathbb{Z})$ is a cyclic group. For any vector v that is non-zero is a single coordinate, say the ith coordinate, the stabilizer of v is $\mathbb{Z}_{b_i}$. Thus, the group Γ generated by all finite stabilizers is the ℓth roots of unity $\mathbb{Z}_\ell\subset S^1$, where $\ell=\operatorname{lcm}(b_0,\dots,b_n)$. Hence, we have a surjection

$$\mathbb{Z}[u,\alpha_0,\alpha_1,\dots,\alpha_{\ell-1}]\twoheadrightarrow H^*_{CR}([\mathbb{C}P^n_{(b)}];\mathbb{Z}). \tag{4.31}$$

In this case, thinking of $e^{\frac{2\pi ik}{\ell}}=\zeta_k\in\mathbb{Z}_\ell\subset S^1$, α_k denotes a generator for

$$H^*_{S^1_{(b)}}((\mathbb{C}^{n+1})^{\zeta_k};\mathbb{Z}). \tag{4.32}$$

To complete the computation, we must determine the orbifold product

$$\alpha_i\smile\alpha_j=\alpha_i\star\alpha_j \tag{4.33}$$

and the kernel of the orbifold Kirwan map (3.5). For any integer $m\in\mathbb{Z}$, we let $[m]$ denote the smallest non-negative integer congruent to m modulo ℓ. For any rational number $q\in\mathbb{Q}$, $\langle q\rangle_f$ denotes its fractional part. Finally, we let

$$a_k(m):=\frac{[b_k\cdot m]}{\ell}=\left\langle\frac{b_k\cdot m}{\ell}\right\rangle_f. \tag{4.34}$$

This is the rational number such that ζ_m acts on the kth coordinate by $e^{2\pi i a_k(m)}$.

THEOREM 4.3. *The Chen-Ruan orbifold cohomology of $\mathbb{C}P^n_{(b)}$ is*

$$H^*_{CR}([\mathbb{C}P^n_{(b)}];\mathbb{Z})\cong\frac{\mathbb{Z}[u,\alpha_0,\alpha_1,\dots,\alpha_{\ell-1}]}{\mathcal{I}+\mathcal{J}}, \tag{4.35}$$

where u is a class in degree 2,

$$\deg(\alpha_j)=2\sum_{k=0}^{n}a_k(j). \tag{4.36}$$

Here, $\mathcal{I}$ is the ideal

$$\mathcal{I}=\left\langle\alpha_i\alpha_j-\left(\prod_{k=0}^{n}(b_ku)^{a_k(i)+a_k(j)-a_k(i+j)}\right)\alpha_{[i+j]}\right\rangle \tag{4.37}$$

generated by the $\star$ product structure, and $\mathcal{J}$ is

$$\mathcal{J}=\sum_{j=0}^{n}\left\langle\left(\prod_{a_k(j)=0}b_ku\right)\alpha_i\right\rangle, \tag{4.38}$$

the kernel of the surjection $\mathcal{K}$ of the orbifold Kirwan map.

REMARK 4.4: The generator u is the generator of S^1-equivariant cohomology and hence has degree 2. The generator α_k is a placeholder for the cohomology of the ζ_k-sector.

REMARK 4.5: Note that the generator α_0 is the placeholder for the identity sector. Indeed, we always have

$$\alpha_0 \star \alpha_0 = \alpha_0 \tag{4.39}$$

as a consequence of the relation in (4.37) where $i = j = 0$, hence we may think of α_0 as 1.

REMARK 4.6: The reader may use this theorem to check that $H^*_{CR}(\ ;\mathbb{Z})$ does distinguish $[\mathbb{C}P^1_{2,2}]$ from $[\mathbb{C}P^1_{4,1}]$.

PROOF. We use the $\star$ product given by Equation (2.1) in [**GHK**]. In the case of a weighted circle action on $\mathbb{C}^{n+1}$, there is exactly one fixed point (the origin), any generator α_i restricted to that fixed point is 1, and the equivariant Euler class for the kth coordinate is precisely $b_k u$, whence

$$\alpha_i \star \alpha_j = \left(\prod_{k=0}^{n} (b_k u)^{a_k(i)+a_k(j)-a_k(i+j)} \right) \alpha_{[i+j]}. \tag{4.40}$$

Turning to the kernel computation, each $(\mathbb{C}^{n+1})^{\zeta_j}$ has a weighted S^1 action, and so we apply Theorem 4.2 to this subspace. Thus, for the ζ_j-sector, the kernel contribution is the equivariant Euler class of $(\mathbb{C}^{n+1})^{\zeta_j}$ times the placeholder α_j. We note that $(\mathbb{C}^{n+1})^{\zeta_j}$ contains the kth coordinate subspace precisely when $a_k(j) = 0$. Hence,

$$e_{S^1_{(b)}}((\mathbb{C}^{n+1})^{\zeta_j}) = \prod_{a_k(j)=0} b_k u, \tag{4.41}$$

and the theorem follows. □

This theorem is an immediate consequence of Theorem 4.2 and [**GHK**]. The importance of this description is its ease in computation, since it avoids any computation of a labeled moment polytope (*á la* [**LT**]) or of a stacky fan (*á la* [**BCS**]). We demonstrate this computational facility in the following concluding example.

EXAMPLE 4.7: Consider the weighted projective space $[\mathbb{C}P^5_{1,2,2,3,3,3}]$. This is a symplectic reduction of $\mathbb{C}^6$, the group Γ is $\mathbb{Z}/6\mathbb{Z}$, and so the Chen-Ruan orbifold cohomology of $[\mathbb{C}P^5_{1,2,2,3,3,3}]$ is a quotient of

$$\mathbb{Z}[u, \alpha_0, \alpha_1, \alpha_2, \alpha_3, \alpha_4, \alpha_5]. \tag{4.42}$$

The following chart contains the data needed to compute the ideals $\mathcal{I}$ and $\mathcal{J}$.

(4.43)

g	ζ_0	ζ_1	ζ_2	ζ_3	ζ_4	ζ_5
$(\mathbb{C}^6)^g$	$\mathbb{C}^6$	$\{0\}$	$3\mathbb{C}_{(3)}$	$2\mathbb{C}_{(2)}$	$3\mathbb{C}_{(3)}$	$\{0\}$
$a_{\mathbb{C}_{(1)}}(g)$	0	$\frac{1}{6}$	$\frac{1}{3}$	$\frac{1}{2}$	$\frac{2}{3}$	$\frac{5}{6}$
$a_{\mathbb{C}_{(2)}}(g)$	0	$\frac{1}{3}$	$\frac{2}{3}$	0	$\frac{1}{3}$	$\frac{2}{3}$
$a_{\mathbb{C}_{(3)}}(g)$	0	$\frac{1}{2}$	0	$\frac{1}{2}$	0	$\frac{1}{2}$
$2 \cdot \mathrm{age}(g)$	0	$\frac{14}{3}$	$\frac{10}{3}$	4	$\frac{8}{3}$	$\frac{22}{3}$
generator of $H^*_{S^1_{(b)}}((\mathbb{C}^6)^g;\mathbb{Z})$	α_0	α_1	α_2	α_3	α_4	α_5
$e_{S^1_{(b)}}((\mathbb{C}^6)^g)$	$108u^6$	1	$27u^3$	$4u^2$	$27u^3$	1

Note that because of the multiplicities,

$$2 \cdot \mathrm{age}(g) = 2 \cdot \big[a_1(g) + 2a_2(g) + 3a_3(g)\big]. \tag{4.44}$$

Since $\alpha_0 = 1$, and α_1 and α_5 are in the kernel ideal $\mathcal{J}$, we only need to compute the products among α_2, α_3 and α_4. For example, we may compute

$$\alpha_2 \star \alpha_2 = (u)^{\frac{1}{3}+\frac{1}{3}-\frac{2}{3}} \left((2u)^{\frac{2}{3}+\frac{2}{3}-\frac{1}{3}}\right)^2 \left((3u)^{0+0-0}\right)^3 \alpha_4 = 4u^2\alpha_4. \tag{4.45}$$

All of the products contributing to $\mathcal{I}$, then, are summarized in the following table.

(4.46)

$\star$	α_2	α_3	α_4
α_2	$4u^2\alpha_4$	$\alpha_5 = 0$	$4u^3$
α_3		$27u^4$	$u\alpha_1 = 0$
α_4			$u\alpha_2$

.

Thus, as a ring,

$$H^*_{CR}([\mathbb{C}P^n_{1,2,2,3,3,3}];\mathbb{Z}) \cong \frac{\mathbb{Z}[u,\alpha_1,\alpha_2,\alpha_3,\alpha_4,\alpha_5]}{\mathcal{I} + \langle 108u^6, \alpha_1, 27u^3\alpha_2, 4u^2\alpha_3, 27u^3\alpha_4, \alpha_5\rangle}. \tag{4.47}$$

This generalizes Jiang's computation [**J**] to a computation over $\mathbb{Z}$.

References

[AGV] D. Abramovich, T. Graber, and A. Vistoli, "Algebraic orbifold quantum products." *Contemp. Math.*, **310** (2002) 1–24. Preprint `math.AG/0112004`.

[A] M. Atiyah, "Convexity and commuting Hamiltonians." *Bull. London Math. Soc.* **14** (1982) no. 1, 1–15.

[BCS] L. Borisov, L. Chen, and G. Smith, "The orbifold Chow ring of toric Deligne-Mumford stacks." *J. Amer. Math. Soc.* **18** (2005) no. 1, 193–215. Preprint `math.AG/0309229`.

[CdS] A. Cannas da Silva, *Lectures on symplectic geometry.* **Lecture Notes in Mathematics**, 1764. Springer-Verlag, Berlin, 2001.

[CS] T. Chang and T. Skjelbred, "Topological Schur lemma and related results." *Bull. Amer. Math. Soc.* **79** (1973) 1036–1038.

[CH] B. Chen and S. Hu, "A deRham model for Chen-Ruan cohomology ring of abelian orbifolds." *Math. Ann.* **336** (2006) no. 1, 51–71. Preprint `math.SG/0408265`.

[CR] W. Chen and Y. Ruan, "A New Cohomology Theory for Orbifold." *Commun.Math.Phys.* **248** (2004) 1-31. Preprint `math.AG/0004129`.

[CCLT] T. Coates, A. Corti, Y.-P. Lee, and H.-H. Tseng, "The quantum orbifold cohomology of weighted projective space." Preprint `math.AG/0608481`.

[D] V. Danilov, "The geometry of toric varieties." *Russian Math. Surveys* **33** (1978) no. 2, 97–154.

[DM] P. Deligne and D. Mumford, "The irreducibility of the space of curves of given genus." *Inst. Hautes tudes Sci. Publ. Math.* **36** (1969) 75–109.

[EHKV] D. Edidin, B. Hassett, A. Kresch, and A. Vistoli, "Brauer groups and quotient stacks." *Amer. J. Math.* **123** (2001) no. 4, 761–777.

[FG] B. Fantechi and L. Göttsche, "Cohomology for global quotients." *Duke Math. J.* **117** (2003) no. 2, 197–227. Preprint `math.AG/0104207`.

[F] T. Frankel, "Fixed points and torsion on Kaḧler manifolds." *Ann. of Math. (2)* **70** 1959 1–8.

[Go] R. Goldin, "An effective algorithm for the cohomology ring of symplectic reduction." *Geom. Funct. Anal.* **12** (2002) 567-583. Preprint `math.SG/0110022`.

[GoH] R. Goldin and T. Holm, "The equivariant cohomology of Hamiltonian G-spaces from residual S^1 actions." *Math. Res. Lett.* **8** (2001) no. 1-2, 67–77. Preprint `math.SG/0107131`.

[GHK] R. Goldin, T. Holm and A. Knutson, "Orbifold cohomology of torus quotients." *Duke Math. Journal*, to appear. Preprint `math.SG/0502429`.

[GKM] M. Goresky, R. Kottwitz, and R. MacPherson, "Equivariant cohomology, Koszul duality, and the localization theorem." *Invent. Math.* **131** (1998) no. 1, 25–83.

[GuH] V. Guillemin and T. Holm, "GKM theory for torus actions with non-isolated fixed points." *Int. Math. Res. Not.* **40** (2004) 2105–2124. Preprint `math.SG/0308008`.

[GuSj] V. Guillemin and R. Sjamaar, *Convexity properties of Hamiltonian group actions.* **CRM Monograph Series 26**. American Mathematical Society, Providence, RI, 2005.

[GuSt] V. Guillemin and S. Sternberg, "Convexity properties of the moment mapping." *Invent. Math.* **67** (1982) no. 3, 491–513.

[GuZ] V. Guillemin and C. Zara, "1-skeleta, Betti numbers, and equivariant cohomology." *Duke Math. J.* **107** (2001) no. 2, 283–349. Preprint `math.DG/9903051`.

[Hæ] A. Haefliger, "Structures feuilletes et cohomologie valeur dans un faisceau de groupodes." **Comment. Math. Helv. 32** (1958) 248–329.

[HHH] M. Harada, A. Henriques, and T. Holm, "Computation of generalized equivariant cohomologies of Kac-Moody flag varieties." *Adv. Math.* **197** (2005) no. 1, 198–221. Preprint `math.AT/0409305`.

[HL] M. Harada and G. Landweber, "Surjectivity for Hamiltonian G-spaces in K-theory." *Trans. AMS* to appear. Preprint `math.SG/0503609`.

[HeMe] A. Henriques and D. Metzler, "Presentations of Noneffective Orbifolds." *Trans. Amer. Math. Soc.* **356** (2004) no. 6 2481–2499. Preprint `math.AT/0302182`.

[HT] T. Holm and S. Tolman, "Integral Kirwan Surjectivity for Hamiltonian T-manifolds," in preparation.

[JK] L. Jeffrey and F. Kirwan, "Localization for nonabelian group actions." *Topology* **34** (1995) no. 2, 291–327.

[J] Y. Jiang, "The Chen-Ruan Cohomology of Weighted Projective Spaces." Preprint `math.AG/0304140`.

[Ka] T. Kawasaki, "Cohomology of twisted projective spaces and lens complexes." *Math. Ann.* **206** (1973) 243–248.

[Ki] F. Kirwan, *Cohomology of quotients in symplectic and algebraic geometry.* Mathematical Notes, **31**. Princeton University Press, Princeton, NJ, 1984.

[Kn] A. Knutson, *Weight varieties*, MIT Ph.D. thesis 1996.

[LT] E. Lerman and S. Tolman, Hamiltonian torus actions on symplectic orbifolds and toric varieties. *Trans. Amer. Math. Soc.* **349** (1997) no. 10, 4201–4230. `dg-ga/9511008`.

[Ma1] E. Mann, *Cohomologie quantique orbifolde des espaces projectifs á poids*, IRMA (Strasbourg) Ph.D. thesis 2005. Available at `math.AG/0510331`.

[Ma2] E. Mann "Orbifold quantum cohomology of weighted projective spaces." *J. of Alg. Geom.*, to appear. Preprint `math.AG/0610617`.

[MaWe] J. Marsden and A. Weinstein, "Reduction of symplectic manifolds with symmetry." *Rep. Mathematical Phys.* **5** (1974) no. 1, 121–130.

[Moe] I. Moerdijk, "Orbifolds as groupoids: an introduction." In *Orbifolds in mathematics and physics (Madison, WI, 2001)* 205–222, *Contemp. Math.*, **310**, Amer. Math. Soc., Providence, RI, 2002.

[MoWi] G. Moore and E. Witten, "Self-duality, Ramond-Ramond fields and K-theory." *J. High Energy Phys.* (2000) no. 5, Paper 32, 32 pp.

[S1] I. Satake, "On a generalization of the notion of manifold." *Procedings of the National Academy of Sciences USA* **42** (1956) 359–363.

[S2] I. Satake, "The Gauss-Bonnet Theorem for V-manifolds." *J. Math. Soc. Japan* **9** (1957) 464–492.

[SL] R. Sjamaar and E. Lerman, "Stratified symplectic spaces and reduction." *Ann. of Math. (2)* **134** (1991) no. 2, 375–422.

[Th] W. Thurston, *Three-dimensional geometry and topology. Vol. 1.* Princeton Mathematical Series, #35. Princeton University Press, 1997. ISBN 0-691-08304-5.

[TW] S. Tolman and J. Weitsman, "The cohomology ring of symplectic quotients." *Communications in Analysis and Geometry* **11** (2003) no. 4, 751–773. Preprint `math.DG/9807173`.

Department of Mathematics, Cornell University, Ithaca, NY 14853-4201 USA
E-mail address: `tsh@math.cornell.edu`

Contemporary Mathematics
Volume **450**, 2008

Deformation of graded Poisson (Batalin-Vilkovisky) Structures

Noriaki IKEDA

ABSTRACT. A Batalin-Vilkovisky formalism is a most general framework to construct consistent quantum field theories. Its mathematical structure is called *a Batalin-Vilkovisky structure*. First we explain the mathematical setting of a Batalin-Vilkovisky formalism. Next, we consider deformation theory of a Batalin-Vilkovisky structure. Especially, we consider deformation of topological sigma models in any dimension, which is closely related to many deformation theories in mathematics, including deformation from commutative geometry to noncommutative geometry. We obtain a series of new nontrivial topological sigma models and we find these models have the Batalin-Vilkovisky structures based on a series of new algebroids.

1. Introduction

Topological field theory is a powerful method to analyze geometry by means of a quantum field theory.[**BBRT**] A Batalin-Vilkovisky formalism [**BV**] is a most general systematic method to treat a consistent quantum field theory. So it is natural to treat topological field theory by means of a Batalin-Vilkovisky formalism.

Deformation theory is one of the main topics in mathematics. On the other hand, deformation theory of a quantum field theory is proposed by [**BBH**][**BH**] in the physical context. Purposes of this deformation theory are to find a new gauge theory, or to prove a no-go theorem to construct a new gauge theory. Here we apply deformation theory to topological field theories, especially to topological sigma models. Our purpose is to construct many geometries by topological field theories, classify geometries as topological field theories, unify many deformation theories as a topological field theories, and analyze deformation theories in mathematics as quantum field theories.

In this article, we consider a *topological sigma model.* Let X and M be two manifolds. We denote by ϕ a (smooth) map from X to M. A *sigma model* is a quantum field theory constructed from a map ϕ (and other auxiliary fields). We analyze structures on M by the structures induced from X. A *topological sigma model* is a sigma model independent of metrics of X and M [**W1**].

The AKSZ formulation [**AKSZ**] of the Batalin-Vilkovisky formalism is a general framework to construct a topological sigma model by the Batalin-Vilkovisky method. This formulation is appropriate to analyze geometry by means of a topological sigma model. The AKSZ formulation is originally formulated in the base manifolds (a worldsheet or a worldvolume) X in two and three dimensions and a target supermanifold with Z_2 gradings[**AKSZ**] . In this article, first we generalize the AKSZ formulation to a general n dimensional base manifold X and a target manifold with general gradings p. We call this genralization of the AKSZ formulation the *Batalin-Vilkovisky structure*. Next we discuss deformation of a Batalin-Vilkovisky structure with general gradings.

We can construct various geometrical structures on M if we consider general n and general grading p. Our purpose is *Poisson geometry realization* of geometries. This formulation provides a clear method to unify and to classify many geometries as Batalin-Vilkovisky structures, and to analyze them as quantum field theories. This construction includes many new topological sigma models as special cases, for examples, the topological sigma model with a Kähler structure (A model), with a complex structure (B model) [**W2**][**AKSZ**], with a symplectic structure [**W1**], with a Poisson structure (the Poisson sigma model) [**II**][**SS**], with a Courant algebroid structure [**R1**][**I4**][**HP**], with a twisted Poisson structure [**KS**][**I5**], with a Dirac structure [**KSS**], with a generalized complex structure [**Z**][**I5**][**Pe**] with a generalized geometry structure [**IT**], and so on.

As explained from the next section, the Batalin-Vilkovisky structure is a generalization of a Schouten bracket formulation of a Poisson structure. Detail discussion appears in the paper [**I6**]. Mathematical rigorous approaches in two and three dimenis ions are described in [**CF2**] and [**R2**].

2. AKSZ formulation of Batalin-Vilkovisky formalism on Graded Bundles

We explain general setting of the AKSZ formulation [**AKSZ**] of the Batalin-Vilkovisky formalism for a general graded bundle in the rather mathematical context.

Let M be a smooth manifold in d dimensions. We consider a *supermanifold* ΠT^*M. Mathematically, ΠT^*M, whose bosonic part is M, is defined as a cotangent bundle with reversed parity of the fiber. That is, the base manifold M has a Grassman even coordinate and the fiber of ΠT^*M has a Grassman odd coordinate. We can introduce a grading. The coordinate on the base manifold has grade zero and the coordinate on the fiber has grade one. This grading is called the *total degrees*. Similarly, we can define ΠTM for a tangent bundle TM.

We can consider more general assignments for the degree of the fibers of T^*M or TM. For a nonnegative integer p, we define $T^*[p]M$, which is called a graded cotangent bundle. $T^*[p]M$ is a cotangent bundle the degree of whose fiber is p. If p is odd, the fiber is Grassman odd, and if p is even, the fiber is Grassman even. The coordinate on the bass manifold has the total degree zero and the coordinate on the fiber has the total degree p. If M itself is graded, we define that $T^*[m]M$ is a cotangent bundle the sum of the degree of whose fiber and base manifold is m. That is, a tangent vector field $\boldsymbol{B}_i$ for a degree p local coordinate $\boldsymbol{\phi}^i$ on M has degree $m-p$.

We define a graded tangent bundle $T[p]M$ in the same way. For a general vector bundle E, a graded vector bundle $E[p]$ is defined in a similar way. That is, $E[p]$ is a vector bundle with degree of the fiber being shifted by p. [1]

[1]Note that only the fiber is shifted and the base space is not shifted.

We consider a Poisson manifold N with a Poisson bracket $\{*, *\}$. If we can construct a graded manifold $\tilde{N}$ from N, then the Poisson structure $\{*, *\}$ shifts to *a graded Poisson structure* by grading of $\tilde{N}$. For example, if $N = T^*M$ and $\tilde{N} = T^*[p]M$, we cosider $N = T^*M$, then we can construct $T^*[p]M$ by shifting the degree of the fiber.

The graded Poisson bracket is called an *antibracket* and denoted by $(*, *)$. $(*, *)$ is graded symmetric and satisfies the graded Leibniz rule and the graded Jacobi identity with respect to the grading of the manifold. The antibracket $(*, *)$ with total degree $-n+1$ satisfies the following identities:

$$\begin{aligned}
&(F, G) = -(-1)^{(|F|+1-n)(|G|+1-n)}(G, F), \\
&(F, GH) = (F, G)H + (-1)^{(|F|+1-n)|G|}G(F, H), \\
&(FG, H) = F(G, H) + (-1)^{|G|(|H|+1-n)}(F, H)G, \\
&(-1)^{(|F|+1-n)(|H|+1-n)}(F, (G, H)) + \text{cyclic permutations} = 0,
\end{aligned} \tag{2.1}$$

where F, G and H are functions on $\tilde{N}$ and $|F|, |G|$ and $|H|$ are the total degrees of functions, respectively. If $n = 1$, the antibacket is nothing but the Scouten bracket. For higher n, the antibracket is equivalent to the Loday bracket [**L**] with the degree $-n+1$. The graded Poisson structure is also called the *P-structure*.

Typical examples of a Poisson manifold N are the cotangent bundle T^*M and the vector bundle $E \oplus E^*$. Both bundles play important roles in this paper. Another example is a vector bundle E with a Poisson structure on the fiber. We consider these three bundles.

First we consider a cotangent bundle T^*M. Since T^*M has a natural symplectic structure, we can define a Poisson bracket induced from the natural symplectic structure. If we take a local coordinate ϕ^i on M and a local coordinate B_i of the fiber, we can define a Poisson bracket as follows: [2]

$$\{F, G\} \equiv F \frac{\overleftarrow{\partial}}{\partial \phi^i} \frac{\overrightarrow{\partial}}{\partial B_i} G - F \frac{\overleftarrow{\partial}}{\partial B_i} \frac{\overrightarrow{\partial}}{\partial \phi^i} G. \tag{2.2}$$

where F and G are a function on T^*M, and $\overleftarrow{\partial}/\partial\varphi$ and $\overrightarrow{\partial}/\partial\varphi$ are the right and left differentiations with respect to φ, respectively. Next we shift the degree of fiber by p, and we consider the space $T^*[p]M$. Then we obtain a graded Poisson structure. $(*, *)$. Let $\boldsymbol{\phi}^i$ be a local coordinate of M and $\boldsymbol{B}_{n-1,i}$ a basis of the fiber of $T^*[p]M$. [3] An *antibracket* $(*, *)$ on a cotangent bundle $T^*[p]M$ is represented as:

$$(F, G) \equiv F \frac{\overleftarrow{\partial}}{\partial \boldsymbol{\phi}^i} \frac{\overrightarrow{\partial}}{\partial \boldsymbol{B}_{p,i}} G - F \frac{\overleftarrow{\partial}}{\partial \boldsymbol{B}_{p,i}} \frac{\overrightarrow{\partial}}{\partial \boldsymbol{\phi}^i} G. \tag{2.3}$$

The total degree of the $(*, *)$ is $-p$.

We consider a vector bundle $E \oplus E^*$. There is a natural Poisson structure on the fiber of $E \oplus E^*$ induced from natural paring between E and E^*. If we take a local coordinate A^a on the fiber of E and B_a on the fiber of E^*, we define

$$\{F, G\} \equiv F \frac{\overleftarrow{\partial}}{\partial A^a} \frac{\overrightarrow{\partial}}{\partial B_a} G - F \frac{\overleftarrow{\partial}}{\partial B_a} \frac{\overrightarrow{\partial}}{\partial A^a} G. \tag{2.4}$$

[2] We take Einstein's summation notation.

[3] We use bold notations for local coordinates of graded (super)vector bundles, while we use nonbold notations for local coordinates of usual vector bundles.

where F and G are functions on $E \oplus E^*$. We shift the degrees of fibers of E and E^* to $E[p] \oplus E^*[q]$, where p and q are positive integers. Then the Poisson structure changes to a graded Poisson structure $(*, *)$. Let $\boldsymbol{A}_p{}^a$ be a basis of the fiber of $E[p]$ and $\boldsymbol{B}_{q,a}$ a basis of the fiber of $E^*[q]$. An antibracket is represented as

$$(F, G) \equiv F \frac{\overleftarrow{\partial}}{\partial \boldsymbol{A}_p{}^a} \frac{\overrightarrow{\partial}}{\partial \boldsymbol{B}_{q,a}} G - (-1)^{pq} F \frac{\overleftarrow{\partial}}{\partial \boldsymbol{B}_{q,a}} \frac{\overrightarrow{\partial}}{\partial \boldsymbol{A}_p{}^a} G. \tag{2.5}$$

The total degree of the $(*, *)$ is $-p-q$.

Next, we consider a vector bundle E on which can be defined a Poisson structure on the fiber. If we shift the degree of the fiber of E to $E[p]$, the Poisson structure changes to a graded Poisson structure $(*, *)$. Let $\boldsymbol{A}_p{}^a$ be a basis of the fiber of $E[p]$, An antibracket is represented as

$$(F, G) \equiv F \frac{\overleftarrow{\partial}}{\partial \boldsymbol{A}_p{}^a} k^{ab} \frac{\overrightarrow{\partial}}{\partial \boldsymbol{A}_p{}^b} G, \tag{2.6}$$

where F and G are a function on $E[p]$ and k^{ab} is a nondegenerate constant bivector induced from a (graded) Poisson structure. The total degree of the antibracket $(*, *)$ is $-2p$.

The another component for a Batalin-Vilkovisky structure is a *Q-structure*. A *Q-structure* is a function S on a supermanifold $\tilde{N}$ which satisfies the classical master equation $(S, S) = 0$. S is called a *Batalin-Vilkovisky action*, or simply an *action*. We require that S satisfies the compatibility condition

$$Q(F, G) = (QF, G) + (-1)^{|F|+1}(F, QG), \tag{2.7}$$

where $Q = (S, *)$, F and G are arbitrary functions, and $|F|$ is the total degree of F. $(S, F) = \delta F$ defines a coboundary operator δ because we can prove $\delta^2 = 0$. We call this operation a *BRST transformation* and a δ cohomology a BRST cohomology.

We define a (classical) *Batalin-Vilkovisky structure* as follows:

DEFINITION 2.1. If a structure on a *supermanifold* has *P-structure* and *Q-structure*, it is called a *Batalin-Vilkovisky structure.*

3. Batalin-Vilkovisky Structures of Abelian Topological Sigma Models

In this section, we construct Batalin-Vilkovisky structures of topological sigma models.

Let X be a base manifold in n dimensions, with or without boundary, and M be a target manifold in d dimensions. We denote ϕ a smooth map from X to M.

We consider a *supermanifold* ΠTX, whose bosonic part is X. ΠTX is defined as a tangent bundle with reversed parity of the fiber. We extend a smooth map ϕ to a smooth map $\boldsymbol{\phi} : \Pi TX \to M$. The *total degree* defined in the previous section is a grading with respect to M. We introduce a nonnegative integer grading on ΠT^*X. A coordinate on a bass manifold is zero and a coordinate on the fiber is one. This grading is called the *form degrees.* We denote $\deg F$ the form degree of a function F. $\mathrm{gh} F = |F| - \deg F$ is called the *ghost number.*

First we consider a *P-structure* on $T^*[p]M$. It is natural to take $p = n-1$ to construct a Batalin-Vilkovisky structure in a topological sigma model. In other words, the dimensions of X labels the total degree of a Batalin-Vilkovisky structure on the supermanifold $T^*[p]M$. We consider $T^*[n-1]M$ for an n-dimensional base manifold X. Let $\boldsymbol{\phi}^i$ be a local coordinate expression of a smooth map from ΠTX

to M, where $i, j, k, \cdots$ are indices of the local coordinate on M. Let $\boldsymbol{B}_{n-1,i}$ be a degree $n-1$ superfield on ΠTX, which takes a value on $\boldsymbol{\phi}^*(T^*[n-1]M)$. According to the discussion in the previous section, we can define an *antibracket* $(*, *)$ on a cotangent bundle $T^*[n-1]M$ as

$$(F, G) \equiv F \frac{\overleftarrow{\partial}}{\partial \boldsymbol{\phi}^i} \frac{\overrightarrow{\partial}}{\partial \boldsymbol{B}_{n-1,i}} G - F \frac{\overleftarrow{\partial}}{\partial \boldsymbol{B}_{n-1,i}} \frac{\overrightarrow{\partial}}{\partial \boldsymbol{\phi}^i} G, \tag{3.1}$$

where F and G are functions of $\boldsymbol{\phi}^i$ and $\boldsymbol{B}_{n-1,i}$. Now we take a Darboux coordinate $\boldsymbol{\phi}^i, \boldsymbol{B}_{n-1,i}$ for simplicity, but we can also take general coordinates. The total degree of the antibracket is $-n+1$.

If F and G are functionals of $\boldsymbol{\phi}^i$ and $\boldsymbol{B}_{n-1,i}$, we define the antibracket on functionals up to total derivative terms in (3.1). This is also written as

$$(F, G) \equiv \int_X F \frac{\overleftarrow{\partial}}{\partial \boldsymbol{\phi}^i} \frac{\overrightarrow{\partial}}{\partial \boldsymbol{B}_{n-1,i}} G - F \frac{\overleftarrow{\partial}}{\partial \boldsymbol{B}_{n-1,i}} \frac{\overrightarrow{\partial}}{\partial \boldsymbol{\phi}^i} G, \tag{3.2}$$

where the integration over X is understood as that on the n-form part of the integrand.

Next we consider a *P-structure* on $E \oplus E^*$. A natural assignment of the total degrees is given by the nonnegative integers p and q with $p+q = n-1$. That is, we consider $E[p] \oplus E^*[n-p-1]$, where $1 \leq p \leq n-2$. We can naturally construct a topological sigma model in this case. Let $\lfloor x \rfloor$ be the floor function which gives the largest integer less than or equal to x. If $\lfloor \frac{n}{2} \rfloor \leq p \leq n-2$, we identify $E[p] \oplus E^*[n-p-1]$ with the dual bundle $E^*[n-p-1] \oplus (E^*)^*[p]$. Therefore $1 \leq p \leq \lfloor \frac{n-1}{2} \rfloor$ provides different structures of grading.

Let $\boldsymbol{A}_p{}^{a_p}$ be a degree p superfield on X, which takes a value on $\boldsymbol{\phi}^*(E[p])$. $\boldsymbol{B}_{n-p-1,a_p}$ a degree $n-p-1$ superfield, which takes a value on $\boldsymbol{\phi}^*(E^*[n-p-1])$. From (2.5), we can define an antibracket as

$$(F, G) \equiv F \frac{\overleftarrow{\partial}}{\partial \boldsymbol{A}_p{}^{a_p}} \frac{\overrightarrow{\partial}}{\partial \boldsymbol{B}_{n-p-1,a_p}} G - (-1)^{np} F \frac{\overleftarrow{\partial}}{\partial \boldsymbol{B}_{n-p-1,a_p}} \frac{\overrightarrow{\partial}}{\partial \boldsymbol{A}_p{}^{a_p}} G. \tag{3.3}$$

We want to consider various grading assignments for $E \oplus E^*$. Because each assignment induces different Batalin-Vilkovisky structures. In order to consider all independent assignments, we define the following bundle. Let E_p be $\lfloor \frac{n-1}{2} \rfloor$ series of vector bundles, where $1 \leq p \leq \lfloor \frac{n-1}{2} \rfloor$. We consider $E_p[p] \oplus E_p^*[n-p-1]$ and consider a direct sum

$$\sum_{p=1}^{\lfloor \frac{n-1}{2} \rfloor} E_p[p] \oplus E_p^*[n-p-1] \tag{3.4}$$

And we consider the graded bundle whose fiber is isomorphic to a direct sum of (3.4) and a cotangent space $T_x^*[n-1]M$,

$$T^*[n-1] \left(\sum_{p=1}^{\lfloor \frac{n-1}{2} \rfloor} E_p[p] \right), \tag{3.5}$$

If $n=2$, there is no $E_p[p]$ and the graded bundle is regarded as

$$T^*[1]M. \tag{3.6}$$

We define a *P-structure* on this graded vector bundle A local (Darboux) coordinate expression of the antibracket $(\cdot,\cdot)$ is a sum of (3.1) and (3.3):

$$(3.7)\quad (F,G) \equiv \sum_{p=0}^{\lfloor\frac{n-1}{2}\rfloor} F\frac{\overleftarrow{\partial}}{\partial \boldsymbol{A}_p{}^{a_p}}\frac{\overrightarrow{\partial}}{\partial \boldsymbol{B}_{n-p-1\ a_p}}G - (-1)^{np}F\frac{\overleftarrow{\partial}}{\partial \boldsymbol{B}_{n-p-1\ a_p}}\frac{\overrightarrow{\partial}}{\partial \boldsymbol{A}_p{}^{a_p}}G.$$

where the $p=0$ component is the antibracket (3.1) on the graded cotangent bundle $T^*[n-1]M$ and $\boldsymbol{A}_0{}^{a_0} = \boldsymbol{\phi}^i$. Note that all terms of the antibracket have the total degree $-n+1$. We can confirm that the antibracket (3.7) satisfies the identity (2.1).

We construct a *Q-structure* on the bundle (3.5). The simplest and natural action for a topological sigma model is

$$(3.8)\qquad S_0 = \sum_{p=0}^{\lfloor\frac{n-1}{2}\rfloor} (-1)^{n-p}\int_X \boldsymbol{B}_{n-p-1\ a_p} d\boldsymbol{A}_p{}^{a_p},$$

where d is an exterior differential on X. The total degree of d is 1 because the form degree is 1 and we assign the ghost number 0. The integration over X is understood as that over the n-form part (the form degree n part) of the integrand. This action is an analogue of the fundamental form $\theta = p_i dq^i$ for a symplectic form ω, which has $\omega = -d\theta$, therefore this action is directly derived from the *P-structure* on the graded bundle. The total degree of S_0 is n. It is called an *abelian BF theory in n dimensions.*

S_0 defines a *Q-structure*, since we can easily confirm that S_0 satisfies the classical master equation:

$$(3.9)\qquad (S_0, S_0) = 0,$$

'Abelian' means that the theory has a $U(1)$ gauge symmetry. The BRST transformation (the gauge symmetry) is defined as

$$(3.10)\qquad \delta_0\boldsymbol{\Phi} \equiv (S_0, \boldsymbol{\Phi}) = d\boldsymbol{\Phi},$$

where $\boldsymbol{\Phi}$ is an arbitrary function of a total bundle $T^*[n-1]\left(\sum_{p=1}^{\lfloor\frac{n-1}{2}\rfloor} E_p[p]\right)$. $\delta_0^2 = 0$ is satisfied from $(S_0, S_0) = 0$, which is consistent with $d^2 = 0$.

For a vector bundle E with a Poisson structure on the fiber, we can construct an another topological sigma model if n is odd. We do not discuss this case in this article. See the paper [**I4**].

4. Deformation

In this section, we consider deformation of Batalin-Vilkovisky structures. Deformation means deformation of the *Q-structure* for a fixed *P-structure*. We consider local deformation from the simplest *Q-structure* (3.8).

The *Q-structure* to begin with is the equation (3.8). We deform this Batalin-Vilkovisky action S_0 to

$$(4.1)\qquad S = S_0 + gS_1,$$

under the condition that S also satisfies the classical master equation:

$$(4.2)\qquad (S, S) = 0,$$

where g is a deformation parameter and S_1 represents all the deformation terms, which are functionals on X and functions on $T^*[n-1]\left(\sum_{p=1}^{\lfloor\frac{n-1}{2}\rfloor} E_p[p]\right)$. We require that S is the total degree n. It is equivalent that the ghost number is zero $\mathrm{gh} S = |S| - \deg S = 0$. This condition is physically necessary, though we can relax this condition mathematically. If two deformations S and S' satisfy $S' = S + (S_0, T) = S + \delta_0 T$ for some functional T, the two Batalin-Vilkovisky structures are equivalent. Therefore the problem is to look for the total degree n cohomology class $H^n\left(T^*[n-1]\left(\sum_{p=1}^{\lfloor\frac{n-1}{2}\rfloor} E_p[p]\right)\right)$.

Substituting $S = S_0 + gS_1$ to (4.2), we obtain g expansion

$$(S_0, S_0) + 2g(S_0, S_1) + g^2(S_1, S_1) = 0, \tag{4.3}$$

The zero-th order of the equation (4.3), $(S_0, S_0) = 0$, is already satisfied.

The first order is $(S_0, S_1) = 0$. Because of the equation (3.10), S_1 is the integration of an arbitrary function F of all fundamental superfields $\boldsymbol{\Phi}$ which are $\boldsymbol{A}_p{}^{a_p}$ or $\boldsymbol{B}_{qb_q}$, and their derivatives $d\boldsymbol{\Phi}$:

$$S_1 = \int_X F(\boldsymbol{\Phi}, d\boldsymbol{\Phi}). \tag{4.4}$$

For simplicity, we assume that there is no boundary contribution to S, that is, integration of total derivative terms on X is always zero. This corresponds to assume that there is no obstruction to deformation. Then we can prove the following theorem:

THEOREM 4.1. *Assume there is no boundary contribution on X, i.e. $\int_X dG(\boldsymbol{\Phi}) = 0$ for any function G. If a monomial of $F(\boldsymbol{\Phi}, d\boldsymbol{\Phi})$ includes at least one $d\boldsymbol{\Phi}$, $\int_X F(\boldsymbol{\Phi}, d\boldsymbol{\Phi})$ is δ_0-exact.*

PROOF. We can assume that $\int_X F(\boldsymbol{\Phi}, d\boldsymbol{\Phi}) = \sum_{p=0}^{n-1} \int_X F_{n-p-1} dG_p$, where F_{n-p-1} are functions with the form degree $n-p-1$ and G_p are functions with the form degree p. From (3.10), we obtain

$$\begin{aligned}
&\delta_0 F_0 = 0, \\
&\delta_0 F_{n-p-1} = dF_{n-p-2} \quad \text{for } -1 \leq p \leq n-2, \\
&dF_n = 0, \\
&\delta_0 G_0 = 0, \\
&\delta_0 G_p = dG_{p-1} \quad \text{for } 1 \leq p \leq n, \\
&dG_n = 0,
\end{aligned} \tag{4.5}$$

For even p, adjoining two terms are combined as

$$\begin{aligned}
&F_{n-p-1} dG_p + F_{n-p-2} dG_{p+1} \\
&= (-1)^{n-p-1} \delta_0 (F_{n-p-1} G_{p+1}) - (-1)^{n-p-1} d(F_{n-p-2} G_{p+1})
\end{aligned} \tag{4.6}$$

using the relations (4.5). Thus integration of these two terms is δ_0-exact.

If n is even, $S_1 = \sum_{p=0}^{n-1} \int_X F_{n-p-1} dG_p$ has even numbers of terms, therefore we combine each two term like (4.6) and we can confirm that S_1 is δ_0-exact.

If n is odd, the last term $F_0 dG_{n-1}$ remains. However this term is δ_0-exact because $F_0 dG_{n-1} = \delta_0(F_0 G_n)$. Therefore S_1 is δ_0-exact. □

Therefore since the nontrivial S_1 cohomological class does not include d, we can take $S_1 = \int_X F(\mathbf{\Phi})$. Concretely we can express

$$
\begin{aligned}
S_1 = & \sum_{p(1),\cdots,p(k),q(1),\cdots,q(l)} \int_X F_{p(1)\cdots p(k),q(1)\cdots q(l)\ a_{p(1)}\cdots a_{p(k)}}{}^{b_{q(1)}\cdots b_{q(l)}}(\boldsymbol{A}_0{}^{a_0}) \\
& \times \boldsymbol{A}_{p_1}{}^{a_{p(1)}} \cdots \boldsymbol{A}_{p_k}{}^{a_{p(k)}} \boldsymbol{B}_{q_1 b_{q(1)}} \cdots \boldsymbol{B}_{q_l b_{q(l)}},
\end{aligned}
\tag{4.7}
$$

where $F_{p(1)\cdots p(k),q(1)\cdots q(l)\ a_{p(1)}\cdots a_{p(k)}}{}^{b_{q(1)}\cdots b_{q(l)}}(\boldsymbol{A}_0{}^{a_0})$ is a function of $\boldsymbol{A}_0{}^{a_0}$ and $p(r) \neq 0, q(s) \neq 0$ for $r = 1, \cdots, k, s = 1, \cdots, l$. From the deg$S = n$ condition, we obtain $\sum_{\alpha=0}^{k} |\boldsymbol{A}_{p_\alpha}{}^{a_{p(\alpha)}}| + \sum_{\beta=0}^{l} |\boldsymbol{B}_{q_\beta b_{q(\beta)}}| = n$.

Second order of the equation (4.3)

$$
(S_1, S_1) = 0, \tag{4.8}
$$

imposes conditions on functions $F_{p(1)\cdots p(k),q(1)\cdots q(l)\ a_{p(1)}\cdots a_{p(k)}}{}^{b_{q(1)}\cdots b_{q(l)}}(\boldsymbol{A}_0{}^{a_0})$. These conditions determine the mathematical structure of a Batalin-Vilkovisky structure.

We call the resulting field theory $S = S_0 + gS_1$ a *nonlinear gauge theory* in n dimensions.

5. Deformation in lower dimensions

In this section, we concretely analyze the algebraic and geometric structure of deformation of a topological sigma model in lower dimensional X.

We can easily find that we cannot obtain nontrivial deformation in case of the total degree $n = 1$, thus we cannot obtain a nontrivial structure.

5.1. $n = 2$. We analyze the algebraic structure of the total degree $n = 2$ topological sigma model. In two dimensions, the total graded bundle (3.5) is $T^*[1]M = \Pi T^*M$.

(4.1) under (4.7) is

$$
\begin{aligned}
& S = S_0 + gS_1, \\
& S_0 = \int_X \boldsymbol{B}_{1i} d\boldsymbol{\phi}^i, \qquad S_1 = \int_\Sigma \frac{1}{2} f^{ij}(\boldsymbol{\phi}) \boldsymbol{B}_{1i} \boldsymbol{B}_{1j},
\end{aligned}
\tag{5.1}
$$

where $i, j, \cdots$ are indices of a local coordinate expressions on $T^*[1]M$. and we rewrite the notations as $\boldsymbol{\phi}^i = \boldsymbol{A}_0{}^i$ and $\frac{1}{2} f^{ij}(\boldsymbol{\phi}) = F_{,11}{}^{ij}(\boldsymbol{A}_0)$. This topological sigma model is known as *the Poisson sigma model* [**II**][**SS**]. If we substitute this S_1 to the condition (4.8), we obtain the geometric structure of the action. We obtain the following identity on f^{ij}:

$$
f^{kl} \frac{\overrightarrow{\partial}}{\partial \boldsymbol{\phi}^l} f^{ij} + f^{il} \frac{\overrightarrow{\partial}}{\partial \boldsymbol{\phi}^l} f^{jk} + f^{jl} \frac{\overrightarrow{\partial}}{\partial \boldsymbol{\phi}^l} f^{ki} = 0. \tag{5.2}
$$

If we restrict the identity on X (i.e. a ghost number zero sector), (5.2) reduces to

$$
f^{kl}(\phi) \frac{\partial f^{ij}(\phi)}{\partial \phi^l} + f^{il}(\phi) \frac{\partial f^{jk}(\phi)}{\partial \phi^l} + f^{jl}(\phi) \frac{\partial f^{ki}(\phi)}{\partial \phi^l} = 0. \tag{5.3}
$$

Under the identity (5.2), $-f^{ij}$ defines a Poisson structure as

$$
\{F(\phi), G(\phi)\} \equiv -f^{ij}(\phi) \frac{\partial F}{\partial \phi^i} \frac{\partial G}{\partial \phi^j}, \tag{5.4}
$$

on the space M. Conversely, if we consider the Poisson structure $-f^{ij}$ on M, which satisfies the identity (5.2), we can define the action (5.1) consistently. This Poisson bracket (5.4) is directly constructed if we restrict the derived bracket [**Kos**] of a Batalin-Vilkovisky structure:

$$\{F(\phi), G(\phi)\} = -F\frac{\overleftarrow{\partial}}{\partial\phi^i}f^{ij}(\phi)\frac{\overrightarrow{\partial}}{\partial\phi^j}G = ((S, F), G), \tag{5.5}$$

to M and X, i.e. to the ghost number zero sector.

The Batalin-Vilkovisky structure of this theory defines the structure of a *Lie algebroid* in $T^*[1]\mathcal{M}$ [**LO**][**O**], where $\mathcal{M}$ is a space of a (smooth) map from ΠTX to a target space M.

A *Lie algebroid* is a generalization of a bundle of a Lie algebra over a base manifold $\mathcal{M}$. A Lie algebroid over a manifold is a vector bundle $\mathcal{E} \to \mathcal{M}$ with a Lie algebra structure on the space of the sections $\Gamma(\mathcal{E})$ defined by the Lie bracket $[e_1, e_2]$, $e_1, e_2 \in \Gamma(\mathcal{E})$ and a bundle map (the anchor) $\rho : \mathcal{E} \to T\mathcal{M}$ satisfying the following properties:

$$\begin{aligned} &1, \quad \text{For any} \quad \mathrm{e}_1, \mathrm{e}_2 \in \Gamma(\mathcal{E}), \quad [\rho(\mathrm{e}_1), \rho(\mathrm{e}_2)] = \rho([\mathrm{e}_1, \mathrm{e}_2]), \\ &2, \quad \text{For any} \quad \mathrm{e}_1, \mathrm{e}_2 \in \Gamma(\mathcal{E}), \ \mathrm{F} \in \mathrm{C}^\infty(\mathcal{M}), \\ &\qquad [e_1, Fe_2] = F[e_1, e_2] + (\rho(e_1)F)e_2, \end{aligned} \tag{5.6}$$

We can derive the bracket of a Lie algebroid from the antibracket and the Batalin-Vilkovisky structure of the $n = 2$ theory. In our case, a vector bundle $\mathcal{E}$ is a cotangent bundle $T^*[1]\mathcal{M}$. The Lie bracket of two sections e_1 and e_2 is defined as a derived bracket of the antibracket by

$$[e_1, e_2] \equiv ((S, e_1), e_2), \tag{5.7}$$

and the anchor is defined by

$$\rho(e)F(\phi) \equiv (e, (S, F(\phi))). \tag{5.8}$$

Then we can confirm $[e_1, e_2] = -[e_2, e_1]$, from $(e_1, e_2) = 0$ and the graded Jacobi identity of the antibracket. Similarly, a Lie algebroid conditions 1 and 2 on the bracket $[\cdot, \cdot]$ and the anchor map ρ are obtained from the classical master equation $(S, S) = 0$. From this derived bracket, we directly obtain the following "noncommutative" relation of the coordinates:

$$[\phi^i, \phi^j] = -f^{ij}(\phi). \tag{5.9}$$

The anchor is a differentiation on functions of ϕ as

$$\rho(\phi^i)F(\phi) = -f^{ij}(\phi)\frac{\overrightarrow{\partial}}{\partial\phi^j}F(\phi). \tag{5.10}$$

5.2. $n = 3$. We analyze a nonlinear gauge theory in three dimensions. In this case, the theory defines the topological open 2-brane as a sigma model [**I2**][**Pa**].

We consider a supermanifold ΠTX whose ghost number zero part is a three-dimensional manifold X. A base space $\mathcal{M}$ is the space of a (smooth) map ϕ from ΠTX to a target space M. In $n = 3$, the total bundle (3.4) is a vector bundle $E[1] \oplus E^*[1]$. From (3.5), we consider $T^*[2]E[1]$. Then we obtain an antibracket $(\cdot, \cdot)$ on the graded bundle $T^*[2]E[1]$ if we set $n = 3$ in (3.7).

In order to write down the Batalin-Vilkovisky action S, we take a local basis of the fiber of $\phi^*(E[1] \oplus E^*[1])$ as $\boldsymbol{A}_1{}^a, \boldsymbol{B}_{1\,a}$, which are 1 forms on X and are Darboux coordinates such that $(\boldsymbol{A}_1{}^a, \boldsymbol{A}_1{}^b) = (\boldsymbol{B}_{1a}, \boldsymbol{B}_{1b}) = 0$ and $(\boldsymbol{A}_1{}^a, \boldsymbol{B}_{1b}) = \delta^a{}_b$ and Moreover we introduce a 2 form $\boldsymbol{B}_{2i}$ on X, a local basis of the fiber of $T^*[2]M$.

The total action (4.1) under (4.7) is

$$
\begin{aligned}
(5.11)\qquad & S = S_0 + gS_1,\\
& S_0 = \int_X [-\boldsymbol{B}_{2i} d\boldsymbol{\phi}^i + \boldsymbol{B}_{1a} d\boldsymbol{A}_1{}^a],\\
& S_1 = \int_X [f_{1a}{}^i(\boldsymbol{\phi})\boldsymbol{A}_1{}^a \boldsymbol{B}_{2i} + f_2^{ib}(\boldsymbol{\phi})\boldsymbol{B}_{2i}\boldsymbol{B}_{1b} + \frac{1}{3!} f_{3abc}(\boldsymbol{\phi})\boldsymbol{A}_1{}^a\boldsymbol{A}_1{}^b\boldsymbol{A}_1{}^c\\
& + \frac{1}{2} f_{4ab}{}^c(\boldsymbol{\phi})\boldsymbol{A}_1{}^a\boldsymbol{A}_1{}^b\boldsymbol{B}_{1c} + \frac{1}{2} f_{5a}{}^{bc}(\boldsymbol{\phi})\boldsymbol{A}_1{}^a\boldsymbol{B}_{1b}\boldsymbol{B}_{1c}\\
& + \frac{1}{3!} f_6{}^{abc}(\boldsymbol{\phi})\boldsymbol{B}_{1a}\boldsymbol{B}_{1b}\boldsymbol{B}_{1c}],
\end{aligned}
$$

where we set $f_{1a}{}^i = F_{1,2a}{}^i$, $f_2^{ib} = F_{,21}{}^{ib}$, $\frac{1}{3!} f_{3abc} = F_{111,abc}$, $\frac{1}{2} f_{4ab}{}^c = F_{11,1ab}{}^c$, $\frac{1}{2} f_{5a}{}^{bc} = F_{1,11a}{}^{bc}$, $\frac{1}{3!} f_6{}^{abc} = F_{,111}{}^{abc}$, for clarity. The condition of the classical master equation (4.8) imposes the following identities on six f_i's, $i = 1, \cdots, 6$[**I2**]:

$$
\begin{aligned}
(5.12)\quad & f_{1e}{}^i f_2{}^{je} + f_2{}^{ie} f_{1e}{}^j = 0,\\
& -\left(\frac{\vec{\partial}}{\partial \phi^j} f_{1c}{}^i\right) f_{1b}{}^j + \left(\frac{\vec{\partial}}{\partial \phi^j} f_{1b}{}^i\right) f_{1c}{}^j + f_{1e}{}^i f_{4bc}{}^e + f_2{}^{ie} f_{3ebc} = 0,\\
& f_{1b}{}^j \left(\frac{\vec{\partial}}{\partial \phi^j} f_2{}^{ic}\right) - f_2{}^{jc}\left(\frac{\vec{\partial}}{\partial \phi^j} f_{1b}{}^i\right) + f_{1e}{}^i f_{5b}{}^{ec} - f_2{}^{ie} f_{4eb}{}^c = 0,\\
& -f_2{}^{jb}\left(\frac{\vec{\partial}}{\partial \phi^j} f_2{}^{ic}\right) + f_2{}^{jc}\left(\frac{\vec{\partial}}{\partial \phi^j} f_2{}^{ib}\right) + f_{1e}{}^i f_6^{ebc} + f_2{}^{ie} f_{5e}{}^{bc} = 0,\\
& -f_{1[a}{}^j\left(\frac{\vec{\partial}}{\partial \phi^j} f_{4bc]}{}^d\right) + f_2{}^{jd}\left(\frac{\vec{\partial}}{\partial \phi^j} f_{3abc}\right) + f_{4e[a}{}^d f_{4bc]}{}^e + f_{3e[ab} f_{5c]}{}^{de} = 0,\\
& -f_{1[a}{}^j\left(\frac{\vec{\partial}}{\partial \phi^j} f_{5b]}{}^{cd}\right) - f_2{}^{j[c}\left(\frac{\vec{\partial}}{\partial \phi^j} f_{4ab}{}^{d]}\right)\\
& + f_{3eab} f_6{}^{ecd} + f_{4e[a}{}^{[d} f_{5b]}{}^{c]e} + f_{4ab}{}^e f_{5e}{}^{cd} = 0,\\
& -f_{1a}{}^j\left(\frac{\vec{\partial}}{\partial \phi^j} f_6{}^{bcd}\right) + f_2{}^{j[b}\left(\frac{\vec{\partial}}{\partial \phi^j} f_{5a}{}^{cd]}\right) + f_{4ea}{}^{[b} f_6{}^{cd]e} + f_{5e}{}^{[bc} f_{5a}{}^{d]e} = 0,\\
& -f_2{}^{j[a}\left(\frac{\vec{\partial}}{\partial \phi^j} f_6{}^{bcd]}\right) + f_6{}^{e[ab} f_{5e}{}^{cd]} = 0,\\
& -f_{1[a}{}^j\left(\frac{\vec{\partial}}{\partial \phi^j} f_{3bcd]}\right) + f_{4[ab}{}^e f_{3cd]e} = 0,
\end{aligned}
$$

where $[\cdots]$ on the indices represents the antisymmetrization of them, e.g., $\Phi_{[ab]} = \Phi_{ab} - \Phi_{ba}$.

If we restrict fields to X, (5.12) reduces to the identities on $E \oplus E^*$. The algebraic structure (5.12) is *a Courant algebroid. A Courant algebroid* has been introduced by Courant in order to analyze the Dirac structure as a generalization of a Lie algebra of vector fields on a vector bundle [**C**][**LWX**]. The Batalin-Vilkovisky structure on a Courant algebroid is first analyzed in [**R1**].

A Courant algebroid is a vector bundle $\mathcal{E} \to \mathcal{M}$ and has a nondegenerate symmetric bilinear form $\langle \cdot , \cdot \rangle$ on the bundle, a bilinear operation $\circ$ on $\Gamma(\mathcal{E})$ (the space of sections on $\mathcal{E}$), and a bundle map (called the anchor) $\rho : \mathcal{E} \to T\mathcal{M}$ satisfying the following properties:

$$\begin{aligned}
&1, \quad e_1 \circ (e_2 \circ e_3) = (e_1 \circ e_2) \circ e_3 + e_2 \circ (e_1 \circ e_3), \\
&2, \quad \rho(e_1 \circ e_2) = [\rho(e_1), \rho(e_2)], \\
&3, \quad e_1 \circ F e_2 = F(e_1 \circ e_2) + (\rho(e_1)F)e_2, \\
&4, \quad e_1 \circ e_2 + e_2 \circ e_1 = \mathcal{D}\langle e_1 \, , e_2 \rangle, \\
&5, \quad \rho(e_1)\langle e_2 \, , e_3 \rangle = \langle e_1 \circ e_2 \, , e_3 \rangle + \langle e_2 \, , e_1 \circ e_3 \rangle,
\end{aligned} \tag{5.13}$$

where e_1, e_2 and e_3 are sections of $\mathcal{E}$, and F is a function on $\mathcal{M}$; $\mathcal{D}$ is a map from functions on $\mathcal{M}$ to $\Gamma(\mathcal{E})$ and is defined by $\langle \mathcal{D}F \, , e \rangle = \rho(e)F$.

In our nonlinear gauge theory, $\mathcal{M}$ is the space of a map $\boldsymbol{\phi}$ from ΠTX to M and $\mathcal{E}$ is the space of a map from ΠTX to $T^*[2]E[1]$. Let e^a be a local basis of the fiber of $\mathcal{E}$. We take $e^a = \boldsymbol{A}_1{}^a$ or $\boldsymbol{B}_1{}_a$, which are a $E[1]$ component or a $E^*[1]$ component respectively. We define a graded symmetric bilinear form $\langle \cdot , \cdot \rangle$, a bilinear operation $\circ$, a bundle map ρ and $\mathcal{D}$ from the antibracket as follows:

$$\begin{aligned}
&e^a \circ e^b \equiv ((S, e^a), e^b), \\
&\langle e^a \, , e^b \rangle \equiv (e^a, e^b), \\
&\rho(e^a)F(\boldsymbol{\phi}) \equiv (e^a, (S, F(\boldsymbol{\phi}))), \\
&\mathcal{D}(*) \equiv (S, *).
\end{aligned} \tag{5.14}$$

Then we can confirm that the classical master equation $(S, S) = 0$, which derive the identity (5.12) on structure functions f's, is equivalent to the conditions 1 to 5 of the equation (5.13). [**R1**][**I4**] We calculate the operations $\circ$ and ρ on the basis as follows:

$$\begin{aligned}
\boldsymbol{A}_1{}^a \circ \boldsymbol{A}_1{}^b &= -f_{5c}{}^{ab}(\boldsymbol{\phi})\boldsymbol{A}_1{}^c - f_6{}^{abc}(\boldsymbol{\phi})\boldsymbol{B}_{1c}, \\
\boldsymbol{A}_1{}^a \circ \boldsymbol{B}_{1b} &= -f_{4bc}{}^a(\boldsymbol{\phi})\boldsymbol{A}_1{}^c + f_{5b}{}^{ac}(\boldsymbol{\phi})\boldsymbol{B}_{1c}, \\
\boldsymbol{B}_{1a} \circ \boldsymbol{B}_{1b} &= -f_{3abc}(\boldsymbol{\phi})\boldsymbol{A}_1{}^c - f_{4ab}{}^c(\boldsymbol{\phi})\boldsymbol{B}_{1c}, \\
\rho(\boldsymbol{A}_1{}^a)\boldsymbol{\phi}^i &= -f_2{}^{ia}(\boldsymbol{\phi}), \\
\rho(\boldsymbol{B}_{1a})\boldsymbol{\phi}^i &= -f_{1a}{}^i(\boldsymbol{\phi}).
\end{aligned} \tag{5.15}$$

This topological sigma model defines a Courant algebroid structure on the space $E[1] \oplus E^*[1]$. We call this model as *the Courant sigma model.*

5.3. General n. In n dimensions, the equation (4.8), $(S_1, S_1) = 0$, impose an algebroid structure on the space (3.5), $T^*[n-1]\left(\sum_{p=1}^{\lfloor \frac{n-1}{2} \rfloor} E_p[p]\right)$. The algebroid structure is derived from the Batalin-Vilkovisky structure $(S_1, S_1) = 0$ of nonlinear gauge theories. Now we obtain an infinite series of algebroids labeled by n. We call this algebroid an *n-algebroid.*

In the previous section, we have found that the $n = 2$ case defines a Lie algebroid on T^*M and the $n = 3$ case defines a Courant algebroid on $E \oplus E^*$. For $n \geq 4$ cases, we can easily calculate algebraic relations but characterization of algebroid structures is still unknown. Higher order generalization has also been discussed in [**Se**].

6. Quantum Version of Deformation

In the previous sections, we have considered a classical BV structure. In this section, we discuss quantum version of deformation of a Batalin-Vilkovisky structure. In this section, we discuss the BF case. We can make a similar discussion in the Chern-Simons with BF case.

In order to quantize a gauge theory, we must fix the gauge. Gauge fixing is carried out by adding a gauge fixing term S_{GF} to the classical action S.[**GPS**] The gauge fixed quantum action is $S_q = S + S_{GF}$.

We need the *BV Laplacian*. The BV Laplacian is defined as follows:

$$\boldsymbol{\Delta} F \equiv \sum_{p=0}^{\lfloor \frac{n-1}{2} \rfloor} \frac{\overrightarrow{\partial}}{\partial \boldsymbol{A}_p{}^{a_p}} \frac{\overrightarrow{\partial}}{\partial \boldsymbol{B}_{n-p-1,a_p}} F \tag{6.1}$$

The BV Laplacian satisfies the following identity:

$$\boldsymbol{\Delta}(F \cdot G) = (\boldsymbol{\Delta} F)G + (-1)^{(n+1)|F|}(F, G) + (-1)^{|F|} F \boldsymbol{\Delta} G, \tag{6.2}$$

In order for the generating functional to be gauge invariant in the quantum sense, the following quantum master equation is required:

$$(S_q, S_q) - 2i\hbar \boldsymbol{\Delta} S_q = 0, \tag{6.3}$$

for the quantum action S_q. In our n-algebroid topological sigma model, two terms are independently satisfied, i.e. $\boldsymbol{\Delta} S_q = 0$ and $(S_q, S_q) = 0$.

$\mathcal{O}$ is called an observable if an operator $\mathcal{O}$ satisfies the following equation:

$$(S_q, \mathcal{O}) - i\hbar \boldsymbol{\Delta} \mathcal{O} = 0. \tag{6.4}$$

Generally, there are two kinds of observables. One is the integration of a local function F on the boundary ∂X. Let $X_r \subset \partial X$ be a r-cycle on the boundary ∂X. If F has the form degree r, the integration of F on the r cycle X_r:

$$\mathcal{O} = \int_{X_r} F(\boldsymbol{\Phi}) \tag{6.5}$$

is nontrivial and satisfies (6.4).

Another observable is constructed from $\boldsymbol{A}_0{}^a$. We consider a function F of $\boldsymbol{A}_0{}^a$ and restrict F on the boundary, $\mathcal{O}_F \equiv F(\boldsymbol{A}_0{}^a)|_{\partial X}$. We can confirm that the form degree zero part $\mathcal{O}_F^{(0)}$ of $\mathcal{O}_F$ is an local observable with ghost number zero on the boundary.

The generating functional is defined by the path integral as

$$Z[\mathcal{O}_k] = \int \prod_{p=0}^{[\frac{n-1}{2}]} \mathcal{D}\boldsymbol{A}_p \mathcal{D}\boldsymbol{B}_{n-p-1} \; e^{\frac{i}{\hbar}(S_q + \sum_r J_r \mathcal{O}_r)}, \tag{6.6}$$

where $\mathcal{D}\boldsymbol{A}_p \mathcal{D}\boldsymbol{B}_{n-p-1}$ is a path integral measure and J_k are source fields and $\mathcal{O}_k$ are observables and $\hbar = g$.

We consider $n = 2$ case. Let X be a two-dimensional disc. Note that classical deformation derives a Poisson structure on T^*M in $n = 2$ case. On the other hand, quantum deformation derives the deformation quantization on a Poisson manifold M. [**Kon**] The correlation function of two local observables $\mathcal{O}_f^{(0)}$ and $\mathcal{O}_g^{(0)}$ derives the Kontsevich's star product formula [**CF1**] on a Poisson manifold:

$$f * g(x) = \int_{\phi(\infty)=x} \mathcal{D}\phi \mathcal{D}\boldsymbol{B}_1 \ \mathcal{O}_f^{(0)}(\phi(1))\mathcal{O}_g^{(0)}(\phi(0)) e^{\frac{i}{\hbar}S_q}, \tag{6.7}$$

where $\phi = \boldsymbol{A}_0$ and $0, 1, \infty$ are three distinct points at the boundary ∂X.

If we calculate the same correlation function for S_0, we obtain the usual product of functions f and g:

$$f(x)g(x) = \int_{\phi(\infty)=x} \mathcal{D}\phi \mathcal{D}\boldsymbol{B}_1 \ \mathcal{O}_f^{(0)}(\phi(1))\mathcal{O}_g^{(0)}(\phi(0)) e^{\frac{i}{\hbar}S_{0q}}, \tag{6.8}$$

where $S_{0q} = S_0 + S_{GF}$. Therefore quantum deformation in $n = 2$ is equivalent to the star deformation on $C^\infty(M)$.

We can generalize this discussion to higher orders. Deformation $S_0 \to S$ derives a generalization of the star deformation to higher dimensions as follows:

$$m_k[\mathcal{O}_1, \mathcal{O}_2, \cdots, \mathcal{O}_k] = \int \prod_{p=0}^{[\frac{n-1}{2}]} \mathcal{D}\boldsymbol{A}_p \mathcal{D}\boldsymbol{B}_{n-p-1} \ \mathcal{O}_1\mathcal{O}_2 \cdots \mathcal{O}_k e^{\frac{i}{\hbar}S_q}, \tag{6.9}$$

under the appropriate regularization and the boundary conditions, where S is the deformation (4.1) of the abelian topological sigma model and $\mathcal{O}_r$'s are two kinds of observables at the boundary. The correlation functions satisfy the Ward-Takahashi identity derived from the gauge symmetry:

$$\int \prod_{p=0}^{[\frac{n-1}{2}]} \mathcal{D}\boldsymbol{A}_p \mathcal{D}\boldsymbol{B}_{n-p-1} \ \boldsymbol{\Delta}\left(\mathcal{O} e^{\frac{i}{\hbar}S_q}\right) = 0, \tag{6.10}$$

which leads a quantum geometric structure on the space of correlation functions.

7. Summary and Outlook

We have discussed deformation of Batalin-Vilkovisky structures of topological sigma models in n dimensions. We have constructed general theory of most general deformation in general n dimensions. We have analyzed structures in the case of $n = 2$ and 3 in detail. In $n = 2$ deformation of a BV structure produces a Lie algebroid structure, and in $n = 3$, deformation produces a Courant algebroid structure. For $n \geq 4$, characterization of n-algebroids obtained by deformation of topological sigma models is still unknown and an open problem.

We have also discussed quantum version of deformation. For $n = 2$ case, the deformation on the disc X is equivalent to the deformation quantization on a Poisson manifold M.

In $n = 2$, there are two special important cases of deformations. They are A-model and B-model. [**W2**] There are many investigations to analyze quantum moduli. For reviews, see [**BCOV**], [**HKKPTVVZ**] and references therein.

$n = 3$ quantum deformation is analyzed in [**HM**]. For general $n \geq 4$, quantum structures are unknown. If we analyze higher n cases, we will obtain interesting mathematical and physical structures.

In this article, we assume the p and $n - p - 1$ are nonnegative integers in $E_p[p] \oplus E_p^*[n - p - 1]$ We will be able to generalize our discussions to negative integers p and $n - p - 1$. A special case has been analyzed in [**BM**][**II2**].

We have to analyze all moduli, i.e. we should consider the Kodaira-Spencer theory of Batalin-Vilkovisky structures.

References

[AKSZ] M. Alexandrov, M. Kontsevich, A. Schwartz and O. Zaboronsky: The Geometry of the master equation and topological quantum field theory, Int. J. Mod. Phys., A **12**, 1405(1997), hep-th/9502010.

[BBH] G. Barnich, F. Brandt and M. Henneaux: Local BRST Cohomology In The Antifield Formalism. 1. General Theorems, Commun. Math. Phys. **174**, 57 (1995), hep-th/9405109; For a review, M. Henneaux: hep-th/9712226.

[BBRT] D. Birmingham, M. Blau, M. Rakowski and G. Thompson: Topological field theory. Phys. Rept., **209**, 129 (1991).

[BCOV] M. Bershadsky, S. Cecotti, H. Ooguri and C. Vafa: Kodaira-Spencer theory of gravity and exact results for quantum string amplitudes, Commun. Math. Phys. **165**, 311 (1994), hep-th/9309140.

[BG] G. Barnich and M. Grigoriev, BRST extension of the non-linear unfolded formalism, hep-th/0504119.

[BH] G. Barnich and M. Henneaux: Consistent Couplings Between Fields With A Gauge Freedom And Deformations Of The Master Equation, Phys. Lett. B **311**, 123 (1993), hep-th/9304057.

[BM] I. Batalin and R. Marnelius: Superfield algorithms for topological field theories, hep-th/0110140.

[BV] I. A. Batalin and G. A. Vilkovisky: Gauge Algebra And Quantization, Phys. Lett. B **102**, 27 (1981); Quantization Of Gauge Theories With Linearly Dependent Generators, Phys. Rev. D **28**, 2567 (1983) , [Erratum-ibid. D **30**, 508 (1984)].

[C] T. Courant, Dirac manifolds: Trans. A. M. S. **319**, 631 (1990).

[CF1] A. S. Cattaneo and G. Felder: A path integral approach to the Kontsevich quantization formula, Commun. Math. Phys. **212**, 591 (2000), math.qa/9902090.

[CF2] A. S. Cattaneo and G. Felder: On the AKSZ formulation of the Poisson sigma model, Lett. Math. Phys. **56**, 163 (2001), arXiv:math/0102108.

[GD] M. A. Grigoriev and P. H. Damgaard, Superfield BRST charge and the master action, Phys. Lett. B **474**, 323 (2000), hep-th/9911092.

[GPS] For a review, J. Gomis, J. Paris and S. Samuel, Antibracket, antifields and gauge theory quantization, Phys. Rept. **259** (1995) 1, hep-th/9412228.

[HM] C. M. Hofman and W. K. Ma: Deformations of closed strings and topological open membranes, JHEP **0106**, 033 (2001), hep-th/0102201.

[HP] C. Hofman and J. S. Park, Topological open membranes, hep-th/0209148.

[HT92] M. Henneaux, C. Teitelboim: Quantization of Gauge Systems, Princeton University (1992).

[HKKPTVVZ] K. Hori, S. Katz, A. Klemm, R. Pandharipande, R. Thomas, C. Vafa, R. Vakil, and E. Zaslow: Mirror symmetry, volume 1 of Clay Mathematics Monographs, American Mathematical Society, Providence, RI, 2003.

[II] N. Ikeda and K. -I. Izawa: Gauge theory based on quadratic Lie algebras and 2-d gravity with dynamical torsion, Prog. Theor. Phys. **89**, 1077(1993); General form of dilaton gravity and nonlinear gauge theory, Prog. Theor. Phys. **90**, 237 (1993) ;
For a review, N. Ikeda: Two-dimensional gravity and nonlinear gauge theory, Ann. Phys., **235**, 435 (1994), hep-th/9312059.

[II2] N. Ikeda and K. I. Izawa: Dimensional reduction of nonlinear gauge theories, JHEP **0409**, 030 (2004), hep-th/0407243.

[I1] N. Ikeda: A deformation of three dimensional BF theory. JHEP **0011**, 009(2000), hep-th/0010096.

[I2] N. Ikeda: Deformation of BF theories, topological open membrane and a generalization of the star deformation, JHEP **0107**, 037(2001) , hep-th/0105286.

[I3] N. Ikeda: Chern-Simons gauge theory coupled with BF theory, Int. J. Mod. Phys. A **18**, 2689 (2003), hep-th/0203043.

[I4] N. Ikeda: Topological field theories and geometry of Batalin-Vilkovisky algebras, JHEP **0210**, 076 (2002), hep-th/0209042.

[I5] N. Ikeda, Three dimensional topological field theory induced from generalized complex structure, hep-th/0412140.

[I6] N. Ikeda, Deformation of Batalin-Vilkovisky Structures, math.sg/0604157.

[IT] N. Ikeda and T. Tokunaga, Topological membranes with 3-form H flux on generalized geometries, hep-th/0609098.

[Kon] M. Kontsevich: Deformation quantization of Poisson manifolds, Lett. Math. Phys., **66**, 157 (2003), q-alg/9709040.

[Kos] Y. Kosmann-Schwarzbach: Derived brackets, Lett. Math. Phys. **69**, 61 (2004), math.dg/0312524.

[KS] C. Klimcik and T. Strobl, WZW-Poisson manifolds, J. Geom. Phys. **43**, 341 (2002), math.sg/0104189.

[KSS] A. Kotov, P. Schaller and T. Strobl, Dirac sigma models, Commun. Math. Phys. **260**, 455 (2005), hep-th/0411112.

[L] J. -L. Loday, Une version non commutative des algebres de Lie: les algebres de Leibniz, Enseign. Math. **39**, 269 (1993).

[LO] A. M. Levin and M. A. Olshanetsky: Hamiltonian algebroid symmetries in W-gravity and Poisson sigma-model, hep-th/0010043.

[LWX] Z. J. Liu, A. Weinstein and P. Xu: Manin Triples for Lie Bialgebroids, J. Differential Geom., **45**, 547-574, (1997), dg-ga/9611001

[O] M. A. Olshanetsky: Lie algebroids as gauge symmetries in topological field theories, hep-th/0201164.

[Pa] J. Park: Topological open p-branes, Symplectic geometry and mirror symmetry, 311-384, Seoul (2000), hep-th/0012141.

[Pe] V. Pestun, Topological strings in generalized complex space, hep-th/0603145.

[R1] D. Roytenberg: Courant algebroids, derived brackets and even symplectic supermanifolds, math.qa/0112152; On the structure of graded symplectic supermanifolds and Courant algebroids, math.sg/0203110.

[R2] D. Roytenberg, AKSZ-BV formalism and Courant algebroid-induced topological field theories, Lett. Math. Phys. **79**, 143 (2007), arXiv:hep-th/0608150.

[Sc] A. Schwarz: Geometry of Batalin-Vilkovisky quantization, Commun. Math. Phys. **155**, 249 (1993), hep-th/9205088

[SS] P. Schaller and T. Strobl: Poisson structure induced (topological) field theories, Mod. Phys. Lett., **A9**, 3129 (1994), hep-th/9405110.

[Se] P. Severa, Some title containing the words "homotopy" and "symplectic", e.g. this one, math.SG/0105080.

[W1] E. Witten. Topological sigma models. Commun. Math. Phys., **118**, 411, (1988).

[W2] E. Witten, Mirror manifolds and topological field theory, hep-th/9112056.

[Z] R. Zucchini, A sigma model field theoretic realization of Hitchin's generalized complex geometry, JHEP **0411**, 045 (2004), hep-th/0409181; Generalized complex geometry, generalized branes and the Hitchin sigma model, JHEP **0503**, 022 (2005), hep-th/0501062; A topological sigma model of biKaehler geometry, JHEP **0601**, 041 (2006), hep-th/0511144; The biHermitian topological sigma model, hep-th/0608145.

Department of Mathematical Sciences, Ritsumeikan University, Kusatsu, Shiga 525-8577, Japan

E-mail address: nikeda@se.ritsumei.ac.jp

Contemporary Mathematics
Volume **450**, 2008

Deformation quantization of a Kähler-Poisson structure vanishing on a Levi nondegenerate hypersurface

Alexander V. Karabegov

Abstract. We give an elementary proof of the result by Leichtnam, Tang, and Weinstein [**LTW**] that there exists a deformation quantization with separation of variables on a complex manifold endowed with a Kähler-Poisson structure vanishing on a Levi nondegenerate hypersurface and nondegenerate on its complement.

1. Introduction

Let V be a vector space and ν a formal parameter. Denote by $V[\nu^{-1},\nu]]$ the space of formal Laurent series of the form

$$v = \sum_{r \geq s} \nu^r v_r,$$

where $s \in \mathbb{Z}$ and $v_r \in V$. The elements of $V[\nu^{-1},\nu]]$ are called formal vectors. Let M be a Poisson manifold with the Poisson structure given by a Poisson bivector field η or, equivalently, by a Poisson bracket $\{\cdot,\cdot\}$. A nondegenerate Poisson structure on M is equivalent to a symplectic structure given by a symplectic form on M. Deformation quantization on (M,η) is an associative algebra structure on $C^\infty(M)[\nu^{-1},\nu]]$ with the product $\star$ (named a star product) given by the formula

$$f \star g = \sum_{r \geq 0} \nu^r C_r(f,g),$$

where C_r are bidifferential operators on M such that $C_0(f,g) = fg$ and $C_1(f,g) - C_1(g,f) = i\{f,g\}$. It is assumed that the unit constant 1 is the unity in the algebra $(C^\infty(M)[\nu^{-1},\nu]],\star)$. A deformation quantization on M can be localized to the open subsets of M. An important feature of the deformation quantizations on symplectic manifolds is the existence of local ν-derivations of the form $\delta = \nu d/d\nu + A$, where A is a formal differential operator (i.e., A does not contain derivatives with respect to ν).

Deformation quantizations were introduced in [**BFFLS**]. The existence of deformation quantizations on the symplectic manifolds was shown by [**DL**], [**F**], and

1991 *Mathematics Subject Classification.* Primary: 53D55.
Key words and phrases. deformation quantization.

[**OMY**]. In the general Poisson case the existence of deformation quantizations was shown by Kontsevich in [**K**].

We call a complex manifold M endowed with a Poisson bivector field of type (1,1) with respect to the complex structure *a Kähler-Poisson manifold.* A nondegenerate Kähler-Poisson structure is equivalent to a pseudo-Kähler structure given by a pseudo-Kähler form. A deformation quantization on a Kähler-Poisson manifold is called deformation quantization with separation of variables (or of the Wick type) if the operators C_r in the definition of the corresponding star product differentiate their first argument in antiholomorphic directions and the second argument in holomorphic ones (or vice versa). Deformation quantizations with separation of variables on an arbitrary pseudo-Kähler manifold were constructed in [**Kar1**], [**BW**], and [**RT**]. It is not yet known whether there exists a deformation quantization with separation of variables on an arbitrary Kähler-Poisson manifold. In [**Kar3**] it was shown that such quantizations cannot be obtained at least by a naive extension of the formula for the star product with separation of variables on a pseudo-Kähler manifold. However, it turns out that there exist deformation quantizations with separation of variables of a Kähler-Poisson structure vanishing on a Levi nondegenerate hypersurface and nondegenerate on its complement. Leichtnam, Tang, and Weinstein proved it in [**LTW**] using para-Kähler Lie algebroids. Their work was motivated by the construction of Berezin-Toeplitz deformation quantization on a complex manifold with strongly pseudoconvex boundary by Engliš in [**E**]. The goal of our paper is to give an elementary proof of this fact.

Given a star product $\star$, we denote by L_f and R_f the corresponding operators of left and right multiplication by an element f, respectively. The standard deformation quantization with separation of variables on a pseudo-Kähler manifold (M, ω) has the following local properties (see [**Kar1**]). On an arbitrary contractible coordinate chart U with holomorphic coordinates $\{z^k\}$

$$(1.1) \quad L_a = a,\ L_{\frac{1}{\nu}\frac{\partial \Phi}{\partial z^k}} = \frac{1}{\nu}\frac{\partial \Phi}{\partial z^k} + \frac{\partial}{\partial z^k},\ R_b = b,\ R_{\frac{1}{\nu}\frac{\partial \Phi}{\partial \bar{z}^l}} = \frac{1}{\nu}\frac{\partial \Phi}{\partial \bar{z}^l} + \frac{\partial}{\partial \bar{z}^l},$$

where a and b are holomorphic and antiholomorphic functions on U, respectively, and Φ is a potential of the pseudo-Kähler form ω on U. These properties determine the standard deformation quantization with separation of variables uniquely and globally on M. It was shown in [**Kar2**] that a local ν-derivation $\delta = \nu d/d\nu + A$ of the standard deformation quantization with separation of variables can be determined from the formula

$$\left[e^{-\frac{1}{\nu}\Phi} \left(\nu \frac{\partial}{\partial \nu} \right) e^{\frac{1}{\nu}\Phi}, L_f \right] = L_{\delta(f)},$$

where Φ is a local potential of the pseudo-Kähler form ω.

Acknowledgments. The author is very grateful to Professor Alan Weinstein for the inspiring discussion which stimulated this paper.

2. A Kähler-Poisson structure vanishing on a Levi nondegenerate hypersurface

We begin with several calculations and reobtain some statements from [**LTW**] in the form convenient for our exposition. Assume that ψ is a real function on a neighborhood $U \subset \mathbb{C}^n$. Denote $\mathbb{C}^\times = \mathbb{C} \backslash \{0\}$ and consider the product $\tilde{U} = U \times \mathbb{C}^\times$. The holomorphic coordinates on U and $\mathbb{C}^\times$ will be denoted by $\{z^k\}$ and

u, respectively. Introduce a function $\rho = \psi(z,\bar{z})u\bar{u}$ and a Hermitian matrix

$$\Gamma = \begin{pmatrix} \frac{\partial^2\psi}{\partial z^k \partial \bar{z}^l} u\bar{u} & \frac{\partial\psi}{\partial z^k} u \\ \frac{\partial\psi}{\partial \bar{z}^l}\bar{u} & \psi \end{pmatrix}$$

on $\tilde{U}$. We will describe the conditions on the function ψ under which the matrix Γ is nondegenerate. If this is the case, let

$$\Pi = \begin{pmatrix} A^{\bar{l}m} & B^{\bar{l}} \\ C^m & D \end{pmatrix}$$

be the inverse matrix of Γ, which is equivalent to the following conditions:

$$\frac{\partial^2\psi}{\partial z^k \partial \bar{z}^l} u\bar{u} A^{\bar{l}m} + \frac{\partial\psi}{\partial z^k} u C^m = \delta_k^m, \quad \frac{\partial\psi}{\partial \bar{z}^l}\bar{u} A^{\bar{l}m} + \psi C^m = 0, \tag{2.1}$$

$$\frac{\partial^2\psi}{\partial z^k \partial \bar{z}^l} u\bar{u} B^{\bar{l}} + \frac{\partial\psi}{\partial z^k} u D = 0, \quad \frac{\partial\psi}{\partial \bar{z}^l}\bar{u} B^{\bar{l}} + \psi D = 1. \tag{2.2}$$

The matrix Γ can be invertible at a fixed point $\tilde{x} = (x,y) \in \tilde{U}$ (with $x \in U$ and $y \in \mathbb{C}^\times$) only in one of the following two cases:

1. $\psi(x) \neq 0$;
2. $\psi(x) = 0$, $\partial\psi(x) \neq 0$, and $\bar{\partial}\psi(x) \neq 0$.

Since the function ψ is real, the conditions $\partial\psi(x) \neq 0$ and $\bar{\partial}\psi(x) \neq 0$ are equivalent to the condition that x is not a critical point of ψ. We will analyse Case 1 assuming that the matrices Γ and Π are inverse to each other. Set

$$g_{k\bar{l}} = \frac{\partial^2}{\partial z^k \partial \bar{z}^l} \log|\psi| = \frac{1}{\psi}\left(\frac{\partial^2\psi}{\partial z^k \partial \bar{z}^l} - \frac{1}{\psi}\frac{\partial\psi}{\partial z^k}\frac{\partial\psi}{\partial \bar{z}^l}\right). \tag{2.3}$$

Solving the second equation in (2.1) for C^m and substituting the resulting expression to the first one, we get

$$\left(\frac{\partial^2\psi}{\partial z^k \partial \bar{z}^l} - \frac{1}{\psi}\frac{\partial\psi}{\partial z^k}\frac{\partial\psi}{\partial \bar{z}^l}\right) u\bar{u} A^{\bar{l}m} = \delta_k^m,$$

which means, according to Eqn. (2.3), that the matrix $(g_{k\bar{l}})$ is invertible and its inverse $(g^{\bar{l}k})$ is

$$g^{\bar{l}k} = \psi u\bar{u} A^{\bar{l}k} = \rho A^{\bar{l}k}. \tag{2.4}$$

Notice that $g^{\bar{l}k}$ does not depend on the variables $u, \bar{u}$ and $A^{\bar{l}k}$ is smooth on $\tilde{U}$. It easily follows from the calculations above and similar calculations applied to Eqn. (2.2) that in Case 1 the matrix Γ is invertible if and only if the matrix $(g_{k\bar{l}})$ is invertible.

Now consider Case 2. Let $x \in U$ be a point where $\psi(x) = 0$ and which is not critical for ψ. Introduce the following $(n-1)$-dimensional subspaces of $\mathbb{C}^n$:

$$V = \left\{ v = (v^1, \ldots, v^n) | \frac{\partial\psi}{\partial z^k}(x) v^k = 0 \right\},$$

$$W = \left\{ w = (w^1, \ldots, w^n) | \frac{\partial\psi}{\partial \bar{z}^l}(x) w^{\bar{l}} = 0 \right\}.$$

The Levi form Q is the bilinear form on $V \times W$ such that

$$Q(v,w) = \frac{\partial^2\psi}{\partial z^k \partial \bar{z}^l}(x) v^k w^{\bar{l}}. \tag{2.5}$$

To determine whether the matrix Γ is invertible, consider its kernel. Assume that a nonzero vector $(w^1, \dots w^n, b) \in \mathbb{C}^{n+1}$ is in the kernel of the matrix Γ, i.e.,

$$\frac{\partial^2 \psi}{\partial z^k \partial \bar{z}^l} u\bar{u}w^{\bar{l}} + \frac{\partial \psi}{\partial z^k} ub = 0 \text{ and } \frac{\partial \psi}{\partial \bar{z}^l} \bar{u}w^{\bar{l}} = 0. \tag{2.6}$$

It follows from the fisrt equation in (2.6) and the assumption that x is not critical for ψ that the vector $w = (w^1, \dots, w^n)$ is nonzero and $Q(v, w) = 0$ for any vector $v \in V$. The second equation in (2.6) means that $w = (w^1, \dots, w^n) \in W$ and therefore the Levi form Q is degenerate. Conversly, if Q is degenerate, there exists a nonzero vector $w = (w^1, \dots, w^n) \in W$ such that $Q(v, w) = 0$ for any vector $v = (v^1, \dots, v^n) \in V$. This means that the vector $a = (a_1, \dots, a_n)$ such that

$$a_k = \frac{\partial^2 \psi}{\partial z^k \partial \bar{z}^l}(x) w^{\bar{l}}$$

is proportional to the vector

$$\left(\frac{\partial \psi}{\partial z^1}(x), \dots, \frac{\partial \psi}{\partial z^n}(x) \right),$$

which implies that the first equation in (2.6) holds for some constant b. Thus in Case 2 the matrix Γ is invertible if and only if the Levi form Q is nondegenerate. Summarizing, we arrive at the following. Assume that $S \subset U$ is a Levi nondegenerate hypersurface given by a defining function ψ. It means that S is the zero set of the function ψ which has no critical points on S and the Levi form (2.5) is nondegenerate on S. The property that S is Levi nondegenerate does not depend on the choice of the defining function. For any point $(x, y) \in \tilde{U}$ such that $x \in S$ the matrix Γ is nondegenerate. Shrinking, if necessary, the neighborhood U around S we will assume from now on that Γ is nondegenerate everywhere on $\tilde{U}$. In this case the function ρ is a potential of the pseudo-Kähler form

$$\Omega = -i\partial_{\tilde{U}} \bar{\partial}_{\tilde{U}} \rho$$

on $\tilde{U}$. Further, the function $\log|\psi|$ is a potential of the pseudo-Kähler form

$$\omega = -i\partial\bar{\partial} \log|\psi|$$

and

$$\eta = ig^{\bar{l}k} \frac{\partial}{\partial z^k} \wedge \frac{\partial}{\partial \bar{z}^l}$$

is the corresponding Poisson bivector field on $U \backslash S$ of type (1,1) with respect to the complex structure. Since S is nowhere dense in U, it follows from Eqn. (2.4) that the Poisson bivector field η has a smooth extension to the whole neighborhood U which vanishes on S. Thus it determines a Kähler-Poisson structure on U which is nondegenerate on the complement of S.

3. Deformation quantization of the Kähler-Poisson structure on U

Denote by $*$ the star product of the standard deformation quantization with separation of variables on $(\tilde{U}, \Omega)$ such that

$$F * G = \sum_{r \geq 0} h^r C_r(F, G).$$

Here h is a formal parameter and $F, G \in C^\infty(\tilde{U})[h^{-1}, h]]$. We will denote by $\tilde{L}_F$ and $\tilde{R}_G$ the left and right multiplication operators in the algebra $(C^\infty(\tilde{U})[h^{-1}, h]], *)$

by the elements F and G, respectively, so that $F * G = \tilde{L}_F G = \tilde{R}_G F$. Adapting Eqns. (1.1) to the star product $*$, we get

$$\tag{3.1} \tilde{L}_{a(z,u)} = a(z,u),\ \tilde{L}_{\frac{1}{h}\frac{\partial\rho}{\partial z^k}} = \frac{1}{h}\frac{\partial\rho}{\partial z^k} + \frac{\partial}{\partial z^k},\ \tilde{L}_{\frac{1}{h}\frac{\partial\rho}{\partial u}} = \frac{1}{h}\frac{\partial\rho}{\partial u} + \frac{\partial}{\partial u},$$
$$\tilde{R}_{b(\bar{z},\bar{u})} = b(\bar{z},\bar{u}),\ \tilde{R}_{\frac{1}{h}\frac{\partial\rho}{\partial\bar{z}^l}} = \frac{1}{h}\frac{\partial\rho}{\partial\bar{z}^l} + \frac{\partial}{\partial\bar{z}^l},\ \tilde{R}_{\frac{1}{h}\frac{\partial\rho}{\partial\bar{u}}} = \frac{1}{h}\frac{\partial\rho}{\partial\bar{u}} + \frac{\partial}{\partial\bar{u}},$$

where $a(z,u)$ and $b(\bar{z},\bar{u})$ are a holomorphic and an antiholomorphic functions on $\tilde{U}$. It follows from Eqns. (3.1) that

$$\tag{3.2} \tilde{L}_{\frac{1}{h}\rho} = \tilde{L}_{\frac{1}{h}u\frac{\partial\rho}{\partial u}} = \tilde{L}_u\tilde{L}_{\frac{1}{h}\frac{\partial\rho}{\partial u}} = \frac{1}{h}\rho + u\frac{\partial}{\partial u}$$

and, similarly,

$$\tag{3.3} \tilde{R}_{\frac{1}{h}\rho} = \frac{1}{h}\rho + \bar{u}\frac{\partial}{\partial\bar{u}}.$$

It follows from Eqns. (3.2) and (3.2) that

$$\tag{3.4} \tilde{L}_{\frac{1}{h}\rho} - \tilde{R}_{\frac{1}{h}\rho} = u\frac{\partial}{\partial u} - \bar{u}\frac{\partial}{\partial\bar{u}}.$$

is an inner derivation of the algebra $(C^\infty(\tilde{U})[h^{-1},h]],*)$. According to [**Kar2**], there exists an h-derivation $\delta = h\frac{d}{dh} + A$ of the star product $*$ (where A is a formal differential operator on $\tilde{U}$) such that for $F \in C^\infty(\tilde{U})[h^{-1},h]]$

$$\tag{3.5} \left[e^{-\frac{1}{h}\rho}\left(h\frac{d}{dh}\right)e^{\frac{1}{h}\rho}, \tilde{L}_F\right] = \left[h\frac{d}{dh} - \frac{1}{h}\rho, \tilde{L}_F\right] = \tilde{L}_{\delta(F)}.$$

Evaluating the operator in Eqn. (3.5) at the unit constant 1 and using Eqn. (3.3), we obtain

$$\delta(F) = \left(h\frac{\partial}{\partial h} - \frac{1}{h}\rho\right)F + F * \left(\frac{1}{h}\rho\right) = \left(h\frac{\partial}{\partial h} + \bar{u}\frac{\partial}{\partial\bar{u}}\right)F.$$

In the rest of the paper we identify the functions on U with their lifts to $\tilde{U}$ and thus we can treat $C^\infty(U)$ as a subspace of $C^\infty(\tilde{U})$. Denote by $\mathcal{F}(\tilde{U})$ the subspace of $C^\infty(\tilde{U})[h^{-1},h]]$ consisting of the elements annihilated by the derivations (3.4) (i.e., commuting with $(1/h)\rho$) and δ. These elements can be written as

$$F = \sum_{r\geq s}\left(\frac{h}{u\bar{u}}\right)^r f_r,$$

where $s \in \mathbb{Z}$ and $f_r \in C^\infty(U)$. The subspace $\mathcal{F}(\tilde{U})$ is a subalgebra of the algebra $(C^\infty(\tilde{U})[h^{-1},h]],*)$ which does not contain the formal parameter h as an element. Notice that

$$\frac{1}{h}\rho = \left(\frac{h}{u\bar{u}}\right)^{-1}\psi$$

is a central element in this subalgebra and $C^\infty(\tilde{U}) \subset \mathcal{F}(\tilde{U})$. Set $\tilde{S} = S \times \mathbb{C}^\times \subset \tilde{U}$. We can define an algebra $(\mathcal{F}(\tilde{U}\backslash\tilde{S}),*)$ by replacing $\tilde{U}$ with $\tilde{U}\backslash\tilde{S}$ in the construction of $\mathcal{F}(\tilde{U})$. Since ρ does not vanish on $\tilde{U}\backslash\tilde{S}$, the element $(1/h)\rho$ has an inverse in the algebra $(\mathcal{F}(\tilde{U}\backslash\tilde{S}),*)$ which is also central. Denote it by κ. Thus $\kappa = h/\rho \pmod{h^2}$ and

$$\tilde{L}_\kappa = \left(\tilde{L}_{\frac{1}{h}\rho}\right)^{-1}.$$

For any integer n, denote by κ^{*n} the nth power of the element κ with respect to the star product $*$. In particular, $\kappa^{*(-1)} = (1/h)\rho$. Since the operator $\tilde{L}_{(1/h)\rho}$ commutes with the pointwise multiplication by the functions from $C^\infty(U\backslash S)$, so does $\tilde{L}_\kappa$, and therefore for any $f \in C^\infty(U\backslash S)$ and $n \in \mathbb{Z}$

$$f * (\kappa^{*n}) = (\kappa^{*n}) * f = \left(\tilde{L}_\kappa\right)^n f = f \cdot \left(\tilde{L}_\kappa\right)^n 1 = (\kappa^{*n}) \cdot f.$$

Let ν be a different formal parameter. Define a mapping $\tau : C^\infty(U\backslash S)[\nu^{-1}, \nu]] \to \mathcal{F}(\tilde{U}\backslash\tilde{S})$ via an h-adically convergent series:

$$\tau\left(\sum_{r\geq s} \nu^r f_r(z,\bar{z})\right) = \sum_{r\geq s} \kappa^{*r} * f_r(z,\bar{z}) = \sum_{r\geq s} \kappa^{*r} \cdot f_r(z,\bar{z}).$$

Notice that

$$(3.6)\quad \tau(f) = f,\ \tau(F\cdot f) = \tau(F)\cdot f,\ \tau(\nu) = \kappa, \text{ and } \tau(\nu F) = \kappa * \tau(F) = \tau(F) * \kappa$$

for any $f \in C^\infty(U\backslash S)$ and $F \in \mathcal{F}(\tilde{U}\backslash\tilde{S})$. Using the fact that $\kappa^{*n} = (h/\psi)^n \pmod{h^{n+1}}$, it is easy to show that τ is a bijection of $C^\infty(U\backslash S)[\nu^{-1}, \nu]]$ onto $\mathcal{F}(\tilde{U}\backslash\tilde{S})$. Denote by $\star$ the pullback of the star-product $*$ to $C^\infty(U\backslash S)[\nu^{-1}, \nu]]$ via the mapping τ and by L_f and R_f the operators of left and right multiplication by $f \in C^\infty(U\backslash S)[\nu^{-1}, \nu]]$ with respect to the product $\star$. One can show with the use of Eqns. (3.6) that $\star$ is actually a star product. Assume that a and b are a holomorphic and an antiholomorphic functions on $U\backslash S$, respectively. Then, using Eqns. (3.6), we get for $f \in C^\infty(U\backslash S)[\nu^{-1}, \nu]]$ that

$$\tau(a \star f) = \tau(a) * \tau(f) = a \cdot \tau(f) = \tau(a \cdot f),$$

which means that $a \star f = af$. Similarly, one can show that $f \star b = f \cdot b$. Thus $\star$ is a star product of a deformation quantization with separation of variables on $U\backslash S$.

Consider the operator

$$A_k = \frac{1}{\nu}\frac{\partial}{\partial z^k}\log|\psi| + \frac{\partial}{\partial z^k} = \frac{1}{\nu\psi}\frac{\partial\psi}{\partial z^k} + \frac{\partial}{\partial z^k}$$

on $C^\infty(U\backslash S)[\nu^{-1}, \nu]]$. We will need the following technical statement.

Lemma 3.1. *Given a function $f \in C^\infty(U\backslash S)$, the formula*

$$\left(\frac{1}{h}\frac{\partial\rho}{\partial z^k}\right) * f = \tau(A_k f).$$

holds.

Proof. Using Eqns (3.1), we get

$$\left(\frac{1}{h}\frac{\partial\rho}{\partial z^k}\right) * f = \left(\frac{1}{h}\frac{\partial\rho}{\partial z^k}\right) f + \frac{\partial f}{\partial z^k} = \left(\frac{1}{h}\rho\frac{\partial}{\partial z^k}\log|\psi|\right) f + \frac{\partial f}{\partial z^k} =$$

$$\frac{1}{h}\rho * \left(\left(\frac{\partial}{\partial z^k}\log|\psi|\right) f\right) + \frac{\partial f}{\partial z^k} = \tau\left(\frac{1}{\nu}\left(\frac{\partial}{\partial z^k}\log|\psi|\right) f + \frac{\partial f}{\partial z^k}\right) = \tau(A_k f),$$

which concludes the proof.

Lemma 3.2. *The operator of left multiplication by $\frac{1}{\nu}\frac{\partial}{\partial z^k}\log|\psi|$ with respect to the product $\star$ coincides with the operator A_k,*

$$L_{\frac{1}{\nu}\frac{\partial}{\partial z^k}\log|\psi|} = A_k.$$

Proof. Given an element $f = \sum_{r \geq s} \nu^f f_r \in C^\infty(U \backslash S)[\nu^{-1}, \nu]]$, we get from Eqns. (3.1), Lemma 3.1, and the fact that κ is central, that

$$\tau\left(\frac{1}{\nu}\frac{\partial}{\partial z^k}\log|\psi| \star f\right) = \left(\frac{1}{h}\rho\frac{\partial}{\partial z^k}\log|\psi|\right) * \tau(f) = \left(\frac{1}{h}\frac{\partial \rho}{\partial z^k}\right) * \tau(f) =$$
$$\sum_{r \geq s} \kappa^{*r} * \left(\frac{1}{h}\frac{\partial \rho}{\partial z^k}\right) * f_r = \sum_{r \geq s} \kappa^{*r} * \tau(A_k f_r) = \sum_{r \geq s} \tau(\nu^r A_k f_r) = \tau(A_k f),$$

which proves the Lemma.

Since $\log|\psi|$ is a potential of the pseudo-Kähler form ω on $U \backslash S$, Lemma 3.2 immediately implies the following

THEOREM 3.3. *The product $\star$ is the star product of the standard deformation quantization with separation of variables on $(U \backslash S, \omega)$.*

We want to show that the star product $\star$ can be extended to the whole neighborhood U. This requires some preparations. For $r \in \mathbb{N}$ denote by $N_r(\nu)$ the following formal number:

$$N_r(\nu) = \prod_{s=1}^{r} \frac{\nu}{1 + \nu s} = \nu^r + \dots,$$

where the division by $1 + \nu s$ is in the field of formal numbers. Set $N_0(\nu) = 1$.

LEMMA 3.4. *Given a function $f \in C^\infty(U \backslash S)$, the following formula holds for any integer $r \geq 0$:*

$$\tau^{-1}\left(\left(\frac{h}{u\bar{u}}\right)^r f\right) = N_r(\nu)\psi^r f.$$

Proof. Using Eqns. (3.1), we have for any $s \in \mathbb{N}$:

$$\frac{1}{\nu}\tau^{-1}\left(\left(\frac{h}{u\bar{u}}\right)^s f\right) = \tau^{-1}\left(\frac{1}{h}\rho\right) \star \tau^{-1}\left(\left(\frac{h}{u\bar{u}}\right)^s f\right) =$$
$$\tau^{-1}\left(\frac{1}{h}\rho * \left(\frac{h}{u\bar{u}}\right)^s f\right) = \tau^{-1}\left(\frac{1}{h}\rho\left(\frac{h}{u\bar{u}}\right)^s f + u\frac{\partial}{\partial u}\left(\left(\frac{h}{u\bar{u}}\right)^s f\right)\right) =$$
$$\tau^{-1}\left(\left(\frac{h}{u\bar{u}}\right)^{s-1} \psi f\right) - s\tau^{-1}\left(\left(\frac{h}{u\bar{u}}\right)^s f\right),$$

whence

$$\frac{1 + \nu s}{\nu}\tau^{-1}\left(\left(\frac{h}{u\bar{u}}\right)^s f\right) = \tau^{-1}\left(\left(\frac{h}{u\bar{u}}\right)^{s-1} \psi f\right).$$

Now the statement of the Lemma follows by induction.

Denote by $\mathcal{F}_{\geq 0}(\tilde{U})$ the subspace of $\mathcal{F}(\tilde{U})$ consisting of the elements

$$f = \sum_{r \geq 0} \left(\frac{h}{u\bar{u}}\right)^r f_r,$$

where $f_r \in C^\infty(U)$. It is a subalgebra of $(\mathcal{F}(\tilde{U}), *)$. Define a mapping $\sigma : \mathcal{F}_{\geq 0}(\tilde{U}) \to C^\infty(U)[[\nu]]$ by the formula

$$\sigma\left(\sum_{r \geq 0} \left(\frac{h}{u\bar{u}}\right)^r f_r\right) = \sum_{r \geq 0} N_r(\nu)\psi^r f_r.$$

Denote $\mathcal{A}(U) = \sigma(\mathcal{F}_{\geq 0}(\tilde{U})) \subset C^\infty(U)[[\nu]]$. Since the mapping σ is injective and local, one can push forward the star product $*$ via σ to a product on $\mathcal{A}(U)$ localizable to the open subsets of U (where it will be defined on the restrictions of the elements from $\mathcal{A}(U)$). It follows from Lemma 3.4 that the restriction of that product to $U \backslash S$ agrees with the product $\star$. Thus it will be denoted by $\star$ as well. Notice that since $C^\infty(U) \subset \mathcal{F}_{\geq 0}(\tilde{U})$, the product $f * g$ of elements $f, g \in C^\infty(U)$ belongs to $\mathcal{F}_{\geq 0}(\tilde{U})$ and thus the term $h^r C_r(f,g)$ of the formal series representing $f * g$ can be rewritten as

$$h^r C_r(f,g) = \left(\frac{h}{u\bar{u}}\right)^r D_r(f,g)$$

for some bidifferential operator D_r on U. Since $C^\infty(U) \subset \mathcal{A}(U)$ and $\sigma(f) = f$ for any $f \in C^\infty(U)$, we see that for $f, g \in C^\infty(U)$

$$f \star g = \sigma(f * g) = \sigma\left(\sum_{r \geq 0} h^r C_r(f,g)\right) =$$

$$\sigma\left(\sum_{r \geq 0} \left(\frac{h}{u\bar{u}}\right)^r D_r(f,g)\right) = \sum_{r \geq 0} N_r(\nu) \psi^r D_r(f,g).$$

This formula gives a smooth extension of the star product $\star$ from $U \backslash S$ to U and shows that $f \star g - fg = 0$ on S. The results of this paper can be globalized as follows. Assume that (M, η) is a Kähler-Poisson manifold such that

- the Kähler-Poisson bivector field η is nondegenerate on the complement of a Levi-nondegenerate hypersurface $S \subset M$, i.e., η determines a pseudo-Kähler form ω on $M \backslash S$; and
- for any point $x \in S$ there exists a local defining function ψ of the hypersurface S in a neighborhood of x such that $\log|\psi|$ is a potential of the form ω on the complement of S.

The results obtained in this paper imply the following theorem

THEOREM 3.5. *The standard deformation quantization with separation of variables on the pseudo-Kähler manifold $(M \backslash S, \omega)$ extends to a deformation quantization with separation of variables on the Kähler-Poisson manifold (M, η).*

EXAMPLE 3.6. Consider a defining function

$$\psi = \sum_{k=1}^{n} |z^k|^2 - 1$$

of the unit sphere $S \subset \mathbb{C}^n$. The unit sphere S is a Levi nondegenerate hypersurface in $\mathbb{C}^n$. On the complement of S the potential $\log|\psi|$ determines the pseudo-Kähler metric

$$g_{kl} = \frac{1}{\psi}\left(\delta_{kl} - \frac{1}{\psi}\bar{z}^k z^l\right).$$

Its inverse

$$g^{lk} = \psi\left(\delta^{kl} - \bar{z}^l z^k\right)$$

is a Kähler-Poisson tensor which gives a global Kähler-Poisson structure on $\mathbb{C}^n$. It extends via the inclusion $\mathbb{C}^n \subset \mathbb{C}P^n$ given by the formula

$$(z^1, \ldots, z^n) \mapsto (z^1 : \ldots : z^n : 1)$$

to a Kähler-Poisson structure on the complex projective space $\mathbb{C}P^n$ invariant with respect to the projective action of the group $SU(n,1)$. There exists a global $SU(n,1)$-invariant star product on $\mathbb{C}P^n$ which coincides with the star product of the standard deformation quantization with separation of variables on the complement of the hypersurface $S \subset \mathbb{C}^n \subset \mathbb{C}P^n$. This invariant star product can be constructed also by the methods developed in [**AL**].

References

[AL] A. Alekseev, A. Lachowska, *Invariant ∗-products on coadjoint orbits and the Shapovalov pairing*, Comment. Math. Helv. **80** (2005), 795-810.

[BFFLS] F. Bayen, M. Flato, C. Fronsdal, A. Lichnerowicz, and D. Sternheimer, *Deformation theory and quantization. I. Deformations of symplectic structures.* Ann. Physics **111** (1978), no. 1, 61 - 110.

[BW] M. Bordemann, S. Waldmann, *A Fedosov star product of the Wick type for Kähler manifolds,* Lett. Math. Phys. **41** (3) (1997), 243-253.

[DL] M. De Wilde, P. B. A. Lecomte, *Existence of star-products and of formal deformations of the Poisson Lie algebra of arbitrary symplectic manifolds,* Lett. Math. Phys. **7** (1983), no. 6, 487-496.

[E] M. Engliš, *Weighted Bergman kernels and quantization,* Commun. Math. Phys. **227** (2002), 211-241.

[F] B. V. Fedosov, *A simple geometrical construction of deformation quantization,* J. Differential Geom. **40** (1994), no. 2, 213-238.

[Kar1] A. Karabegov, *Deformation quantizations with separation of variables on a Kähler manifold,* Commun. Math. Phys. **180** (1996), 745-755.

[Kar2] A. Karabegov, *Cohomological classification of deformation quantizations with separation of variables*, Lett. Math. Phys. **43** (1998), 347-357.

[Kar3] A. Karabegov, *A covariant Poisson deformation quantization with separation of variables of the third order*, Lett. Math. Phys. **61** (2002), 255-261.

[K] M. Kontsevich, *Deformation quantization of Poisson manifolds, I,* Lett. Math. Phys. **66** (2003), 157 – 216.

[LTW] E. Leichtnam, X. Tang, and A. Weinstein, *Poisson geometry and deformation quantization near a strictly pseudoconvex boundary*, preprint math.SG/0603350, 26 pp.

[OMY] H. Omori, Y. Maeda, and A. Yoshioka, *Weyl manifolds and deformation quantization,* Adv. Math. **85** (1991), 224–255.

[RT] N. Reshetikhin, L. Takhtajan, *Deformation quantization of Kähler manifolds,* L. D. Faddeev's Seminar on Mathematical Physics, Amer. Math. Soc. Transl. Ser. 2, **201**, Amer. Math. Soc., Providence, RI, (2000), 257-276.

Department of Mathematics, Abilene Christian University, ACU Box 28012, 215 Foster Science Building, Abilene, TX 79699-8012

E-mail address: alexander.karabegov@math.acu.edu

Contemporary Mathematics
Volume **450**, 2008

A note on Poisson homogeneous spaces

Jiang-Hua Lu

Abstract. We identify the cotangent bundle Lie algebroid of a Poisson homogeneous space G/H of a Poisson Lie group G as a quotient of a transformation Lie algebroid over G. As applications, we describe the modular vector fields of G/H, and we identify the Poisson cohomology of G/H with coefficients in powers of its canonical line bundle with relative Lie algebra cohomology of the Drinfeld Lie algebra associated to G/H. We also construct a Poisson groupoid over $(G/H, \pi)$ which is symplectic near the identity section. This note serves as preparation for forthcoming papers, in which we will compute explicitly the Poisson cohomology and study their symplectic groupoids for certain examples of Poisson homogeneous spaces related to semi-simple Lie groups.

1. Introduction

The cotangent bundle of a Poisson manifold (P, π) is naturally a Lie algebroid [**21**] called the cotangent bundle Lie algebroid of (P, π) and denoted by $T^*(P, \pi)$. Let $K_P = \wedge^{\mathrm{top}} T^*P$ be the canonical line bundle over P. Then the Lie algebroid $T^*(P, \pi)$ has a natural representation on K_P. The Poisson cohomology of (P, π) as defined in [**13**], the Poisson homology of (P, π) as defined in [**3**], and the twisted Poisson cohomology of (P, π) as defined in [**7**], can be regarded as the Lie algebroid cohomology of $T^*(P, \pi)$ with coefficients in, respectively, the trivial line bundle, K_P and K_P^2 (see [**7, 21, 26**]). In general, one can consider the Lie algebroid cohomology of $T^*(P, \pi)$ with coefficients in K_P^N for any integer N, which we will denote by $H^\bullet(P, \pi; K_P^N)$ and refer to as *generalized Poisson cohomology* of (P, π). A symplectic groupoid of (P, π) is a Lie groupoid over P with Lie algebroid $T^*(P, \pi)$ and a compatible symplectic structure [**24**].

This note concerns the cotangent bundle Lie algebroids of Poisson homogeneous spaces of a Poisson Lie group (G, π_G). More precisely, by a theorem of Drinfeld [**6**], each Poisson homogeneous space $(G/H, \pi)$ of (G, π_G) corresponds to a Lie subalgebra $\mathfrak{l}$ of the double Lie algebra $\mathfrak{d}$ of (G, π_G). In this note, we identify the cotangent bundle Lie algebroid of $(G/H, \pi)$ with a *quotient* of the transformation Lie algebroid $G \rtimes_\lambda \mathfrak{l}$ over G associated to an infinitesimal action λ of $\mathfrak{l}$ on G. We also identify the representation of $T^*(G/H, \pi)$ on $K_{G/H}$ with a *quotient representation* of $G \rtimes_\lambda \mathfrak{l}$ (see §2.3 for the detail).

2000 *Mathematics Subject Classification.* Primary 53D17; Secondary 18B40.

Research partially supported by HKRGC grants 701603, 703304, and 703405.

We give two applications. First, for any integer N, we identify the generalized Poisson cohomology $H^\bullet\left(G/H, \pi; K_{G/H}^N\right)$ with Lie algebra cohomology of $\mathfrak{l}$ relative to H with coefficients in $C^\infty(G)_N$, the space of smooth functions on G together with an $(\mathfrak{l}, H)$-module structure that depends on N (see Corollary 4.12 for detail). We also discuss the canonical pairing between $H^\bullet\left(G/H, \pi; K_{G/H}^N\right)$ and $H^\bullet\left(G/H, \pi; K_{G/H}^{2-N}\right)$ as a pairing on relative Lie algebra cohomology of $\mathfrak{l}$, and we compute the modular vector fields of $(G/H, \pi)$. The identifications of the Poisson cohomology and homology (i.e., when $N=0$ and $N=1$) with relative Lie algebra cohomology of $\mathfrak{l}$ have been established in [**15**] and [**19**] but by different methods.

As a second application, we construct a Poisson groupoid Γ over $(G/H, \pi)$ that is symplectic near the identity section, and we give conditions and examples when it is symplectic. The groupoid structure on Γ is a quotient of a transformation groupoid over G (see Mackenzie's book [**18**] for a general treatment of quotients of groupoids), while the Poisson structure on Γ is obtained by reduction of a quasi-Poisson manifold by an action of a quasi-Poisson Lie group, a theory developed by Alekseev and Kosmann-Schwarzbach in [**1**]. In the special case when (G, π_G) is complete and when H is a Poisson Lie subgroup of (G, π_G) with π being the projection of π_G to G/H, a symplectic groupoid of $(G/H, \pi)$ was constructed by P. Xu in [**25**].

There are many examples of Poisson homogeneous spaces associated to semi-simple Lie groups, and they are in general not of the type G/H with H being a Poisson Lie subgroup. See [**8, 9, 16**] for studies of certain varieties which can serve as moduli spaces of Poisson homogeneous spaces. In forthcoming papers, we will use results from this note to compute explicitly the Poisson cohomology and study their symplectic groupoids for certain examples of Poisson homogeneous spaces treated in [**8, 9, 16**]. Such examples included flag varieties of complex semi-simple groups [**8**] and semi-simple Riemannian symmetric spaces [**10**] (see Example 5.14).

1.1. Notation. For a smooth manifold P, the tangent and cotangent bundles of P are denoted by TP and T^*P respectively. For an integer $0 \leq k \leq \dim P$, $\mathcal{V}^k(P)$ and $\Omega^k(P)$ will denote respectively the spaces of smooth k-vector fields and smooth k-forms on P, and

$$\mathcal{V}(P) = \oplus_{k=0}^{\dim P} \mathcal{V}^k(P) \quad \text{and} \quad \Omega(P) = \oplus_{k=0}^{\dim P} \Omega^k(P).$$

If P and Q are smooth manifolds and $F: P \to Q$ is a smooth map, F_* will denote the induced map $TP \to TQ$.

For a vector bundle A over P, $\Gamma(A)$ will denote the space of smooth sections of A. If V is an n-dimensional vector space, $\wedge^{\mathrm{top}} V$ always denotes $\wedge^n V$. Let V^* be the dual space of V. For $x \in \wedge^k V$ and $\xi \in \wedge^j V^*$ with $k \leq j$, $\iota_x \xi \in \wedge^{j-k} V^*$ is defined by $(\iota_x \xi, y) = (\xi, x \wedge y)$ for all $y \in \wedge^{j-k} V$. Unless otherwise specified, all vector spaces are real.

For a Lie group G and $g \in G$, l_g and r_g denote respectively the left and right translation on G by g. The identity element of a group is always denoted by e.

1.2. Acknowledgement. We thank K. Mackenzie for references on quotients of Lie algebroids and groupoids and Bing-Kwan So for helpful discussions.

2. Some basic facts on Lie algebroids

We refer to [**17, 18**] for details on the facts reviewed in this section.

2.1. Lie algebroids and Lie algebroid cohomology. Recall that a Lie algebroid over a manifold P is a vector bundle A over P together with a vector bundle homomorphism $\rho_A : A \to TP$ and a Lie bracket $[\,,\,]$ on $\Gamma(A)$ such that

1) $[fa_1, a_2] = f[a_1, a_2] - \rho_A(a_2)(f)a_1$ for all $f \in C^\infty(P)$ and $a_1, a_2 \in \Gamma(A)$;

2) $\rho_A[a_1, a_2] = [\rho_A(a_1), \rho_A(a_2)]$ for all $a_1, a_2 \in \Gamma(A)$.

Let A be a Lie algebroid over P. A *representation* of A on a vector bundle E over P is an $\mathbb{R}$-bilinear map $D : \Gamma(A) \times \Gamma(E) \to \Gamma(E) : (a, s) \mapsto D_a s$, such that for any $a, b \in \Gamma(A)$, $s \in \Gamma(E)$, and $f \in C^\infty(P)$,

$$\begin{aligned} &1) \quad D_{fa}s = fD_a s;\\ &2) \quad D_a(fs) = fD_a s + (\rho(a)f)s;\\ &3) \quad D_a(D_b s) - D_b(D_a s) = D_{[a,b]}s. \end{aligned}$$

The *trivial representation* of A is the one on the trivial line bundle $E = P \times \mathbb{R}$ given by $D_a f = \rho(a)(f)$ for $a \in \Gamma(A)$ and $f \in \Gamma(E) \cong C^\infty(P)$. One has the natural notion of tensor products and duals of representations of A. In particular, a representation D of A on a line bundle L gives rise to a representation of A on the N-th power L^N of L for any integer $N \geq 0$. For a negative integer N, we use the natural identification between L^N and$(L^{-N})^*$ and thus have a representation of A on L^N as well.

For a representation D of A on E, and for $k \geq 0$, define

$$\begin{aligned} d_{A,E} : \Gamma(\mathrm{Hom}(\wedge^k A, E)) \quad &\longrightarrow \quad \Gamma(\mathrm{Hom}(\wedge^{k+1} A, E))\\ (d_{A,E}\phi)(a_1, a_2, \cdots, a_{k+1}) \quad &= \quad \sum_{j=1}^{k+1} (-1)^{j+1} D_{a_j}\phi(a_1, \cdots, \hat{a}_j, \cdots, a_{k+1})\\ &\quad + \sum_{i<j} (-1)^{i+j} \phi([a_i, a_j], \cdots, \hat{a}_i, \cdots \hat{a}_j, \cdots, a_{k+1}) \end{aligned}$$

for $a_1, \ldots, a_{k+1} \in \Gamma(A)$. Then $d_{A,E}^2 = 0$. The cohomology of the cochain complex

$$(\Gamma(\mathrm{Hom}(\wedge A, E)),\ d_{A,E}),$$

which will be denoted by $H^\bullet_{\mathrm{Lie}}(A; E)$, is called the *Lie algebroid cohomology of A with coefficients in E*. When E is the trivial representation, we set $H^\bullet(A; E) = H^\bullet_{\mathrm{Lie}}(A)$.

2.2. Relative Lie algebra cohomology. Our reference for this section is [**2**]. A Lie algebra $\mathfrak{l}$ can be regarded as a Lie algebroid over a one point space, so for every $\mathfrak{l}$-module V, we have the coboundary operators

$$d_{\mathfrak{l},V} : \mathrm{Hom}(\wedge^k \mathfrak{l}, V) \longrightarrow \mathrm{Hom}(\wedge^{k+1} \mathfrak{l}, V), \qquad k \geq 0.$$

Let $\mathfrak{h} \subset \mathfrak{l}$ be a Lie subalgebra, H a Lie group with Lie algebra $\mathfrak{h}$, and $H \to \mathrm{Aut}(\mathfrak{l}) : h \mapsto \mathrm{Ad}_h$ a group homomorphism integrating the adjoint action of $\mathfrak{h}$ on $\mathfrak{l}$.

DEFINITION 2.1. An $(\mathfrak{l}, H)$*-module* is a topological vector space V which is both an $\mathfrak{l}$-module and an H-module such that

1) for every $v \in V$, the map $H \to V : h \mapsto hv$ is smooth, and that the restriction to $\mathfrak{h}$ of the action of $\mathfrak{l}$ on V coincides with the one induced from the H-action;

2) for every $v \in V, x \in \mathfrak{l}$, and $h \in H$, $h(x(h^{-1}(v))) = (\mathrm{Ad}_h x)(v)$.

Let V be an $(\mathfrak{l}, H)$-module. For $k \geq 0$, let

$$C^k_{\mathfrak{l},H;V} = \left(\wedge^k(\mathfrak{l}/\mathfrak{h})^* \otimes V\right)^H,$$

where the superscript H denotes the subspace of H-invariants. Identify $(\mathfrak{l}/\mathfrak{h})^* \cong \{\xi \in \mathfrak{l}^* \mid \xi|_{\mathfrak{h}} = 0\} \subset \mathfrak{l}^*$ and regard $C^k_{\mathfrak{l},H;V}$ as in $\wedge^k\mathfrak{l}^* \otimes V \cong \mathrm{Hom}(\wedge^k\mathfrak{l}, V)$. Then

$$\bigoplus_{k\geq 0} C^k_{\mathfrak{l},H;V} \subset \bigoplus_{k\geq 0} \mathrm{Hom}(\wedge^k\mathfrak{l}, V)$$

is invariant under $d_{\mathfrak{l},V}$. The cohomology of the cochain complex $(C^\bullet_{\mathfrak{l},H;V}, d_{\mathfrak{l},V})$, which will be denoted by $H^\bullet_{\mathrm{Lie}}(\mathfrak{l}, H; V)$, is called the *Lie algebra cohomology of* $\mathfrak{l}$ *relative to* H *with coefficients in* V.

Suppose that U and V are two $(\mathfrak{l}, H)$-modules. Then $U \otimes V$ is naturally an $(\mathfrak{l}, H)$-module. For any $0 \leq j, k \leq n = \dim(\mathfrak{l}/\mathfrak{h})$, define

$$C^j_{\mathfrak{l},H;U} \times C^k_{\mathfrak{l},H;V} \longrightarrow C^{j+k}_{\mathfrak{l},H;\,U\otimes V}: \quad (c_1, c_2) \longmapsto c_1 \otimes c_2 := \phi \wedge \psi \otimes u \otimes v,$$

where $c_1 = \phi \otimes u, c_2 = \psi \otimes v$ with $\phi \in \wedge^j(\mathfrak{l}/\mathfrak{h})^*$, $\psi \in \wedge^k(\mathfrak{l}/\mathfrak{h})^*$, $u \in U$, and $v \in V$. It is easy to check that

$$d_{\mathfrak{l},\,U\otimes V}(c_1 \otimes c_2) = d_{\mathfrak{l},U}(c_1) \otimes c_2 + (-1)^j c_1 \otimes d_{\mathfrak{l},V}(c_2) \tag{2.1}$$

if $c_1 \in C^j_{\mathfrak{l},H;U}$. Assume that $\nu \in \left(C^n_{\mathfrak{l},H;\,U\otimes V}\right)^*$ is such that

$$\nu\left(d_{\mathfrak{l},\,U\otimes V}(C^{n-1}_{\mathfrak{l},\,H;\,U\otimes V})\right) = 0. \tag{2.2}$$

For $0 \leq k \leq n$, define the pairing $(\,,\,)_\nu$ between $C^k_{\mathfrak{l},H;\,U}$ and $C^{n-k}_{\mathfrak{l},H;\,V}$ by

$$(c_1,\ c_2)_\nu = \nu(c_1 \otimes c_2).$$

It follows from (2.1) that

$$(d_{\mathfrak{l},U}(c_1),\ c_2)_\nu + (-1)^{k-1}(c_1,\ d_{\mathfrak{l},V}(c_2))_\nu = 0$$

for all $c_1 \in C^{k-1}_{\mathfrak{l},H;U}$ and $c_2 \in C^{n-k}_{\mathfrak{l},H;V}$. Thus $(\,,\,)_\nu$ induces a well-defined pairing, still denoted by $(\,,\,)_\nu$, between $H^k_{\mathrm{Lie}}(\mathfrak{l}, H; U)$ and $H^{n-k}_{\mathrm{Lie}}(\mathfrak{l}, H; V)$ for every $0 \leq k \leq n$.

2.3. Quotients of transformation Lie algebroids. Let again $\mathfrak{l}$ be a Lie algebra, $\mathfrak{h} \subset \mathfrak{l}$ a Lie subalgebra, H a Lie group with Lie algebra $\mathfrak{h}$, and $H \to \mathrm{Aut}(\mathfrak{l}) : h \to \mathrm{Ad}_h$ a group homomorphism integrating the adjoint action of $\mathfrak{h}$ on $\mathfrak{l}$.

DEFINITION 2.2. An $(\mathfrak{l}, H)$-*space* is a smooth manifold M together with a Lie algebra homomorphism $\lambda : \mathfrak{l} \to \mathcal{V}^1(M)$ and a right action of H on M such that

1) the restriction of λ on $\mathfrak{h}$ coincides with the infinitesimal action of $\mathfrak{h}$ on M induced by the right H-action, and

2) for all $m \in M, x \in \mathfrak{l}$ and $h \in H$, $\lambda_x(mh) = h_*\lambda_{\mathrm{Ad}_h x}(m)$, where h_* is the differential of the map $h : M \to M : m_1 \mapsto m_1 h$ for $m_1 \in M$.

We will sometimes denote an $(\mathfrak{l}, H)$-space by the pair (M, λ) without explicitly mentioning the action of H on M.

Let (M, λ) be an $(\mathfrak{l}, H)$-space. Using the action λ of $\mathfrak{l}$ on M, one can form the *transformation Lie algebroid* $M \rtimes_\lambda \mathfrak{l}$ over M, which is the trivial vector bundle $M \times \mathfrak{l}$ over M with the anchor map

$$M \times \mathfrak{l} \longrightarrow TM : \ (m, x) \longmapsto \lambda_x(m), \ m \in M, x \in \mathfrak{l},$$

and the Lie bracket $[\,,\,]_{M\rtimes_\lambda \mathfrak{l}}$ on $\Gamma(M\times\mathfrak{l})\cong C^\infty(M,\mathfrak{l})$ determined by

$$[\bar{x}_1,\bar{x}_2]_{M\rtimes_\lambda \mathfrak{l}}=\overline{[x_1,x_2]},$$

where for $x\in\mathfrak{l}$, $\bar{x}$ is the constant function on M with value x.

Assume in addition that the H-action on M is free and proper so that the quotient M/H is a smooth manifold. Consider the associated vector bundle $A=M\times_H(\mathfrak{l}/\mathfrak{h})$ over M/H, where $h\in H$ acts on $\mathfrak{l}/\mathfrak{h}$ by Ad_h. Points in A will be denoted by $[m,x+\mathfrak{h}]$, where $m\in M$ and $x\in\mathfrak{l}$. Note that

$$\rho_A: A\longrightarrow T(M/H):\ [m,\ x+\mathfrak{h}]\longmapsto q_*\lambda_x(m)$$

is a well-defined bundle map, where $q: M\to M/H$ is the natural projection, and

$$\begin{aligned}\Gamma(A) &= C^\infty(M,\mathfrak{l}/\mathfrak{h})^H\\ &= \{a\in C^\infty(M,\mathfrak{l}/\mathfrak{h})\mid a(mh)=\mathrm{Ad}_{h^{-1}}a(m),\ \forall m\in M,h\in H\}.\end{aligned}$$

Let

$$\Gamma(M\rtimes_\lambda\mathfrak{l})^H=\{a\in C^\infty(M,\mathfrak{l})\mid a(mh)=\mathrm{Ad}_{h^{-1}}a(m),\ \forall m\in M,h\in H\}.$$

For $a_1,a_2\in\Gamma(A)$, let $\tilde{a}_1,\tilde{a}_2\in\Gamma(M\rtimes_\lambda\mathfrak{l})^H$ be such that $\mathfrak{p}(\tilde{a}_1)=a_1$ and $\mathfrak{p}(\tilde{a}_2)=a_2$, where $\mathfrak{p}: M\rtimes_\lambda\mathfrak{l}\to A:(m,x)\mapsto[m,x+\mathfrak{h}]$ is the natural vector bundle projection. Define $[a_1,a_2]\in\Gamma(A)$ by

$$[a_1,a_2]=\mathfrak{p}([\tilde{a}_1,\tilde{a}_2]_{M\rtimes_\lambda\mathfrak{l}}). \tag{2.3}$$

The proof of the following lemma is omitted since it is straightforward.

LEMMA 2.3. *Formula* (2.3) *is a well-defined Lie bracket on* $\Gamma(A)$. *With the Lie bracket in* (2.3) *on* $\Gamma(A)$ *and* ρ_A *as the anchor map,* A *is a Lie algebroid over* M/H. *Moreover, the bundle map* $\mathfrak{p}: M\rtimes_\lambda\mathfrak{l}\to A$ *is a Lie algebroid morphism.*

DEFINITION 2.4. The Lie algebroid A in Lemma 2.3 is called the *H-quotient* of the transformation Lie algebroid $M\rtimes_\lambda\mathfrak{l}$ and will be denoted by $M\rtimes_{\lambda,H}(\mathfrak{l}/\mathfrak{h})$.

EXAMPLE 2.5. If G is a Lie group and $H\subset G$ a closed subgroup, the tangent bundle Lie algebroid $T(G/H)$ is a quotient by H of the tangent bundle Lie algebroid TG. A more general discussion on quotients of Lie algebroid can be found in [**18**, Chap. 4].

We now turn to a special class of representations of $M\rtimes_{\lambda,H}(\mathfrak{l}/\mathfrak{h})$ that arise from representations of $M\rtimes_\lambda\mathfrak{l}$.

DEFINITION 2.6. An *$(\mathfrak{l},H)$-vector bundle* is an H-equivariant vector bundle E over an $(\mathfrak{l},H)$-space (M,λ) together with a representation of $\mathfrak{l}$ on $\Gamma(E)$ such that

1) $x\cdot(fs)=\lambda_x(f)s+f(x\cdot s)$ for all $x\in\mathfrak{l}$, $f\in C^\infty(M)$, and $s\in\Gamma(E)$;

2) the $\mathfrak{l}$-action and the H-action on $\Gamma(E)$ induced from the H-action on E make $\Gamma(E)$ into an $(\mathfrak{l},H)$-module (see Definition 2.1).

Let E be an $(\mathfrak{l},H)$-vector bundle over M such that the H-action on M is free and proper. One then has the representation $\widetilde{D}$ of $M\rtimes_\lambda\mathfrak{l}$ on E given by

$$(\widetilde{D}_b s)(m)=(b(m)\cdot s)(m),\qquad b\in\Gamma(M\rtimes_\lambda\mathfrak{l})=C^\infty(M,\mathfrak{l}),\ m\in M,s\in\Gamma(E).$$

Let E/H be the quotient bundle over M/H with $\Gamma(E/H)=\Gamma(E)^H$, the space of H-invariant smooth sections of E. For $a\in\Gamma(A)$, let $\tilde{a}\in\Gamma(M\rtimes_\lambda\mathfrak{l})^H$ be such that $\mathfrak{p}(\tilde{a}) = a$. It is easy to see that $\widetilde{D}_{\tilde{a}}s \in \Gamma(E)$ is H-invariant for any $s \in \Gamma(E/H) \cong \Gamma(E)^H$, so we can regard $\widetilde{D}_{\tilde{a}}s$ as in $\Gamma(E/H)$. Define

$$D_a s = \widetilde{D}_{\tilde{a}} s, \qquad s \in \Gamma(E/H) \cong \Gamma(E)^H. \tag{2.4}$$

The proof of the following Lemma 2.7 is straightforward.

LEMMA 2.7. *Formula (2.4) is a well-defined representation of the quotient Lie algebroid $M \bowtie_{\lambda,H} (\mathfrak{l}/\mathfrak{h})$ on E/H, and we call it the H-quotient of the representation $\widetilde{D}$ of $M \bowtie_\lambda \mathfrak{l}$ on E.*

LEMMA 2.8. *The Lie algebroid cohomology of $A = M \bowtie_{\lambda,H} (\mathfrak{l}/\mathfrak{h})$ with coefficient in E/H is isomorphic to the Lie algebra cohomology of $\mathfrak{l}$ relative to H with coefficients in $\Gamma(E)$, i.e.,*

$$H^k_{\mathrm{Lie}}(A; E/H) \cong H^k_{\mathrm{Lie}}(\mathfrak{l}, H; \Gamma(E)), \qquad \forall k \geq 0.$$

PROOF. Let $\mathcal{T}$ be the trivial vector bundle over M with fiber $\mathfrak{l}/\mathfrak{h}$. Then for every $k \geq 0$, the vector bundle $\mathrm{Hom}(\wedge^k A, E/H)$ over M/H is the quotient by H of the H-equivariant vector bundle $\mathrm{Hom}(\wedge^k \mathcal{T}, E)$, so

$$\Gamma(\mathrm{Hom}(\wedge^k A, E/H)) \cong \left(\wedge^k(\mathfrak{l}/\mathfrak{h})^* \otimes \Gamma(E)\right)^H \cong C^k_{\mathfrak{l},H;\Gamma(E)}. \tag{2.5}$$

By following the definitions of the Lie algebroid structure on A and the representation of A on E/H, it is straightforward to check that the identifications in (2.5) give an isomorphism of cochains

$$\left(\bigoplus_{k\geq 0} \Gamma(\mathrm{Hom}(\wedge^k A, E/H)),\ d_{A,E/H}\right) \longrightarrow \left(\bigoplus_{k\geq 0} C^k_{\mathfrak{l},H;\Gamma(E)},\ d_{\mathfrak{l},H}\right).$$

□

REMARK 2.9. Suppose that F is an $(\mathfrak{l}, H)$-line bundle over an $(\mathfrak{l}, H)$-space (M, λ) and that E is an H-equivariant square root of F, i.e., $E^2 \cong F$. Then E is naturally an $(\mathfrak{l}, H)$-line bundle with the $\mathfrak{l}$-action on $\Gamma(E)$ uniquely defined as follows: if t is a nowhere vanishing local section of E, then $x \cdot t = \frac{1}{2}\frac{x \cdot t^2}{t}$ for any $x \in \mathfrak{l}$ (see [7]). Consequently, one has the quotient representation of the quotient Lie algebroid $A = M \bowtie_{\lambda,H} (\mathfrak{l}/\mathfrak{h})$ on E/H.

3. Poisson cohomology and modular vector fields

3.1. The cotangent bundle Lie algebroid and Poisson cohomology. The cotangent bundle Lie algebroid $T^*(P, \pi)$ of a Poisson manifold (P, π) is by definition the cotangent bundle T^*P of P with the anchor map

$$\tilde{\pi}: \quad T^*P \longrightarrow TP: \quad \tilde{\pi}(\alpha)(\beta) = \pi(\alpha, \beta), \qquad \alpha, \beta \in \Omega^1(P),$$

and the Lie bracket $\{\,,\,\}_\pi$ on $\Omega^1(P)$ given by

$$\{\alpha, \beta\}_\pi = d(\pi(\alpha, \beta)) + \iota_{\tilde{\pi}(\alpha)} d\beta - \iota_{\tilde{\pi}(\beta)} d\alpha, \qquad \alpha, \beta \in \Omega^1(P). \tag{3.1}$$

Let

$$K_P = \wedge^{\mathrm{top}} T^*P$$

be the canonical line bundle over P. It is shown in [7, 26] that there is a representation of the Lie algebroid $T^*(P, \pi)$ on K_P given by

$$D_\alpha \mu = L_{\tilde{\pi}(\alpha)}\mu + (\pi, d\alpha)\mu = \{\alpha, \mu\}_\pi - (\pi, d\alpha)\mu = \alpha \wedge d(i_\pi \mu), \quad \mu \in \Omega^{\mathrm{top}}(P), \tag{3.2}$$

where $\{\,,\,\}_\pi$ is the Schouten bracket on the space $\Omega(P)$ induced from the bracket in (3.1) on $\Omega^1(P)$.

DEFINITION 3.1. The representation of $T^*(P,\pi)$ on K_P is called the *canonical representation* of $T^*(P,\pi)$ on K_P.

For any integer N, let K_P^N be the N-th power of K_P, equipped with the natural extension of the representation of $T^*(P,\pi)$. When N is negative, we will understand K_P^N as $(K_P^{-N})^*$.

DEFINITION 3.2. For a Poisson manifold (P,π) and for any integer N, we define the *Poisson cohomology of* (P,π) *with coefficients* K_P^N to be the Lie algebroid cohomology of $T^*(P,\pi)$ with coefficients in K_P^N, and we denote it by $H^\bullet(P,\pi;K_P^N)$. When $N=0$, we simply write $H^\bullet(P,\pi;K_P^N)$ as $H^\bullet(P,\pi)$. The totality of $H^\bullet(P,\pi;K_P^N)$ for all integers N is called the *generalized Poisson cohomology* of (P,π).

REMARK 3.3. The Poisson cohomology of (P,π) defined in [**13**] is $H^\bullet(P,\pi)$. It is shown in [**7, 26**] that the Poisson homology of (P,π) defined in [**3**] is isomorphic to $H^\bullet(P,\pi;K_P)$. In [**7**], the cohomology $H^\bullet(P,\pi;K_P^2)$ is called the twisted Poisson cohomology of (P,π).

3.2. The canonical pairing on Poisson cohomology. Suppose that P is compact and oriented. For $0\le k\le n=\dim P$ and an integer N, set

$$C^k_{P,N}=\Gamma(\mathrm{Hom}(\wedge^k T^*P,\ K_P^N))\cong\Gamma(\wedge^k TP\otimes K_P^N).$$

The natural identifications of bundles

$$\wedge^k TP\otimes\wedge^{n-k}TP\cong\wedge^n TP,\qquad K_P^N\otimes K_P^{2-N}\cong K_P^2,\qquad \wedge^n TP\otimes K_P^2\cong K_P$$

give rise to an identification

$$J:\ \left(\wedge^k TP\otimes K_P^N\right)\otimes\left(\wedge^{n-k}TP\otimes K_P^{2-N}\right)\longrightarrow K_P$$

and thus an $\mathbb{R}$-bilinear pairing

$$(c_1,\ c_2):=\int_P J(c_1,c_2),\qquad c_1\in C^k_{P,N},\ c_2\in C^{n-k}_{P,2-N}.$$

A proof similar to that of Theorem 5.1 of [**7**] shows that $(\,,\,)$ induces a well-defined pairing between $H^k(P,\pi;K_P^N)$ and $H^{n-k}(P,\pi;K_P^{2-N})$. We will refer to $(\,,\,)$ the *canonical pairing* on the generalized Poisson cohomology of (P,π).

3.3. Modular vector fields. Let (P,π) be an orientable Poisson manifold, and let μ be a volume form of P. The *modular vector field of* π *with respect to* μ (see [**23**]) is defined to be the vector field θ_μ on P such that

$$D_\alpha\mu=(\theta_\mu,\alpha)\mu,\quad \forall\alpha\in\Omega^1(P),$$

where $D_\alpha\mu\in\Omega^{\mathrm{top}}(P)$ is given in (3.2). For an integer N, set $d_N=d_{T^*P,K_P^N}\in\mathrm{End}(C^\bullet_{P,N})$.

PROPOSITION 3.4. *Let N be any integer. For any volume form μ, the action of the modular vector field θ_μ on $C^\bullet_{P,N}=\oplus_{k\ge0}C^k_{P,N}$ by Lie derivative commutes with the operator d_N. When $N\neq1$, the induced action of θ_μ on $H^\bullet(P,\pi;K_P^N)$ is trivial.*

PROOF. Consider the identification

$$\mathcal{I}: \quad \mathcal{V}^k(P) \longrightarrow C^k_{P,N}: \quad V \longmapsto V \otimes \mu^N.$$

Since $L_{\theta_\mu}\mu = 0$, $L_{\theta_\mu} \circ \mathcal{I} = \mathcal{I} \circ L_{\theta_\mu}$. It is also easy to show (see [**7**, Lemma 4.4]) that the operator $\delta_N := \mathcal{I}^{-1} \circ d_N \circ \mathcal{I}$ is given by

$$\delta_N: \mathcal{V}^k(P) \longrightarrow \mathcal{V}^{k+1}(P): \quad V \longmapsto [\pi, V] + N\theta_\mu \wedge V.$$

Since $L_{\theta_\mu}\pi = 0$, it is clear that L_{θ_μ} commutes with δ_N. Consider the operator

$$b_\mu: \quad \mathcal{V}^k(P) \longrightarrow \mathcal{V}^{k-1}(P): \quad \iota_{b_\mu V}\mu = (-1)^k d(\iota_V \mu).$$

It is easy to see that $b_\mu^2 = 0$ and that $\theta_\mu = b_\mu \pi$. Moreover, for $V_1 \in \mathcal{V}^k(P)$ and $V_2 \in \mathcal{V}(P)$,

$$\begin{aligned} b_\mu(V_1 \wedge V_2) &= b_\mu(V_1) \wedge V_2 + (-1)^k V_1 \wedge b_\mu(V_2) + (-1)^k [V_1, V_2] \\ b_\mu[V_1, V_2] &= [b_\mu(V_1), V_2] + (-1)^{k-1}[V_1, b_\mu(V_2)] \end{aligned}$$

It follows that $b_\mu \delta_N + \delta_N b_\mu = (1-N)L_{\theta_\mu}$. Thus θ_μ acts trivially on $H^\bullet(P, \pi; K_P^N)$ when $N \neq 1$. □

4. Poisson homogeneous spaces

4.1. Review on Poisson Lie groups. Recall that [**5, 11, 14, 20**] a *Poisson Lie group* is a Lie group G with a Poisson structure π_G such that the group multiplication map $(G, \pi_G) \times (G, \pi_G) \to (G, \pi_G): (g, h) \mapsto gh$ is Poisson. Let (G, π_G) be a Poisson Lie group. Then π_G necessarily vanishes at the identity element e of G. Let $\delta: \mathfrak{g} \to \wedge^2 \mathfrak{g}$ be the linearization of π_G at e. Then the dual map

$$\delta^*: \wedge^2 \mathfrak{g}^* \longrightarrow \mathfrak{g}^*: \quad \xi \wedge \eta \longmapsto [\xi, \eta]$$

of δ defines a Lie bracket on $\mathfrak{g}^*$, and the pair $(\mathfrak{g}, \delta)$ becomes a *Lie bialgebra* [**5**]. For $x \in \mathfrak{g}$ and $\xi \in \mathfrak{g}^*$, define $\mathrm{ad}_x^* \xi \in \mathfrak{g}^*$ and $\mathrm{ad}_\xi^* x \in \mathfrak{g}$ by

$$(ad_x^* \xi, y) = (\xi, [y, x]) \quad \text{and} \quad (\mathrm{ad}_\xi^* x, \eta) = (x, [\eta, \xi]), \quad \text{where } y \in \mathfrak{g}, \eta \in \mathfrak{g}^*.$$

Let $\mathfrak{d} = \mathfrak{g} \oplus \mathfrak{g}^*$. Then the bracket on $\mathfrak{d}$ given by

$$[x + \xi, y + \eta] = [x, y] + \mathrm{ad}_\xi^* y - \mathrm{ad}_\eta^* x + [\xi, \eta] + \mathrm{ad}_x^* \eta - \mathrm{ad}_y^* \xi, \quad x, y \in \mathfrak{g}, \xi, \eta \in \mathfrak{g}^*,$$

is a Lie bracket, and the bilinear form $\langle\, , \rangle$ on $\mathfrak{d}$ given by

$$\langle x + \xi, y + \eta \rangle = (x, \eta) + (y, \xi), \qquad x, y \in \mathfrak{g}, \ \xi, \eta \in \mathfrak{g}^*,$$

is ad-invariant with respect to $[\, ,]$. The pair $(\mathfrak{d}, \langle\, , \rangle)$ is called the double of the Lie bialgebra $(\mathfrak{g}, \delta)$. The adjoint action of $\mathfrak{g}$ on $\mathfrak{d}$ integrates to an action of G on $\mathfrak{d}$, still denoted by $\mathrm{Ad}_g: \mathfrak{d} \to \mathfrak{d}$ for $g \in G$, which is given by [**6**]

$$\mathrm{Ad}_g(x + \xi) =: \mathrm{Ad}_g x + \iota_{\mathrm{Ad}^*_{g^{-1}}\xi}(r_{g^{-1}} \pi_G(g)) + \mathrm{Ad}^*_{g^{-1}}\xi, \tag{4.1}$$

where $\mathrm{Ad}_g: \mathfrak{g} \to \mathfrak{g}$ and $\mathrm{Ad}^*_{g^{-1}}: \mathfrak{g}^* \to \mathfrak{g}^*$ are the adjoint and co-adjoint actions of $g \in G$ on $\mathfrak{g}$ and on $\mathfrak{g}^*$ respectively. A subspace $\mathfrak{l}$ of $\mathfrak{d}$ is said to be Lagrangian if $\langle x, y \rangle = 0$ for all $x, y \in \mathfrak{l}$ and if $\dim \mathfrak{l} = \dim \mathfrak{g}$.

4.2. Drinfeld Lagrangian subalgebras. Let H be a closed subgroup of G.

DEFINITION 4.1. [**6**] A (G, π_G)-homogeneous Poisson structure on G/H is a bivector field π on G/H such that 1) π is Poisson, and 2) the map

$$(4.2) \qquad \sigma: \ (G, \pi_G) \times (G/H, \pi) \longrightarrow (G/H, \pi): \quad (g_1, g_2H) \longmapsto g_1g_2H$$

is Poisson.

By definition, the map σ in (4.2) is Poisson if and only if

$$(4.3) \qquad \pi(gH) = (\sigma_g)_*\pi(eH) + q_*\pi_G(g), \qquad \forall g \in G,$$

where $q : G \to G/H$ is the projection, and for $g \in G$, $\sigma_g : G/H \to G/H$ is defined by $g_1H \to gg_1H$ for $g_1 \in G$. Thus, π is uniquely determined by $\pi(eH) \in \wedge^2 T_{eH}(G/H)$, and Conditions 1) and 2) on π in Definition 4.1 become the following two conditions on $\pi(eH) \in \wedge^2 T_{eH}(G/H)$:

(i) $\pi(eH) = (\sigma_h)_*\pi(eH) + q_*\pi_G(h)$ for all $h \in H$ (so that π given by (4.3) is well-defined); and

(ii) the bi-vector field π on G/H determined by $\pi(eH)$ via (4.3) is Poisson.

Let $\mathfrak{h}$ be the Lie algebra of H. Simple linear algebra arguments show that there is a one to one correspondence between $\wedge^2(\mathfrak{g}/\mathfrak{h})$ and the set of Lagrangian subspaces $\mathfrak{l}$ of $\mathfrak{d}$ such that $\mathfrak{l} \cap \mathfrak{g} = \mathfrak{h}$. The explicit correspondence is given by

$$(4.4) \qquad \wedge^2(\mathfrak{g}/\mathfrak{h}) \ni r \mapsto \mathfrak{l}_r := \{x + \xi \mid x \in \mathfrak{g}, \xi \in \mathfrak{g}^*, \xi|_{\mathfrak{h}} = 0, \iota_\xi r = x + \mathfrak{h}\}.$$

Identify $T_{eH}(G/H) \cong \mathfrak{g}/\mathfrak{h}$. Then an element $\pi(eH) \in \wedge^2 T_{eH}(G/H) \cong \wedge^2(\mathfrak{g}/\mathfrak{h})$ corresponds to the Lagrangian subspace $\mathfrak{l}_{\pi(eH)}$ of $\mathfrak{d}$. Drinfeld showed [**6**] that Conditions (i) and (ii) on $\pi(eH) \in \wedge^2 T_{eH}(G/H)$ are respectively equivalent to

(a) $\mathrm{Ad}_h \mathfrak{l}_{\pi(eH)} = \mathfrak{l}_{\pi(eH)}$ for all $h \in H$, where $\mathrm{Ad}_h : \mathfrak{d} \to \mathfrak{d}$ is given in (4.1), and

(b) $\mathfrak{l}_{\pi(eH)}$ is a Lie subalgebra of $\mathfrak{d}$.

DEFINITION 4.2. When $(G/H, \pi)$ is a Poisson homogeneous space of (G, π_G), the Lie subalgebra $\mathfrak{l}_{\pi(eH)}$ of $\mathfrak{d}$ is called the Drinfeld Lagrangian subalgebra associated to $\pi(eH)$.

Let $(G/H, \pi)$ be a Poisson homogeneous space of (G, π_G). Let q also denote the projection $\mathfrak{g} \to \mathfrak{g}/\mathfrak{h}$. Let $\Lambda \in \wedge^2\mathfrak{g}$ be any element such that

$$(4.5) \qquad q(\Lambda) = \pi(eH) \in \wedge^2 T_{eH}(G/H) \cong \wedge^2(\mathfrak{g}/\mathfrak{h}).$$

The following Lemma 4.3 is straightforward to prove [**4**].

LEMMA 4.3. *Conditions (i) and (ii) on $\pi(eH)$ are equivalent to*

1) $\mathrm{Ad}_h\Lambda - \Lambda + (r_{h^{-1}})_*\pi_G(h) \in \mathfrak{h} \wedge \mathfrak{g}$ *for all* $h \in H$*;*

2) $[\Lambda, \Lambda] + 2\delta(\Lambda) \in \mathfrak{h} \wedge \mathfrak{g} \wedge \mathfrak{g}$,

where $[\,,\,]$ *is the Schouten bracket on* $\wedge\mathfrak{g}$ *and* $\delta : \mathfrak{g} \to \wedge^2\mathfrak{g}$ *is the linearization of* π_G *at* e *as well as its extension* $\delta : \wedge^2\mathfrak{g} \to \wedge^3\mathfrak{g}$ *given by*

$$\delta(x \wedge y \wedge z) = \delta(x) \wedge y \wedge z - x \wedge \delta(y) \wedge z + x \wedge y \wedge \delta(z), \qquad x, y, z \in \mathfrak{g}.$$

For $\Lambda \in \wedge^2\mathfrak{g}$ as in (4.5), define the bi-vector field π_Λ on G by

$$(4.6) \qquad \pi_\Lambda = \Lambda^l + \pi_G,$$

where Λ^l is the left invariant bi-vector field on G with value Λ at e. Condition 1) on Λ in Lemma 4.3 implies that $q_*\pi_\Lambda$ is a well-defined bi-vector field on G/H. In fact,

$$q_*\pi_\Lambda = \pi.$$

Let $\Lambda\xi = \iota_\xi \Lambda$ for $\xi \in \mathfrak{g}^*$. The Drinfeld Lagrangian subalgebra $\mathfrak{l}_{\pi(eH)}$ is also given by

$$\mathfrak{l}_{\pi(eH)} = \{x + \Lambda\xi + \xi \mid x \in \mathfrak{h}, \xi \in \mathfrak{g}^*, \xi|_{\mathfrak{h}} = 0\}. \tag{4.7}$$

REMARK 4.4. Although the bi-vector field π_Λ on G is not necessarily Poisson, we can still define the skew-symmetric bracket $\{\, , \,\}_{\pi_\Lambda}$ on $\Omega^1(G)$ by replacing π by π_Λ in (3.1). Moreover, the space of left invariant 1-forms on G is invariant under $\{\, , \,\}_{\pi_\Lambda}$. In fact, it is easy to show that

$$\{\xi^l, \eta^l\}_{\pi_\Lambda} = ([\xi, \eta]_\Lambda)^l, \qquad \xi, \eta \in \mathfrak{l},$$

where for $\zeta \in \mathfrak{g}^*$, ζ^l is the left invariant 1-form on G with value ζ at e, and

$$[\xi, \eta]_\Lambda \stackrel{\text{def}}{=} [\xi, \eta] + \mathrm{ad}^*_{\Lambda\xi}\, \eta - \mathrm{ad}^*_{\Lambda\eta}\, \xi, \qquad \xi, \eta \in \mathfrak{g}^*. \tag{4.8}$$

LEMMA 4.5. *Let* $\mathfrak{h}^0 = \{\xi \in \mathfrak{g}^* \mid \xi|_{\mathfrak{h}} = 0\}$. *Then* $[\xi, \eta]_\Lambda \in \mathfrak{h}^0$ *for all* $\xi, \eta \in \mathfrak{h}^0$.

PROOF. The condition $\mathrm{Ad}_h \mathfrak{l}_{\pi(eH)} = \mathfrak{l}_{\pi(eH)}$ for all $h \in H$ implies that

$$[x, \mathfrak{l}_{\pi(eH)}] \subset \mathfrak{l}_{\pi(eH)}$$

for all $x \in \mathfrak{h}$, so $[x, \Lambda\xi + \xi] \in \mathfrak{l}_{\pi(eH)}$ for all $x \in \mathfrak{h}$ and $\xi \in \mathfrak{h}^0$, from which it follows that $[\xi, \eta]_\Lambda \in \mathfrak{h}^0$ for all $\xi, \eta \in \mathfrak{h}^0$. See also [**4**]. □

Let $\chi_{\mathfrak{h}^0, \Lambda} \in (\mathfrak{h}^0)^*$ be defined by

$$\chi_{\mathfrak{h}^0, \Lambda}(\xi) = \mathrm{tr}(T_\xi), \qquad \xi \in \mathfrak{h}^0,$$

where $T_\xi \in \mathrm{End}(\mathfrak{h}^0) : \eta \mapsto [\xi, \eta]_\Lambda$ for $\xi, \eta \in \mathfrak{h}^0$. Let $\chi_{\mathfrak{l}} \in \mathfrak{l}^*, \chi_{\mathfrak{g}} \in \mathfrak{g}^*$, and $\chi_{\mathfrak{g}^*} \in \mathfrak{g}$ be the adjoint characters of $\mathfrak{l}, \mathfrak{g}$ and $\mathfrak{g}^*$ respectively. Let $b\Lambda = \sum_i [x_i, y_i] \in \mathfrak{g}$ if $\Lambda = \sum_i x_i \wedge y_i$. We now prove a fact that will be used in the proof of Theorem 4.7.

LEMMA 4.6. *For every* $\xi \in \mathfrak{h}^0$,

$$\chi_{\mathfrak{h}^0, \Lambda}(\xi) + (b\Lambda,\ \xi) = \frac{1}{2}\left(\chi_{\mathfrak{l}}(\Lambda\xi + \xi) - \chi_{\mathfrak{g}}(\Lambda\xi) + \chi_{\mathfrak{g}^*}(\xi)\right). \tag{4.9}$$

PROOF. For $\xi \in \mathfrak{g}^*$, consider the operator $T_\xi \in \mathrm{End}(\mathfrak{g}^*) : T_\xi(\eta) = [\xi,\ \eta]_\Lambda$, and define $\chi_{\mathfrak{g}^*, \Lambda}(\xi) = \mathrm{tr}(T_\xi \in \mathrm{End}(\mathfrak{g}^*))$. It is easy to see that

$$\chi_{\mathfrak{g}^*, \Lambda}(\xi) = \chi_{\mathfrak{g}^*}(\xi) - \chi_{\mathfrak{g}}(\Lambda\xi) - 2(b\Lambda, \xi), \qquad \xi \in \mathfrak{g}^*.$$

For $\xi \in \mathfrak{h}^0$, since $T_\xi(\mathfrak{h}^0) \subset \mathfrak{h}^0$, we have an induced map $T_\xi \in \mathrm{End}(\mathfrak{g}^*/\mathfrak{h}^0)$. Define $\chi(\xi) = \mathrm{tr}(T_\xi \in \mathrm{End}(\mathfrak{g}^*/\mathfrak{h}^0))$ for $\xi \in \mathfrak{h}^0$. Then

$$\chi_{\mathfrak{h}^0, \Lambda}(\xi) = \chi_{\mathfrak{g}^*, \Lambda}(\xi) - \chi(\xi) = \chi_{\mathfrak{g}^*}(\xi) - \chi_{\mathfrak{g}}(\Lambda\xi) - 2(b\Lambda, \xi) - \chi(\xi) \tag{4.10}$$

for all $\xi \in \mathfrak{h}^0$. On the other hand, consider the embedding $\kappa : \mathfrak{h}^0 \hookrightarrow \mathfrak{l}$ by $\xi \mapsto \Lambda\xi + \xi$, and let $\mathfrak{p}_{\mathfrak{h}} : \mathfrak{l} \to \mathfrak{h}$ be the projection with respect to the decomposition $\mathfrak{l} = \mathfrak{h} + \kappa(\mathfrak{h}^0)$. For $\xi \in \mathfrak{h}^0$, let $S_\xi \in \mathrm{End}(\mathfrak{h})$ be the operator $S_\xi(x) = \mathfrak{p}_{\mathfrak{h}}[\Lambda\xi + \xi,\ x]$ for $x \in \mathfrak{h}$. Then

$$\chi_{\mathfrak{h}^0, \Lambda}(\xi) = \chi_{\mathfrak{l}}(\Lambda\xi + \xi) - \mathrm{tr}(S_\xi \in \mathrm{End}(\mathfrak{h})), \qquad \forall \xi \in \mathfrak{h}^0.$$

By identifying $\mathfrak{g}^*/\mathfrak{h}^0 \cong \mathfrak{h}^*$, one can show that $-S_\xi^* = T_\xi \in \mathrm{End}(\mathfrak{g}^*/\mathfrak{h}^0)$, and so $\mathrm{tr}(S_\xi \in \mathrm{End}(\mathfrak{h})) = -\chi(\xi)$ for all $\xi \in \mathfrak{h}^0$. Thus

$$\chi_{\mathfrak{h}^0, \Lambda}(\xi) = \chi_{\mathfrak{l}}(\Lambda\xi + \xi) + \chi(\xi), \qquad \forall \xi \in \mathfrak{h}^0. \tag{4.11}$$

Adding (4.10) and (4.11), we get (4.9). □

4.3. The cotangent bundle Lie algebroid of $(G/H, \pi)$. Let (G, π_G) be a Poisson Lie group. For $x \in \mathfrak{g}$ and $\xi \in \mathfrak{g}^*$, let x^l (resp. ξ^l) be the left invariant vector field (resp. 1-form) on G with value x (resp. ξ) at e. Then [**15**] the map

$$\lambda : \mathfrak{d} \longrightarrow \mathcal{V}^1(G) : x + \xi \longmapsto \lambda_{x+\xi} := x^l + \tilde{\pi}_G(\xi^l) \tag{4.12}$$

is a Lie algebra homomorphism from $\mathfrak{d}$ to the space $\mathcal{V}^1(G)$ of vector fields on G. Let $p_{\mathfrak{g}} : \mathfrak{d} \to \mathfrak{g}$ be the projection along $\mathfrak{g}^*$. By (4.1), we also have

$$\lambda_{x+\xi}(g) = (r_g)_* p_{\mathfrak{g}} \mathrm{Ad}_g(x + \xi), \qquad g \in G,\ x \in \mathfrak{g},\ \xi \in \mathfrak{g}^*. \tag{4.13}$$

Let now $(G/H, \pi)$ be a (G, π_G)-homogeneous Poisson space, and let $\mathfrak{l} = \mathfrak{l}_{\pi(eH)}$ be the Drinfeld Lagrangian subalgebra of $\mathfrak{d}$ as in Definition 4.2. Then G, with the right action of H by right translations and the infinitesimal action of $\mathfrak{l}$ by λ, becomes an $(\mathfrak{l}, H)$-space in the sense of Definition 2.2. Let $G \rtimes_\lambda \mathfrak{l}$ be the corresponding transformation Lie algebroid over G.

THEOREM 4.7. *The cotangent bundle Lie algebroid of $(G/H, \pi)$ is isomorphic to the H-quotient $A = G \rtimes_{\lambda, H} (\mathfrak{l}/\mathfrak{h})$ of the transformation Lie algebroid $G \rtimes_\lambda \mathfrak{l}$.*

PROOF. Let $\Lambda \in \wedge^2 \mathfrak{g}$ be any element with $q(\Lambda) = \pi(eH) \in \wedge^2 T_{eH}(G/H) \cong \wedge^2(\mathfrak{g}/\mathfrak{h})$. Recall that $\mathfrak{h}^0 = \{\xi \in \mathfrak{g}^* \mid \xi|_{\mathfrak{h}} = 0\}$. The projection $\mathfrak{l} \to \mathfrak{g}^* : x + \xi \to \xi$ gives an H-equivariant isomorphism $\mathfrak{l}/\mathfrak{h} \to \mathfrak{h}^0$ whose inverse is $\mathfrak{h}^0 \to \mathfrak{l}/\mathfrak{h} : \xi \mapsto \xi + \Lambda\xi + \mathfrak{h}$.

Using left translations by elements in G and the identification $T^*_{eH}(G/H) \cong \mathfrak{h}^0$, we have the vector bundle isomorphism

$$I : T^*(G/H) \longrightarrow G \times_H \mathfrak{h}^0 \cong G \times_H (\mathfrak{l}/\mathfrak{h}). \tag{4.14}$$

It remains to show that I is a Lie algebroid isomorphism. Recall that $\pi = q_* \pi_\Lambda$, where π_Λ is the bi-vector field on G given by $\pi_\Lambda = \Lambda^l + \pi_G$.

Let $n = \dim \mathfrak{h}^0$, and let $\xi_1, \xi_2, \ldots, \xi_n$ be a basis of $\mathfrak{h}^0$. For $\alpha \in \Omega^1(G/H)$, write

$$q^*\alpha = \sum_{j=1}^n f_{\alpha,j} \xi_j^l \in \Omega^1(G), \qquad \text{where} \qquad f_{\alpha,j} \in C^\infty(G),\ j = 1, \ldots, n. \tag{4.15}$$

Then $I(\alpha) = \sum_{j=1}^n f_{\alpha,j} \xi_j \in C^\infty(G, \mathfrak{h}^0)^H \cong \Gamma(A)$, and $b_\alpha = \sum_{j=1}^n f_{\alpha,j}(\Lambda\xi_j + \xi_j) \in C^\infty(G, \mathfrak{l})^H$ is an H-invariant lifting of $I(a)$. Using $q_* \pi_\Lambda = \pi$, one has

$$\begin{aligned}
\tilde{\pi}(\alpha) &= q_* \tilde{\pi}_\Lambda(q^*\alpha) = q_* \left(\sum_{j=1}^n f_{\alpha,j} \tilde{\pi}_\Lambda(\xi_j^l) \right) = q_* \left(\sum_{j=1}^n f_{\alpha,j} ((\Lambda\xi_j)^l + \tilde{\pi}_G(\xi_j^l)) \right) \\
&= q_* \left(\sum_{j=1}^n f_{\alpha,j} \lambda_{\Lambda\xi_j + \xi_j} \right) = \rho_A(I(\alpha)).
\end{aligned}$$

Thus I maps the anchor map of $T^*(G/H)$ to the anchor map ρ_A of A.

It remains to show that $I\{\alpha, \beta\}_\pi = [I(\alpha), I(\beta)]$ for any $\alpha, \beta \in \Omega^1(G/H)$. Let $\{\,,\,\}_{\pi_\Lambda}$ be the skew-symmetric bracket on $\Omega^1(G)$ defined by replacing π by π_Λ in

(3.1). Using again the fact that $\pi = q_* \pi_\Lambda$, we have

$$\begin{aligned} q^*\{\alpha, \beta\}_\pi &= \{q^*\alpha, q^*\beta\}_{\pi_\Lambda} = \sum_{j,k} \{f_{\alpha,j}\xi_j^l,\, f_{\beta,k}\xi_k^l\}_{\pi_\Lambda} \\ &= \sum_{j,k} \left(f_{\alpha,j}\tilde{\pi}_\Lambda(\xi_j^l)(f_{\beta,k})\xi_k^l\right) - \sum_{j,k} \left(f_{\beta,k}\tilde{\pi}_\Lambda(\xi_k^l)(f_{\alpha,j})\xi_j^l\right) \\ &\quad + \sum_{j,k} \left(f_{\alpha,j} f_{\beta,k} \{\xi_j^l, \xi_k^l\}_{\pi_\Lambda}\right). \end{aligned}$$

Thus, by Remark 4.4,

$$\begin{aligned} I\{\alpha, \beta\}_\pi &= \sum_{j,k} \left(f_{\alpha,j}\lambda_{\Lambda\xi_j+\xi_j}(f_{\beta,k})\xi_k\right) - \sum_{j,k} \left(f_{\beta,k}\lambda_{\Lambda\xi_k+\xi_k}(f_{\alpha,j})\xi_j\right) \\ &\quad + \sum_{j,k} \left(f_{\alpha,j} f_{\beta,k}[\xi_j, \xi_k]_\Lambda\right), \end{aligned}$$

where the bracket $[\, ,\,]_\Lambda$ on $\mathfrak{g}^*$ is defined in (4.8). On the other hand, using

$$b_\alpha = \sum_{j=1}^n f_{\alpha,j}(\Lambda\xi_j + \xi_j) \quad \text{and} \quad b_\beta = \sum_{k=1}^n f_{\beta,k}(\Lambda\xi_k + \xi_k)$$

as H-invariant liftings of $I(\alpha)$ and $I(\beta)$ to smooth sections of $G \rtimes_\lambda \mathfrak{l}$, one can compute $[I(\alpha), I(\beta)] \in \Gamma(A)$ and see that $I\{\alpha, \beta\} = [I(\alpha), I(\beta)]$. This completes the proof that I is a Lie algebroid isomorphism. □

4.4. The canonical representation of $T^*(G/H, \pi)$ on $K_{G/H}$. Let $(G/H, \pi)$ be a Poisson homogeneous space of (G, π_G). Let

$$E = G \times \wedge^{\mathrm{top}}\mathfrak{h}^0$$

be the trivial H-equivariant line bundle over G, where

$$(g,\, Y)\cdot h = (gh,\, \mathrm{Ad}_h^* Y), \qquad g \in G,\, Y \in \wedge^{\mathrm{top}}\mathfrak{h}^0.$$

Then the identification $I : T^*(G/H) \to G \times_H \mathfrak{h}^0$ by left translation induces an identification $I : K_{G/H} \to E/H$. In this section, we show that E is naturally an $(\mathfrak{l}, H)$-line bundle and that the canonical representation of $T^*(G/H)$ on $K_{G/H} \cong E/H$ can be identified with the H-quotient of the representation of $G \rtimes_\lambda \mathfrak{l}$ on E, where $\mathfrak{l}$ is the Drinfeld Lagrangian subalgebra of $\mathfrak{d}$ associated to $\pi(eH)$, and λ is the infinitesimal action of $\mathfrak{l}$ on G given in (4.12) (see Definition 3.1 and Lemma 2.7).

Let $\wedge^{\mathrm{top}}\mathfrak{l}$ be the 1-dimensional $(\mathfrak{l}, H)$-module, on which $\mathfrak{l}$ acts by the adjoint character $\chi_\mathfrak{l}$ and $h \in H$ acts by $\mathrm{Ad}_h \in \mathrm{Aut}(\mathfrak{l})$. The trivial line bundle over G with fiber $\wedge^{\mathrm{top}}\mathfrak{l}$, still denoted by $\wedge^{\mathrm{top}}\mathfrak{l}$, is then an $(\mathfrak{l}, H)$-line bundle. Regard $\wedge^{\mathrm{top}}T^*G$ as an (l, H)-line bundle, on which H acts by right translation and $\mathfrak{l}$ acts by Lie derivatives via λ. Set

$$F = \wedge^{\mathrm{top}}\mathfrak{l} \otimes \wedge^{\mathrm{top}}T^*G.$$

Then F is an $(\mathfrak{l}, H)$-line bundle. Clearly, left translation in G gives rise to an H-equivariant trivialization

$$F \xrightarrow{\cong} G \times (\wedge^{\mathrm{top}}\mathfrak{l} \otimes \wedge^{\mathrm{top}}\mathfrak{g}^*),$$

where $h \in H$ acts on $G \times (\wedge^{\mathrm{top}}\mathfrak{l} \otimes \wedge^{\mathrm{top}}\mathfrak{g}^*)$ by

$$(g,\, X \otimes \mu)\cdot h = (gh,\, (\mathrm{Ad}_{h^{-1}}X) \otimes (\mathrm{Ad}_h^*\mu)), \qquad X \in \wedge^{\mathrm{top}}\mathfrak{l},\, \mu \in \wedge^{\mathrm{top}}\mathfrak{g}^*.$$

LEMMA 4.8. $\wedge^{\mathrm{top}}\mathfrak{l} \otimes \wedge^{\mathrm{top}}\mathfrak{g}^* \cong (\wedge^{\mathrm{top}}\mathfrak{h}^0)^2$ *as* H*-modules, so* $E^2 \cong F$ *as* H*-equivariant line bundles over* G.

PROOF. For $V \in \{\mathfrak{h}, \mathfrak{l}, \mathfrak{h}^0, \mathfrak{g}^*\}$, let $\chi_{H,V}$ be the character of the H-action on $\wedge^{\mathrm{top}}V$ induced from the adjoint and co-adjoint actions. It is easy to see that

$$\chi_{H,\mathfrak{l}} = \chi_{H,\mathfrak{h}}\chi_{H,\mathfrak{h}^0} \quad \text{and} \quad \chi_{H,\mathfrak{g}^*} = \chi_{H,\mathfrak{h}}^{-1}\chi_{H,\mathfrak{h}^0}.$$

Thus $\chi_{H,\mathfrak{l}}\chi_{H,\mathfrak{g}^*} = \chi_{H,\mathfrak{h}^0}^2$. □

Since F is an (l, H)-line bundle, so is E as a square root of F by Remark 2.9. In the next Lemma 4.9, we determine the l-module structure on $\Gamma(E)$. Recall that $\chi_{\mathfrak{l}} \in \mathfrak{l}^*, \chi_{\mathfrak{g}} \in \mathfrak{g}^*$ and $\chi_{\mathfrak{g}^*}$ are the adjoint characters of $\mathfrak{l}$, $\mathfrak{g}$, and $\mathfrak{g}^*$ respectively.

LEMMA 4.9. *Fix* $Y_0 \in \wedge^{\mathrm{top}}\mathfrak{h}^0$, $Y_0 \neq 0$, *and write elements in*

$$\Gamma(E) = C^\infty(G, \wedge^{\mathrm{top}}\mathfrak{h}^0)$$

as fY_0 *for* $f \in C^\infty(G)$. *Then the* l*-module structure on* $\Gamma(E)$ *is given by*

$$(x+\xi)\cdot(fY_0) = \left(\lambda_{x+\xi}(f) + \frac{1}{2}\left(\chi_l(x+\xi) - \chi_{\mathfrak{g}}(x) + \chi_{\mathfrak{g}^*}(\xi) - 2(\pi_G, d\xi^l\right) f\right) Y_0$$

for any $x+\xi \in \mathfrak{l}$ *and* $f \in C^\infty(G)$.

PROOF. Fix non-zero elements $X_0 \in \wedge^{\mathrm{top}}l$ and $\mu_0 \in \wedge^{\mathrm{top}}\mathfrak{g}^*$, and let μ_0^l be the left invariant volume form on G with $\mu_0^l(e) = \mu_0$. Then $X_0 \otimes \mu_0^l$ is a nowhere vanishing section of F. For $x+\xi \in \mathfrak{l}$, one has

$$\begin{aligned}(x+\xi)\cdot(X_0 \otimes \mu_0^l) &= \chi_{\mathfrak{l}}(x+\xi)X_0 \otimes \mu_0^l + X_0 \otimes L_{\lambda_{x+\xi}}\mu_0^l \\ &= (\chi_l(x+\xi) - \chi_{\mathfrak{g}}(x))X_0 \otimes \mu_0^l + X_0 \otimes L_{\tilde{\pi}_G(\xi^l)}\mu_0^l.\end{aligned}$$

By (3.2),

$$L_{\tilde{\pi}_G(\xi^l)}\mu_0^l = \{\xi^l, \mu_0^l\}_{\pi_G} - 2(\pi_G, d\xi^l)\mu_0^l = (\chi_{\mathfrak{g}^*}(\xi) - 2(\pi_G, d\xi^l))\mu_0^l,$$

from which the formula in Lemma 4.9 follows. □

By §2.3, the (l, H)-line bundle structure on E gives rise to a representation of the transformation Lie algebroid $G \rtimes_\lambda \mathfrak{l}$ on E and a representation of the H-quotient Lie algebroid $A = G \rtimes_{\lambda,H} (\mathfrak{l}/\mathfrak{h})$ on E/H.

THEOREM 4.10. *Under the identification* $I : T^*(G/H, \pi) \cong A = G \rtimes_{\lambda,H} (\mathfrak{l}/\mathfrak{h})$ *of Lie algebroids and the identification* $I : K_{G/H} \cong E/H$ *of line bundles, the canonical representation of* $T^*(G/H, \pi)$ *on* $K_{G/H}$ *becomes the* H*-quotient representation of* A *on* E/H.

PROOF. Denote by D both the canonical representation of $T^*(G/H, \pi)$ on $K_{G/H}$ and the quotient representation of A on E/H. We need to show that

$$D_{I(\alpha)}I(\mu) = I(D_\alpha\mu), \qquad \forall\, \alpha \in \Omega^1(G/H),\ \mu \in \Omega^{\mathrm{top}}(G/H). \tag{4.16}$$

Let $Y_0 = \xi_1 \wedge \cdots \wedge \xi_n \in \wedge^{\mathrm{top}}\mathfrak{h}^0$, where $\xi_1, \ldots, \xi_n$ is a basis for $\mathfrak{h}^0$, and write

$$q^*\alpha = \sum_{j=1}^n f_{\alpha,j}\xi_j^l \in \Omega^1(G) \quad \text{and} \quad q^*\mu = \phi\xi_1^l \wedge \cdots \xi_n^l \in \Omega^n(G),$$

where $f_{\alpha,j} \in C^\infty(G)$ for $j = 1, \ldots, n$, and $\phi \in C^\infty(G)$. Then

$$I(\alpha) = \sum_{j+1}^{n} f_{\alpha,j}\xi_j \in C^\infty(G, \mathfrak{h}^0)^H \quad \text{and} \quad I(\mu) = \phi Y_0 \in \Gamma(E)^H.$$

Moreover $b_\alpha := \sum_{j=1}^n f_{\alpha,j}(\Lambda\xi_j + \xi_j) \in \Gamma(G \bowtie_\lambda \mathfrak{l})^H$ is an H-invariant lifting of $I(\alpha)$ to a section of $G \bowtie_\lambda \mathfrak{l}$. Let $\widetilde{D}$ be the representation of $G \bowtie_\lambda \mathfrak{l}$ on E. By Lemma 4.9,

$$\begin{aligned} \widetilde{D}_{b_\alpha} f_\mu &= \sum_{j=1}^{n} f_{\alpha,j} \left(\lambda_{\Lambda\xi_j+\xi_j}(\phi) - (\pi_G, d\xi_j^l)\phi\right) Y_0 \\ &\quad + \frac{1}{2}\sum_{j=1}^{n} f_{\alpha,j} \left(\chi_{\mathfrak{l}}(\Lambda\xi_j + \xi_j) - \chi_{\mathfrak{g}}(\Lambda\xi_j) + \chi_{\mathfrak{g}^*}(\xi_j)\right) \phi Y_0. \end{aligned} \tag{4.17}$$

On the other hand, let Y_0^l be the left invariant n-form on G with $Y_0^l(e) = Y_0$. Then

$$\begin{aligned} q^* D_\alpha \mu &= q^*\left(\{\alpha, \mu\}_\pi - (\pi, d\alpha)\mu\right) = \{q^*\alpha, q^*\mu\}_{\pi_\Lambda} - (\pi_\Lambda, dq^*\alpha)q^*\mu \\ &= \sum_{j=1}^{n} \left(\{f_{\alpha,j}\xi_j^l, \phi Y_0^l\}_{\pi_\Lambda} - (\pi_\Lambda, d(f_{\alpha,j}\xi_j^l))\phi Y_0^l\right) \\ &= \sum_{j=1}^{n} \left(f_{\alpha,j}\tilde{\pi}_\Lambda(\xi_j^l)(\phi)Y_0^l + \{f_{\alpha,j}\xi_j^l, Y_0^l\}_{\pi_\Lambda}\phi\right) \\ &\quad - \sum_{j=1}^{n} (\pi_\Lambda, df_{\alpha,j} \wedge \xi_j^l + f_{\alpha,j} d\xi_j^l)\phi Y_0^l \\ &= \sum_{j=1}^{n} f_{\alpha,j} \left(\lambda_{\Lambda\xi_j+\xi_j}(\phi) - (\pi_G, d\xi_j^l)\phi\right) Y_0^l + \{f_{\alpha,j}\xi_j^l, Y_0^l\}_{\pi_\Lambda}\phi \\ &\quad + \sum_{j=1}^{n} \left(\tilde{\pi}_\Lambda(\xi_j^l)(f_{\alpha,j}) - f_{\alpha,j}(\Lambda^l, d\xi_j^l)\right) \phi Y_0^l. \end{aligned}$$

Using the properties of the Schouten bracket $\{\,,\,\}_{\pi_\Lambda}$ on $\Omega(G)$, one has

$$\sum_{j=1}^{n} \{f_{\alpha,j}\xi_j^l, Y_0^l\}_{\pi_\Lambda} = \sum_{j=1}^{n} \left(f_{\alpha,j}\{\xi_j^l, Y_0^l\}_{\pi_\Lambda} - \tilde{\pi}_\Lambda(\xi_j^l)(f_{\alpha,j})Y_0^l\right).$$

Thus

$$\begin{aligned} q^* D_\alpha \mu &= \sum_{j=1}^{n} f_{\alpha,j} \left(\lambda_{\Lambda\xi_j+\xi_j}(\phi) - (\pi_G, d\xi_j^l)\phi\right) Y_0^l \\ &\quad + \sum_{j=1}^{n} f_{\alpha,j} \left(\{\xi_j^l, Y_0^l\}_{\pi_\Lambda}\phi - (\Lambda^l, d\xi_j^l)\phi Y_0^l\right). \end{aligned} \tag{4.18}$$

By Lemma 4.6,

$$\{\xi, Y_0^l\}_{\pi_\Lambda} - (\Lambda^l, d\xi^l)Y_0^l = \frac{1}{2}\left(\chi_{\mathfrak{l}}(\Lambda\xi + \xi) - \chi_{\mathfrak{g}}(\Lambda\xi) + \chi_{\mathfrak{g}^*}(\xi)\right) Y_0^l, \qquad \forall \xi \in \mathfrak{h}^0.$$

Comparing with (4.17) and (4.18), we see that (4.16) holds. □

4.5. Poisson cohomology of $(G/H,\pi)$. Let the notation be as in §4.3. For any integer N, since E is a trivial line bundle over G, $\Gamma(E^N)\cong C^\infty(G)$ as vector spaces. The induced (l,H)-module structure on $C^\infty(G)$ is specified as follows.

NOTATION 4.11. For an integer N, denote $C^\infty(G)_N$ the vector space $C^\infty(G)$ with the following $(\mathfrak{l},H)$-module structure: for $x+\xi\in\mathfrak{l}, h\in H$ and $f\in C^\infty(G)$,

$$\begin{aligned}(x+\xi)\cdot_N f &= \lambda_{x+\xi}(f)+\frac{N}{2}\left(\chi_{\mathfrak{l}}(x+\xi)-\chi_{\mathfrak{g}}(x)+\chi_{\mathfrak{g}^*}(\xi)-2(\pi_G,d\xi^l)\right)f,\\ h\cdot_N f &= (\chi_{H,\mathfrak{h}^0}(h))^N\,(f\circ r_h),\end{aligned}$$

where $\chi_{H,\mathfrak{h}^0}(h)=\det(\mathrm{Ad}^*_{h^{-1}}:\mathfrak{h}^0\to\mathfrak{h}^0)$ and r_h is the right translation by h.

We can now identify the Poisson cohomology of G/H with relative Lie algebra cohomology. Corollary 4.12 follows directly from Lemma 2.8, Theorem 4.7, and Theorem 4.10.

COROLLARY 4.12. *For any integer N,*

$$H^\bullet\left(G/H,\pi;K^N_{G/H}\right)\cong H^\bullet_{\mathrm{Lie}}(\mathfrak{l},H;C^\infty(G)_N).$$

where the left hand side is the generalized Poisson cohomology of $(G/H,\pi)$ and the right hand side is the Lie algebra cohomology of $\mathfrak{l}$ relative to H with coefficients in $C^\infty(G)_N$.

The special case of Corollary 4.12 when $N=0$ was proved in [**15**].

4.6. The pairing on the Poisson cohomology. Assume that G/H is compact and orientable with a fixed orientation, so one has the map

$$\Omega^{\mathrm{top}}(G/H)\longrightarrow\mathbb{R}:\ \omega\longmapsto\int_{G/H}\omega. \tag{4.19}$$

By §3.2, for any integer N and any $0\le k\le n=\dim(G/H)$, there is a well-defined pairing $(\ ,\ )$ between $H^k\left(G/H,\pi;K^N_{G/H}\right)$ and $H^{n-k}\left(G/H,\pi;K^{2-N}_{G/H}\right)$. In view of Corollary 4.12, we now identify this pairing with a pairing on the corresponding relative Lie algebra cohomology spaces. Let the notation be as in §4.4. Then we have the identifications of H-modules:

$$\begin{aligned}\wedge^{\mathrm{top}}(\mathfrak{l}/\mathfrak{h})^*\otimes\Gamma(E^N)\otimes\Gamma(E^{2-N}) &\cong \wedge^{\mathrm{top}}(\mathfrak{l}/\mathfrak{h})^*\otimes\Gamma(E^2)\\ &\cong \wedge^{\mathrm{top}}(\mathfrak{l}/\mathfrak{h})^*\otimes\Gamma(F)\\ &\cong \wedge^{\mathrm{top}}(\mathfrak{l}/\mathfrak{h})^*\otimes\wedge^{\mathrm{top}}\mathfrak{l}\otimes\Omega^{\mathrm{top}}(G)\\ &\cong \wedge^{\mathrm{top}}(\mathfrak{l}/\mathfrak{h})^*\otimes\wedge^{\mathrm{top}}\mathfrak{l}\otimes\wedge^{\mathrm{top}}\mathfrak{g}^*\otimes C^\infty(G)\\ &\cong \wedge^{\mathrm{top}}(\mathfrak{l}/\mathfrak{h})^*\otimes(\wedge^{\mathrm{top}}\mathfrak{h}^0)^2\otimes C^\infty(G)\\ &\cong \wedge^{\mathrm{top}}\mathfrak{h}^0\otimes C^\infty(G),\end{aligned}$$

where we used Lemma 4.8 to identify $\wedge^{\mathrm{top}}\mathfrak{l}\otimes\wedge^{\mathrm{top}}\mathfrak{g}^*\cong(\wedge^{\mathrm{top}}\mathfrak{h}^0)^2$ and left translation in G to identify $\Omega^{\mathrm{top}}(G)\cong\wedge^{\mathrm{top}}\mathfrak{g}^*\otimes C^\infty(G)$. Thus we have an identification

$$(\wedge^{\mathrm{top}}(\mathfrak{l}/\mathfrak{h})^*\otimes\Gamma(E^N)\otimes\Gamma(E^{2-N}))^H\cong(\wedge^{\mathrm{top}}(\mathfrak{l}/\mathfrak{h})\otimes C^\infty(G))^H\cong\Omega^{\mathrm{top}}(G/H). \tag{4.20}$$

Let $\nu:(\wedge^{\mathrm{top}}(\mathfrak{l}/\mathfrak{h})^*\otimes\Gamma(E^N)\otimes\Gamma(E^{2-N}))^H\to\mathbb{R}$ be the composition of the identification in (4.20) with the integration map in (4.19). One checks directly that (2.2) holds and that, under the identifications in Corollary 4.12, the canonical pairing between $H^k\left(G/H,\pi;K_{G/H}^N\right)$ and $H^{n-k}\left(G/H,\pi;K_{G/H}^{2-N}\right)$ coincides with the pairing between $H^k_{\mathrm{Lie}}(\mathfrak{l},H;C^\infty(G)_N)$ and $H^{n-k}_{\mathrm{Lie}}(\mathfrak{l},H;C^\infty(G)_{2-N})$ induced by ν (see §2.2).

4.7. Modular vector fields of $(G/H,\pi)$. Assume again that G/H is orientable and let μ be a fixed volume form on G/H. Fix a non-zero $Y_0\in\wedge^{\mathrm{top}}\mathfrak{h}^0$, and let Y_0^l be the corresponding left invariant form on G. Write $q^*\mu=\phi Y_0^l$, with $\phi\in C^\infty(G)$ everywhere non-zero. Let $\Lambda\in\wedge^2\mathfrak{g}$ be any element such that $q(\Lambda)=\pi(eH)\in\wedge^2T_eG/H\cong\wedge^2\mathfrak{g}/\mathfrak{h}$, and let $\pi_\Lambda=\Lambda^l+\pi_G$ so that $q_*\pi_\Lambda=\pi$. Recall that $\chi_{\mathfrak{l}}\in\mathfrak{l}^*,\chi_{\mathfrak{g}}\in\mathfrak{g}^*$ and $\chi_{\mathfrak{g}^*}\in\mathfrak{g}$ are the adjoint characters of $\mathfrak{l},\mathfrak{g}$ and $\mathfrak{g}^*$ respectively. Write $x_0=\chi_{\mathfrak{g}^*}\in\mathfrak{g}$, $\xi_0=\chi_{\mathfrak{g}}\in\mathfrak{g}^*$, and let $x_{\mathfrak{l}}$ be any element in $\mathfrak{g}^*$ such that $x_{\mathfrak{l}}(\xi)=\chi_{\mathfrak{l}}(\Lambda\xi+\xi)$ for $\xi\in\mathfrak{h}^0$. Recall that for $x\in\mathfrak{g}$ and $\xi\in\mathfrak{g}$, x^l (resp. x^r and ξ^l) is the left (resp. right) invariant vector field and one form on G with values x and ξ at $e\in G$.

LEMMA 4.13. *Let the notation be as above. Let X be the vector field on G given by*

$$X=-\tilde{\pi}_\Lambda(d\log|\phi|)+\frac{1}{2}\left(x_{\mathfrak{l}}^l+x_0^r+\tilde{\pi}_\Lambda(\xi_0^l)\right).$$

Then q_*X is a well-defined vector field on G/H, and it is the modular vector field of π with respect to μ.

PROOF. Let $\xi_1,\ldots,\xi_n$ be a basis of $\mathfrak{h}^0$ such that $\xi_1\wedge\cdots\wedge\xi_n=Y_0$. Let $\alpha\in\Omega^1(G/H)$, and write $q^*\alpha=\sum_{j=1}^n f_{\alpha,j}\xi_j^l\in\Omega^1(G)$. As in the proof of Theorem 4.10,

$$\begin{aligned}q^*D_\alpha\mu &= -\sum_{j=1}^n f_{\alpha,j}((\xi_j^l,\,\tilde{\pi}_\Lambda(d\log|\phi|)+(\pi_G,\,d\xi_j^l))Y_0^l\\ &\quad+\frac{1}{2}\sum_{j=1}^n f_{\alpha,j}(\chi_{\mathfrak{l}}(\Lambda\xi_j+\xi_j)-\chi_{\mathfrak{g}}(\Lambda\xi_j)+\chi_{\mathfrak{g}^*}(\xi_j))Y_0^l\\ &= \left(q^*\alpha,\;-\tilde{\pi}_\Lambda(d\log|\phi|)-F_0+\frac{1}{2}(x_{\mathfrak{l}}^l+(\Lambda\xi_0)^l+x_0^l)\right),\end{aligned}$$

where F_0 is the vector field on G such that $(F_0,\xi^l)=(\pi_G,d\xi^l)$ for all $\xi\in\mathfrak{g}^*$. It is shown in Proposition 4.7 of [7] that $F_0=\frac{1}{2}(x_0^l-x_0^r-\tilde{\pi}_G(\xi_0^l))$. Thus we have

$$q^*D_\alpha\mu=(q^*\alpha,\;X).$$

It follows that q_*X is a well-defined vector field on G/H and it is the modular vector field of π with respect to μ. □

REMARK 4.14. Note that if μ is a G-invariant volume form on G/H, the modular vector field of π with respect to μ is

$$q_*X=\frac{1}{2}\left(x_{\mathfrak{l}}^l+x_0^r+\tilde{\pi}_\Lambda(\xi_0^l)\right).$$

This formula for the special case when $\mathfrak{h}^0$ is an ideal of $\mathfrak{g}^*$ has been obtained in [7].

5. A Poisson groupoid over $(G/H, \pi)$

When H is a Poisson Lie subgroup of (G, π_G) and $\pi = q_*\pi_G$, where $q : G \to G/H$ is the projection, a symplectic groupoid of $(G/H, \pi)$ was constructed in [**25**] (under the additional assumption that (G, π_G) is complete). In this section, let $(G/H, \pi)$ be an arbitrary Poisson homogeneous space of (G, π_G) with the Drinfeld Lagrangian subalgebra $\mathfrak{l} = \mathfrak{l}_{\pi(eH)}$. We assume that G is a closed subgroup of a connected Lie group D with Lie algebra $\mathfrak{d}$, $H = G \cap L$, where L is the connected subgroup of D with Lie algebra $\mathfrak{l}$, and that the infinitesimal action λ of $\mathfrak{l}$ on G in (4.12) integrates to an action of L on G. We will show that the associated space $\Gamma = G \times_H (L/H)$ is a Poisson groupoid over $(G/H, \pi)$. We also give conditions for Γ to be symplectic. The Poisson structure on Γ is obtained from reduction of a quasi-Poisson manifold by an action of a quasi-Poisson Lie group [**1**].

5.1. The quasi-Poisson Lie group $(G, \pi_{G,\Lambda}, \varphi)$. Let (G, π_G) be a Poisson Lie group corresponding to Manin triple $(\mathfrak{d}, \mathfrak{g}, \mathfrak{g}^*)$. Then any $\Lambda \in \wedge^2\mathfrak{g}$ (not necessarily related to any Poisson homogeneous space of (G, π_G) as in §4.2) can be used to twist the Manin triple $(\mathfrak{d}, \mathfrak{g}, \mathfrak{g}^*)$ to a Manin quasi-triple $(\mathfrak{d}, \mathfrak{g}, \mathfrak{g}')$ [**1**], where

$$\mathfrak{g}' = \{\Lambda\xi + \xi \,|\, \xi \in \mathfrak{g}^*\}, \tag{5.1}$$

and thus defines a quasi-Poisson Lie group structure on G. More precisely, let $p_1 : \mathfrak{d} \to \mathfrak{g} : x + \xi \mapsto x - \Lambda\xi$, where $x \in \mathfrak{g}$ and $\xi \in \mathfrak{g}^*$, be the projection from $\mathfrak{d} = \mathfrak{g} + \mathfrak{g}'$ to $\mathfrak{g}$ along $\mathfrak{g}'$, and define $\varphi \in \wedge^3\mathfrak{g}$ by

$$\varphi(\xi \wedge \eta \wedge \zeta) = \langle p_1[\Lambda\xi + \xi,\, \Lambda\eta + \eta],\, \Lambda\zeta + \zeta\rangle, \qquad \xi, \eta, \zeta \in \mathfrak{g}^*.$$

It is straightforward to check that, for any $\xi, \eta, \zeta \in \mathfrak{g}^*$,

$$\begin{aligned}\varphi(\xi \wedge \eta \wedge \zeta) = {} & \langle[\Lambda\xi, \Lambda\eta],\, \zeta\rangle + \langle[\Lambda\eta, \Lambda\zeta],\, \xi\rangle + \langle[\Lambda\zeta, \Lambda\xi],\, \eta\rangle \\ & + \langle\Lambda\xi,\, [\eta, \zeta]\rangle + \langle\Lambda\eta,\, [\zeta, \xi]\rangle + \langle\Lambda\zeta,\, [\xi, \eta]\rangle.\end{aligned}$$

In fact $\varphi = \frac{1}{2}[\Lambda, \Lambda] + \delta(\Lambda)$. Let Λ^l and Λ^r be respectively the left and right invariant bi-vector fields on G with value Λ at e, and define

$$\pi_{G,\Lambda} = \Lambda^l - \Lambda^r + \pi_G, \tag{5.2}$$

LEMMA 5.1. [**1**] *$(G, \pi_{G,\Lambda}, \varphi)$ is a quasi-Poisson Lie group corresponding to the Manin quasi-triple $(\mathfrak{d}, \mathfrak{g}, \mathfrak{g}')$ in the sense that $\pi_{G,\Lambda}$ is multiplicative,*

$$\frac{1}{2}[\pi_{G,\Lambda},\, \pi_{G,\Lambda}] = \varphi^l - \varphi^r, \qquad \text{and} \qquad [\pi_{G,\Lambda},\, \varphi^l] = 0,$$

where φ^r (resp. φ^l) is the right (resp. left) invariant tri-vector field on G with value φ at e.

Recall from [**1**] that a (right) quasi-Poisson action of $(G, \pi_{G,\Lambda}, \varphi)$ on a manifold P with a bi-vector field π_P is a right action $\rho : P \times G \to P$ of G on P such that

1) $[\pi_P, \pi_P] = 2\rho_\varphi$ and

2) $\rho : (P, \pi_P) \times (G, \pi_{G,\Lambda}) \to (P, \pi_P)$ is a bi-vector map,

where $\rho : x \mapsto \rho_x$ also denotes the Lie algebra homomorphism $\mathfrak{g} \to \mathcal{V}^1(P)$ given by

$$\rho_x(p) = \frac{d}{dt}|_{t=0}\, p\exp(tx), \qquad x \in \mathfrak{g},\, p \in P \tag{5.3}$$

as well as its multi-linear extension $\wedge^k\mathfrak{g} \to \mathcal{V}^k(P) : X \mapsto \rho_X$ for any integer $k \geq 1$. Left quasi-Poisson actions of $(G, \pi_{G,\Lambda}, \varphi)$ are similarly defined.

EXAMPLE 5.2. Let $\pi_\Lambda = \Lambda^l + \pi_G$. It is easy to see that the action of $(G, \pi_{G,\Lambda}, \varphi)$ on (G, π_Λ) by right multiplication is a right quasi-Poisson action. For another example, assume that D is a connected Lie group with Lie algebra $\mathfrak{d}$ and that G is a closed subgroup of D. For $d \in D$, let $\underline{d} = dG \in D/G$, and for $x + \xi \in \mathfrak{d}$ with $x \in \mathfrak{g}$ and $\xi \in \mathfrak{g}^*$, let $\sigma_{x+\xi}$ be the vector field on D/G given by

$$\sigma_{x+\xi}(\underline{d}) = \frac{d}{dt}|_{t=0} \exp(t(x+\xi))\underline{d} \in T_{\underline{d}}(D/G), \qquad d \in D. \tag{5.4}$$

Let $\sigma : \wedge^k \mathfrak{d} \to \mathcal{V}^k(D/G) : X \mapsto \sigma_X$ also denote the multi-linear extension of σ. Let $\{x_i\}_{n=1}^n$ be a basis of $\mathfrak{g}$ and let $\{\xi_i\}_{i=1}^n$ be its dual basis of $\mathfrak{g}^*$. Define the bi-vector fields $\pi_{D/G}$ and $\pi_{D/G,\Lambda}$ on D/G respectively by

$$\pi_{D/G} = \frac{1}{2}\sum_i \sigma_{\xi_i} \wedge \sigma_{x_i} \quad \text{and} \quad \pi_{D/G,\Lambda} = \frac{1}{2}\sum_i \sigma_{\Lambda\xi_i + \xi_i} \wedge \sigma_{x_i} = \pi_{D/G} - \sigma_\Lambda. \tag{5.5}$$

Then [**1**] $\pi_{D/G}$ is Poisson and the action

$$(G, \pi_{G,\Lambda}, \varphi) \times (D/G, \pi_{D/G,\Lambda}) \longrightarrow (D/G, \pi_{D/G,\Lambda}) : \ (g, \underline{d}) \longmapsto g\underline{d}, \quad g \in G, d \in D,$$

is a left quasi-Poisson action of $(G, \pi_{G,\Lambda}, \varphi)$. In particular,

$$[\pi_{D/G,\Lambda}, \ \pi_{D/G,\Lambda}] = -2\sigma_\varphi. \tag{5.6}$$

Moreover, let $\delta_{\mathfrak{g}'} : \mathfrak{g}' \to \wedge^2 \mathfrak{g}'$ be defined by

$$\langle \delta_{\mathfrak{g}'}(\Lambda\xi + \xi), \ x \wedge y\rangle = \langle \Lambda\xi + \xi, \ [x, y]\rangle = \langle \xi, \ [x, y]\rangle, \qquad \xi \in \mathfrak{g}^*, \ x, y \in \mathfrak{g}.$$

Then one can check that

$$[\sigma_{\Lambda\xi+\xi}, \ \pi_{D/G,\Lambda}] = -\sigma_{\delta_{\mathfrak{g}'}(\Lambda\xi+\xi)} + \sigma_{\iota_\xi \varphi}, \qquad \forall \xi \in \mathfrak{g}^*. \tag{5.7}$$

5.2. The bivector field π_P on $P = G \times (D/G)$. Let the assumptions be as in §5.1. In particular, assume that D is a connected Lie group with Lie algebra $\mathfrak{d}$ and that G is a closed subgroup of D. Let $P = G \times (D/G)$. For any integer $k \geq 1$ and for a k-vector field V on G, let $(V, 0)$ be the corresponding k-vector field on P. Similarly a k-vector field U on D/G gives rise to the k-vector field $(0, U)$ on P. For $x \in \mathfrak{g}$, recall that x^l is the left invariant vector field on G with $x^l(e) = x$. Define the bi-vector field π_P on P by

$$\pi_P = (\pi_\Lambda, 0) - (0, \pi_{D/G,\Lambda}) + \sum_{i=1}^n (0, \sigma_{\Lambda\xi_i + \xi_i}) \wedge (x_i^l, 0). \tag{5.8}$$

LEMMA 5.3. *The right action*

$$\rho : \ (P, \pi_P) \times (G, \pi_{G,\Lambda}, \varphi) \longrightarrow (P, \pi_P) : \ (g, \underline{d}) \cdot g_1 = (gg_1, \ g_1^{-1}\underline{d}), \qquad g, g_1 \in G, d \in D,$$

is a quasi-Poisson action of $(G, \pi_{G,\Lambda}, \varphi)$.

PROOF. To show that $[\pi_P, \pi_P] = 2\rho_\varphi$, let $\varphi = \sum_k a_k \wedge b_k \wedge c_k \in \wedge^3 \mathfrak{g}$, and let

$$\rho'_\varphi = \sum_k \left((0, \sigma_{a_k}) \wedge (b_k^l \wedge c_k^l, 0) + (0, \sigma_{b_k}) \wedge (c_k^l \wedge a_k^l, 0) + (0, \sigma_{c_k}) \wedge (a_k^l \wedge b_k^l, 0)\right)$$

$$\rho''_\varphi = \sum_k \left((a_k^l, 0) \wedge (0, \sigma_{b_k \wedge c_k}) + (b_k^l, 0) \wedge (0, \sigma_{c_k \wedge a_k}) + (c_k^l, 0) \wedge (0, \sigma_{a_k \wedge b_k})\right).$$

It is easy to see that $\rho_\varphi = (\varphi^l, 0) - (0, \sigma_\varphi) - \rho'_\varphi + \rho''_\varphi$. On the other hand, let $\pi_0 = \sum_{i=1}^n (0, \sigma_{\Lambda\xi_i+\xi_i}) \wedge (x_i^l, 0)$, so that $\pi_P = (\pi_\Lambda, 0) - (0, \pi_{D/G,\Lambda}) + \pi_0$, and

$$\begin{aligned}[\pi_P,\ \pi_P] &= ([\pi_\Lambda, \pi_\Lambda], 0) + (0, [\pi_{D/G,\Lambda}, \pi_{D/G,\Lambda}]) \\ &\quad + 2[\pi_0,\ (\pi_\Lambda, 0) - (0, \pi_{D/G,\Lambda})] + [\pi_0,\ \pi_0] \\ &= 2(\varphi^l, 0) - 2(0, \sigma_\varphi) + 2[\pi_0,\ (\pi_\Lambda, 0) - (0, \pi_{D/G,\Lambda})] + [\pi_0,\ \pi_0].\end{aligned}$$

It is easy to see that $[\pi_0, \pi_0] = \pi_1 + \pi_2$, where

$$\pi_1 = -\sum_{i,j=1}^n (0, \sigma_{[\Lambda\xi_i+\xi_i, \Lambda\xi_j+\xi_j]}) \wedge (x_i^l \wedge x_j^l, 0),$$

$$\pi_2 = \sum_{i,j=1}^n ([x_i, x_j]^l, 0) \wedge (0, \sigma_{(\Lambda\xi_i+\xi_i)\wedge\Lambda\xi_j+\xi_j)}).$$

Thus $[\pi_P, \pi_P] = 2(\varphi^l, 0) - 2(0, \sigma_\varphi) + 2[\pi_0, (\pi_\Lambda, 0)] + \pi_1 - 2[\pi_0, (0, \pi_{D/G,\Lambda})] + \pi_2$. Now

$$2[\pi_0, (\pi_\Lambda, 0)] = 2\sum_{i=1}^n (0, \sigma_{\Lambda\xi_i+\xi_i}) \wedge (([x_i, \Lambda] + \delta(x_i))^l, 0).$$

Recall that $p_1 : \mathfrak{d} \to \mathfrak{g}$ is the projection along $\mathfrak{g}'$. Let $p' : \mathfrak{d} \to \mathfrak{g}'$ be the projection along $\mathfrak{g}$. It is easy to check that

$$\sum_{i,j=1}^n p'[\Lambda\xi_i + \xi_i,\ \Lambda\xi_j + \xi_j] \otimes x_i \wedge x_j = 2\sum_{i=1}^n (\Lambda\xi_i + \xi_i) \otimes ([x_i, \Lambda] + \delta(x_i))$$

$$\sum_{i,j=1}^n p_1[\Lambda\xi_i + \xi_i,\ \Lambda\xi_j + \xi_j] \otimes x_i \wedge x_j = 2\tilde{\varphi},$$

where $\tilde{\varphi} = \sum_k (a_k \otimes b_k \wedge c_k + b_k \otimes c_k \wedge a_k + c_k \otimes a_k \wedge b_k)$. Thus $2[\pi_0, (\pi_\Lambda, 0)] + \pi_1 = -2\rho'_\varphi$. Similarly, by (5.7),

$$[\pi_0, \pi_{D/G,\Lambda}] = \sum_{i=1}^n (x_i^l, 0) \wedge (0, -[\sigma_{\Lambda\xi_i+\xi_i},\ \pi_{D/G,\Lambda}]) = \sum_{i=1}^n (x_i^l, 0) \wedge (0, \sigma_{\delta_{\mathfrak{g}'}(\Lambda\xi_i+\xi_i)} - \sigma_{\iota_{\xi_i}\varphi}).$$

It is easy to check that $\sum_{i=1}^n x_i \otimes \iota_{\xi_i}\varphi = \tilde{\varphi}$ and that

$$2\sum_{i=1}^n x_i \otimes \delta_{\mathfrak{g}'}(\Lambda\xi_i + \xi_i) = \sum_{i,j=1}^n [x_i, x_j] \otimes (\Lambda\xi_i + \xi_i) \wedge (\Lambda\xi_j + \xi_j).$$

Thus $-2[\pi_0, \pi_{D/G,\Lambda}] + \pi_2 = 2\rho''_\varphi$. Hence $[\pi_P, \pi_P] = 2\rho_\varphi$. The proof that ρ is a bi-vector map is straightforward and we omit the details. □

We now study when π_P on $P = G \times (D/G)$ is nondegenerate. For $d \in D$, the linear map $\mathfrak{d} \to T_{\underline{d}}(D/G) : x + \xi \mapsto \sigma_{x+\xi}(\underline{d})$, where $x \in \mathfrak{g}$ and $\xi \in \mathfrak{g}^*$, induces an isomorphism $\mathfrak{d}/\mathrm{Ad}_d\mathfrak{g} \to T_{\underline{d}}(D/G)$. For $y \in \mathfrak{g}$ and $\eta \in \mathfrak{g}^*$, let $\alpha_{y+\eta}(\underline{d}) \in T^*_{\underline{d}}(D/G)$ be such that

(5.9) $$(\alpha_{y+\eta}(\underline{d}),\ \sigma_{x+\xi}(\underline{d})) = \langle y + \eta,\ x + \xi\rangle, \qquad x \in \mathfrak{g}, \xi \in \mathfrak{g}^*.$$

Then we have the isomorphism

(5.10) $$\mathrm{Ad}_d\mathfrak{g} \longrightarrow T^*_{\underline{d}}(D/G) :\ y + \eta \longmapsto \alpha_{y+\eta}(\underline{d}), \qquad y \in \mathfrak{g},\ \eta \in \mathfrak{g}^*,\ y + \eta \in \mathrm{Ad}_d\mathfrak{g}.$$

Note that when $y+\eta \in \mathrm{Ad}_d\mathfrak{g}$, $\sigma_{y+\eta}(\underline{d}) = 0$, so $\sigma_y(\underline{d}) = -\sigma_\eta(\underline{d})$. The proof of the first identity in the following Lemma 5.4 is straightforward and is omitted. The second identity follows from (4.13).

LEMMA 5.4. *For $g \in G, d \in D, \xi \in \mathfrak{g}^*$ and $y+\eta \in \mathrm{Ad}_d\mathfrak{g}$ with $y \in \mathfrak{g}$ and $\eta \in \mathfrak{g}^*$,*

$$\begin{aligned}\tilde{\pi}_P(l_{g^{-1}}^*\xi,\, \alpha_{y+\eta}(\underline{d})) &= \left(\tilde{\pi}_G(l_{g^{-1}}^*\xi) + (l_g)_*(y+\Lambda\xi - \Lambda\eta),\ \ \sigma_{\Lambda\eta+\eta-\Lambda\xi-\xi}(\underline{d})\right)\\ &= (\lambda_{y-\Lambda\eta+\Lambda\xi+\xi}(g),\ \ -\sigma_{y-\Lambda\eta+\Lambda\xi+\xi}(\underline{d}))\end{aligned}$$

LEMMA 5.5. *The bi-vector field π_P on $P = G \times (D/G)$ is nondegenerate at $(g, \underline{d})$, where $g \in G$ and $d \in D$, if*

$$\mathfrak{g}' \cap \mathrm{Ad}_d\mathfrak{g} = 0 \quad \textit{and} \quad \mathfrak{g}^* \cap \mathrm{Ad}_{gd}\mathfrak{g} = 0. \tag{5.11}$$

In particular, π_P is nondegenerate at $(g, \underline{e})$ for any $g \in G$, where $e \in D$ is the identity.

PROOF. Assume that (5.11) holds at $(g, \underline{d}) \in P$. Suppose that $\xi \in \mathfrak{g}^*$ and $y + \eta \in \mathrm{Ad}_d\mathfrak{g}$ are such that $\tilde{\pi}_P(l_{g^{-1}}^*\xi,\, \alpha_{y+\eta}(\underline{d})) = 0$. Then $\sigma_{\Lambda\eta+\eta-\Lambda\xi-\xi}(\underline{d}) = 0$ by Lemma 5.4, so $\Lambda\eta + \eta - \Lambda\xi - \xi \in \mathfrak{g}' \cap \mathrm{Ad}_d\mathfrak{g} = 0$. Thus $\xi = \eta$. By (4.1), $\tilde{\pi}_G(l_{g^{-1}}^*\xi) = (r_g)_* p_{\mathfrak{g}} \mathrm{Ad}_g \xi$, where $p_{\mathfrak{g}} : \mathfrak{d} \to \mathfrak{g}$ is the projection along $\mathfrak{g}^*$. Thus Lemma 5.4 implies that $p_{\mathfrak{g}}\mathrm{Ad}_g(y+\xi) = 0$, so $\mathrm{Ad}_g(y+\xi) \in \mathfrak{g}^* \cap \mathrm{Ad}_{gd}\mathfrak{g} = 0$. Thus $y = 0$ and $\xi = \eta = 0$. □

REMARK 5.6. Let $N(\mathfrak{g}^*)$ be the normalizer subgroup of $\mathfrak{g}^*$ in D. Suppose that $D = N(\mathfrak{g}^*)G$ and that $\Lambda = 0$ (so $\pi(eH) = 0$). Then (5.11) holds for all $(g, d) \in G \times D$, and π_P is nondegenerate everywhere on P. See Example 5.14 for an example.

5.3. The Poisson structure Π on $G \times_H (L/H)$. Let the notation be as in §5.1 and §5.2, but assume now that $(G/H, \pi)$ is a Poisson homogeneous space of (G, π) and that $\Lambda \in \wedge^2\mathfrak{g}$ is such that $q(\Lambda) = \pi(eH) \in \wedge^2 T_{eH}(G/H) \cong \wedge^2(\mathfrak{g}/\mathfrak{h})$, where q denotes both projections $G \to G/H$ and $\mathfrak{g} \to \mathfrak{g}/\mathfrak{h}$. Let $(G, \pi_{G,\Lambda}, \varphi)$ be the quasi-Poisson Lie group defined using Λ as in §5.1.

LEMMA 5.7. *Let P be any manifold with a bi-vector field π_P. Suppose that $\rho : (P, \pi_P) \times (G, \pi_{G,\Lambda}, \varphi) \to (P, \pi_P)$ is a right quasi-Poisson action of $(G, \pi_{G,\Lambda}, \varphi)$ and that ρ restricts to a free and proper action of H. Let $j : P \to P/H$ be the projection. Then $j_*\pi_P$ is a well-defined Poisson structure on P/H.*

PROOF. By 1) in Lemma 4.3, $q_*\pi_{G,\Lambda}(h) = 0$ for all $h \in H$. It follows from the fact that ρ is a bi-vector map that $j_*\pi_P$ is well-defined. Since $\varphi \in \mathfrak{h} \wedge \mathfrak{g} \wedge \mathfrak{g}$ by 2) of Lemma 4.3, $[j_*\pi_P,\, j_*\pi_P] = j_*[\pi_P,\, \pi_P] = 2j_*\rho_\varphi = 0$, so $j_*\pi_P$ is Poisson. □

We now state a lemma from linear algebra.

LEMMA 5.8. *Let (V, π) be a Poisson vector space. Suppose that U and W are subspaces of V such that $\tilde{\pi}(U^0) \subset W \subset U$, where $U^0 = \{\xi \in V^* \,|\, \xi|_U = 0\}$. Let $\phi : V \to V/W$ be the projection. Then U/W is a Poisson subspace of $(V/W, \phi(\pi))$.*

The following Lemma 5.9 follows immediately from Lemma 5.8.

LEMMA 5.9. *Let the notation be as in Lemma 5.7. Suppose that Q is an H-invariant submanifold of P such that $\tilde{\pi}_P(T_q^0 Q) \subset T_q(qH)$ for every $q \in Q$, where $T_q^0 Q = \{\alpha \in T_q^* P \,|\, \alpha|_{T_q Q} = 0\}$ and qH is the H-orbit through q. Then Q/H is a Poisson submanifold of $(P/H,\, j_*\pi_P)$.*

We now apply Lemma 5.7 to $P = G \times (D/G)$ as in §5.2, π_P as in (5.8), and the action ρ as in Lemma 5.3. Denote by $G \times_H (D/G)$ the quotient of P by H with the projection $j : P \to G \times_H (D/G)$. By Lemma 5.7, $j_* \pi_P$ is a well-defined Poisson structure on $G \times_H (D/G)$. Set $[g, \underline{d}] = j(g, \underline{d})$ for $g \in G$ and $d \in D$.

NOTATION 5.10. The Poisson structure $j_*\pi_P$ on $G \times_H (D/G)$ will be denoted by Π.

Recall that $\mathfrak{l} = \mathfrak{l}_{\pi(eH)}$ is the Drinfeld Lie subalgebra of $\mathfrak{d}$ associated to $\pi(eH)$. Let L be the connected Lie subgroup of D with Lie algebra $\mathfrak{l}$ and assume that $H = G \cap L$. Let $\mathcal{O}$ be the L-orbit in D/G through $\underline{e} \in D/G$, where e is the identity element of D. Identify L/H with $\mathcal{O}$ and regard $G \times_H (L/H)$ as a submanifold of $G \times_H (D/G)$.

LEMMA 5.11. *$G \times_H (L/H)$ is a Poisson submanifold of $(G \times_H (D/G),\ \Pi)$, and Π is nondegenerate at $[g, \underline{d}]$ for all $g \in G$ and $d \in L$ such that* (5.11) *holds.*

PROOF. Let $Q = G \times \mathcal{O} \subset P$. Then Q is H-invariant. To see that Q/H is a Poisson submanifold of $(P/H, \Pi)$, it suffices, by Lemma 5.9, to show that $\tilde{\pi}_P(T_q^0 Q) \subset T_q(qH)$ for every $q = (g, \underline{d}) \in Q$, where $g \in G$ and $d \in L$. Using the isomorphism in (5.10), $T_q^0 Q = \{(0, \alpha_{y+\eta}(\underline{d})) \,|\, y + \eta \in \mathfrak{l} \cap \mathrm{Ad}_d \mathfrak{g}\}$, and by Lemma 5.4,

$$\tilde{\pi}_P(T_q^0 Q) = \{((l_g)_*(y - \Lambda\eta),\ \sigma_{\Lambda\eta - y}(\underline{d})) \,|\, y + \eta \in \mathfrak{l} \cap \mathrm{Ad}_d \mathfrak{g}\}.$$

By (4.7), $y + \eta \in \mathfrak{l}$ implies that $y - \Lambda\eta \in \mathfrak{h}$. Thus $\tilde{\pi}_P(T_q^0 Q) \subset T_q(qH)$.

By Lemma 5.5, π_P is nondegenerate at $(g, \underline{d})$ for all $g \in G$ and $d \in D$ such that (5.11) holds. At such a point $(g, \underline{d})$ where $d \in L$, $\mathfrak{l} \cap \mathrm{Ad}_d \mathfrak{g} = \mathrm{Ad}_d(\mathfrak{l} \cap \mathfrak{g}) = \mathrm{Ad}_d \mathfrak{h}$, and the map $\mathfrak{l} \cap \mathrm{Ad}_d \mathfrak{g} \to \mathfrak{h} : y + \eta \mapsto y - \Lambda\eta$ is an isomorphism, so $\tilde{\pi}_P(T_q^0 Q) = T_q(qH)$. It follows from a linear algebra argument that Π is nondegenerate at $[g, \underline{d}]$. □

5.4. The Poisson groupoid $(G \times_H (L/H), \Pi)$**.** Let the notation be as in §5.3. Recall that $\lambda : \mathfrak{d} \to \mathcal{V}^1(G)$ is the infinitesimal action of $\mathfrak{d}$ on G given in (4.12). Assume in addition that the restriction of λ to $\mathfrak{l}$ integrates to a right action of L on G, denoted by $(g, l) \mapsto g^l$ for $g \in G$ and $l \in L$, such that $g^h = gh$ for $g \in G$ and $h \in H$. Then G is a $(\mathfrak{d}, L)$-space (see Definition 2.2).

Let $\Gamma = G \times_H (L/H)$. It is straightforward to show (we omit the proof) that the following is a groupoid structure on Γ over G/H: for $g, g_1, g_2 \in G$ and $l, l_1, l_2 \in L$,

1) source map $s : \Gamma \to G/H :\ [g,\ lH] \mapsto gH$;

2) target map $t : \Gamma \to G/H :\ [g,\ lH] \mapsto g^l H$;

3) multiplication $\cdot_\Gamma$: $[g_1,\ l_1 H] \cdot_\Gamma [g_2,\ l_2 H] = [g_1,\ l_1 h l_2 H]$ when $g_1^{l_1} H = g_2 H$, where $h = (g_1^{l_1})^{-1} g_2$;

4) inverse $\tau : \Gamma \to \Gamma :\ [g,\ lH] \mapsto [g^l,\ l^{-1} H]$;

5) identity section $\epsilon : G/H \to \Gamma :\ gH \mapsto [g,\ eH]$.

THEOREM 5.12. *With the groupoid structure described above, $(G \times_H (L/H), \Pi)$ is a Poisson groupoid over $(G/H, \pi)$.*

The proof of Theorem 5.12 will be given in §5.5.

REMARK 5.13. Assume that $\pi(eH) = 0$, so that we can take $\Lambda = 0$. Recall that G^* is the connected subgroup of D with Lie algebra $\mathfrak{g}^*$. Assume further that the map $G^* \times G \to D : (u, g) \mapsto ug$ is a diffeomorphism. Identify G with $G^* \backslash D$. Then the restriction to L of the right action of D on $G \cong G^* \backslash D$ integrates the infinitesimal action λ of $\mathfrak{l}$ on G. By Remark 5.6 and Lemma 5.11, Π is nondegenerate everywhere on $G\times_H(L/H)$. Thus $(G\times_H(L/H),\Pi)$ is a symplectic groupoid over $(G/H,\pi)$. Note that in this case, the bi-vector field π_P on $P\cong D$ is Poisson by Lemma 5.3 and everywhere nondegenerate by Lemma 5.5. Moreover, by Lemma 4.5, $\mathfrak{h}^0$ is a subalgebra of $\mathfrak{g}^*$. Let H^0 be the connected subgroup of G^* with Lie algebra $\mathfrak{h}^0$. Then $L=HH^0$ is a coisotropic submanifold (D,π_P) and $L/H\cong H^0$. Our construction of the symplectic structure Π on $G\times_H(L/H)\cong G\times_H H^0$ is a special case of coisotropic reduction for symplectic manifolds. In the special case when H is a Poisson subgroup of (G,π_G) and when $\pi=q_*\pi_G$, this construction was carried out in [**26**]

EXAMPLE 5.14. Let G be a connected and simply connected Lie group and let X be the variety of Borel subgroups of G. Let G_0 be a real form of G and K a compact real form of G such that $K_0:=G\cap K$ is a maximal compact subgroup of G_0. Choose an Iwasawa decomposition $G=KAN$ of G such that the Borel subgroup B of G containing AN lies in the unique closed G_0-orbit in X. This choice of B gives rise to a Poisson Lie group (K,π_K) with AN as a dual Poisson Lie group. Although K_0 is not a Poisson Lie subgroup of (K,π_K), it is shown in [**10**] that the projection π of π_K is a well-defined Poisson structure on K/K_0, making $(K/K_0,\pi)$ a Poisson homogeneous space of (K,π_K), and the Drinfeld Lagrangian subalgebra associated to $\pi(eK_0)$ is $\mathfrak{g}_0$, the Lie algebra of G_0. Let $T=K\cap B$, a maximal torus of K. The set of T-orbits of symplectic leaves of π in K/K_0 is shown in [**10**] to be in one to one correspondence with the set of G_0-orbits in X. Due to the importance in representation theory of G_0-orbits in X, the Poisson geometrical properties of $(K/K_0,\pi)$ are worth further study. Since $\pi(eK_0)=0$ and since $G=KAN=ANK$, the conditions in Remark 5.6 are satisfied. By Remark 5.13 and Theorem 5.12, $K\times_{K_0}(G_0/K_0)$ has the structure of a symplectic groupoid over K/K_0. More details of this example, in particular, the generalized Poisson cohomology of $(K/K_0,\pi)$, will be studied in a future paper.

5.5. Proof of Theorem 5.12. Let the assumptions be as in §5.4. We need two lemmas. Recall that G^* is the connected subgroup of D with Lie algebra $\mathfrak{g}^*$.

LEMMA 5.15. *For any $g\in G$ and $l\in L$, $g^l l^{-1}g^{-1}\in G^*$.*

PROOF. Fix $g\in G$ and $l\in L$. To avoid confusion with the notation set up in §1.1 for left and right translations on G, if $v\in T_gG$, we let $g^{-1}v\in\mathfrak{g}$ and $vg^{-1}\in\mathfrak{g}$ be the left and right translation of v by g^{-1}.

Let $l(t)$ be a smooth curve in L such that $l(0)=e$ and $l(1)=l$, and let $u(t)=g^{l(t)}l(t)^{-1}g^{-1}\in D$. Let $u'(t)\in T_{u(t)}D$ and $l'(t)\in T_{l(t)}L$ be respectively the derivatives of $u(t)$ and $l(t)$ at t. Let $x(t)=l(t)^{-1}l'(t)\in\mathfrak{l}$. Then, for every t,

$$u'(t)=\lambda_{x(t)}(g^{l(t)})l(t)^{-1}g^{-1}-g^{l(t)}x(t)l(t)^{-1}g^{-1}$$

so by (4.13), $u'(t)u(t)^{-1}=-\mathrm{Ad}_{g^{l(t)}}x(t)+p_{\mathfrak{g}}\mathrm{Ad}_{g^{l(t)}}x(t)=-p_{\mathfrak{g}^*}\mathrm{Ad}_{g^{l(t)}}x(t)\in\mathfrak{g}^*$, where $p_{\mathfrak{g}}$ and $p_{\mathfrak{g}^*}$ are projections from $\mathfrak{d}$ to $\mathfrak{g}$ and $\mathfrak{g}^*$ with respect to the decomposition $\mathfrak{d}=\mathfrak{g}+\mathfrak{g}^*$. It follows from $u(0)=e$ that $u(t)\in G^*$ for all t. In particular, $g^l l^{-1}g^{-1}=u(1)\in G^*$. □

The following Lemma 5.16 is equivalent to 1) in Lemma 4.3, and we omit its proof.

LEMMA 5.16. *One has* $\mathrm{Ad}_h\Lambda\xi + p_{\mathfrak{g}}\mathrm{Ad}_h\xi - \Lambda\mathrm{Ad}^*_{h^{-1}}\xi \in \mathfrak{h}$ *for all* $\xi \in \mathfrak{h}^0$ *and* $h \in H$.

We can now start the proof of Theorem 5.12.

Let $\mathcal{G}_\Gamma = \{(\gamma_1, \gamma_2, \gamma_3) \in \Gamma\times\Gamma\times\Gamma \,|\, t(\gamma_1) = s(\gamma_2),\ \gamma_3 = \gamma_1\cdot_\Gamma\gamma_2\}$. By the definition of Poisson groupoids [**24**], we need to show that $\mathcal{G}_\Gamma$ is coisotropic in $\Gamma\times\Gamma\times\Gamma$ with the Poisson structure $\Pi\oplus\Pi\oplus(-\Pi)$. Let (P, π_P) be as in §5.2 and recall that $j : P \to P/H$ is the projection. Since (Γ, Π) is a Poisson submanifold of $(P/H, \Pi)$, and since $j : (P, \pi_P) \to (P/H, \Pi)$ is a bi-vector map, it is enough [**22**, Corollary 2.2.5] to show that $\mathcal{G}_P := (j\times j\times j)^{-1}(\mathcal{G}_\Gamma)$ is coisotropic in $(P\times P\times P,\ \pi_P\oplus\pi_P\oplus(-\pi_P))$. Recall that $\mathcal{O}$ is the L-orbit in D/G through $\underline{e} \in D/G$ and that $Q = G\times\mathcal{O} \subset P$. Identify$L/H$ with $\mathcal{O}$ by identifying $lH \in L/H$ with $\underline{l} = lG \in D/G$ for $l \in L$. Then

$$\begin{aligned}\mathcal{G}_P =&\{((g_1,\, l_1H),\, (g_2,\, l_2H),\, (g_1h_3,\, h_3^{-1}l_1hl_2H))\ |\\ &g_1, g_2, g_2 \in G,\, l_1, l_2 \in L,\, h_3 \in H,\, g_1^{l_1}H = g_2H,\, h = (g_1^{l_1})^{-1}g_2\} \subset Q\times Q\times Q.\end{aligned}$$

We will first describe the tangent bundle of $\mathcal{G}_P$ and then the co-normal bundle of $\mathcal{G}_P$ in $P\times P\times P$.

Let $\mathcal{G}_2 = \{((g_1,\, l_1H),\, (g_2,\, l_2H))\,|\, g_1, g_2 \in G,\, l_1, l_2 \in L,\, g_1^{l_1}H = g_2H\} \subset Q\times Q$. We now compute $T_{(q_1,q_2)}\mathcal{G}_2$ for $(q_1, q_2) = ((g_1,\, l_1H),\, (g_2,\, l_2H)) \in \mathcal{G}_2$. Define $\tilde{t}, \tilde{s} : Q \to G/H$ by $\tilde{t}(g, lH) = g^lH$ and $\tilde{s}(g, lH) = gH$ for $g \in G$ and $l \in L$. Then

$$T_{(q_1,q_2)}\mathcal{G}_2 = \{(v_1, v_2)\,|\, v_1 \in T_{q_1}Q,\ v_2 \in T_{q_2}Q,\ \tilde{t}_*(v_1) = \tilde{s}_*(v_2)\}.$$

Recall that $\sigma : \mathfrak{d} \to \mathcal{V}^1(D/G)$ is given in (5.4). Let $\kappa : \mathfrak{g} \to \mathcal{V}^1(G/H) : x \to \kappa_x$ be the Lie algebra anti-homomorphism given by

$$\kappa_x(gH) = \frac{d}{dt}|_{t=0}\exp(tx)gH, \qquad x \in \mathfrak{g},\ g \in G. \tag{5.12}$$

For $x, z \in \mathfrak{g}$, $\zeta \in \mathfrak{g}^*$ with $z+\zeta \in \mathfrak{l}$, and $q = (g, lH) \in Q$, let

$$v_{x,\, z+\zeta}(q) = ((l_g)_*x,\ \sigma_{z+\zeta}(lH)) \in T_qQ.$$

(Recall from §1.1 that the "l" in l_g denotes the left translation by g. This is not to be confused with an element in L.) Recall that $p_{\mathfrak{g}} : \mathfrak{d} \to \mathfrak{g}$ is the projection along $\mathfrak{g}^*$. Since G is a $(\mathfrak{d}, L)$-space via the infinitesimal action λ of $\mathfrak{d}$ and the action of L, one has

$$\tilde{t}_*v_{x_1,\, z_1+\zeta_1}(q_1) = \kappa_{p_{\mathfrak{g}}\mathrm{Ad}_{g_1^{l_1}l_1^{-1}}(x_1+z_1+\zeta_1)}(g_1^{l_1}H), \qquad \text{for } x_1 \in \mathfrak{g},\ z_1+\zeta_1 \in \mathfrak{l}.$$

Since $\tilde{s}_*v_{x_2,\, z_2+\zeta_2}(q_2) = \kappa_{\mathrm{Ad}_{g_2}x_2}(g_2H)$ for $x_2 \in \mathfrak{g}$, $z_2+\zeta_2 \in \mathfrak{l}$, we get

(5.13)
$$\begin{aligned}T_{(q_1,q_2)}\mathcal{G}_2 = \{&(v_{x_1,\, z_1+\zeta_1}(q_1),\ v_{x_2,\, z_2+\zeta_2}(q_2))\,|\, x_1, x_2 \in \mathfrak{g},\ z_1+\zeta_1, z_2+\zeta_2 \in \mathfrak{l},\\ &x_2 = \bar{x}_2 + \mathrm{Ad}_{g_2^{-1}}p_{\mathfrak{g}}\mathrm{Ad}_{g_1^{l_1}l_1^{-1}}(x_1+z_1+\zeta_1) \text{ for some } \bar{x}_2 \in \mathfrak{h}\}.\end{aligned}$$

Define

$$f : \mathcal{G}_2 \longrightarrow Q : ((g_1,\, l_1H),\, (g_2,\, l_2H)) \longmapsto (g_1,\, l_1(g_1^{l_1})^{-1}g_2l_2H).$$

Fix $g_i \in G, l_i \in L$, for $i = 1, 2$, such that $(q_1, q_2) = ((g_1,\, l_1H),\, (g_2,\, l_2H)) \in \mathcal{G}_2$. Let $x_i \in \mathfrak{g}$ and $z_i+\zeta_i \in \mathfrak{l}$ be such that $(v_{x_1,\, z_1+\zeta_1}(q_1),\ v_{x_2,\, z_2+\zeta_2}(q_2)) \in T_{(q_1,q_2)}Q$ as in (5.13). Let $g_1(t), l_1(t)$, and $l_2(t)$ be smooth curves in G and L respectively such that $g_1(0) = g_1$, $g_1'(0) = (l_{g_1})_*x_1$, and $l_i(0) = l_i$, $l_i'(0) = (r_{l_i})_*(z_i+\zeta_i)$ for $i = 1, 2$, where the superscript $'$ denotes derivative at 0. Let $g_2(t) = g_1(t)^{l_1(t)} h \exp t\bar{x}_2$, where $h = (g_1^{l_1})^{-1} g_2 \in H$. It is easy to see that $g_2'(0) = (l_{g_2})_* x_2$. Let

$$c(t) = ((g_1(t),\ l_1(t)H),\ (g_2(t),\ l_2(t)H)) \in Q \times Q.$$

Then $c(t) \in \mathcal{G}_2$ for all t, $c(0) = (q_1, q_2)$, and $c'(0) = (v_{x_1,\, z_1+\zeta_1}(q_1),\ v_{x_2,\, z_2+\zeta_2}(q_2))$. Since $f(c(t)) = (g_1(t),\ l_1(t) h \exp t\bar{x}_2 l_2(t)H)$, we have

$$f_*(v_{x_1,\, z_1+\zeta_1}(q_1),\ v_{x_2,\, z_2+\zeta_2}(q_2)) = \frac{d}{dt}|_{t=0} f(c(t)) = v_{x_1,\, z_3+\zeta_3}(g_1,\ l_1 h l_2 H)),$$

where $z_3 \in \mathfrak{g}$ and $\zeta_3 \in \mathfrak{g}^*$ are such that

$$z_3 + \zeta_3 = z_1 + \zeta_1 + \mathrm{Ad}_{l_1 h}(\bar{x}_2 + \mathrm{Ad}_{l_2}\bar{x}_3 + z_2 + \zeta_2) \qquad \text{for some } \bar{x}_3 \in \mathfrak{h}. \tag{5.14}$$

Thus for $(q_1, q_2, q_3) = ((g_1, l_1 H), (g_2, l_2 H), (g_1 h_3,\ h_3^{-1} l_1 h l_2 H)) \in \mathcal{G}_P$, where $h_3 \in H$,

$$\begin{aligned} T_{(q_1,q_2,q_3)}\mathcal{G}_P = \Big\{ \Big(& v_{x_1,\, z_1+\zeta_1}(q_1),\ v_{x_2,\, z_2+\zeta_2}(q_2),\ v_{x_3 + \mathrm{Ad}_{h_3}^{-1} x_1,\, -x_3 + \mathrm{Ad}_{h_3}^{-1}(z_3+\zeta_3)}(q_3) \Big) \mid \\ & x_1, x_2 \in \mathfrak{g},\ x_3 \in \mathfrak{h},\ z_i \in \mathfrak{g}, \zeta_i \in \mathfrak{g}^*,\ z_i + \zeta_i \in \mathfrak{l},\ i = 1,2,3, \\ & x_2 = \bar{x}_2 + \mathrm{Ad}_{g_2^{-1}} p_{\mathfrak{g}} \mathrm{Ad}_{g_1^{l_1} l_1^{-1}}(x_1 + z_1 + \zeta_1) \text{ for some } \bar{x}_2 \in \mathfrak{h}, \\ & z_3 + \zeta_3 = z_1 + \zeta_1 + \mathrm{Ad}_{l_1 h}(\bar{x}_2 + \mathrm{Ad}_{l_2}\bar{x}_3 + z_2 + \zeta_2) \text{ for some } \bar{x}_3 \in \mathfrak{h} \Big\}. \end{aligned}$$

Let $T^0_{(q_1,q_2,q_3)}\mathcal{G}_P$ be the co-normal subspace of $T_{(q_1,q_2,q_3)}\mathcal{G}_P$ in $T^*_{(q_1,q_2,q_3)}(P \times P \times P)$. Recall that for $\underline{d} \in D/G$, $\alpha_{y+\eta}(\underline{d}) \in T^*_{\underline{d}}(D/G)$ is given in (5.9). For $y \in \mathfrak{g}$, $\xi, \eta \in \mathfrak{g}^*$, and $q = (g, lH) \in Q$, let $\alpha_{\xi,\, y+\eta}(q) = (l^*_{g^{-1}}\xi,\ \alpha_{y+\eta}(lH)) \in T^*_q P$. Then for $x, z \in \mathfrak{g}$ and $\zeta \in \mathfrak{g}^*$ with $z + \zeta \in \mathfrak{l}$,

$$(\alpha_{\xi,\, y+\eta}(q),\ v_{x,\, z+\zeta}(q)) = (x, \xi) + \langle y + \eta,\ z + \zeta \rangle = (x, \xi) + (y, \zeta) + (z, \eta), \tag{5.15}$$

and the map $\mathfrak{g}^* \times \mathrm{Ad}_l \mathfrak{g} \to T^*_q P : (\xi,\ y + \eta) \mapsto \alpha_{\xi,\, y+\eta}(q)$ is an isomorphism. Let $l_3 = h_3 l_1 h l_2 \in L$, where $h = (g_1^{l_1})^{-1} g_2$. It follows from (5.15) that $T^0_{(q_1,q_2,q_3)}\mathcal{G}_P$ consists of all triples

$$(\alpha_{\xi_1,\, y_1+\eta_1}(q_1),\ \alpha_{\xi_2,\, y_2+\eta_2}(q_2),\ \alpha_{\xi_3,\, y_3+\eta_3}(q_3)) \in T^*_{(q_1,q_2,q_3)}(P \times P \times P), \tag{5.16}$$

where $\xi_i, \eta_i \in \mathfrak{g}^*$, $y_i \in \mathfrak{g}$ and $y_i + \eta_i \in \mathrm{Ad}_{l_i}\mathfrak{g}$ for $i = 1, 2, 3$, such that

$$\begin{aligned} 0 = {} & (\xi_1, x_1) + (\xi_2,\ \mathrm{Ad}_{g_2^{-1}} p_{\mathfrak{g}} \mathrm{Ad}_{g_1^{l_1} l_1^{-1}} x_1) + (\xi_3,\ \mathrm{Ad}_{h_3^{-1}} x_1) \\ & + \langle y_1 + \eta_1, z_1 + \zeta_1 \rangle + (\xi_2, \mathrm{Ad}_{g_2^{-1}} p_{\mathfrak{g}} \mathrm{Ad}_{g_1^{l_1} l_1^{-1}}(z_1 + \zeta_1)) + \langle y_3 + \eta_3,\ \mathrm{Ad}_{h_3^{-1}}(z_1 + \zeta_1) \rangle \\ & + (\xi_2, \bar{x}_2) + \langle y_3 + \eta_3,\ \mathrm{Ad}_{h_3^{-1} l_1 h} \bar{x}_2 \rangle \\ & + \langle y_2 + \eta_2,\ z_2 + \zeta_2 \rangle + \langle y_3 + \eta_3,\ \mathrm{Ad}_{h_3^{-1} l_1 h}(z_2 + \zeta_2) \rangle \\ & + (\xi_3, x_3) - \langle y_3 + \eta_3,\ x_3 \rangle \end{aligned}$$

for all $x_1 \in \mathfrak{g}$, $\bar{x}_2, x_3 \in \mathfrak{h}$, $\mathfrak{z}_1 + \zeta_1 \in \mathfrak{l}$ and $z_2 + \zeta_2 \in \mathfrak{l}$, which is equivalent to

$$\xi_1 + \mathrm{Ad}_{l_1 (g_1^{l_1})^{-1}} \mathrm{Ad}^*_{g_2^{-1}} \xi_2 + \mathrm{Ad}_{h_3} \xi_3 \in \mathfrak{g}; \tag{5.17}$$

$$y_1 + \eta_1 + \mathrm{Ad}_{l_1 (g_1^{l_1})^{-1}} \mathrm{Ad}^*_{g_2^{-1}} \xi_2 + \mathrm{Ad}_{h_3}(y_3 + \eta_3) \in \mathfrak{l}; \tag{5.18}$$

$$\xi_2 + \mathrm{Ad}_{h^{-1} l_1^{-1} h_3}(y_3 + \eta_3) \in \mathfrak{g} + \mathfrak{l}; \tag{5.19}$$

$$y_2 + \eta_2 + \mathrm{Ad}_{h^{-1} l_1^{-1} h_3}(y_3 + \eta_3) \in \mathfrak{l}; \tag{5.20}$$

$$\xi_3 - \eta_3 \in \mathfrak{h}^0, \tag{5.21}$$

where recall that $\mathfrak{h}^0 = \{\xi \in \mathfrak{g}^* | (\xi, \mathfrak{h}) = 0\}$. Since $\mathfrak{g} + \mathfrak{l} = \mathfrak{g} + \mathfrak{h}^0$ by (4.7), (5.19) and (5.20) imply that $\eta_2 - \xi_2 \in \mathfrak{h}^0$. Similarly, (5.17), (5.18), and (5.21) imply that $\eta_1 - \xi_1 \in \mathfrak{h}^0$. Thus

$$\Lambda(\eta_i - \xi_i) + \eta_i - \xi_i \in \mathfrak{l} \quad \text{for } i = 1, 2, 3. \tag{5.22}$$

It remains to show that for any triple in (5.16) satisfying (5.17) - (5.21),

$$(\tilde{\pi}_P(\alpha_{\xi_1, y_1+\eta_1}(q_1)),\ \tilde{\pi}_P(\alpha_{\xi_2, y_2+\eta_1}(q_2)),\ -\tilde{\pi}_P(\alpha_{\xi_3, y_3+\eta_3}(q_3))) \in T_{(q_1,q_2,q_3)}\mathcal{G}_P. \tag{5.23}$$

Let $g_3 = g_1 h_3$. By Lemma 5.4 and (5.22),

$$\tilde{\pi}_P(\alpha_{\xi_i, y_i+\eta_i}(q_i)) = v_{x_i, z_i+\zeta_i}(q_i) \in T_{q_i}Q$$

for $i = 1, 2, 3$, and

$$x_i = y_i + \Lambda\xi_i - \Lambda\eta_i + \mathrm{Ad}_{g_i^{-1}} p_{\mathfrak{g}} \mathrm{Ad}_{g_i}\xi_i, \quad z_i + \zeta_i = \Lambda(\eta_i - \xi_i) + \eta_i - \xi_i. \tag{5.24}$$

Thus by our description of $T_{(q_1,q_2,q_3)}\mathcal{G}_P$, to show (5.23), it suffices to show

$$\bar{x}_2 := x_2 - \mathrm{Ad}_{g_2^{-1}} p_{\mathfrak{g}} \mathrm{Ad}_{g_1^{l_1} l_1^{-1}}(x_1 + z_1 + \zeta_1) \in \mathfrak{h}, \tag{5.25}$$

$$x_1 + \mathrm{Ad}_{h_3} x_3 \in \mathfrak{h} \tag{5.26}$$

$$\mathrm{Ad}_{h_3}(x_3 + z_3 + \zeta_3) + x_1 + z_1 + \zeta_1 + \mathrm{Ad}_{l_1 h}(\bar{x}_2 + z_2 + \zeta_2) \in \mathrm{Ad}_{l_1 h l_2}\mathfrak{h}. \tag{5.27}$$

By (5.24),

$$x_i + z_i + \zeta_i = y_i + \eta_i - \mathrm{Ad}_{g_i^{-1}} \mathrm{Ad}^*_{g_i^{-1}}\xi_i, \qquad i = 1, 2, 3. \tag{5.28}$$

We first prove (5.25). Since $\mathrm{Ad}_{l_1^{-1}}(y_1 + \eta_1) \in \mathfrak{g}$ and $g_1^{l_1} l_1^{-1} g_1^{-1} \in G^*$ by Lemma 5.15,

$$\bar{x}_2 = y_2 - \Lambda\eta_2 + \Lambda\xi_2 + \mathrm{Ad}_{g_2^{-1}} p_{\mathfrak{g}} \mathrm{Ad}_{g_2}\xi_2 - \mathrm{Ad}_{h^{-1} l_1^{-1}}(y_1 + \eta_1).$$

Note from (5.18) that

$$\mathrm{Ad}_{h^{-1} l_1^{-1}}(y_1 + \eta_1) + \xi_2 - \mathrm{Ad}_{g_2^{-1}} p_{\mathfrak{g}} \mathrm{Ad}_{g_2}\xi_2 + \mathrm{Ad}_{h^{-1} l_1^{-1} h_3}(y_3 + \eta_3) \in \mathfrak{l},$$

so by (5.20), $\mathrm{Ad}_{g_2^{-1}} p_{\mathfrak{g}} \mathrm{Ad}_{g_2}\xi_2 - \mathrm{Ad}_{h^{-1} l_1^{-1}}(y_1 + \eta_1) - \xi_2 + y_2 + \eta_2 \in \mathfrak{l}$. It follows from (4.7) that $\bar{x}_2 \in \mathfrak{h}$, so (5.25) holds. To prove (5.26), note that (5.17) implies that $p_{\mathfrak{g}^*} \mathrm{Ad}_{l_1 (g_1^{l_1})^{-1}} \mathrm{Ad}^*_{g_2^{-1}}\xi_2 = -\xi_1 - \mathrm{Ad}^*_{h_3^{-1}}\xi_3$, so by (5.18) and (4.7),

$$y_1 + p_{\mathfrak{g}} \mathrm{Ad}_{l_1 (g_1^{l_1})^{-1}} \mathrm{Ad}^*_{g_2^{-1}}\xi_2 + \mathrm{Ad}_{h_3} y_3 + p_{\mathfrak{g}} \mathrm{Ad}_{h_3}\eta_3 + \Lambda(\xi_1 - \eta_1) + \Lambda \mathrm{Ad}^*_{h_3^{-1}}(\xi_3 - \eta_3) \in \mathfrak{h}.$$

Thus by (5.17), (5.26) is equivalent to

$$\mathrm{Ad}_{h_3}\Lambda(\xi_3 - \eta_3) + p_{\mathfrak{g}} \mathrm{Ad}_{h_3}(\xi_3 - \eta_3) - \Lambda \mathrm{Ad}^*_{h_3^1}(\xi_3 - \eta_3) \in \mathfrak{h}$$

which holds because of Lemma 5.16. This proves (5.26). It remains to prove (5.27). Using Lemma 5.15 and (5.28), one sees that the left hand side of (5.27) is equal to

$$\mathrm{Ad}_{h_3}(y_3 + \eta_3) + \mathrm{Ad}_{l_1 h}(y_2 + \eta_2) - \mathrm{Ad}_{g_1^{-1}} p_{\mathfrak{g}^*} \mathrm{Ad}_{g_1}(\xi_1 + \mathrm{Ad}_{h_3}\xi_3) - \mathrm{Ad}_{l_1 (g_1^{l_1})^{-1}} \mathrm{Ad}^*_{g_2^{-1}}\xi_2.$$

By (5.17), $\mathrm{Ad}_{g_1^{-1}} p_{\mathfrak{g}^*} \mathrm{Ad}_{g_1}(\xi_1 + \mathrm{Ad}_{h_3}\xi_3) + \mathrm{Ad}_{l_1 (g_1^{l_1})^{-1}} \mathrm{Ad}^*_{g_2^{-1}}\xi_2 = 0$. Moreover, since $y_2 + \eta_2 \in \mathrm{Ad}_{l_2}\mathfrak{g}$ and $y_3 + \eta_3 \in \mathrm{Ad}_{h_3^1 l_1 h l_2}\mathfrak{g}$,

$$y_2 + \eta_2 + \mathrm{Ad}_{h^{-1} l_1^{-1} h_3}(y_3 + \eta_3) \in \mathrm{Ad}_{l_2}\mathfrak{g} \cap l = \mathrm{Ad}_{l_2}\mathfrak{h}$$

by (5.20). Thus the left hand side of (5.27) is in $\mathrm{Ad}_{l_1 h l_2}\mathfrak{h}$, so (5.27) holds.

This finishes the proof of Theorem 5.12.

References

[1] Alekseev, A. and Kosmann-Schwarzbach, Y., Manin pairs and moment maps, *J. Differential Geom.* **56** (1) (2000), 133–165.

[2] Borel, A. and Wallach, N., Continuous cohomology, discrete subgroups, and representations of reductive groups, *Mathematical Surveys and Monographs* **67**, second edition.

[3] Brylinski, J-L., A differential complex for Poisson manifolds, *J. Diff. Geom.* **28** (1) (1988), 93 - 114.

[4] Diatta, A. and Medina, A., Espaces de Poisson homogènes d'un groupe de Lie-Poisson, *C. R. Acad. Sci. Paris. Sér. I Math.* **328** (8) (1999), 687 - 690.

[5] Drinfeld, V. G., Hamiltonian structures on Lie groups, Lie bialgebras and the geometric meaning of the classical Yang - Baxter equations, *Soviet Math. Dokl.* **27 (1)** (1983), 68 - 71.

[6] Drinfel'd, V. G., On Poisson homogeneous spaces of Poisson-Lie groups, *Theo. Math. Phys.* **95** (2) (1993), 226 - 227.

[7] Evens, S. and Lu, J.-H., and Weinstein, A., Transversal measures, the modular class, and cohomology pairing for Lie algebroids, *Quart. J. Math. Oxford* **50** (2) (1999), 417 - 436.

[8] Evens, S. and Lu, J.-H., On the variety of Lagrangian subalgebras, I. *Ann. Sci. Ecole Norm. Sup.* (4) **34** (5) (2001), 631–668.

[9] Evens, S. and Lu, J.-H., On the variety of Lagrangian subalgebras, II. *Ann. Sci. Ecole Norm. Sup.* (4) **39**(2) (2006), 347–379.

[10] Foth, P. and Lu, J.-H., A Poisson structure on compact symmetric spaces, *Comm. Math. Phys.* **251**(3) (2004), 557–566.

[11] Korogodski, L. and Soibelman Y., *Algebras of functions on quantum groups, Part I*, Mathematical Surveys and Monographs, **56**, A.M.S., Providence, 1998.

[12] Koszul, J. L., Crochet de Schouten-Nijenhuis et cohomologie, *Astérisque, hors série, Soc. Math. France,* Paris (1985), 257 - 271.

[13] Lichnerowicz, A., Les variétés de Poisson et leurs algèbres de Lie associées, *J. Diff. Geom.* **12** (1977), 253-300.

[14] Lu, J. H. and Weinstein, A., Poisson Lie groups, dressing transformations, and Bruhat decompositions, *Journal of Differential Geometry* **31** (1990), 501 - 526.

[15] Lu, J. H., Poisson homogeneous spaces and Lie algebroids associated to Poisson actions, *Duke. Math. J.* **86** (2) (1997), 261 - 304.

[16] Lu, J. H. and Yakimov, M., Group orbits and regular partitions of Poisson manifolds, math.SG/0609732.

[17] Mackenzie, K., *Lie groupoids and Lie algebroids in differential geometry*, London Mathematical Society Lecture Note Series, **124**, Cambridge University Press, Cambridge, 1987.

[18] Mackenzie, K., *General theory of Lie groupoids and Lie algebroids*, London Mathematical Society Lecture Note Series, **213**, Cambridge University Press, Cambridge, 2005.

[19] So, B.-K., On some examples of Poisson homology and cohomology - Analytic and Lie theoretic approaches, Mphil thesis of the University of Hong Kong, 2005.

[20] Semenov-Tian-Shansky, M. A., Dressing transformations and Poisson Lie group actions, *Publ. RIMS, Kyoto University* **21** (1985), 1237 - 1260.

[21] Vaisman, I., *Lectures on the geometry of Poisson manifolds*, Progress in Mathematics, **118**, Birkhauser, Boston, 1994.

[22] Weinstein, A., Coisotropic calculus and Poisson groupoids, *J. Math. Soc. Japan* **40** (4) (1988), 705 - 726.

[23] Weinstein, A., The modular automorphism group of a Poisson manifold, *J. Geom. Phys.* **23** (1997), 379 - 394.

[24] Weinstein, A., Symplectic groupoids and Poisson manifolds, *Bull. Amer. Math. Soc.* **16** (1) (1987), 101–104.

[25] Xu, P., Symplectic groupoids of reduced Poisson spaces. *C. R. Aca d. Sci. Paris Serie I* 314:457-461 (1992).

[26] Xu, P, Gerstenhaber algebras and BV-algebras in Poisson geometry, *Comm. Math. Phys.* **200** (1999), 545 - 560.

Department of Mathematics, the University of Hong Kong, Pokfulam Rd., Hong Kong, SAR, China

E-mail address: `jhlu@maths.hku.hk`

Contemporary Mathematics
Volume **450**, 2008

Lectures on Pure Spinors and moment maps

E. Meinrenken

Contents

1. Introduction

This article is an expanded version of notes for my lectures at the summer school on 'Poisson geometry in mathematics and physics' at Keio University, Yokohama, June 5–9 2006. The plan of these lectures was to give an elementary introduction to the theory of Dirac structures, with applications to Lie group valued moment maps. Special emphasis was given to the *pure spinor* approach to Dirac structures, developed in Alekseev-Xu [**7**] and Gualtieri [**20**]. (See [**11, 12, 16**] for the more standard approach.) The connection to moment maps was made in the work of Bursztyn-Crainic [**10**]. Parts of these lecture notes are based on a forthcoming joint paper [**1**] with Anton Alekseev and Henrique Bursztyn.

I would like to thank the organizers of the school, Yoshi Maeda and Guiseppe Dito, for the opportunity to deliver these lectures, and for a greatly enjoyable meeting. I also thank Yvette Kosmann-Schwarzbach, Shlomo Sternberg, and the referee for a number of helpful comments.

2. Volume forms on conjugacy classes

We will begin with the following FACT, which at first sight may seem quite unrelated to the theme of these lectures:

FACT. Let G be a simply connected semi-simple real Lie group. Then every conjugacy class in G carries a canonical invariant volume form.

By definition, a conjugacy class $\mathcal{C}$ is an orbit for the conjugation action,

$$\mathrm{Ad}\colon G \to \mathrm{Diff}(G),\ \mathrm{Ad}(g).a = gag^{-1}.$$

It is thus a smooth Ad-invariant submanifold of G. By the existence of a 'canonical' volume form, we mean that there exists an explicit construction, not depending on any further choices.

More generally, the above FACT holds for simply connected Lie groups with a bi-invariant pseudo-Riemannian metric. In the semi-simple case, such a metric is provided by the Killing form. The assumption that G is simply connected may be relaxed as well – the precise condition will be given below. Without any assumption on $\pi_1(G)$, the conjugacy classes may be non-orientable, but they still carry canonical invariant measures.

EXERCISE 2.1. (a) Show that SO(3) has a conjugacy class diffeomorphic to $\mathbb{R}P(2)$. This is the simplest example of a non-orientable conjugacy class.
(b) Let G be the conformal group of the real line $\mathbb{R}$ (the group generated by dilations and translations). Show that G does not admit a bi-invariant pseudo-Riemannian metric, and that G has conjugacy classes not admitting invariant measures.

The above FACT does not appear to be well-known. Indeed, it is not obvious how to use the pseudo-Riemannian metric to produce a measure on $\mathcal{C}$, since the restriction of this metric to $\mathcal{C}$ may be degenerate or even zero. Recall on the other hand that any co-adjoint orbit $\mathcal{O} \subset \mathfrak{g}^*$ carries a canonical volume form, the *Liouville form* for the Kirillov-Kostant-Souriau symplectic form ω on $\mathcal{O}$:

$$\omega(\xi_1^\sharp, \xi_2^\sharp)\big|_\mu = \langle \mu, [\xi_1, \xi_2] \rangle, \quad \mu \in \mathcal{O}. \tag{1}$$

Here $\xi^\sharp \in \mathfrak{X}(\mathcal{O})$ denotes the vector field generated by $\xi \in \mathfrak{g}$ under the co-adjoint action. Letting $n = \frac{1}{2} \dim \mathcal{O}$, the Liouville form is $\frac{1}{n!}\omega^n$, or equivalently the top degree part of the differential form $\exp \omega$. One is tempted to try something similar for conjugacy classes. Unfortunately, conjugacy classes need not admit symplectic forms, in general:

EXERCISE 2.2. Show that the group Spin(5) (the connected double cover of SO(5)) has a conjugacy class isomorphic to S^4. The 4-sphere does not admit an almost complex structure, hence also no non-degenerate 2-form.

Nevertheless, our construction of the volume form on $\mathcal{C}$ will be similar to that of the Liouville form on coadjoint orbits. Let

$$B\colon \mathfrak{g} \times \mathfrak{g} \to \mathbb{R}$$

be the Ad-invariant inner product on $\mathfrak{g}$, corresponding to the bi-invariant pseudo-Riemannian metric on G. There is an Ad-invariant 2-form $\omega \in \Omega^2(\mathcal{C})$, given by the formula

$$\omega(\xi_1^\sharp, \xi_2^\sharp)\big|_g = B\Big(\frac{\mathrm{Ad}_g - \mathrm{Ad}_{g^{-1}}}{2}\xi_1, \xi_2\Big), \quad g \in \mathcal{C}. \tag{2}$$

This 2-form was introduced by Guruprasad-Huebschmann-Jeffrey-Weinstein in their paper [**21**] on moduli spaces of flat connections, and plays a key role in the theory of group valued moment maps [**4**]. Its similarity to the KKS formula (1) becomes evident if we use B to identify $\mathfrak{g}^*$ with $\mathfrak{g}$: The KKS 2-form is defined by the skew-adjoint operator $\mathrm{ad}_\mu = [\mu, \cdot]$, while the GHJW 2-form is defined by the skew-adjoint operator $\frac{1}{2}(\mathrm{Ad}_g - \mathrm{Ad}_{g^{-1}})$. An important difference is that the GHJW 2-form may well be degenerate. It can even be zero:

EXERCISE 2.3. Show that the GHJW 2-form vanishes on the conjugacy class $\mathcal{C}$ if and only if the elements of $\mathcal{C}$ square to elements of the center of G. For $G = \mathrm{SU}(2)$, there is one such conjugacy class (besides the central elements themselves): $\mathcal{C} = \{A \in \mathrm{SU}(2) |\ \mathrm{tr}(A) = 0\}$.

To proceed we need a certain differential form on the group G. Let $\theta^L, \theta^R \in \Omega^1(G) \otimes \mathfrak{g}$ denote the left-, right-invariant Maurer-Cartan forms. Thus $\theta^L = g^{-1}\mathrm{d}g$ and $\theta^R = \mathrm{d}gg^{-1}$ in matrix representations of G.

THEOREM 2.4. *Suppose G is a simply connected Lie group with a bi-invariant pseudo Riemannian metric, corresponding to the scalar product B on $\mathfrak{g}$. Then there is a well-defined smooth,* Ad*-invariant differential form $\psi \in \Omega(G)$ such that*

$$\psi_g = \det{}^{1/2}\big(\tfrac{\mathrm{Ad}_g + 1}{2}\big)\ \exp\Big(\tfrac{1}{4} B\big(\tfrac{1-\mathrm{Ad}_g}{1+\mathrm{Ad}_g}\theta^L, \theta^L\big)\Big) \tag{3}$$

at elements $g \in G$ such that $\mathrm{Ad}_g + 1$ *is invertible.*

Note that the 2-form in the exponential becomes singular at points where $\mathrm{Ad}_g + 1$ fails to be invertible. The Theorem ensures that these singularities are compensated by the zeroes of the determinant factor. We can now write down our formula for the volume form on conjugacy classes.

THEOREM 2.5. *With the assumptions of Theorem 2.4, the top degree part of the differential form*

$$e^{\omega}\iota_{\mathcal{C}}^*\psi \tag{4}$$

defines a volume form on $\mathcal{C}$. Here $\omega \in \Omega^2(\mathcal{C})$ is the GHJW 2-form on $\mathcal{C}$, and $\iota_{\mathcal{C}} \colon \mathcal{C} \hookrightarrow G$ denotes the inclusion.

Since ψ is an even form, the Theorem says in particular that $\dim \mathcal{C}$ is even. Although Formula (4) is very explicit, it is not very easy to evaluate in practice. In particular, it is a non-trivial task to work out the top degree part 'by hand', and to verify that it is indeed non-vanishing! Also, the complicated formula for ψ may seem rather mysterious at this point.

What I would like to explain, in the first part of these lectures, is that the differential form ψ is a *pure spinor* on G, and that Theorem 2.5 may be understood as the non-degeneracy of a pairing between two pure spinors, $e^{-\omega}$ and $\iota_{\mathcal{C}}^*\psi$ on $\mathcal{C}$.

REMARK 2.6. For the case of a *compact* Lie group, with B positive definite, Theorem 2.5 was first proved in [**6**], using a cumbersome evaluation of the top degree part of (4). The general case was obtained in [**1**].

The volume forms on conjugacy classes are not only similar to the Liouville volume form on coadjoint orbits, but are actually generalizations of the latter:

EXERCISE 2.7. Let K be any Lie group. The semi-direct product $G = \mathfrak{k}^* \rtimes K$ (where K acts on $\mathfrak{k}^*$ by the co-adjoint action) carries a bi-invariant pseudo-Riemannian metric, with associated bilinear form B on $\mathfrak{g} = \mathfrak{k}^* \rtimes \mathfrak{k}$ given by the pairing between $\mathfrak{k}$ and $\mathfrak{k}^*$. Show that the inclusion $\mathfrak{k}^* \hookrightarrow G$ restricts to a diffeomorphism from any K-coadjoint orbit $\mathcal{O}$ onto a G-conjugacy class $\mathcal{C}$. Furthermore, the GHJW 2-form on $\mathcal{C}$ equals the KKS 2-form on $\mathcal{O}$, and the volume form on $\mathcal{C}$ constructed above is just the ordinary Liouville form on $\mathcal{O}$.

3. Clifford algebras and spinors

This Section summarizes a number of standard facts about Clifford algebras and spinors. Further details may be found in the classic monograph [**15**].

3.1. The Clifford algebra. Let W be a vector space, equipped with a non-degenerate symmetric bilinear form $\langle \cdot, \cdot \rangle$. Let $m = \dim W$. The *Clifford algebra* over $(W, \langle \cdot, \cdot \rangle)$ is the associative unital algebra, linearly generated by the elements $w \in W$ subject to relations

$$w_1 w_2 + w_2 w_1 = \langle w_1, w_2 \rangle \, 1, \quad w_i \in W.$$

Elements of the Clifford algebra $\mathrm{Cl}(W)$ may be written as linear combinations of products of elements $w_i \in W$. There is a canonical filtration,

$$\mathrm{Cl}(W) = \mathrm{Cl}^{(m)}(W) \supset \cdots \mathrm{Cl}^{(1)}(W) \supset \mathrm{Cl}^{(0)}(W) = \mathbb{R}$$

with $\mathrm{Cl}^{(k)}(W)$ the subspace spanned by products of $\leq k$ generators. The associated graded algebra $\mathrm{gr}(\mathrm{Cl}(W))$ is the exterior algebra $\wedge(W)$.

The Clifford algebra has a $\mathbb{Z}_2$-grading compatible with the algebra structure, in such a way that the generators $w \in W$ are odd. With the usual sign conventions for $\mathbb{Z}_2$-graded ('super') algebras, the defining relations may be written $[w_1, w_2] = \langle w_1, w_2 \rangle \, 1$ where $[\cdot, \cdot]$ denotes the super-commutator. A *module over the Clifford algebra* $\mathrm{Cl}(W)$ is a vector space $\mathcal{S}$ together with an algebra homomorphism

$$\varrho \colon \mathrm{Cl}(W) \to \mathrm{End}(\mathcal{S}).$$

Equivalently, the module structure is described by a linear map $\varrho \colon W \to \mathrm{End}(\mathcal{S})$ such that

$$\varrho(w)\varrho(w') + \varrho(w')\varrho(w) = \langle w, w' \rangle \, 1$$

for all $w, w' \in W$. A Clifford module $\mathcal{S}$ is called a *spinor module* if it is irreducible, i.e. if there are no non-trivial sub-modules.

3.2. The Pin group. Let $\Pi \colon \mathrm{Cl}(W) \to \mathrm{Cl}(W)$ be the parity automorphism of $\mathrm{Cl}(W)$, equal to $+1$ on the even part and to -1 on the odd part. The *Clifford group* $\Gamma(W)$ is the subgroup of the group $\mathrm{Cl}(W)^\times$ of invertible elements, consisting of all x such that the transformation

$$y \mapsto \Pi(x) y x^{-1} \tag{5}$$

of $\mathrm{Cl}(W)$ preserves the subspace $W \subset \mathrm{Cl}(W)$. Let $A_x \in \mathrm{GL}(W)$ denote the induced transformation of W.

PROPOSITION 3.1. *The homomorphism $A \colon \Gamma(W) \to \mathrm{GL}(W)$ has kernel $\mathbb{R}^\times$ and range $\mathrm{O}(W)$. Thus, the Clifford group fits into an exact sequence,*

$$1 \longrightarrow \mathbb{R}^\times \longrightarrow \Gamma(W) \longrightarrow \mathrm{O}(W) \longrightarrow 1.$$

EXERCISE 3.2. Show that any $w \in W$ with $B(w,w) \neq 0$ lies in $\Gamma(W)$, with A_w the reflection defined by w. Since any element of $\mathrm{O}(W)$ may be written as a product of reflections (E.Cartan-Dieudonné theorem), conclude that any element in $\Gamma(W)$ is a product $g = w_1 \cdots w_k$ with $B(w_i, w_i) \neq 0$. Use this to prove the above Proposition.

Let $x \mapsto x^\top$ denote the canonical anti-homomorphism of $\mathrm{Cl}(W)$, i.e. $(w_1 \cdots w_k)^\top = w_k \cdots w_1$ for $w_i \in W$. Then $g^\top g \in \mathbb{R}^\times$ for all $g \in \Gamma(W)$. Letting

$$\mathrm{Pin}(W) = \{g \in \Gamma(W) |\ g^\top g = \pm 1\}$$

one obtains an exact sequence,

$$1 \longrightarrow \mathbb{Z}_2 \longrightarrow \mathrm{Pin}(W) \longrightarrow \mathrm{O}(W) \longrightarrow 1.$$

Thus $\mathrm{Pin}(W)$ is a double cover of $\mathrm{O}(W)$. Its restriction to $\mathrm{SO}(W)$ is denoted $\mathrm{Spin}(W)$.

3.3. Lagrangian subspaces. For any subspace $E \subset W$, we denote by $E^\perp$ the space of vectors orthogonal to E. The subspace E is called *Lagrangian* if $E = E^\perp$. Let $\mathrm{Lag}(W)$ denote the *Lagrangian Grassmannian*, i.e. the set of Lagrangian subspaces. If $\mathrm{Lag}(W) \neq \emptyset$, the bilinear form $\langle \cdot, \cdot \rangle$ is called *split*. Since we are working over $\mathbb{R}$, the non-degenerate symmetric bilinear forms are classified by their signature, and $\langle \cdot, \cdot \rangle$ is split if and only if the signature is (n, n). That is, $(W, \langle \cdot, \cdot \rangle)$ is isometric to $\mathbb{R}^{n,n}$, the vector space $\mathbb{R}^{2n}$ with the bilinear form

$$\langle e_i, e_j \rangle = \pm \delta_{ij}, \quad i, j = 1, \ldots, 2n$$

with a $+$ sign for $i = j \leq n$ and a $-$ sign for $i = j > n$.

EXERCISE 3.3. For any invertible matrix $A \in \mathrm{GL}(n)$ let

$$E_A = \{(Av, v) |\ v \in \mathbb{R}^n\} \subset \mathbb{R}^{n,n}.$$

(a) Show that E_A is Lagrangian if and only if $A \in \mathrm{O}(n)$.
(b) Show that every Lagrangian subspace E is of the form E_A for a unique $A \in \mathrm{O}(n)$.

We will assume for the rest of this Section that the bilinear form $\langle \cdot, \cdot \rangle$ on W is split, of signature (n, n). The exercise shows that the Lagrangian Grassmannian $\mathrm{Lag}(W)$ is diffeomorphic to $\mathrm{O}(n)$. In particular, it is a manifold of dimension $n(n-1)/2$, with *two connected components*.

EXERCISE 3.4. Show that $E, F \in \mathrm{Lag}(W)$ are in the same component of $\mathrm{Lag}(W)$ if and only if $n + \dim(E \cap F)$ is even.

The orthogonal group $\mathrm{O}(W) \cong \mathrm{O}(n, n)$ acts transitively on $\mathrm{Lag}(W)$, as does its maximal compact subgroup $\mathrm{O}(n) \times \mathrm{O}(n)$. (By Exercise 3.3, already the subgroup $\mathrm{O}(n, 0)$ acts transitively.)

REMARK 3.5. Compare with the situation in symplectic geometry: If (Z, ω) is a *symplectic* vector space (thus $Z \cong \mathbb{R}^{2n}$ with the standard symplectic form ω), a subspace E is called Lagrangian if it coincides with its ω-orthogonal space E^ω. The symplectic group $\mathrm{Sp}(Z, \omega)$ acts transitively on the set $\mathrm{Lag}(Z)$ of Lagrangian subspaces, as does its maximal compact subgroup $\mathrm{U}(n)$, and $\mathrm{Lag}(Z) = \mathrm{U}(n)/\mathrm{O}(n)$. Thus $\mathrm{Lag}(Z)$ is connected and has dimension $n(n+1)/2$.

For any pair of transverse Lagrangian subspaces E, F, the pairing $\langle \cdot, \cdot \rangle$ defines an isomorphism $F \cong E^*$. Equivalently, one obtains an isometric isomorphism

$$W \cong E \oplus E^*$$

where the bilinear form on the right hand side is defined by extension of the pairing between E and E^*.

EXERCISE 3.6. Show that for any given $E \in \mathrm{Lag}(W)$, the open subset $\{F \in \mathrm{Lag}(W) \mid E \cap F = 0\}$ is (canonically) an affine space, with $\wedge^2 E$ its space of motions. Show that the closure of this subset is a connected component of $\mathrm{Lag}(W)$. Which of the two components is it?

The sub-algebra of $\mathrm{Cl}(W)$ generated by a Lagrangian subspace $E \in \mathrm{Lag}(W)$ is just the exterior algebra $\wedge E$. Given a Lagrangian subspace F transverse to E, and using the commutator relations to 'write elements of F to the left', we see that

$$\mathrm{Cl}(W) = \wedge(F) \otimes \wedge(E), \tag{6}$$

thus $\mathrm{Cl}(W) = \wedge(W)$ as a $\mathbb{Z}_2$-graded vector space, and also as a filtered vector space (but not as an algebra). If $w \in W$, the isomorphism intertwines the operator $[w, \cdot]$ (graded commutator) on $\mathrm{Cl}(W)$ with the contraction operators $\iota(w)$ on $\wedge(W)$.

LEMMA 3.7. (a) *The Clifford algebra* $\mathrm{Cl}(W)$ *has no non-trivial two-sided ideals.*
(b) *For* $E \in \mathrm{Lag}(W)$, *the left-ideal* $\mathrm{Cl}(W)E$ *is maximal.*

PROOF. We use the following simple fact: If a non-zero subspace of an exterior algebra $\wedge(S)$ is stable under all contraction operators $\iota(u)$, $u \in S^*$, then the subspace contains the scalars.

a) Suppose $\mathcal{I}$ is a proper 2-sided ideal in $\mathrm{Cl}(W)$. Then $\mathcal{I}$ is invariant under all $[w, \cdot]$ with $w \in W$. The above isomorphism (6) takes scalars to scalars, and intertwines $[w, \cdot]$ with contractions. Hence $\mathcal{I} = 0$.

b) Let $\mathcal{I}$ be a proper left-ideal containing $\mathrm{Cl}(W)E$. By the isomorphism (6), we have a direct sum decomposition

$$\mathrm{Cl}(W) = \wedge(F) \oplus \mathrm{Cl}(W)E.$$

On $\wedge(F)$, the operators $[w, \cdot]$ for $w \in E \cong F^*$ coincide with the contractions $\iota(w)$. Since $\mathcal{I} \cap \wedge(F)$ is stable under these operators, it must be zero (or else it would contain the scalars). Thus $\mathcal{I} = \mathrm{Cl}(W)E$. □

COROLLARY 3.8. *For any non-zero Clifford module* $\mathcal{S}$ *over* $\mathrm{Cl}(W)$, *the action map* $\varrho\colon \mathrm{Cl}(W) \to \mathrm{End}(\mathcal{S})$ *is injective.*

PROOF. The kernel of the map ϱ is a 2-sided ideal in $\mathrm{Cl}(W)$, hence it must be zero. □

3.4. The spinor module. A Clifford module $\mathcal{S}$ over $\mathrm{Cl}(W)$ is called a *spinor module* if it is irreducible, i.e. if there are no non-trivial sub-modules.

EXAMPLE 3.9. If $E \in \mathrm{Lag}(W)$ is Lagrangian, the quotient $\mathcal{S} := \mathrm{Cl}(W)/\mathrm{Cl}(W)E$ is a spinor module. The irreducibility is immediate from the fact that $\mathrm{Cl}(W)E$ is a maximal left-ideal.

We will see below that all spinor modules over $\mathrm{Cl}(W)$ are isomorphic.

PROPOSITION 3.10. *Let $\mathcal{S}$ be a spinor module, and $E \in \mathrm{Lag}(W)$. Then the subspace*

$$\mathcal{S}^E = \{\phi \in \mathcal{S} |\, \varrho(w)\phi = 0 \ \forall w \in E\}$$

of elements fixed by E is 1-dimensional.

PROOF. Let F be a complementary Lagrangian subspace, and choose bases $e_1, \ldots, e_n$ of E and $f^1, \ldots f^n$ of F with $B(e_i, f^j) = \delta_i^j$. Define $p \in \mathrm{Cl}(W)$ as a product

$$p = \prod_{i=1}^n e_i f^i = \prod_{i=1}^n (1 - f^i e_i)$$

One easily verifies that p has the following properties:

$$p^2 = p, \quad Ep = 0, \quad pF = 0, \quad p - 1 \in \mathrm{Cl}(W)E. \tag{7}$$

Since $p^2 = p$ the operator $\varrho(p) \in \mathrm{End}(\mathcal{S})$ is a projection operator. By $Ep = 0$ its range lies in $\mathcal{S}^E$, and by $p - 1 \in \mathrm{Cl}(W)E$ it acts as the identity on $\mathcal{S}^E$. Hence

$$\mathcal{S}^E = \varrho(p)\, \mathcal{S}.$$

Since ϱ is injective, we have $\varrho(p) \neq 0$, hence $\mathcal{S}^E \neq 0$. Pick a non-zero element $\phi \in \mathcal{S}^E$. Then $\mathcal{S} = \varrho(\mathrm{Cl}(W))\phi = \varrho(\wedge(F))\phi$ by irreducibility, and since the left ideal $\mathrm{Cl}(W)E$ acts trivially on ϕ. Since $pF = 0$, and hence $p \wedge(F) = \mathbb{R}p$ we obtain

$$\mathcal{S}^E = \varrho(p)\, \mathcal{S} = \varrho(p)\, \varrho(\wedge(F))\, \phi = \mathbb{R}\varrho(p)\phi = \mathbb{R}\phi. \tag{8}$$

Equation (8) proves that $\mathcal{S}^E$ is 1-dimensional. □

The kernel and range of any homomorphism of Clifford modules are submodules. Hence, any non-zero homomorphism of spinor modules is an isomorphism. In particular, this applies to the action map

$$\mathrm{Cl}(W)/\mathrm{Cl}(W)E \otimes \mathcal{S}^E \to \mathcal{S}, \ x \otimes \phi \mapsto \varrho(x)\, \phi. \tag{9}$$

for any spinor module $\mathcal{S}$. Thus:

COROLLARY 3.11. *Any two spinor modules $\mathcal{S}_1, \mathcal{S}_2$ over $\mathrm{Cl}(W)$ are isomorphic. Furthermore, the isomorphism is unique up to non-zero scalar, i.e. the space $\mathrm{Hom}_{\mathrm{Cl}(W)}(\mathcal{S}_1, \mathcal{S}_2)$ is 1-dimensional.*

As a consequence, the projectivization $\mathbb{P}(\mathcal{S})$ of the spinor module is canonically defined (i.e. up to a *unique* isomorphism). In other words, the Clifford algebra $\mathrm{Cl}(W)$ has a unique irreducible *projective* module. The map taking E to $\mathcal{S}^E$ defines a canonical equivariant embedding

$$\mathrm{Lag}(W) \to \mathbb{P}(\mathcal{S})$$

as an orbit for the action of $\mathrm{O}(W)$. The image can be characterized as follows. Given $\phi \in \mathcal{S}$, let $N_\phi \subset W$ be its 'null space',

$$N_\phi = \{w \in W |\, \varrho(w)\, \phi = 0\}.$$

If $w_1, w_2 \in N_\phi$ then

$$0 = \varrho(w_1)\varrho(w_2)\phi + \varrho(w_2)\varrho(w_1)\phi = \varrho([w_1, w_2])\, \phi = B(w_1, w_2)\phi.$$

Hence, if $\phi \neq 0$ the subspace N_ϕ is isotropic.

DEFINITION 3.12. [**14, 13**] A non-zero spinor $\phi \in \mathcal{S}$ is called a *pure spinor* if N_ϕ is Lagrangian. Let $\mathrm{Pure}(\mathcal{S})$ denote the set of pure spinors of $\mathcal{S}$.

Note that the pure spinors defining a given Lagrangian subspace E are exactly the non-zero elements of the line $\mathcal{S}^E$. We can summarize the discussion in the following commutative diagram, equivariant for the action of $\mathrm{Pin}(W)$:

$$\begin{array}{ccc} \mathrm{Pure}(\mathcal{S}) & \longrightarrow & \mathcal{S}^\times \\ \downarrow & & \downarrow \\ \mathrm{Lag}(W) & \longrightarrow & \mathbb{P}(\mathcal{S}) \end{array}$$

EXERCISE 3.13. Show that for any spinor module, the map $\varrho\colon \mathrm{Cl}(W) \to \mathrm{End}(\mathcal{S})$ is an isomorphism.

EXERCISE 3.14. Any maximal left ideal $\mathcal{I} \subset \mathrm{Cl}(W)$ defines a spinor module $\mathrm{Cl}(W)/I$. Prove that the set of maximal left ideals is canonically isomorphic to $\mathbb{P}(\mathcal{S})$, and that the inclusion of $\mathrm{Lag}(W)$ is just the map $E \to \mathcal{I} = \mathrm{Cl}(W)E$.

3.5. The bilinear pairing on spinors. Let $\mathcal{S}$ be a spinor module over $\mathrm{Cl}(W)$. Then the dual space $\mathcal{S}^*$ is again a spinor module, with Clifford action given as

$$\varrho_{\mathcal{S}^*}(x) = \varrho_{\mathcal{S}}(x^\top)^*.$$

We obtain a 1-dimensional line $K_{\mathcal{S}} = \mathrm{Hom}_{\mathrm{Cl}(W)}(\mathcal{S}^*, \mathcal{S})$. The evaluation map defines an isomorphism of Clifford modules,

$$\mathcal{S} \cong \mathcal{S}^* \otimes K_{\mathcal{S}}.$$

Tensoring with $\mathcal{S}$, and composing with the duality pairing $\mathcal{S} \otimes \mathcal{S}^* \to \mathbb{R}$, we obtain a pairing

$$\mathcal{S} \otimes \mathcal{S} \to K_{\mathcal{S}}, \quad \phi \otimes \psi \mapsto (\phi, \psi).$$

This pairing was introduced by E. Cartan [**14, 13**], and more transparently by Chevalley [**15**]. By construction, the pairing satisfies

$$(\phi, \varrho(x)\psi) = (\varrho(x^\top)\phi, \psi)$$

for all $x \in \mathrm{Cl}(W)$. In particular,

$$(g.\phi, g.\psi) = \pm(\phi, \psi) \tag{10}$$

for all $g \in \mathrm{Pin}(W)$, where the dot indicates the Clifford action. This just follows since $g^\top g = \pm 1$ by definition of the Pin group.

EXERCISE 3.15. Let $\mathcal{S} = \mathrm{Cl}(W)/\mathrm{Cl}(W)E$. Show that there is a canonical isomorphism $K_{\mathcal{S}} = \wedge^n(E^*)$, where $n = \frac{1}{2}\dim W$. (Hint: $\mathcal{S}^*$ is identified with the submodule of $\mathrm{Cl}(W)$ generated by the line $\wedge^n(E) \subset \mathrm{Cl}(W)$.) Choose a Lagrangian subspace complementary to E to identify $\mathcal{S} = \wedge E^*$. Show that that $(\phi, \psi) = (\phi^\top \wedge \psi)_{[n]}$ using the wedge product in $\wedge E^*$.

PROPOSITION 3.16 (E. Cartan). [**13**, Section 111] *and* [**15**, Theorem III.2.4]. *If ϕ, ψ are pure spinors, then the Lagrangian subspaces N_ϕ, N_ψ are transverse if and only if $(\phi, \psi) \neq 0$.*

PROOF. Let $E = N_\phi$. For any Lagrangian complement $F \cong E^*$ to E, there is a unique isomorphism of spinor modules $\mathcal{S} \cong \wedge E^*$ taking ϕ to the pure spinor $1 \in \wedge E^*$. In this model, $K_{\mathcal{S}} = \wedge^n E^*$, and $(\phi, \psi) = \psi_{[n]}$. Suppose that $(\phi, \psi) \neq 0$, thus $\psi_{[n]} \neq 0$. For $w \in E - \{0\}$ we have

$$(\varrho(w)\psi)_{[n-1]} = \iota_w(\psi_{[n]}) \neq 0.$$

It follows that $N_\psi \cap E = \{0\}$. Conversely, if N_ψ is transverse to N_ϕ, we may take $F = N_\psi$. Then $\psi \in \wedge^n F^* - \{0\}$, and in particular $(\phi, \psi) = \psi_{[n]} \neq 0$. □

In particular, pure spinors in $\mathcal{S} = \mathrm{Cl}(W)/\mathrm{Cl}(W)E$, for any pair of transverse Lagrangian subspaces define a non-zero element of $\wedge^n E^*$, i.e. a volume form on E.

4. Linear Dirac geometry

The field of Dirac geometry was initiated by T. Courant in [**16**]. One of the original motivations of this theory was to describe manifolds with 'pre-symplectic foliations', arising for instance as submanifolds of Poisson manifolds. The term 'Dirac geometry' stems from its relation with the Dirac brackets arising in this context. One of the key features of Dirac geometry is that it treats Poisson geometry and pre-symplectic geometry on an equal footing. More recently, it was observed by Hitchin [**22**] that complex geometry can be understood in this framework as well, leading to the new field of generalized complex geometry [**22, 20**].

As in Courant's original paper, we will first discuss the linear case.

4.1. Linear Dirac structures. Let V be any vector space, and $\mathbb{V} = V \oplus V^*$ equipped with the bilinear form $\langle \cdot, \cdot \rangle$ given by the pairing between V and V^*:

$$\langle v_1 \oplus \alpha_1,\ v_2 \oplus \alpha_2 \rangle = \langle \alpha_1, v_2 \rangle + \langle \alpha_2, v_1 \rangle$$

for $v_i \in V$ and $\alpha_i \in V^*$. Specializing the constructions from the last section to the case $W = \mathbb{V}$, we note that $\mathbb{V}$ has two distinguished Lagrangian subspaces, V and V^*. We will call the corresponding spinor modules over $\mathrm{Cl}(\mathbb{V})$,

$$\mathrm{Cl}(\mathbb{V})/\mathrm{Cl}(\mathbb{V})V \cong \wedge V^*, \qquad \mathrm{Cl}(\mathbb{V})/\mathrm{Cl}(\mathbb{V})V^* \cong \wedge V$$

the contravariant and covariant spinor modules, respectively. The star operator for any volume form on V defines an isomorphism between these two spinor modules.

DEFINITION 4.1. A linear Dirac structure on a vector space V is a Lagrangian subspace $E \in \mathrm{Lag}(\mathbb{V})$.

As we have seen, a linear Dirac structure E may be described by a line $\mathcal{S}^E$ of pure spinors, using e.g. the covariant or contravariant spinor module.

EXAMPLES 4.2. Consider the contravariant spinor representation $\wedge V^*$. Here are some examples of pure spinors ϕ and associated Lagrangian subspaces N_ϕ:

(a) $\phi = 1$ corresponds to $N_\phi = V$.

(b) For any 2-form $\omega \in \wedge^2 V^*$, the exponential $\phi = e^{-\omega}$ is a pure spinor, with N_ϕ the graph
$$\mathrm{Gr}_\omega = \{v \oplus \alpha |\ \alpha = \omega(v, \cdot)\}.$$

(c) Any volume form $\mu \in \wedge^{\mathrm{top}} V^* \backslash 0$ defines a pure spinor $\phi = \mu$, with $N_\phi = V^*$.

(d) If μ is a volume form and $\pi \in \wedge^2 V$, the element $\phi = e^{-\iota(\pi)}\mu$ (where $\iota\colon \wedge V \to \mathrm{End}(\wedge V^*)$ is the algebra homomorphism extending the contraction operators $v \mapsto \iota(v)$) is a pure spinor, with N_ϕ the graph
$$\mathrm{Gr}_\pi = \{v \oplus \alpha |\ v = \pi(\alpha, \cdot)\}.$$

For any Lagrangian subspace $E \subset \mathbb{V} = V \oplus V^*$, define its *range* $\mathrm{ran}(E)$ to be the projection onto V. One observes that $S = \mathrm{ran}(E)$ carries a well-defined 2-form,

$$\omega_S(v_1, v_2) = \langle \alpha_1, v_2 \rangle = -\langle \alpha_2, v_1 \rangle \tag{11}$$

where $v_i \oplus \alpha_i \in E$ are lifts of $v_i \in \mathrm{ran}(E)$. The kernel of this 2-form is $\ker \omega_S = \{v \in V |\ (v, 0) \in E\}$. Conversely, E may be recovered from S together with the 2-form $\omega_S \in \wedge^2 S^*$, as $E = \{(v, \alpha) |\ v \in S,\ \alpha|_S = \omega_S(v, \cdot)\}$.

Let $\mathrm{ann}(S) \subset V^*$ be the annihilator of S, and choose a non-zero element $\mu \in \wedge^{\mathrm{top}} \mathrm{ann}(S) \subset \wedge V^*$.

EXERCISE 4.3. Show that

$$\phi = e^{-\omega_S} \mu \in \wedge V^*$$

is a pure spinor with $N_\phi = E$. (Here we have chosen an arbitrary extension of ω_S to a 2-form on V. Note that the element ϕ does not depend on this choice.) Conversely, show that every contravariant pure spinor has the form $\phi = e^{-\omega} \mu$, for uniquely given $S, \mu \in \wedge^{\mathrm{top}} \mathrm{ann}(S), \omega \in \wedge^2 S^*$.

Put differently, a contravariant pure spinor is equivalent to a Lagrangian subspace E together with a volume form on $V / \mathrm{ran}(E)$.

EXERCISE 4.4. Work out a similar description for covariant pure spinors.

4.2. Dirac maps. Let $A \colon V \to V'$ be a linear map. We say that two elements $w = v \oplus \alpha \in \mathbb{V}$ and $w' = v' \oplus \alpha' \in \mathbb{V}'$ are *A-related*, and write

$$w \sim_A w'$$

if $v' = A(v)$ and $\alpha = A^*(\alpha')$. Then the pull-back map of contravariant spinors has the property,

$$\varrho(w)(A^* \phi') = A^*(\varrho(w')\phi')$$

for $w \sim_A w'$ and $\phi' \in \wedge(V')^*$. From this, we see that the pull-back of a contravariant pure spinor ϕ' is again a pure spinor *unless* $A^*\phi' = 0$. Hence, if $F' \subset \mathbb{V}'$ is a Lagrangian subspace, and $\mathcal{S}^{F'} \subset \wedge(V')^*$ the pure spinor line in the contravariant spinor module, then $A^* \mathcal{S}^{F'}$ is either zero, or is a pure spinor line corresponding to some Lagrangian subspace $F \subset \mathbb{V}$.

EXERCISE 4.5. Suppose $\mathcal{S}^F = A^* \mathcal{S}^{F'}$. Show that

$$F = \{w \in \mathbb{V} |\ \exists w' \in F' \colon w \sim_A w'\}. \tag{12}$$

Similarly, in the covariant spinor representation we have

$$\varrho(w')(A_* \chi) = A_*(\varrho(w)\chi)$$

for $w \sim_A w'$ and $\chi \in \wedge V$. Hence, if $E \subset \mathbb{V}$ is a Lagrangian subspace, and $\mathcal{S}^E$ is the pure spinor line in the covariant spinor representation, then $A_*(\mathcal{S}^E)$ is either zero, or is the pure spinor line for a Lagrangian subspace $E' \subset \mathbb{V}'$.

EXERCISE 4.6. Suppose $\mathcal{S}^{E'} = A_*(\mathcal{S}^E)$. Show that

$$E' = \{w' \in \mathbb{V}' |\ \exists w \in E \colon w \sim_A w'\}. \tag{13}$$

DEFINITION 4.7. Let V, V' be vector spaces with linear Dirac structures E, E'. A linear map $A \colon V \to V'$ is called a *Dirac map* if the spaces E, E' are related by (13). It is a *strong Dirac map* if the induced map $A_* \colon \mathcal{S} = \wedge(V) \to \mathcal{S}' = \wedge(V')$ satisfies $A(\mathcal{S}^E) = (\mathcal{S}')^{E'}$.

Strong Dirac maps are also called *Dirac realizations* in the literature.

EXERCISE 4.8. Show that a Dirac map A is a strong Dirac map if and only if $E \cap (\ker(A) \oplus 0) = 0$.

EXAMPLE 4.9. Let $\pi \in \wedge^2 V$, $\pi' \in \wedge^2 V'$ be 2-forms, with $\pi' = A_*\pi$. Then the map A is a strong Dirac map relative to $E' = \mathrm{Gr}_{\pi'}$ and $E = \mathrm{Gr}_\pi$.

EXAMPLE 4.10. Let E be a linear Dirac structure on V. Recall that $S = \mathrm{ran}(E)$ carries a unique 2-form ω, with

$$E = \{v \oplus \alpha |\ v \in S,\ \alpha|_S = \omega(v, \cdot)\}$$

View (S, ω) as a Dirac space, with Dirac structure defined by the graph $\mathrm{Gr}_\omega \subset S \oplus S^*$. Then the inclusion map $\iota_S \colon S \to V$ is a strong Dirac map.

EXERCISE 4.11. Let V carry the Dirac structure E. Then the collapsing map $V \to \{0\}$ is (trivially) a Dirac map. Show that it is a strong Dirac map if and only if the 2-form ω on $S = \mathrm{ran}(E)$ is non-degenerate, if and only if $E = \mathrm{Gr}_\pi$ for a bi-vector π.

Suppose $E, F \in \mathrm{Lag}(\mathbb{V})$ are Lagrangian subspaces. Then E, F are transverse if and only if $\langle \psi, \chi \rangle \neq 0$, where $\psi \in \wedge V^*$ is a contravariant pure spinor defining F and $\chi \in \wedge(V)$ is a covariant pure spinor defining E. (This is equivalent to Proposition 3.16.) The following result says that Lagrangian complements may be 'pulled back' under strong Dirac maps.

PROPOSITION 4.12. *Suppose $A \colon V \to V'$ is a strong Dirac map relative to Lagrangian subspaces $E \subset \mathbb{V}$, $E' \subset \mathbb{V}'$. Let $\psi' \in \wedge(V')^*$ be a covariant pure spinor, with $N_{\psi'} = F'$ transverse to E'. Then $\psi = A^*\psi'$ is non-zero, and $N_\psi = F$ is transverse to E. Equivalently, if ϕ is a contravariant pure spinor with $N_\phi = E$ we have*

$$(\phi, A^*\psi') \neq 0.$$

PROOF. Let $\chi \in \wedge V$ be a *covariant* pure spinor defining E. Then $A_*(\chi) \in \wedge(V')$ is a pure spinor defining E'. Since E', F' are transverse,

$$0 \neq \langle \psi',\ A_*\chi \rangle = \langle A^*\psi', \chi \rangle.$$

This shows that $\psi = A^*\psi'$ is a pure spinor, with $N_\psi = F$ transverse to E. $\square$

4.3. The map $\mathrm{O}(V) \to \mathrm{Lag}(\mathbb{V})$. Suppose now that V itself carries a non-degenerate symmetric bilinear form B. Let $\overline{V}$ denote the same vector space, but with the bilinear form $-B$. There is an isometric isomorphism

$$\kappa \colon V \oplus \overline{V} \to \mathbb{V} = V \oplus V^*, \quad v \oplus w \mapsto (v - w) \oplus \tfrac{1}{2} B(v + w, \cdot).$$

This identifies $\mathrm{O}(V \oplus \overline{V}) \cong \mathrm{O}(\mathbb{V})$, and in particular yields an inclusion of the subgroup $\mathrm{O}(V)$:

$$\mathrm{O}(V) \hookrightarrow \mathrm{O}(\mathbb{V}),\ A \mapsto A^\kappa = \kappa \circ \begin{pmatrix} A & 0 \\ 0 & I \end{pmatrix} \circ \kappa^{-1}.$$

If we use B to identify V and V^*, the matrix on the right is easily computed to be,

$$A^\kappa = \begin{pmatrix} (A+I)/2 & (A-I) \\ (A-I)/4 & (A+I)/2 \end{pmatrix}. \tag{14}$$

Its action on V describes a new Lagrangian subspace,

$$F = A^\kappa(V).$$

Let $\Gamma(V) \to \Gamma(\mathbb{V})$ be the inclusion of Clifford groups defined by the homomorphism $\mathrm{Cl}(V) \subset \mathrm{Cl}(V \oplus \bar{V}) \cong \mathrm{Cl}(\mathbb{V})$. This lifts the map $\mathrm{O}(V) \to \mathrm{O}(\mathbb{V})$, and restricts to a homomorphism of Pin groups. For any lift $\tilde{A} \in \Gamma(V)$ of A, we obtain a lift $\tilde{A}^\kappa \in \Gamma(\mathbb{V})$ of A^κ. The pure spinor $1 \in \wedge(V^*)$ defines the Lagrangian subspace $V \subset \mathbb{V}$, hence

$$\psi = \varrho(\tilde{A}^\kappa)\, 1$$

represents $F = A^\kappa(V)$. The situation is described in the following diagram:

$$\begin{array}{ccc} \Gamma(V) & \longrightarrow & \mathrm{Pure}(\mathbb{V}) \\ \downarrow & & \downarrow \\ \mathrm{O}(V) & \longrightarrow & \mathrm{Lag}(\mathbb{V}) \end{array}$$

where the lower map is $A \mapsto A^\kappa(v)$ and the upper map is $\tilde{A} \mapsto \varrho(\tilde{A}^\kappa)(1)$. Since $\mathrm{Pin}(V) \subset \Gamma(V)$ is a double cover of $\mathrm{O}(V)$, there is a lift $\tilde{A} \in \mathrm{Pin}(V)$ that is unique up to sign. One has an explicit formula for the resulting ψ, valid for $\det(A+I) \neq 0$:

$$\psi = \det{}^{1/2}\big(\tfrac{A+I}{2}\big) \exp\big(\tfrac{1}{4} \textstyle\sum_i (\tfrac{I-A}{I+A} v_i) \wedge v^i\big). \tag{15}$$

See [**5**] for a proof. Here we have used B to identify $V^* \cong V$, and v_i, v^i are bases with $B(v_i, v^j) = \delta_i^j$. The sign of the square root depends on the choice of lift $\tilde{A}$.

Similarly, the action of A^κ on V^* defines a Lagrangian subspace transverse to F,

$$E = A^\kappa(V^*).$$

Given an orientation on V, the associated Riemannian volume form μ is a pure spinor defining V^*, hence

$$\phi = \varrho(\tilde{A}^\kappa)\, \mu$$

is a pure spinor defining E.

REMARK 4.13. If the scalar product B on V is definite, the inclusion $\mathrm{O}(V) \to \mathrm{Lag}(V)$ is a bijection: This is exactly the diffeomorphism $\mathrm{Lag}(V) \cong \mathrm{O}(n)$ mentioned earlier. (This isomorphism is described in the paper [**16**] under the name 'generalized Cayley transform'.) Similarly, the map $\Gamma(V) \to \mathrm{Pure}(\mathbb{V})$, $g \mapsto \varrho(g)\, 1$ defines a bijection of the set of pure spinors with the Clifford group:

$$\mathrm{Pure}(\mathbb{V}) \cong \Gamma(n) := \Gamma(\mathbb{R}^n).$$

5. The Cartan-Dirac structure

5.1. Almost Dirac structures. It is straightforward to generalize the above considerations from vector spaces to vector bundles, and in particular to the tangent bundle of a manifold. Thus, let

$$\mathbb{T}M = TM \oplus T^*M$$

be the generalized tangent bundle, with fiberwise inner product $\langle \cdot, \cdot \rangle$ given by the pairing of 1-forms with vector fields, and $\mathrm{Cl}(\mathbb{T}M)$ the corresponding bundle of Clifford algebras. Covariant spinors are multi-vector fields, $\chi \in \mathfrak{X}^\bullet(M) = \Gamma(M, \wedge TM)$, while contravariant spinors are differential forms, $\phi \in \Omega(M) = \Gamma(M, \wedge(T^*M))$.

An *almost Dirac structure* on M is a Lagrangian sub-bundle $E \subset \mathbb{T}M$. (In Section 6, we will discuss the integrability condition turning an almost Dirac structure to a Dirac structure.) A smooth map between almost Dirac manifolds $f\colon M \to M'$ is called a (strong) Dirac map if each tangent map $(\mathrm{d}f)_x$ is a (strong) Dirac map.

Any almost Dirac structure may be described (at least locally) by a contravariant pure spinor $\phi \in \Omega(M)$, or by a covariant pure spinor $\chi \in \mathfrak{X}(M)$. Our basic examples for vector spaces carry over to manifolds: Any 2-form on a manifold defines an almost Dirac structure, as does any bi-vector field. If E is an almost Dirac structure, described (locally) by a pure spinor ϕ, and $\tau \in \Omega^2(M)$ any 2-form, one may define a new almost Dirac structure E^τ described locally by pure spinor $e^{-\tau}\phi$. One calls E^τ the *gauge transformation* of E by the 2-form τ. For instance, taking $E = TM$, one obtains the graph of τ:

$$(TM)^\tau = \mathrm{Gr}_\tau .$$

EXERCISE 5.1. In general, show that E^τ is the image of E under the automorphism $v \oplus \alpha \mapsto v \oplus (\alpha + \iota_v \tau)$ of $\mathbb{T}M$.

Given a pseudo-Riemannian metric on M, any section A of the group bundle $\mathrm{O}(TM)$ defines a pair of transverse Lagrangian sub-bundles

$$E = A^\kappa(T^*M), \;\; F = A^\kappa(TM)$$

of $\mathbb{T}M$. A lift $\tilde{A}$ to a section of $\mathrm{Pin}(TM)$ defines pure spinors ϕ, ψ corresponding to E, F, where ϕ depends on the choice of an orientation on M.

5.2. The case $M = G$. Let G be a Lie group. For any $\xi \in \mathfrak{g}$, let $\xi^L, \xi^R \in \mathfrak{X}^1(G)$ the corresponding left-,right-invariant vector fields. The bundle $\mathrm{GL}(TG)$ has a unique section A with the property

$$A(\xi^L) = \xi^R.$$

Suppose G carries a bi-invariant pseudo-Riemannian metric, and let B the corresponding inner product on the Lie algebra $\mathfrak{g}$. Then A is an $\mathrm{Ad}(G)$-invariant section of $\mathrm{O}(TG) \subset \mathrm{GL}(TG)$. Hence, it determines transverse $\mathrm{Ad}(G)$-invariant Lagrangian sub-bundles $E, F \subset \mathbb{T}G$. Recall that $\theta^L, \theta^R \in \Omega^1(G) \otimes \mathfrak{g}$ denote the Maurer-Cartan forms.

PROPOSITION 5.2. *Define bundle maps $e, f\colon \mathfrak{g} \to \mathbb{T}G$ by*

$$e(\xi) = (\xi^L - \xi^R) \oplus B(\frac{\theta^L + \theta^R}{2}, \xi)$$

$$f(\xi) = (\frac{\xi^L + \xi^R}{2}) \oplus B(\frac{\theta^L - \theta^R}{4}, \xi)$$

The maps e, f are injective, and have range E, F.

PROOF. Under left-trivialization of the tangent bundle $TG = G \times \mathfrak{g}$, the section A is just the adjoint action, $g \to \mathrm{Ad}_g$. Hence, the section A^κ is given by (14), with A replaced by Ad_g. Writing elements $\xi_1 \oplus \xi_2 \in \mathfrak{g} \oplus \mathfrak{g} \cong T_gG \oplus T_g^*G$ as column vectors, we see that

$$A^\kappa(\xi_1 \oplus \xi_2) = f(\xi_1) + e(\xi_2). \qquad \square$$

The sub-bundle E is called the *Cartan-Dirac structure.* (It satisfies the integrability condition discussed below.) Since $\xi^\sharp = \xi^L - \xi^R$ are the generating vector fields for the conjugation action, the generalized distribution $\mathrm{ran}(E) = \mathrm{pr}_{TM}(E)$ is just the distribution tangent to the conjugacy classes $\mathcal{C}$ of G. Hence, by (11) the conjugacy classes $\mathcal{C} \subset G$ acquire $\mathrm{Ad}(G)$-invariant 2-forms ω.

PROPOSITION 5.3. *The 2-forms ω on conjugacy classes $\mathcal{C}$ are exactly the GHJW 2-forms.*

We leave the proof as an exercise. Equivalently, the inclusion maps

$$\iota_{\mathcal{C}} \colon \mathcal{C} \hookrightarrow G$$

are strong Dirac maps, relative to the (almost) Dirac structures given by the GHJW 2-form ω on $\mathcal{C}$ and the Lagrangian sub-bundle $E \subset \mathbb{T}G$.

Let us now *assume* that the adjoint action $\mathrm{Ad}\colon G \to \mathrm{O}(\mathfrak{g})$ lifts to a group homomorphism $\widetilde{\mathrm{Ad}}\colon G \to \mathrm{Pin}(\mathfrak{g})$ into the Pin group. (This is automatic if G is simply connected.) The lift $\widetilde{\mathrm{Ad}}$ determines lifts $\tilde{A}, \tilde{A}^\kappa$. Hence it defines an invariant pure spinor $\psi = \varrho(\tilde{A}^\kappa)\,1$ with $N_\psi = F$, and given an invariant volume form μ it also defines a pure spinor $\phi = \varrho(\tilde{A}^\kappa)\,\mu$ with $N_\phi = E$.

Consider now a conjugacy class $\mathcal{C}$, with GHJW 2-form ω. By Lemma 4.12, the pull-back $\iota_{\mathcal{C}}^*\psi$ is a pure spinor defining a Lagrangian sub-bundle transverse to Gr_ω. Equivalently, the pairing between the two pure spinors $e^{-\omega}, \iota_{\mathcal{C}}^*\psi$ is non-vanishing, that is

$$0 \neq (e^{-\omega}, \iota_{\mathcal{C}}^*\psi) = (e^{\omega}\iota_{\mathcal{C}}^*\psi)_{[\mathrm{top}]}$$

is a volume form on $\mathcal{C}$. We have shown the following more precise version of the FACT:

THEOREM 5.4. *Suppose that the adjoint action $\mathrm{Ad}\colon G \to \mathrm{O}(\mathfrak{g})$ lifts to a homomorphism $\widetilde{\mathrm{Ad}}\colon G \to \mathrm{Pin}(\mathfrak{g})$, and let $\psi \in \Omega(G)$ be the pure spinor defined by such a lift. Then, for any conjugacy class $\mathcal{C}$ in G the top degree part of*

$$e^{\omega}\iota_{\mathcal{C}}^*\psi$$

defines an invariant volume form on $\mathcal{C}$.

The explicit formula (3) for $\psi = \tilde{A}^\kappa.1$ is obtained as a special case from (15):

PROPOSITION 5.5. *If G is connected and the adjoint action $G \to \mathrm{SO}(\mathfrak{g})$ lifts to a group homomorphism $G \to \mathrm{Spin}(\mathfrak{g})$, the formula (3) defines a pure spinor ψ with $N_\psi = F$.*

Up to a scalar function the expression (3) for ψ can be directly obtained, as follows:

EXERCISE 5.6. Over the set where $\mathrm{Ad}_g + 1$ is invertible, the vector fields $\frac{\xi^L+\xi^R}{2}$ span the tangent space. Hence, there is a unique 2-form ς on this set, with

$$\iota(\tfrac{\xi^L+\xi^R}{2})\varsigma + B(\tfrac{\theta^L-\theta^R}{4}, \xi) = 0.$$

Deduce that e^ς is a pure spinor defining F, hence coincides with ψ up to a scalar function. Next, check that

$$\varsigma = \tfrac{1}{4} B\Big(\tfrac{1-\mathrm{Ad}_g}{1+\mathrm{Ad}_g}\theta^L, \theta^L\Big)$$

is the unique solution of the defining equation for ς.

6. Dirac structures

6.1. Courant's integrability condition. One of Courant's main discoveries in [**16**] was the existence of a natural integrability condition for almost Dirac structures $E \subset \mathbb{T}M$. Following Alekseev-Xu [**7**] and Gualtieri [**20**], we will express the Courant integrability condition in terms of the spinor representation. The Lagrangian sub-bundle E defines a filtration on the spinor module $\Omega(M)$:

$$\Omega(M) = \Omega^{(n)}(M) \supset \cdots \Omega^{(1)}(M) \supset \Omega^{(0)}(M).$$

Here $\Omega^{(k)}(M)$ consists of differential forms γ with $\rho(w_0)\cdots\rho(w_k)\gamma = 0$ for all $w_i \in \Gamma(E)$.

Let us fix a closed 3-form $\eta \in \Omega^3(M)$ (possibly zero). Note that $\mathrm{d}+\eta$ is again a differential.

LEMMA 6.1. *Let $\phi \in \Omega(M)$ be a (locally defined) pure spinor with $N_\phi = E$. Then*

$$(d+\eta)\phi \in \Omega^{(3)}(M).$$

PROOF. Let $w_i \in \Gamma(E)$. Since $\rho(w_i)$ annihilates ϕ, we have

$$\rho(w_1)\rho(w_2)\rho(w_3)(\mathrm{d}+\eta)\phi = [\rho(w_1),[\rho(w_2),[\rho(w_3),\mathrm{d}+\eta]]]\phi, \tag{16}$$

using graded commutators of operators on $\Omega(M)$. A calculation (cf. Exercise 6.8 below) shows that the triple commutator of operators is multiplication by a smooth function. Thus $\rho(w_1)\rho(w_2)\rho(w_3)(\mathrm{d}+\eta)\phi$ is a function times ϕ, and hence is annihilated by $\rho(w_0)$. □

We may now state the Courant integrability condition.

DEFINITION 6.2. An almost Dirac structure $E \subset \mathbb{T}M$ is called integrable relative to the closed 3-form η if, for any (locally defined) pure spinor ϕ with $N_\phi = E$,

$$\mathrm{gr}^3\left((\mathrm{d}+\eta)\phi\right) = 0. \tag{17}$$

Note that this condition does not depend on the choice of ϕ, since

$$\mathrm{gr}^3\left((\mathrm{d}+\eta)(f\phi)\right) = f\,\mathrm{gr}^3((\mathrm{d}+\eta)(\phi)).$$

By (17), E is integrable if and only if $(\mathrm{d}+\eta)\phi \in \Omega^{(2)}(M)$. Since ϕ and $(\mathrm{d}+\eta)\phi$ have opposite parity, this is in fact equivalent to the condition

$$(\mathrm{d}+\eta)\phi \in \Omega^{(1)}(M). \tag{18}$$

DEFINITION 6.3. A *Dirac manifold* is a triple (M, E_M, η_M), consisting of a manifold M, an almost Dirac structure E_M, and a closed 3-form η_M such that E_M is integrable relative to η_M. A smooth map $\Phi\colon M \to M'$ between two Dirac manifolds is called a *(strong) Dirac map* if each $\mathrm{d}_x\Phi\colon T_xM \to T_{\Phi(x)}M'$ is a linear (strong) Dirac map, and in addition

$$\Phi^*\eta_{M'} = \eta_M.$$

REMARK 6.4. The integrability condition may be rephrased as $(\mathrm{d}+\eta)\phi = \rho(w)\phi$ for *some* section $w \in \Gamma(\mathbb{T}M)$. It is not always possible to choose ϕ in such a way that $(\mathrm{d}+\eta)\phi = 0$. As shown by Alekseev-Xu [**7**], the obstruction is the 'modular class' of E.

6.2. Examples.

EXAMPLES 6.5. (a) Let ω be a 2-form and $\phi = e^{-\omega}$. Then $(\mathrm{d}+\eta)\phi = (-\mathrm{d}\omega + \eta)\wedge\phi$ lies in $\Omega^{(1)}(M)$ if and only if $\mathrm{d}\omega = \eta$. From now on, we will view any manifold M with 2-form ω as a Dirac manifold, taking $E_M = \mathrm{Gr}_\omega$. Observe that $\Phi\colon M \to \mathrm{pt}$ is a strong Dirac map if and only if ω is symplectic (closed and non-degenerate).

(b) More generally, if E is integrable with respect to η, and τ is any 2-form, then E^τ is integrable with respect to $\eta + \mathrm{d}\tau$.

(c) Let π be a bi-vector field and μ_M a volume form on M. Then $\phi = e^{-\iota(\pi)}\mu_M$ satisfies

$$(\mathrm{d}+\eta)\phi = \iota\big(-\tfrac{1}{2}[\pi,\pi]_{\mathrm{Sch}} - \pi^\sharp(\eta) + X_\pi + Y_{\pi,\eta}\big)\phi.$$

Here $[\cdot,\cdot]_{\mathrm{Sch}}$ is the Schouten bracket on multi-vector fields, $X_\pi \in \mathcal{X}^1(M)$ is the vector field defined by $\mathrm{d}\iota(\pi)\mu_M = -\iota(X_\pi)\mu_M$, $\pi^\sharp$ is the bundle map from $\wedge T^*M$ to $\wedge TM$, and $Y_{\pi,\eta}$ is the vector field $Y_{\pi,\eta} = \pi^\sharp(\iota(\pi)\eta)$. The Courant integrability condition reduces to the condition

$$\tfrac{1}{2}[\pi,\pi]_{\mathrm{Sch}} + \pi^\sharp(\eta) = 0,$$

defining a *twisted Poisson structure*. These structures were introduced by Klimcik-Strobl [**24**] and further studied by Ševera-Weinstein [**30**]. It was argued by Kosmann-Schwarzbach-Laurent-Gengoux [**26**, Theorem 6.1] (see also [**7**, Example 6.2]) that the sum $X_\pi + Y_{\pi,\eta}$ plays the role of the *modular vector field* for a twisted Poisson structure.

(d) Take $\eta = 0$, and let $\alpha_1,\ldots,\alpha_k \in \Omega^1(M)$ be a collection of pointwise linearly independent 1-forms, and $K \subset TM$ be the codimension k distribution given as the intersection of their kernels. Then $\phi = \alpha_1\wedge\cdots\wedge\alpha_k$ is pure spinor defining $E = K \oplus \mathrm{ann}(K)$. The integrability condition $\mathrm{d}\phi \in \Omega^{(1)}(M)$ holds if and only if $\mathrm{d}\phi = \beta\wedge\phi$ for a 1-form β. This is one version of the standard (Frobenius) integrability condition for distributions.

(e) The Courant integrability condition has an obvious generalization to complex almost Dirac structures $E \subset \mathbb{T}M \otimes \mathbb{C}$. Given an almost complex structure on M, i.e. a linear complex structure on the tangent bundle, $J \in \Gamma(\mathrm{End}(TM))$, $J^2 = -\mathrm{Id}_{TM}$, one obtains a linear complex structure $\mathbb{J} = J \oplus (-J^*) \in \Gamma(\mathrm{End}(\mathbb{T}M))$, $\mathbb{J}^2 = -\mathrm{Id}_{\mathbb{T}M}$. Let $E \subset \mathbb{T}M \otimes \mathbb{C}$ be the $+i$ eigenbundle of $\mathbb{J}$. It turns out that E is Courant integrable if and only if the almost complex structure J is integrable, i.e comes from complex coordinate charts with holomorphic transition functions. This is the motivating example for the *generalized complex geometry*, developed by Hitchin [**22**] and Gualtieri [**20**].

EXERCISE 6.6. In Example 6.5(e), give a pure spinor $\phi \in \Omega(M)\otimes\mathbb{C}$ defining E.

EXERCISE 6.7. (See [**18**].) Work out a formula for

$$e^{\iota(\pi)}\circ\mathrm{d}\circ e^{-\iota(\pi)} = \mathrm{d} + [\iota(\pi),\mathrm{d}] + \tfrac{1}{2}[\iota(\pi),[\iota(\pi),\mathrm{d}]] + \cdots.$$

(The resulting expression contains terms at most quadratic in π.) Use this to show

$$\mathrm{d}(e^{-\iota(\pi)}\mu_M) = \iota(-\tfrac{1}{2}[\pi,\pi]_{\mathrm{Sch}} + X_\pi)\mu_M$$

for any volume form μ_M. Similarly show that

$$\eta \wedge (e^{-\iota(\pi)}\mu_M) = \iota\big(-\pi^\sharp(\eta) + Y_{\pi,\eta}\big)(e^{-\iota(\pi)}\mu_M).$$

EXERCISE 6.8. (a) Verify that the following formula defines a bilinear map $[\![\cdot,\cdot]\!]\colon \Gamma(\mathbb{T}M) \times \Gamma(\mathbb{T}M) \to \Gamma(\mathbb{T}M)$:

$$\rho([\![w_1, w_2]\!]) = [\rho(w_1), [\rho(w_2), \mathrm{d} + \eta]].$$

This is the definition of the (non skew-symmetric) *Courant bracket* $[\![\cdot,\cdot]\!]$ on $\Gamma(\mathbb{T}M)$ as a *derived bracket*. See Roytenberg [**29**], Alekseev-Xu [**7**] and Kosmann-Schwarzbach [**25**].

(b) Conclude that for any $w_1, w_2, w_3 \in \Gamma(\mathbb{T}M)$, the operator

$$[\rho(w_1), [\rho(w_2), [\rho(w_3), \mathrm{d} + \eta]]]$$

on $\Omega(M)$ is multiplication by the smooth function $Y(w_1, w_2, w_3)$.

(c) Show that for any almost Dirac structure $E \subset \mathbb{T}M$, the restriction of Y to sections of E defines an anti-symmetric tensor $Y_E \in \wedge^3 E^*$.

PROPOSITION 6.9. *The almost Dirac structure E is integrable if and only if $\Gamma(E)$ is closed under Courant bracket $[\![\cdot,\cdot]\!]$. In this case, the restriction $[\cdot,\cdot]_E$ of the Courant bracket to $\Gamma(E)$ defines a Lie algebroid structure on E: That is, it is a Lie bracket, the projection map $a\colon \Gamma(E) \to \mathfrak{X}(M)$ is a Lie algebra homomorphism, and*

$$[w_1, f w_2]_E = f[w_1, w_2]_E + v_1(f)\, w_2, \quad w_i \in \Gamma(E)$$

where $v_1 = a(w_1)$.

PROOF. Since E is Lagrangian, we have $[\![w_2, w_3]\!] \in \Gamma(E)$ for all $w_2, w_3 \in \Gamma(E)$ if and only if

$$Y(w_1, w_2, w_3) = [\rho(w_1), \rho([\![w_2, w_3]\!])] = \langle w_1, [\![w_2, w_3]\!]\rangle = 0$$

for all $w_1, w_2, w_3 \in \Gamma(E)$. The remaining claims are left as an exercise. □

The theory of Lie algebroids [**17, 28**] shows that the generalized distribution $\mathrm{ran}(E) = \mathrm{pr}_{TM}(E)$ is *integrable*, i.e. defines a generalized foliation. Moreover, the leaves $S \subset M$ of this foliation carry *2-forms* $\omega_S \in \Omega^2(S)$, defined pointwise by (11).

For $E = \mathrm{Gr}_\pi$ the graph of a Poisson bi-vector field (i.e. $\eta = 0$), this is just the usual foliation by symplectic leaves $S \subset M$, with ω_S the symplectic 2-forms. More generally, in the twisted Poisson case $\frac{1}{2}[\pi,\pi]_{\mathrm{Sch}} + \pi^\sharp\eta = 0$ one still obtains a foliation. The 2-forms on the leaves are again non-degenerate (since $E \cap TM = \{0\}$), but are not closed in general:

PROPOSITION 6.10. *Let $E \subset \mathbb{T}M$ be a Dirac structure (relative to a closed 3-form $\eta \in \Omega^3(M)$). Then the 2-forms ω_S on the leaves $S \subset M$ satisfy $d\omega_S = \iota_S^*\eta$.*

PROOF. Given any point $x \in S$, we may pass to a neighbourhood of x to reduce to the case $M = S \times N$, with ι_S the inclusion as $S \times \{y\}$ for some $y \in N$. View ω_S as a form on $S \times N$, and define $\gamma := e^{\omega_S}\phi$. Then $\gamma|_S$ is a nowhere vanishing section of the top exterior power of $T^*N|_S \cong \mathrm{ann}(S) \subset T^*M|_S$. By assumption, there exists a vector field v and a 1-form α such that $(\mathrm{d} + \eta)\phi = \iota(v)\phi + \alpha\phi$. This yields:

$$0 = e^{\omega_S}(\mathrm{d} + \eta - \iota(v) - \alpha)\phi = \big(\eta - \mathrm{d}\omega_S + \iota(v)\omega - \alpha\big)\gamma + (\mathrm{d}\gamma - \iota(v)\gamma).$$

Restricting to S, and taking the component in $\Gamma(\wedge^3 T^*S \otimes \wedge^{\mathrm{top}} T^*N|_S)$ we find $(\iota_S^*\eta - \mathrm{d}\omega_S)\gamma|_S = 0$. Hence $\iota_S^*\eta = \mathrm{d}\omega_S$. □

6.3. Integrability of the Cartan-Dirac structure. Let us now return to the example of a Lie group G with an invariant inner product B on $\mathfrak{g}$. Suppose that G admits an invariant orientation and that $\mathrm{Ad}\colon G \to \mathrm{O}(\mathfrak{g})$ lifts to the Pin group. Let $\phi, \psi \in \Omega(G)$ be the pure spinors defining the almost Dirac structures E, F. By construction, both ϕ and ψ are Ad-invariant differential forms.

Now let $\eta \in \Omega^3(G)$ be the left-invariant 3-form

$$\eta = \frac{1}{12} B(\theta^L, [\theta^L, \theta^L]).$$

Since B is invariant, one may replace θ^L with θ^R in this formula, thus η is also right-invariant. In particular, η is closed (since any bi-invariant differential form on a Lie group is closed). Letting $\xi^\sharp = \xi^L - \xi^R$ be the generating vector fields for the conjugation action, one finds

$$\iota(\xi^\sharp)\eta = -\mathrm{d}\ B(\tfrac{\theta^L+\theta^R}{2}, \xi)$$

As a consequence, we see that the commutator of $\mathrm{d}+\eta$ with the generating sections $e(\xi)$ of E are,

$$[\rho(e(\xi)), \mathrm{d} + \eta] = [\iota(\xi^\sharp) + B(\tfrac{\theta^L+\theta^R}{2}, \xi), \mathrm{d} + \eta] = L(\xi^\sharp).$$

(Here $L(X) = [\iota(X), \mathrm{d}]$ denotes the Lie derivative in the direction of a vector field X.) It hence follows that

$$\rho(e(\xi))(\mathrm{d} + \eta)\phi = [\rho(e(\xi)), \mathrm{d} + \eta]\phi = L(\xi^\sharp)\phi = 0.$$

Thus $(\mathrm{d} + \eta)\phi \in \Omega^{(0)}(M)$. Since the parity of $(\mathrm{d} + \eta)\phi$ is opposite to that of ϕ, we obtain:

THEOREM 6.11. *The pure spinor ϕ satisfies*

$$(d + \eta)\phi = 0.$$

In particular, we see that E is a Dirac structure.

DEFINITION 6.12. The Dirac structure E on G is called the *Cartan-Dirac structure.*

The integrability of E explains our earlier observation that the distribution $\mathrm{ran}(E)$ is just the tangent distribution for the generalized foliation by conjugacy classes. Furthermore, Proposition 6.10 tells us that the GHJW 2-form $\omega_{\mathcal{C}}$ on the conjugacy classes satisfies,

$$\mathrm{d}\omega_{\mathcal{C}} = \iota_{\mathcal{C}}^* \eta.$$

REMARK 6.13. The Cartan-Dirac structure was discovered independently by Anton Alekseev, Pavol Ševera and Thomas Strobl, around the end of the last century.

REMARK 6.14. By contrast, the almost Dirac structure F is not integrable. Instead, one has [**1**]

$$(\mathrm{d} + \eta)\psi = \rho(e(\Xi))\psi$$

where $\Xi \in \wedge^3\mathfrak{g}$ is the 'structure constants tensor', and $e(\Xi) \in \Gamma(\wedge^3 E)$ is defined using the extension of $e\colon \mathfrak{g} \to \Gamma(E)$ to an algebra homomorphism $\wedge\mathfrak{g} \to \Gamma(\wedge(E))$.

7. Group-valued moment maps

The theory of G-valued moment maps was introduced in the paper [**4**]. One of its main applications is that it provides a natural framework for the construction of symplectic forms on moduli spaces of flat connections.

7.1. Definition of q-Hamiltonian G-spaces. Let G be a connected Lie group with a bi-invariant pseudo-Riemannian metric, and let B be the corresponding invariant inner product on $\mathfrak{g}$. Let M be a manifold. A *G-action* on M is a group homomorphism $\mathcal{A}\colon G \to \mathrm{Diff}(M)$ such that the action map $G \times M \to M$, $(g,x) \mapsto \mathcal{A}(g).x$ is smooth. Similarly, a *$\mathfrak{g}$-action* is a Lie algebra homomorphism $\mathcal{A}\colon \mathfrak{g} \to \mathfrak{X}(M)$ such that the map $\mathfrak{g} \times M \to TM$, $(\xi, x) \mapsto \mathcal{A}(\xi)_x$ is smooth. We will write $\xi^\sharp = \mathcal{A}(\xi)$. For any G-action, the generating vector fields (defined with the appropriate sign) give a $\mathfrak{g}$-action. Conversely, if M is compact and G is simply connected, any $\mathfrak{g}$-action integrates to a G-action.

DEFINITION 7.1. [**4**] A Hamiltonian $\mathfrak{g}$-space with G-valued moment map is a $\mathfrak{g}$-manifold M, together with a $\mathfrak{g}$-invariant 2-form $\omega \in \Omega^2(M)$ and a $\mathfrak{g}$-equivariant map $\Phi \in C^\infty(M,G)$ such that

(a) $\mathrm{d}\omega = \Phi^*\eta$,
(b) $\iota(\xi^\sharp)\omega = \Phi^* B(\frac{\theta^L+\theta^R}{2}, \xi)$, $\quad \xi \in \mathfrak{g}$ (Moment map condition.)
(c) $\ker(\omega_x) = \{\xi^\sharp(x) |\ \mathrm{Ad}_{\Phi(x)}\,\xi = -\xi\}$, $x \in M$ (Minimal degeneracy condition.)

REMARK 7.2. As pointed out in [**4**], (b) is the simplest G-valued analogue to the defining property for $\mathfrak{g}^*$-valued moment maps $\Phi_0\colon M \to \mathfrak{g}^*$, $\iota(\xi^\sharp)\omega_0 = -\mathrm{d}\langle \Phi_0, \xi\rangle$. It follows from the work of Bursztyn-Crainic [**10**] and Xu [**32**] (see also [**1**]) that (c) may be replaced by the more elegant condition,

$$\ker(\omega_x) \cap \ker(\mathrm{d}_x \Phi) = 0.$$

The theory of G-valued moment maps was developed in [**4**], and subsequent papers, in full analogy to the familiar theory of $\mathfrak{g}^*$-valued moment maps. However, the proofs were much more complicated than in the $\mathfrak{g}^*$-valued theory, and for technical reasons it was necessary to assume that B is positive definite. Unfortunately, this restriction excludes several interesting examples, such as representation varieties for non-compact semi-simple Lie groups. (The Killing form of such groups is indefinite.) In the following approach to group-valued moment maps via Dirac structures these difficulties are no longer present.

THEOREM 7.3 (Bursztyn-Crainic). *Definition 7.1 is equivalent to the following Definition 7.4.*

DEFINITION 7.4. A Hamiltonian $\mathfrak{g}$-space with G-valued moment map is a manifold M with a 2-form ω, together with a strong Dirac map $\Phi\colon M \to G$.

Here M is viewed as a Dirac manifold with $E_M = \mathrm{Gr}_\omega$ and 3-form $\eta_M = \mathrm{d}\omega$, while G carries the Cartan-Dirac structure. Recall that $\Phi^*\eta = \eta_M$ as part of the definition of a Dirac map from $(M, \mathrm{Gr}_\omega, \mathrm{d}\omega)$ to (G, E, η).

Note that Definition 7.4 no longer mentions the $\mathfrak{g}$-action on M, the equivariance of ω and Φ, or the minimal degeneracy property: as shown by Bursztyn-Crainic, all of this comes for free!

REMARKS 7.5. One is immediately led to consider arbitrary Dirac manifolds (M, E_M, η_M) together with strong Dirac maps $M \to G$. As shown by Bursztyn-Crainic, one recovers the theory of q-Poisson manifolds [**2, 3**]. Definition 7.4 is parallel to the definition of Hamiltonian $\mathfrak{g}$-spaces with $\mathfrak{g}^*$-valued moment maps: These may be defined as manifolds M with closed 2-forms ω and strong Dirac maps $\Phi\colon M \to \mathfrak{g}^*$. Here $\mathfrak{g}^*$ carries the Dirac structure coming from its Kirillov-Poisson structure. Similarly, Lu's notion [**27**] of moment maps $\Phi\colon M \to G^*$ for Poisson G-actions on symplectic manifolds (where G, G^* are dual Poisson Lie groups) can be phrased in this way.

In most cases of interest, the $\mathfrak{g}$-action on M exponentiates to an action of G:

DEFINITION 7.6. Let M be a G-manifold, together with a G-invariant 2-form ω and a G-equivariant map $\Phi\colon M \to G$. Then (M, ω, Φ) is called a Hamiltonian G-space with G-valued moment map, or simply a *q-Hamiltonian G-space*, if Φ is a Dirac map, and the $\mathfrak{g}$-action generated by Φ is the infinitesimal G-action.

EXAMPLE 7.7. Every conjugacy class $\mathcal{C} \subset G$, equipped with the GHJW 2-form, is a q-Hamiltonian G-space, with moment map the inclusion.

7.2. Volume forms. Definition 7.4 greatly simplifies many of the constructions with G-valued moment maps. For instance, generalizing our arguments for the FACT about conjugacy classes, one obtains the following

THEOREM 7.8. [**1**] *Assume that the homomorphism* $\mathrm{Ad}\colon G \to \mathrm{O}(\mathfrak{g})$ *lifts to the group* $\mathrm{Pin}(\mathfrak{g})$, *and let* $\psi \in \Omega(G)$ *be the pure spinor with* $N_\psi = F$, *defined by this lift* $\widetilde{\mathrm{Ad}}$. *Let* (M, ω, Φ) *be a q-Hamiltonian G-space. Then*

$$(e^\omega \Phi^* \psi)_{[\mathrm{top}]} \in \Omega(M)$$

is a G-invariant volume form.

If G is connected, we see that $\dim M$ must be even (since ψ is an even form in this case).

7.3. Products. Let us next consider products of q-Hamiltonian G-spaces.

For ordinary Hamiltonian G-spaces (M_i, ω_i) with moment maps $\Phi_i\colon M_i \to \mathfrak{g}^*$, the product is simply the direct product $M_1 \times M_2$ with the diagonal action, the sum of the 2-forms $\omega_1 + \omega_2$ and the sum $\Phi_1 + \Phi_2$ of the moment maps. Similarly, if G is a Poisson Lie group with dual Poisson-Lie group G^*, the product operation for Lu's Hamiltonian G-spaces with G^*-valued moment maps takes the sum of the 2-forms and the pointwise product of the moment maps. In this case, the G-action is a certain twist of the diagonal action, see [**19, 27**].

These two constructions work because the addition map $\mathrm{Add}\colon \mathfrak{g}^* \times \mathfrak{g}^* \to \mathfrak{g}^*$, respectively the product map $\mathrm{Mult}\colon G^* \times G^* \to G^*$, are Poisson. For G-valued moment maps, the situation is slightly different since the multiplication map $\mathrm{Mult}\colon G \times G \to G$, *as it stands*, is not a strong Dirac map if one simply takes the direct product Dirac structure on $G \times G$. Instead, the product operation involves a *gauge transformation*.

DEFINITION 7.9. Let E_M be an almost Dirac structure on M, defined (locally) by a pure spinor ϕ_M, and $\tau \in \Omega^2(M)$ a 2-form. Then the *gauge transformation* E_M^τ is the almost Dirac structure defined (locally) by the pure spinor $e^{-\tau}\phi_M$.

Note that if E_M is integrable with respect to a closed 3-form η_M, then E_M^τ is integrable with respect to $\eta_M + \mathrm{d}\tau$. In our case, we need a suitable gauge transformation of $E_{G\times G} := E_G \times E_G$. Let

$$\tau := \tfrac{1}{2}B(\mathrm{pr}_1^*\,\theta^L, \mathrm{pr}_2^*\,\theta^R) \in \Omega^2(G \times G).$$

This 2-form has the property [**31**],

$$\mathrm{Mult}^*\,\eta = \mathrm{pr}_1^*\,\eta + \mathrm{pr}_2^*\,\eta + \mathrm{d}\tau.$$

THEOREM 7.10. [**1**] *The multiplication map* Mult: $G\times G \to G$ *is a strong Dirac map from* $(G \times G, E^\tau_{G\times G})$ *to* (G, E_G).

One may use this result to define the *fusion product* of two q-Hamiltonian G-spaces M_1, M_2, or more generally to pass to the diagonal action in a q-Hamiltonian $G \times G$-space M (e.g. $M = M_1 \times M_2$). Indeed let (M, ω, Φ) be a q-Hamiltonian $G \times G$-space with moment map $\Phi = \Phi_1 \times \Phi_2$. Put

$$\Phi^{\mathrm{fus}} = \Phi_1\Phi_2, \ \ \omega^{\mathrm{fus}} = \omega + (\Phi_1, \Phi_2)^*\tau.$$

Then $(M, \omega^{\mathrm{fus}}, \Phi^{\mathrm{fus}})$, with diagonal G-action, is a q-Hamiltonian G-space. This follows rather easily from Theorem 7.10, since the composition of two strong Dirac maps is again a strong Dirac map.

REMARK 7.11. For the case of compact Lie groups, and working with the Definition 7.1, this result was obtained in [**4**] by a fairly complicated argument. The main difficulty in this approach was to show that ω^{fus} is again minimally degenerate: It is not easy to compute the kernel of ω^{fus} by 'direct calculation'!

7.4. Exponentials. Let the dual of the Lie algebra $\mathfrak{g}^*$ be equipped with the Kirillov-Poisson structure π. Its graph Gr_π defines a Dirac structure. Use the inner product B to identify $\mathfrak{g}^* \cong \mathfrak{g}$. Let $\varpi \in \Omega^2(\mathfrak{g})$ be the 2-form, obtained by applying the de Rham homotopy operator to $\exp^*\eta$. Thus $(\mathrm{Gr}_\pi)^\varpi$ is a Dirac structure relative to the closed 3-form $\mathrm{d}\varpi = \exp^*\eta$. Now let $\mathfrak{g}_\natural \subset \mathfrak{g}$ be the open subset where exp is a local diffeomorphism.

THEOREM 7.12. [**1**] *The restriction of* exp *to the subset* $\mathfrak{g}_\natural$ *is a strong Dirac map, relative to the Dirac structures* $(\mathrm{Gr}_\pi)^\varpi$ *on* $\mathfrak{g}$ *and the Cartan-Dirac structure on* G.

Suppose now that (M, ω_0, Φ_0) is an ordinary Hamiltonian G-space (thus ω_0 is a symplectic form, and $\Phi_0 \colon M \to \mathfrak{g}^* \cong \mathfrak{g}$ a moment map in the usual sense). Let $\omega = \omega_0 + \Phi_0^*\varpi$ and $\Phi = \exp\circ\Phi_0$. Then (M, ω, Φ) is a q-Hamiltonian G-space *provided that* $\Phi_0(M) \subset \mathfrak{g}_\natural$. Again, this just follows from the fact that the composition of two strong Dirac maps is again a strong Dirac map. Conversely, suppose $U \subset \mathfrak{g}_\natural$ is an open subset where exp is a diffeomorphism (with inverse denoted log), and (M, ω, Φ) is a q-Hamiltonian G-space. Put $\Phi_0 = \log(\Phi)$ and $\omega_0 = \omega - \Phi_0^*\varpi$. Then ω_0 is symplectic, and (M, ω_0, Φ_0) is a Hamiltonian G-space in the usual sense. For (M, ω_0, Φ_0) one has all the standard results from symplectic geometry, which one may then translate back to the q-Hamiltonian setting. For instance, the Meyer-Marsden-Weinstein reduction theorem for Hamiltonian manifolds (see Bates-Lerman [**9**] for a very general version) yields:

PROPOSITION 7.13 (Symplectic reduction of q-Hamiltonian manifolds). *Let* (M, ω, Φ) *be a q-Hamiltonian* G*-space, with proper moment map. Suppose the action of* G *is proper, and that* e *is a regular value of the moment map. Then* G *acts* *locally free on* $\Phi^{-1}(e)$*, and the reduced space*

$$M/\!/G = \Phi^{-1}(e)/G$$

is a symplectic orbifold. (If e is not a regular value, $M/\!/G$ *is a stratified symplectic space.)*

7.5. Examples.

7.5.1. *Homogeneous spaces.* Let G be a Lie group with involution $\sigma \in \mathrm{Aut}(G)$, $\sigma^2 = 1$, and consider the symmetric space $M = G/G^\sigma$. Let $\widehat{G} = \mathbb{Z}_2 \ltimes G$ be the semi-direct product, defined using the action of $\mathbb{Z}_2 = \{1, \sigma\}$ on G. Then M may be viewed as the conjugacy class of the element $(\sigma, e) \in \widehat{G}$. Hence, if $\mathfrak{g}$ carries an invariant scalar product which is preserved under the involution, the space M becomes a q-Hamiltonian $\widehat{G}$-space, with moment map

$$\Phi\colon M \to \widehat{G}, \quad gG^\sigma \mapsto (\sigma, g\sigma(g)^{-1}).$$

Since (σ, e) squares to the group unit, Exercise 2.3 shows that the 2-form ω on M is identically zero. Note also that the action of $\mathbb{Z}_2 \subset \widehat{G}$ is trivial, so that the action of $\widehat{G}$ descends to G.

Consider now the fusion product of M with itself. Letting $\Phi_i = \Phi \circ \mathrm{pr}_i\colon M \times M \to \widehat{G}$, the map $\Phi = \Phi_1\Phi_2$ takes values in the subgroup $G \subset \widehat{G}$:

$$\Phi(g_1G^\sigma,\ g_2G^\sigma) = \sigma(g_1)g_1^{-1}g_2\sigma(g_2^{-1}).$$

Hence $M \times M$ is a q-Hamiltonian G-space.

7.5.2. *The double.* Any Lie group G may be viewed as a symmetric space for the group $G \times G$, with action $(g_1, g_2).a = g_1 a g_2^{-1}$. Here $\sigma \in \mathrm{Aut}(G \times G)$ is the involutive automorphism $\sigma(g_1, g_2) = (g_2, g_1)$, fixing the diagonal subgroup, and the inclusion of the first factor identifies the quotient $(G \times G)/G$ with G. Hence, given an invariant scalar product on $\mathfrak{g}$, the example in Section 7.5.1 shows that G is a q-Hamiltonian $\mathbb{Z}_2 \ltimes (G \times G)$-space. Taking a fusion product of G with itself, we find that $D(G) := G \times G$ is a q-Hamiltonian $G \times G$-space, with action

$$(g_1, g_2).(a, b) = (g_1 a g_2^{-1},\ g_2 b g_1^{-1})$$

and moment map $(a, b) \mapsto (ab, a^{-1}b^{-1})$. The space $D(G)$ is called the *double* of G.

REMARK 7.14. The double $D(G)$ is the counterpart, in the q-Hamiltonian category, of the cotangent bundle T^*G in the usual Hamiltonian category. In fact, as observed in Bursztyn-Crainic-Weinstein-Zhu [**11**] the double $D(G) \rightrightarrows G$ (viewed as a groupoid over G, with source and target maps the two components of the moment map) 'integrates' the Dirac manifold G in a similar sense as $T^*G \rightrightarrows \mathfrak{g}^*$ integrates the Poisson manifold $\mathfrak{g}^*$. Ping Xu [**32**] presents $D(G) \rightrightarrows G$ as an example of a *quasi-symplectic groupoid.*

Passing to the diagonal action, $G \times G$ becomes a q-Hamiltonian G-space with moment map the group commutator:

$$(a, b) \mapsto aba^{-1}b^{-1}.$$

This is called the *fused double*, denoted $\tilde{D}(G)$. Taking a fusion product of several copies of $\tilde{D}(G)$ with itself, the space G^{2h} becomes a q-Hamiltonian G-space with moment map

$$\Phi\colon (a_1, b_1, \ldots, a_h, b_h) \mapsto \prod_{i=1}^{h} a_i b_i a_i^{-1} b_i^{-1}.$$

The symplectic quotient $M/\!/G$ is just the representation variety for a closed oriented surface of genus h:

$$M/\!/G = \mathrm{Hom}(\pi_1(\Sigma), G)/G$$

Equivalently, $M/\!/G$ is the moduli space of flat principal G-bundles on Σ. It was shown in [**4**] that the symplectic structure obtained by this finite-dimensional reduction, coincides with that coming from Atiyah-Bott's [**8**] gauge theory construction. More generally, if $\mathcal{C}_1, \ldots, \mathcal{C}_r$ are conjugacy classes in G, the symplectic quotient

$$(G^{2s} \times \mathcal{C}_1 \times \cdots \times \mathcal{C}_r)/\!/G$$

is the moduli space of flat G-bundles over an oriented surface Σ of genus h with r boundary components, with restrictions to the jth boundary component $(\partial\Sigma)_j \cong S^1$ in the given conjugacy classes. (Note $\mathrm{Hom}(\pi_1(S^1), G)/G) = G/\mathrm{Ad}(G)$ is the set of conjugacy classes.)

REMARK 7.15. We stress that no compactness assumption is needed for these results. In fact, one could even work over the complex numbers, and obtain a complex symplectic structure over the representation variety for a complex Lie group.

7.5.3. *Spheres.* There are other examples of q-Hamiltonian spaces which are unrelated to moduli spaces, such as various examples of *multiplicity-free* q-Hamiltonian spaces. Let $\mathrm{SU}(n)$ act on $\mathbb{C}^n$ in the standard way, and consider the unit sphere $S^{2n} \subset \mathbb{C}^n \times \mathbb{R}$ with the restricted action. By a result of Hurtubise-Jeffrey-Sjamaar [**23**], there exists an invariant 2-form ω and an equivariant map $\Phi\colon S^{2n} \to \mathrm{SU}(n)$ for which (S^{2n}, ω, Φ) is a q-Hamiltonian $\mathrm{SU}(n)$-space. (The special case $n = 2$ was discussed in [**6**].)

References

1. A. Alekseev, H. Bursztyn, and E. Meinrenken, *Pure spinors on Lie groups*, in preparation.
2. A. Alekseev and Y. Kosmann-Schwarzbach, *Manin pairs and moment maps*, J. Differential Geom. **56** (2000), no. 1, 133–165.
3. A. Alekseev, Y. Kosmann-Schwarzbach, and E. Meinrenken, *Quasi-Poisson manifolds*, Canad. J. Math. **54** (2002), no. 1, 3–29.
4. A. Alekseev, A. Malkin, and E. Meinrenken, *Lie group valued moment maps*, J. Differential Geom. **48** (1998), no. 3, 445–495.
5. A. Alekseev and E. Meinrenken, *Clifford algebras and the classical dynamical Yang-Baxter equation*, Mathematical Research Letters **10** (2003), no. 2-3, 253–268.
6. A. Alekseev, E. Meinrenken, and C. Woodward, *Duistermaat-Heckman measures and moduli spaces of flat bundles over surfaces*, Geom. and Funct. Anal. **12** (2002), 1–31.
7. A. Alekseev and P. Xu, *Derived brackets and Courant algebroids*, Unfinished manuscript (2002).
8. M. F. Atiyah, *The geometry and physics of knots*, Cambridge University Press, Cambridge, 1990.
9. L. Bates and E. Lerman, *Proper group actions and symplectic stratified spaces*, Pacific J. Math. **181** (1997), no. 2, 201–229.
10. H. Bursztyn and M. Crainic, *Dirac structures, momentum maps, and quasi-Poisson manifolds*, The breadth of symplectic and Poisson geometry, Progr. Math., vol. 232, Birkhäuser Boston, Boston, MA, 2005, pp. 1–40.
11. H. Bursztyn, M. Crainic, A. Weinstein, and C. Zhu, *Integration of twisted Dirac brackets*, Duke Math. J. **123** (2004), no. 3, 549–607.
12. H. Bursztyn and A. Weinstein, *Poisson geometry and Morita equivalence*, Poisson geometry, deformation quantisation and group representations, London Math. Soc. Lecture Note Ser., vol. 323, Cambridge Univ. Press, Cambridge, 2005, pp. 1–78.
13. Élie Cartan, *Lecons sur la théorie des spineurs*, Paris: Hermann & Cie., 1938.

14. ______, *The theory of spinors*, The M.I.T. Press, Cambridge, Mass., 1967, translation of 1938 French original.
15. C. Chevalley, *The algebraic theory of spinors*, Columbia University Press, New York, 1954.
16. T. Courant, *Dirac manifolds*, Trans. Amer. Math. Soc. **319** (1990), no. 2, 631–661.
17. A. Cannas da Silva and A. Weinstein, *Geometric models for noncommutative algebras*, American Mathematical Society, Providence, RI, 1999.
18. S. Evens and J.-H. Lu, *Poisson harmonic forms, Kostant harmonic forms, and the S^1-equivariant cohomology of K/T*, Adv. Math. **142** (1999), no. 2, 171–220.
19. H. Flaschka and T. Ratiu, *A convexity theorem for Poisson actions of compact Lie groups*, Ann. Sci. Ecole Norm. Sup. **29** (1996), no. 6, 787–809.
20. M. Gualtieri, *Generalized complex geometry*, Ph.D. thesis, Oxford, 2004, arXiv:math.DG/0401221.
21. K. Guruprasad, J. Huebschmann, L. Jeffrey, and A. Weinstein, *Group systems, groupoids, and moduli spaces of parabolic bundles*, Duke Math. J. **89** (1997), no. 2, 377–412.
22. N. Hitchin, *Generalized Calabi-Yau manifolds*, Q. J. Math. **54** (2003), no. 3, 281–308.
23. J. Hurtubise, L. Jeffrey, and R. Sjamaar, *Group-valued implosion and parabolic structures*, Amer. J. Math. **128** (2006), no. 1, 167–214.
24. C. Klimčík and T. Strobl, *WZW-Poisson manifolds*, J. Geom. Phys. **43** (2002), no. 4, 341–344.
25. Y. Kosmann-Schwarzbach, *Derived brackets*, Lett. Math. Phys. **69** (2004), 61–87.
26. Y. Kosmann-Schwarzbach and C. Laurent-Gengoux, *The modular class of a twisted Poisson structure*, Travaux Mathmatiques **16** (2005), 315–339.
27. J.-H. Lu, *Momentum mappings and reduction of Poisson actions*, Symplectic geometry, groupoids, and integrable systems (Berkeley, CA, 1989), Springer, New York, 1991, pp. 209–226.
28. K. Mackenzie, *General theory of Lie groupoids and Lie algebroids*, London Mathematical Society Lecture Note Series, vol. 213, Cambridge University Press, Cambridge, 2005.
29. D. Roytenberg, *Courant algebroids, derived brackets and even symplectic supermanifolds*, Thesis, Berkeley 1999. arXiv:math.DG/9910078.
30. P. Ševera and A. Weinstein, *Poisson geometry with a 3-form background*, Progr. Theoret. Phys. Suppl. (2001), no. 144, 145–154, Noncommutative geometry and string theory (Yokohama, 2001).
31. A. Weinstein, *The symplectic structure on moduli space*, The Floer memorial volume, Progr. Math., vol. 133, Birkhäuser, Basel, 1995, pp. 627–635.
32. P. Xu, *Momentum maps and Morita equivalence*, J. Differential Geom. **67** (2004), no. 2, 289–333.

University of Toronto, Department of Mathematics, 40 St George Street, Toronto, Ontario M4S2E4, Canada

E-mail address: `mein@math.toronto.edu`

Contemporary Mathematics
Volume **450**, 2008

Lectures on props, Poisson geometry and deformation quantization

S.A. Merkulov

§1 Introduction

One of our purposes in this paper is to explain a curious fact (together with a mathematical language needed to formulate it precisely) that some local geometric structures are resolutions of very compact graphical data called their *genome*, and that knowing such a genome can be useful. For example, genome of species "local Poisson structures" consists of two *genes*

with the following engineering rules,

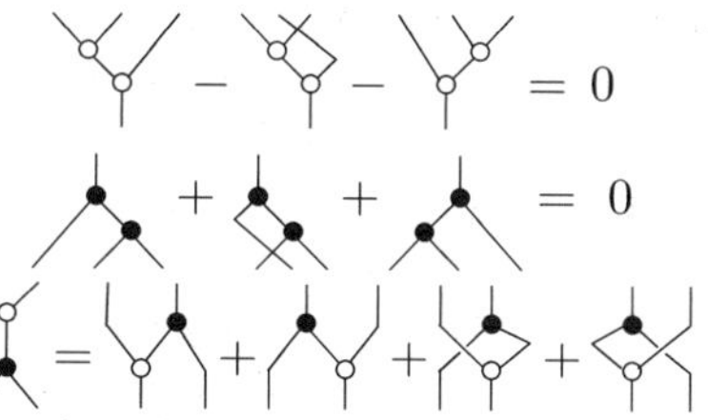

We shall explain below how the above pictures are related to the *prop*, $\mathsf{Lie^1B}$, of Lie 1-bialgebras, and how resolution of the above graph relations leads us to another differential graded (dg, for short) prop, $\mathsf{Lie^1B}_\infty$, whose representations in an arbitrary dg space V are, rather surprisingly, in one-to-one correspondence with the Maurer-Cartan elements in the Schouten Lie algebra, $\wedge^\bullet \mathcal{T}_V$, of polyvector fields, where $\mathcal{T}_V := \mathrm{Der}\mathcal{O}_V$, the module of derivations of the ring, $\mathcal{O}_V := \widehat{\odot^\bullet} V^*$, of formal smooth functions on V; we call such elements *Poisson structures* on a formal graded manifold V[1] ; if V is $\mathbb{R}^n$ concentrated in degree 0, then this notion coincides with the ordinary notion of Poisson structure on $\mathbb{R}^n$.

Kontsevich's famous universal deformation quantization formulae involve graphs with *wheels* — closed directed paths — which encode *traces* of tensor powers of polyvector fields and their partial derivatives and hence make, in general, sense

This work was partially supported by the Göran Gustafsson foundation.

[1]Strictly speaking, a representation of $\mathsf{Lie^1B}_\infty$ in V gives a degree 1 element $\nu \in \wedge^\bullet \mathcal{T}_V$ which satisfies not only the Maurer-Cartan condition, $[\nu, \nu] = 0$, but also the vanishing condition, $\nu|_{x=0} = 0$, at $0 \in V$ (see §3.4.1); given, however, $\nu \in \wedge^\bullet \mathcal{T}_V$ with $\nu|_{x=0} \neq 0$, then, for a formal parameter h viewed as a coordinate on $\mathbb{R}$, the element $\bar{\nu} := h\nu \in \wedge^\bullet \mathcal{T}_{\hat{V}}$, $\hat{V} := V \times \mathbb{R}$, satisfies $\bar{\nu}|_{x=0} = 0$ and hence comes from a representation of the prop $\mathsf{Lie^1B}_\infty$; thus such representations encompass *arbitrary* formal Poisson structures.

only in finite-dimension. We show in this paper that every universal deformation quantization of a *generic* Poisson structure on a graded manifold *must* involve such traces, i.e. graphs with wheels are unavoidable. This fact prompts us to generalize in §4 the notion of a prop to include graphs with wheels and construct wheeled extensions, $\mathsf{Lie^1B}^{\circlearrowright}$ and $\mathsf{Lie^1B}^{\circlearrowright}_\infty$, of the props related to Poisson geometry. Moreover, we show that a universal quantization of Poisson structures (in particular, the Kontsevich quantization) can be understood as a morphism of dg prop,

$$Q : \mathsf{DefQ} \to \widehat{\mathsf{Lie^1B}^{\circlearrowright}_\infty},$$

where $\widehat{\mathsf{Lie^1B}^{\circlearrowright}_\infty}$ is the completion of $\mathsf{Lie^1B}^{\circlearrowright}_\infty$ with respect to the number of vertices, and DefQ is the dg free prop whose representations in a vector space V describe Maurer-Cartan elements of the Hochschild dg Lie algebra, $\mathcal{D}_V$, of polydifferential operators on $\mathcal{O}_V$ (see §4.7 for its precise definition).

We show that the cohomology of the dg prop $\mathsf{Lie^1B}^{\circlearrowright}_\infty$ is strictly larger than $\mathsf{Lie^1B}^{\circlearrowright}$, and that there exists a natural dg free prop $[\mathsf{Lie^1B}^{\circlearrowright}]_\infty$ which extends the above prop, $\mathsf{Lie^1B}^{\circlearrowright}_\infty$, of polyvector fields and fits into a commutative diagram

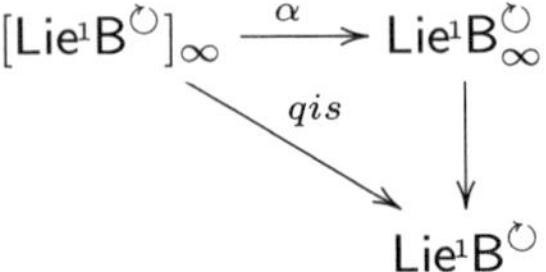

where α is an epimorphism of nondifferential props, and qis a quasi-isomorphism of differential props. It is called a *quasi-minimal prop resolution* of $\mathsf{Lie^1B}^{\circlearrowright}$.
As we mentioned above, representations of $\mathsf{Lie^1B}_\infty$ in a dg space V are precisely Poisson structures on V. This prompts us to call representations of $[\mathsf{Lie^1B}^{\circlearrowright}]_\infty$ in a finite dimensional dg space V *wheeled Poisson structures* on the formal graded manifold V. The geometric meaning of such wheeled Poisson structures is not clear at present — the differential equations behind these new structures involve not only Schouten brackets but also *traces* of derivatives of tensor products of polyvector fields[2] (and hence makes sense only in finite dimensions). Moreover, polyvector fields are only part of the data — there are other tensors (including new ones in degree 0) in the content list of a wheeled Poisson structure. Remarkably, one can construct "star products" out of wheeled Poisson structures, i.e. they can be deformation quantized:

Theorem. *There exists a morphism of dg props,*

$$\hat{Q} : \mathsf{DefQ} \to \widehat{[\mathsf{Lie^1B}^{\circlearrowright}]}_\infty,$$

where $\widehat{[\mathsf{Lie^1B}^{\circlearrowright}]}_\infty$ *is the completion of the prop* $[\mathsf{Lie^1B}^{\circlearrowright}]_\infty$ *with respect to the number of vertices. Moreover, this morphism exists in the category of props over the field,* $\mathbb{Q}$*, of rational numbers.*

Corollary. *Every wheeled Poisson structure on a finite dimensional formal manifold can be deformation quantized, i.e. there exists an associated Maurer-Cartan element in the Hochschild dg Lie algebra* $\mathsf{D}_V[[\hbar]]$.

[2]It is proven in [**Me4**] that this system of differential equations can *not* contain equations involving just a *single* trace of some tensorial expression built from the Poisson tensor and its derivatives (which correspond to graphs with genus *one* wheels).

The quantization morphism $\hat{Q}$ is very non-trivial: the proof of Theorem §5.2 below implies that $\hat{Q}$ involves, e.g., *infinite* jets of the polyvector fields constituent of a wheeled Poisson structure. There is a canonical monomorphism of dg props, $i : \mathsf{Lie^1B}^{\circlearrowright}_\infty \rightarrow [\mathsf{Lie^1B}^{\circlearrowright}]_\infty$, so that those quantization morphisms $\hat{Q}$ which factor though i provide us with quantizations of ordinary Poisson structures; we shall discuss a purely propic construction of such morphisms elsewhere.

One can get some intuition into the geometric meaning of wheeled Poisson structures from their simpler analogues — representations of quasi-minimal prop resolutions, $[\mathsf{Ass}^{\circlearrowright}]_\infty$ and $[\mathsf{Com}^{\circlearrowright}]_\infty$, of the operads of associative and, respectively, commutative algebras which have been computed in [**SMM**]. We show some details in §4.6: all the new fields in $[\mathsf{Ass}^{\circlearrowright}]_\infty$-structure in comparison with the well-known Ass_∞-structure as well as all the associated new (involving trace) equations can be explicitly described. Unfortunately, such a detailed picture of wheeled Poisson structures is out of reach at present: this problem appears to have complexity level comparable with that of the problem of computing homologies of the directed versions of famous Kontsevich's graph complexes [**Ko1**] (in fact, the graph complex behind wheeled Poisson structures is closely related to the directed version of Kontsevich's "commutative" graph complex) .

Analogous "genetic engineering" ideas can be used to get a strongly homotopy generalization of the Drinfeld-Etingof-Kazhdan theory of quantization of Lie bialgebras, and in fact, a new formality theorem behind such quantizations [**Me5**]. Genome of another important equation in geometry, the Nijenhuis integrability conditions for almost complex structures, is described in [**Me3**].

The paper is organized as follows. In §§2-4 we give an introduction to the theory of operads and props with emphasis on examples which are relevant to Poisson geometry and deformation quantization. In §5 we apply the theory of wheeled props to prove the theorem formulated above.

A few words about our notations. The cardinality of a finite set I is denoted by $|I|$. The degree of a homogeneous element, a, of a graded vector space is denoted by $|a|$ (this should never lead to a confusion). $\mathbb{S}_n$ stands for the group of all bijections, $[n] \rightarrow [n]$, where $[n]$ denotes (here and everywhere) the set $\{1, 2, \ldots, n\}$. The set of positive integers is denoted by $\mathbb{N}^*$. If $V = \oplus_{i\in\mathbb{Z}} V^i$ is a graded vector space, then $V[k]$ is a graded vector space with $V[k]^i := V^{i+k}$.

We work throughout over a field $\mathbb{K}$ of characteristic 0.

§2 Algebraic structures ⇒ Graphs complexes

Our purpose in this section is to create a rather non-trivial graph complex out of such a commonplace notion as associative algebra, and motivate, thereby, introduction of operads and props.

2.1. Definition. An *associative algebra* is the data
(i) a graded vector space V over a field $\mathbb{K}$ (always of characteristic 0 in this paper),
(ii) a linear map $\circ : V \otimes V \rightarrow V$ of degree 0,
(iii) the associativity condition: $(a \circ b) \circ c = a \circ (b \circ c) \quad \forall a, b, c \in V$.

We construct below a graph complex out of species *associative algebras* in three movements, the first of which, §2.2, is an attempt to get rid of the input data (i) above.

2.2. Smile of the Cheshire cat. What is left of an associative algebra when the underlying vector space V has gone? Of course, the multiplication, $\circ : V \otimes V \to V$, as a linear map of vector spaces is also gone, but its calculational meaning — a "machine" with two inputs and one output — may linger in the air as a directed (i.e. equipped with a flow which we implicitly assume to go from bottom to top) degree 0 graph,

As the multiplication in V is not assumed to be graded commutative, we have to find a simple pictorial way to distinguish between $a \circ b$ and $b \circ a$ after V departed, that is to equip input legs with different labels, say with 1 and 2. Thus the smile of an associative algebra consists in fact of two different graphs,

1 2 and 2 1.

The vector space

$\mathsf{E} :=$ span ⟨ 1 2 , 2 1 ⟩

is 2-dimensional and caries a natural representation of the group $\mathbb{S}_2$. In fact, as a representation space, $\mathsf{E} \simeq \mathbb{K}[\mathbb{S}_2]$. What is then left of the associativity condition? The answer is obvious: an equation of the form,

1 2 3 − 1 2 3 = 0

or, to be more precise, a system of 3! equations,

(1) $\sigma(1)\ \sigma(2)\ \sigma(3)$ − $\sigma(1)\ \sigma(2)\ \sigma(3)$ $= 0,\ \forall\, \sigma \in \mathbb{S}_3.$

obtained from the first one simply by changing input labels. (For shortness, we shall be showing in other examples only the "first" generating equation and tacitly assuming its closure with respect to the action of permutation groups).

2.3. Ass. The smile (1) of the associativity condition, which is a linear combination of graphs with two vertices, persuades us to introduce a vector space,

$\mathsf{Free}\langle \mathsf{E} \rangle =$ span ⟨ 1 4 2 3 , … ⟩

spanned by all possible calculational schemes — graphs — on one binary operation. This space is infinite-dimensional but is naturally decomposed into a direct sum,

$$\mathsf{Free}\langle \mathsf{E} \rangle = \bigoplus_{n\geq 2} \mathsf{Free}\langle E \rangle(n) \tag{2}$$

of finite-dimensional subspaces with the subspace $\mathsf{Free}\langle \mathsf{E} \rangle(n)$ spanned by graphs with precisely n inputs. Clearly, each $\mathsf{Free}\langle \mathsf{E} \rangle(n)$ is a right $\mathbb{S}_n$-module with respect to the relabeling of input legs.

Then the l.h.s. of (1) are 6 vectors in $\mathsf{Free}\langle \mathsf{E} \rangle(3)$. We denote by R the vector subspace of $\mathsf{Free}\langle \mathsf{E} \rangle(3)$ spanned by these 6 vectors, and then define an equivalence relation, $\sim$, in $\mathsf{Free}\langle\mathsf{E}\rangle$ by identifying graphs which have pairs of neighboring vertices related to each other in accordance with (1). For example,

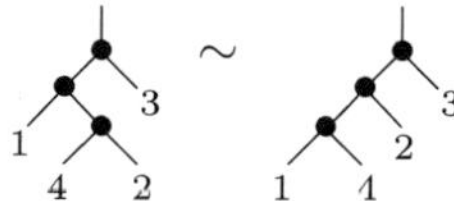

where we applied (1) to the first and second (from the bottom) vertices; we could equally well apply (1) to the second and third vertices and get another equivalent configuration. The associated vector space of equivalence classes,

$$\mathsf{Ass} := \mathsf{Free}\langle\mathsf{E}\rangle/\sim,$$

is called an *operad of associative algebras*. The gradation (2) induces an associated gradation,

$$\mathsf{Ass} := \bigoplus_{n\geq 2} \mathsf{Ass}(n).$$

As equivalence relation $\sim$ is generated by vectors spanning the subspace R, the quotient space, $\mathsf{Free}\langle\mathsf{E}\rangle/\sim$, is often denoted by $\mathsf{Free}\langle\mathsf{E}\rangle/(R)$.

2.3.1. Exercise. Show that as an $\mathbb{S}_n$-module $\mathsf{Ass}(n)$ is isomorphic to $\mathbb{K}[\mathbb{S}_n]$.

2.4. Operads. For any $i \in \{1, 2, \ldots, n\}$ there is an obvious composition,

$$\circ_i : \mathsf{Ass}(n) \otimes \mathsf{Ass}(m) \longrightarrow \mathsf{Ass}(n+m-1),$$

which is given by gluing the root leg of a graph from $\mathsf{Ass}(m)$ into the i-th input leg of a graph from $\mathsf{Ass}(n)$ and the natural renumbering of input legs of the resulting graph as in the following example

$$\circ_2 : \quad \mathsf{Ass}(3) \quad \otimes \quad \mathsf{Ass}(2) \quad \longrightarrow \quad \mathsf{Ass}(4)$$

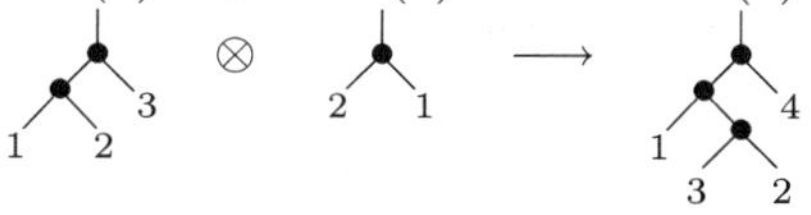

It is easy to draw pictures verifying the following (awkwardly looking when projected onto the plane of paper) associativity type identities,

$$(3) \qquad (f_1 \circ_i f_2) \circ_{j+m-1} f_3 = (-1)^{|f_1||f_2|} (f_1 \circ_j f_3) \circ_i f_2, \quad \forall\, 1 \leq i < j \leq n,$$

$$(4) \qquad f_1 \circ_i (f_2 \circ_j f_3) = (f_1 \circ_i f_2) \circ_{i+j-1} f_3, \quad \forall\, 1 \leq i \leq n, 1 \leq j \leq m.$$

which hold for any $f_1 \in \mathsf{Ass}(n)$, $f_2 \in \mathsf{Ass}(m)$ and $f_3 \in \mathsf{Ass}(k)$. Operads are no more than a formalization of the above obvious properties of $\mathsf{Ass} = \{\mathsf{Ass}(n)\}$.

2.4.1. Definitions [**M, SMM**]. An *operad* (without a unit), P, is a family of $\mathbb{S}_n$-modules, $\mathsf{P} = \{\mathsf{P}(n)\}_{n\geq 1}$, together with a family of equivariant compositions, $\{\circ_i : \mathsf{P}(n) \otimes \mathsf{P}(m) \to \mathsf{P}(n+m-1)\}_{1\leq i\leq n}$, satisfying the axioms (3) and (4) for any $f_1 \in \mathsf{P}(n)$, $f_2 \in \mathsf{P}(m)$ and $f_3 \in \mathsf{P}(k)$.

A *morphism* of operads, $\mathsf{F} : \mathsf{P}_1 \to \mathsf{P}_2$, is a family of $\mathbb{S}_n$-equivariants maps, $\{\mathsf{F}_n : \mathsf{P}_1(n) \to \mathsf{P}_2(n)\}_{n\geq 1}$, such that the diagram,

$$\begin{array}{ccc} \mathsf{P}_1(n) \otimes \mathsf{P}_1(m) & \xrightarrow{\circ_i} & \mathsf{P}_1(m+n-1) \\ \downarrow {\scriptstyle \mathsf{F}_m \otimes \mathsf{F}_n} & & \downarrow {\scriptstyle \mathsf{F}_{m+n-1}} \\ \mathsf{P}_2(n) \otimes \mathsf{P}_2(m) & \xrightarrow[\circ_i]{} & \mathsf{P}_2(n) \otimes \mathsf{P}_2(m) \end{array}$$

commutes for any $i \in \{1,\dots,n\}$.

A *differential graded* (or *dg*, for short) operad is an operad, $\mathsf{P} = \{\mathsf{P}(n)\}_{n\geq 1}$, such that every $\mathbb{S}_n$-module $\mathsf{P}(n)$ is a $\mathbb{Z}$-graded complex, $\mathsf{P}(n) = \oplus_{j\in\mathbb{Z}}\mathsf{P}(n)^j$, with the differential, $\delta : \mathsf{P}(n)^j \to \mathsf{P}(n)^{j+1}$, satisfying the equation,

$$\delta(f\circ_i g) = (\delta f)\circ_i g + (-1)^{|f|} f\circ_i \delta g, \quad i\in[n],$$

for arbitrary $f\in\mathsf{P}(n)$, $g\in\mathsf{P}(m)$.

An *ideal* in an operad P is a collection, $\mathsf{J} = \{\mathsf{J}(n)\}$ of $\mathbb{S}_n$-submodules, $\mathsf{J}(n)\subset\mathsf{P}(n)$ such that $f\circ_i g\in\mathsf{J}(n+m-1)$ whenever $f\in\mathsf{J}(n)$ or $g\in\mathsf{J}(m)$; in particular, J is a suboperad of P. It is clear that the quotient $\mathbb{S}$-module $\{\mathsf{P}(n)/\mathsf{J}(n)\}$ has a naturally induced structure of an operad called the *quotient operad*. For example, Ass is the quotient of the free operad $\mathsf{Free}\langle\mathsf{E}\rangle$ generated by the $\mathbb{S}_2$-module E by the ideal generated by the $\mathbb{S}_3$-submodule $R\subset\mathsf{Free}\langle\mathsf{E}\rangle(3)$. Such operads are often called *quadratic*.

2.4.2. Examples. **(i)** Let V be a graded vector space and $\mathsf{End}\langle V\rangle := \{\mathrm{Hom}(V^{\otimes n}, V)\}$ a collection of $\mathbb{S}_n$-modules with the action of $\mathbb{S}_n$ given just by permuting inputs of linear maps from $V^{\otimes n}$ to V. It is an operad with the composition,

$$\begin{array}{cccc}\circ_i: & \mathrm{Hom}(V^{\otimes n}, V)\otimes\mathrm{Hom}(V^{\otimes m}, V) & \longrightarrow & \mathrm{Hom}(V^{m+n-1}, V)\\ & f\otimes g & \longrightarrow & f\circ_i g\end{array}$$

defined simply as the substitution,

$$f\circ_i g(v_1,\dots,v_{m+n-1}) := (-1)^{|g|(|v_1|+\dots+|v_{i-1}|)} f(v_1,\dots,v_{i-1}, g(v_i,\dots,v_{i+m-1}),\dots,v_{m+n-1}),$$

for any $v_1,\dots,v_{m+n-1}\in V$. The operad $\mathsf{End}\langle V\rangle$ is called the *endomorphism operad* of the vector space V.

(ii) The collection, $\mathsf{Free}\langle\mathsf{E}\rangle = \{\mathsf{Free}\langle\mathsf{E}\rangle(n)\}$, of $\mathbb{S}_n$-modules introduced in §2.3 is an operad with $\circ_i : \mathsf{Free}\langle\mathsf{E}\rangle(n)\otimes\mathsf{Free}\langle\mathsf{E}\rangle(m)\to\mathsf{Free}\langle\mathsf{E}\rangle(m+n-1)$ defined as grafting of the root of a graph from $\mathsf{Free}\langle\mathsf{E}\rangle(m)$ into the i-th input leg of a graph from $\mathsf{Free}\langle\mathsf{E}\rangle(m)$. It is called the *free* operad generated by the $\mathbb{S}_2$-module

$\mathsf{E} := \mathrm{span}\left\langle \underset{1\quad 2}{\curlywedge}, \underset{2\quad 1}{\curlywedge}\right\rangle$.

(iii) Example (ii) can be generalized to an arbitrary $\mathbb{S}$-*module*, that is, a collection, $\mathsf{E} = \{\mathsf{E}(n)\}$, of $\mathbb{S}_n$-modules for all $n\geq 1$. The associated *free operad* , $\mathsf{Free}\langle\mathsf{E}\rangle$, generated by E is, by definition, a vector space spanned by *decorated graphs*, that is, pairs,

$$\left(G, g\in\bigotimes_{v\in Vert(G)}\mathsf{E}(|in_v|)\right),$$

where G is a directed (i.e. every edge is directed) connected graph of genus 0 whose every vertex v has $|in_v|$ incoming edges and precisely one outgoing edge, $Vert(G)$ is the set of vertices, and

$$\bigotimes_{v\in Vert(G)}\mathsf{E}(|in_v|) := \left(\bigoplus_{f:[k]\to Vert(G)}\mathsf{E}\left(|in_{f(1)}|\right)\otimes\dots\otimes\mathsf{E}\left(|in_{f(k)}|\right)\right)_{\mathbb{S}_k},$$

is the *unordered tensor product* over the set of vertices of G. Here $k := |Vert(G)|$, and $(\ )_{\mathbb{S}_k}$ stands for the space of invariants with respect to the natural action of $\mathbb{S}_k$ on the direct sum over bijective enumerations, $f : [k] \to Vert(G)$, of vertices.

Compositions $\circ_i$ in the operad $\mathsf{Free}\langle \mathsf{E}\rangle$ are obvious graftings and tensor products. Thus every element in $\mathsf{Free}\langle \mathsf{E}\rangle$ is an iterated composition of basic decorated graphs,

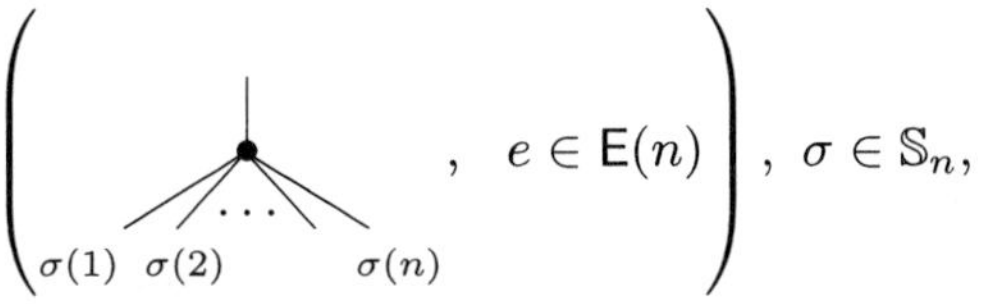

$$\left(\begin{array}{c} \text{corolla with inputs } \sigma(1)\ \sigma(2) \ \cdots\ \sigma(n) \end{array}, \ e \in \mathsf{E}(n) \right), \ \sigma \in \mathbb{S}_n,$$

called *generating corollas*. From now on we abbreviate elements, (G, e), of $\mathsf{Free}\langle \mathsf{E}\rangle$ into just G assuming implicitly that the vertices of G are decorated with elements of E.

As in a free algebra, any differential, δ, in a free operad is completely determined by its values on the generators, i.e. on the generating corollas. Thus any such a differential δ is completely determined by a sum of the form,

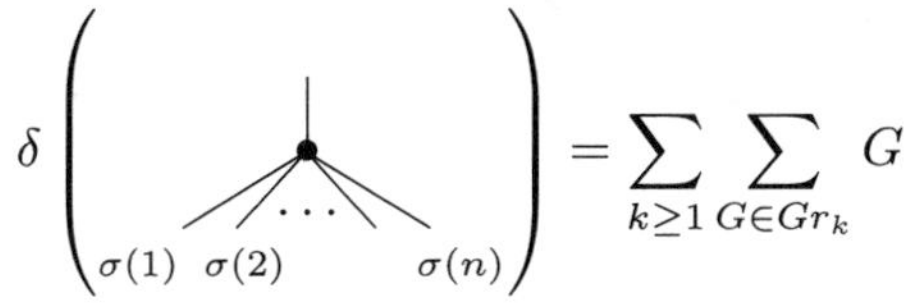

$$\delta \left(\begin{array}{c} \text{corolla with inputs } \sigma(1)\ \sigma(2) \ \cdots\ \sigma(n) \end{array} \right) = \sum_{k\geq 1} \sum_{G \in Gr_k} G$$

where Gr_k is some family of graphs with precisely k vertices. If $Gr_1 = \emptyset$, then the differential δ is called *minimal*. If all $Gr_k = \emptyset$ except $k = 2$, the differential is called *quadratic*.

2.4.3. Representations of operads. Given a (differential) operad $\mathsf{P} = \{\mathsf{P}(n)\}$. Its *representation* in a (differential) vector space V is a morphism,

$$\rho : \mathsf{P} \longrightarrow \mathsf{End}\langle V\rangle,$$

of (differential) operads. Such a morphism is also called a P-*structure* on V.

2.4.4. Proposition. *There is a one-to-one correspondence between morphisms of operads,*

$$\rho : \mathsf{Ass} \longrightarrow \mathsf{End}\langle V\rangle,$$

and associative algebra structures on V.

Proof. Such a morphism is completely determined by its value on the generating corolla of $\mathsf{Free}\langle E\rangle/(R)$,

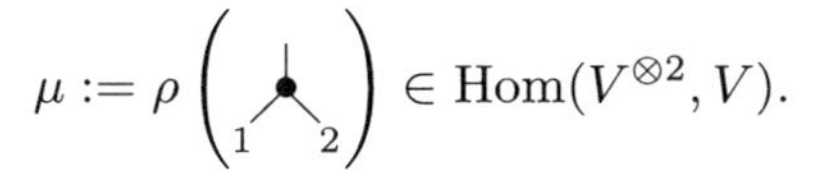

$$\mu := \rho\left(\begin{array}{c} \text{corolla with inputs } 1\ 2 \end{array} \right) \in \mathrm{Hom}(V^{\otimes 2}, V).$$

Then equation $\rho(R) = 0$ immediately implies associativity of the multiplication μ. □

2.5. Final movement: resolution of the smile of an associative algebra. Now, after learning some language, we can return back to our example of $\mathsf{Ass} = \mathsf{Free}\langle E\rangle/(R)$ and try to get rid of relations R, that is, try to find a *free resolution* of Ass which is, by definition, a dg *free* operad, $(\mathsf{Free}\langle \mathsf{E}\rangle, \delta)$ together with an epimorphism,

$$\pi : (\mathsf{Free}\langle \mathsf{E}\rangle, \delta) \longrightarrow \mathsf{Ass},$$

which is a quasi-isomorphism, i.e. induces an isomorphism of operads, $[\pi]$: $\mathsf{H}(\mathsf{Free}\langle \mathsf{E}\rangle, \delta) \to \mathsf{Ass}$, where $\mathsf{H}(\mathsf{Free}\langle \mathsf{E}\rangle, \delta)$ is the associated cohomology operad. A general theorem says that for any operad P there exists its minimal free resolution, $(\mathsf{P}_\infty, \delta)$, which is unique up to isomorphism.

What could be $(\mathsf{Ass}_\infty, \delta)$? It must be a free operad on a family of generators which must contain the generators of Ass,

$$\mathsf{Ass}_\infty := \mathsf{Free}(\ , \ldots),$$

and the corresponding map

$$\pi : \mathsf{Free}(\ , \ldots) \longrightarrow \mathsf{Free}(\ )/(R) = \mathsf{Ass},$$

must send the generator to its equivalence class in Ass, while all other (yet unknown) generators to zero. As π must be a quasi-isomorphism, the generators must be cycles in $(\mathsf{Ass}_\infty, \delta)$, i.e. we have to set

$$\delta \ \ {}_{1 \quad 2} = 0.$$

Next we must handle the 3!-dimensional space of relations (1) in Ass. The r.h.s. of (1) are cycles with respect to δ. As they must be zero in cohomology, we have to introduce new corollas with 3 input legs,

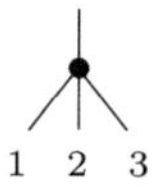

1 2 3 (plus all possible permutations of (1,2,3)),

and set

(5) 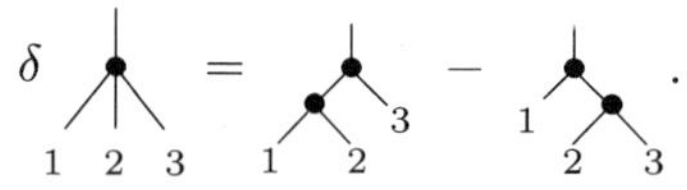 $$\delta \ {}_{1\ 2\ 3} = {}_{1\ 2\ 3} - {}_{1\ 2\ 3}.$$

Note that we have to assign to these new generators degree -1 (as δ has degree $+1$ by our conventions). In this way we get a well-defined differential free operad together a natural epimorphism,

$$\Big(\mathsf{Free}(\ , \ ), \ \delta\Big) \longrightarrow \mathsf{Ass},$$

which, however, is *not* a quasi-isomorphism as the following equality,

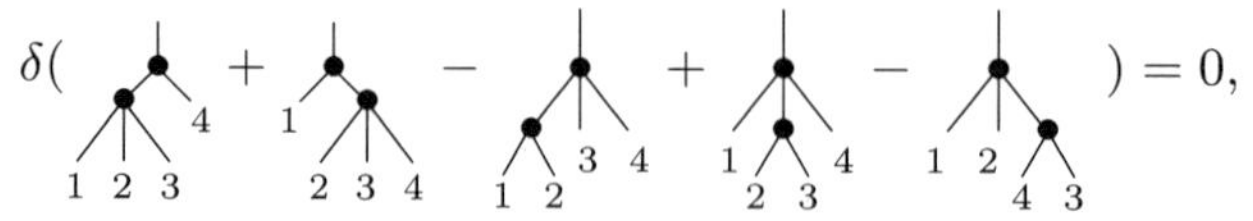

$$\delta({}_{1\,2\,3\,4} + {}_{1\,2\,3\,4} - {}_{1\,2\,3\,4} + {}_{1\,2\,3\,4} - {}_{1\,2\,4\,3}) = 0,$$

provides us with a nontrivial cohomology class in $\mathsf{H}^{-1}(\mathsf{Free}(\ , \ ), \delta)$ which is mapped under π to zero. To get rid of this class, we have to to make the corresponding cycle into a coboundary, i.e. introduce a new generating corolla of degree -2 with with 4 input legs and set

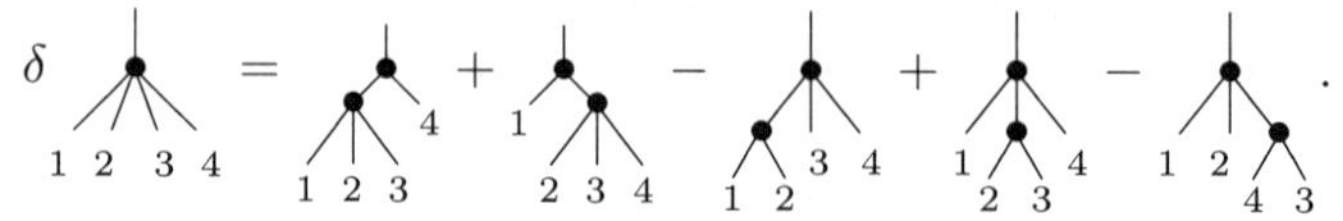

$$\delta \ {}_{1\,2\,3\,4} = {}_{1\,2\,3\,4} + {}_{1\,2\,3\,4} - {}_{1\,2\,3\,4} + {}_{1\,2\,3\,4} - {}_{1\,2\,4\,3}.$$

Again we get a well-defined dg free operad together with a natural homomorphism,

$$\left(\mathsf{Free}\langle \ \cdot \ , \ \cdot \ , \ \cdot \ \rangle, \ \delta\right) \longrightarrow \mathsf{Ass},$$

which, again, fails to be a quasi-isomorphism. To treat the new problem one has to introduce a new generating corolla of degree -3 with 5 input legs and so on.

2.5.1. Theorem [**St**]. *The minimal resolution of* Ass *is a dg free operad,* $\mathsf{Ass}_\infty := (\mathsf{Free}\langle\mathsf{E}\rangle, \delta)$ *generated by the* $\mathbb{S}$*-module* $\mathsf{E} = \{\mathsf{E}(n)\}$,

$$E(n) := \mathbb{K}[\mathbb{S}_n] = span \left(\ \sigma(1) \ \sigma(2) \ \ldots \ \sigma(n) \ \right)_{\sigma\in\mathbb{S}_n},$$

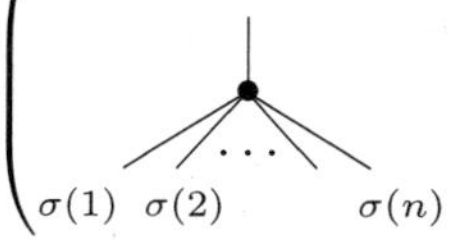

and with the differential given on the generators as

$$\delta \quad \sigma(1) \ \ldots \ \sigma(n) \quad = \sum_{k=0}^{n-2}\sum_{l=2}^{n-k}(-1)^{k+l(n-k-l)+1} \quad \sigma(1)\ldots\sigma(k) \quad \sigma(k+1)\ldots\sigma(k+l) \quad \sigma(k+l+1)\ldots\sigma(n) \ .$$

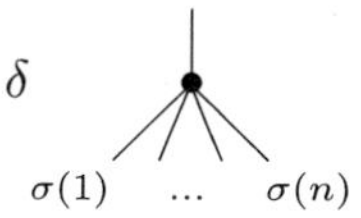

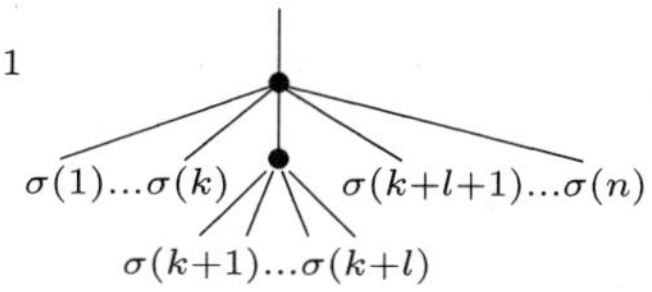

Our arguments above are no more than a motivation for this theorem. Its rigorous proof is by no means easy. The most beautiful proof is due to Stasheff [**St**] who applied methods of algebraic topology to his family of polytopes called nowadays *associahedra*.

This is the promised graph complex we are left with after "the vector space V" has gone. Now let us have the vector space back:

2.5.2. A_∞-algebras. A structure of A_∞*-algebra* on a dg vector space V is a morphism of dg operads,

$$\rho : (\mathsf{Ass}_\infty, \delta) \longrightarrow (\mathsf{End}\langle V\rangle, d).$$

The latter is the same as a family of linear maps,

$$\mu_n = \rho \left(\ 1 \ \ 2 \ \ldots \ n \ \right) : V^{\otimes n} \longrightarrow V, \quad n \geq 2,$$

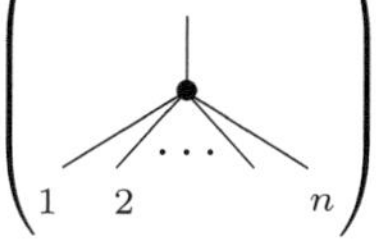

of degree $2-n$ which satisfy, for each $n \geq 1$, a system of equations,

$$\sum_{k=0}^{n-1}\sum_{l=1}^{n-k}(-1)^{\varepsilon}\mu_{n-l+1}(v_1, \ldots, v_k, \mu_l(v_{k+1}, \ldots, v_{k+l}), v_{k+l+1}, \ldots, v_v) = 0,$$

for any $v_1, \ldots, v_n \in V$. Here we set $\mu_1 := d$ and $\varepsilon := k+l(n-k-l)+l(|v_1|+\ldots+|v_k|)$.

Exercise. Show that an A_∞-algebra, $(V, \mu_\bullet)$, with all operations μ_n for $n \geq 3$ being zero is nothing but the ordinary dg associative algebra.

A_∞-algebras are homotopy perfect objects in the sense that given two homotopy equivalent dg vector spaces (V, d) and (W, d), then any A_∞-algebra structure on V induces via the homotopy equivalence an A_∞-structure on W; notice that if the original A_∞-structure is just an ordinary dg associative algebra structure on V, the induced one on W will have, in general, higher operations $\mu_{\geq 3}$ non-zero; explicit formulae for such an induced A_∞-structure via homotopy equivalence are given in [**Me1**].

§3 Algebraic structures ⇒ Graphs complexes ⇒ Geometric structures

3.1. In the previous section we started with an associative algebra structure on a vector space V. Then we got rid of V by encoding algebraic operations in terms of graphs with relations and obtained the operad Ass. Then we got rid of relations by resolving Ass into a dg free operad Ass_∞ which is a vector space spanned by graphs with no relations but with the differential. And finally, when looking at representations of Ass_∞ in vector spaces, we discovered a *new* notion of A_∞-algebra structure.

In this section we follow the same pattern starting with another class of simple algebraic structures — Lie 1-bialgebras, but now, after playing with graphs and representing them finally in a vector space, we recover a *well-known* geometric structure — the Poisson structure. Thus the final result is not new, but the way we get it is unprecedented. Let us describe this path to Poisson geometry in more detail.

3.2. Definition. A *Lie n-bialgebra* is the data
(i) a graded vector space V,
(ii) a linear map $\Delta : V \to V \wedge V$ making V into Lie coalgebra,
(iii) a linear map $[\ \bullet\ ] : \wedge^2(V[-n]) \to V[-n]$ making $V[-n]$ into a Lie algebra,
(iv) the compatibility condition:

$$\Delta[a\bullet b] = \sum a_1\otimes[a_2\bullet b]+[a\bullet b_1]\otimes b_2+(-1)^{|a||b|+n|a|+n|b|}([b\bullet a_1]\otimes a_2+b_1\otimes[b_2\bullet a]),$$

$\forall a,b \in V$. Here $\Delta a =: \sum a_1 \otimes a_2$, $\Delta b =: \sum b_1 \otimes b_2$.

The case $n = 0$ gives us the ordinary definition of Lie bialgebra. The case $n = 1$, as we shall see below, is relevant to Poisson geometry. Note that in the $n = 1$ case $\wedge^2(V[-1]) = (\odot^2 V)[-2]$ so that the brackets $[\ \bullet\ ]$ give us a degree 1 linear map $\odot^2 V \to V$.

3.3. Graph complex behind species "Lie 1-bialgebras". We attempt to formulate the notion of Lie 1-bialgebra without using the vector space V by coding two basic operations into corollas, "genes",

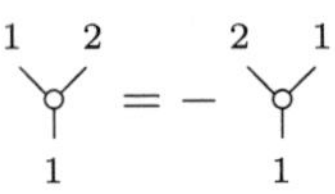

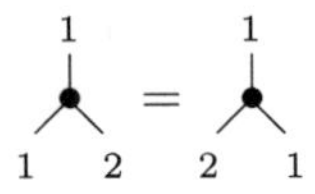

$\Delta \leftrightarrow$ [graph] $= -$ [graph] , $[\bullet] \Leftrightarrow$ [graph] $=$ [graph]

and also coding relations between basic operations as follows,

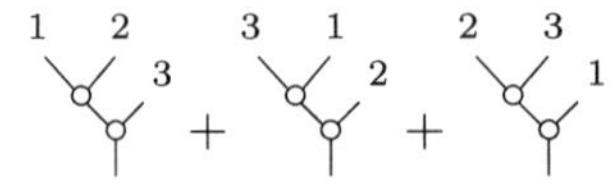

[graph] + [graph] + [graph] $=$ 0 (co-Jacobi identity for Δ)

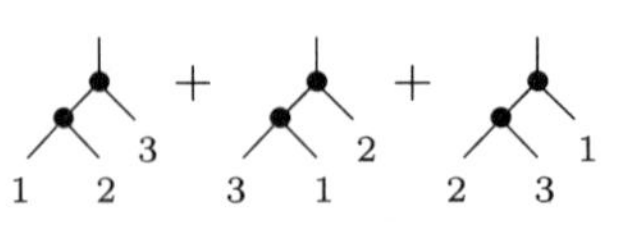

[graph] + [graph] + [graph] $=$ 0 (Jacobi identity for $[\bullet]$)

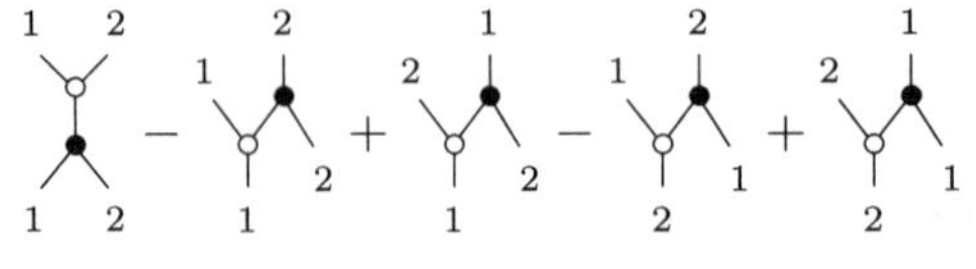

[graph] − [graph] + [graph] − [graph] + [graph] $=$ 0(compatibility condition).

Next we consider a vector space,

$$\mathsf{Free}\langle \ldots , \ldots \rangle := \mathrm{span}\langle \ldots , \ldots \rangle,$$

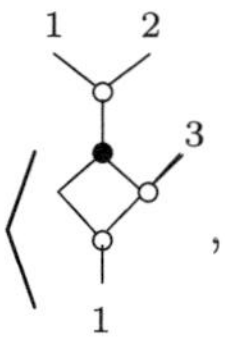

spanned by all possible graphs generated by the two genes (and with input and output legs labeled with integers), and then consider its quotient,

$$\mathsf{Lie^1B} = \frac{\mathsf{Free}\langle \ldots , \ldots \rangle}{\langle \mathsf{Relations} \rangle},$$

with respect to an equivalence relation generated by the above three families of equations. The resulting infinite dimensional vector space is naturally bigraded,

$$\mathsf{Lie^1B} = \bigoplus_{m+n\geq 3} \mathsf{Lie^1B}(m,n),$$

with $\mathsf{Lie^1B}(m,n)$ being the subspace spanned by (equivalence classes of) graphs with precisely n input legs and m output legs. For example,

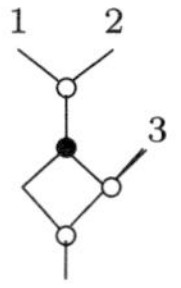

$$\ldots \in \mathsf{Lie^1B}(3,1).$$

There is a natural action of the group $\mathbb{S}_m \times \mathbb{S}_n$ on $\mathsf{Lie^1B}(m,n)$ via permutations of output and input labels. For any $1 \leq k \leq \min(n_1, m_2)$ and a pair of injections, $\alpha : [k] \to [n_1]$ and $\beta : [k] \to [m_2]$ there is an obvious composition,

$$\begin{array}{rccc} {}_\alpha\circ_\beta : & \mathsf{Lie^1B}(m_1,n_1) \otimes \mathsf{Lie^1B}(m_2,n_2) & \longrightarrow & \mathsf{Lie^1B}(m_1+m_2-k, n_1+n_2-k) \\ & G_1 \otimes G_2 & \longrightarrow & G_1 \,{}_\alpha\circ_\beta\, G_2 \end{array}$$

given by gluing the labeled by $\beta(i)$, $i \in [k]$, output leg of a graph G_2 to the labeled by $\alpha(i)$ input leg of a graph G_1. These compositions satisfy obvious "associativity" conditions which, when formalized, lead us to the notion of *properad* (see §4.4 and 4.5 below for rigorous definitions). The vector space $\mathsf{Lie^1B}$ together with this family of compositions, ${}_\alpha\circ_\beta$, is a typical example of a properad. Another typical example is the so called *endomorphism properad* of a vector space V,

$$\mathsf{End}\langle V\rangle = \bigoplus_{m,n\geq 0} \mathsf{End}\langle V\rangle(m,n),$$

with $\mathsf{End}\langle V\rangle(m,n) := \mathrm{Hom}(V^{\otimes n}, V^{\otimes m})$, and the compositions,

$$\begin{array}{rcl} {}_\alpha\circ_\beta : \mathrm{Hom}(V^{\otimes n_1}, V^{\otimes m_1}) \otimes \mathrm{Hom}(V^{\otimes n_2}, V^{\otimes m_2}) & \to & \mathrm{Hom}(V^{\otimes n_1+n_2-k}, V^{\otimes m_1+m_2-k}) \\ f_1 \otimes f_2 & \to & f_1 \,{}_\alpha\circ_\beta\, f_2 \end{array}$$

given by plugging in k outputs of a linear map f_2 into the k inputs of a linear map f_1 in accordance with the rule specified by injections α and β.

Next we resolve the properad $\mathsf{Lie^1B}$ in a way completely analogous to the resolution of Ass into Ass_∞ in §2.5 and get a *free* properad,

$$\mathsf{Lie^1B}_\infty := \mathsf{Free}\left\langle \begin{array}{c} 1\ 2 \ldots\ m \\ \times \\ 1\ 2 \ldots\ n \end{array} \right\rangle,$$

which is a graded vector space spanned by all possible connected graphs built on (m,n)-corollas which have degree $2-m$ and the following symmetries,

$$\begin{array}{c} 1\ 2 \ldots\ m \\ \times \\ \sigma(1)\ \sigma(2) \ldots\ \sigma(n) \end{array} = \begin{array}{c} 1\ 2 \ldots\ m \\ \times \\ 1\ 2 \ldots\ n \end{array}, \qquad \begin{array}{c} \tau(1)\ \tau(2) \ldots\ \tau(m) \\ \times \\ 1\ 2 \ldots\ n \end{array} = (-1)^{sgn(\tau)} \begin{array}{c} 1\ 2 \ldots\ m \\ \times \\ 1\ 2 \ldots\ n \end{array}$$

for arbitrary $\sigma \in \mathbb{S}_n$ and $\tau \in \mathbb{S}_m$. The differential in $\mathsf{Lie^1Bi}_\infty$ is given by

$$d \begin{array}{c} 1\ 2 \ldots\ m \\ \times \\ 1\ 2 \ldots\ n \end{array} = \sum_{\substack{I_1 \sqcup I_2 = (1,\ldots,m) \\ J_1 \sqcup J_2 = (1,\ldots,n) \\ |I_1|\geq 0, |I_2|\geq 1 \\ |J_1|\geq 1, |J_2|\geq 0}} (-1)^{\sigma(I_1\sqcup I_2)+|I_1||I_2|} \begin{array}{c} \overbrace{\ldots}^{I_1} \quad \overbrace{\ldots}^{I_2} \\ \times \!-\! \times \\ \underbrace{\ldots}_{J_1} \quad \underbrace{\ldots}_{J_2} \end{array}$$

where $\sigma(I_1 \sqcup I_2)$ is the sign of the shuffle $I_1 \sqcup I_2 = (1,\ldots,m)$.

What does this graph complex give us when we return back to a vector space V?

3.4. $\mathsf{Lie^1B}_\infty$-algebras and Poisson structures. A *representation* of the dg properad $(\mathsf{Lie^1B}_\infty, \delta)$ in a dg space (V,d) is a morphism of properads,

$$\rho : (\mathsf{Lie^1B}_\infty, \delta) \longrightarrow (\mathsf{End}\langle V\rangle, d),$$

which, by definition, is a morphism of complexes which respects the compositions, that is, satisfies $\rho(x\, {}_\alpha\circ_\beta\, y) = \rho(x){}_\alpha\circ_\beta\, \rho(y)$ for any $x, y \in \mathsf{Lie^1B}_\infty$. As any graph in $\mathsf{Lie^1B}_\infty$ is composed from the generating (m,n)-corollas by repeated application of compositions ${}_\alpha\circ_\beta$, the morphism ρ is completely defined by its values on these corollas,

$$\mu_{m,n} := \rho\left(\begin{array}{c} 1\ 2 \ldots\ m \\ \times \\ 1\ 2 \ldots\ n \end{array} \right) \in \mathrm{Hom}(\odot^n V, \wedge^m V) \subset \mathsf{End}\langle V\rangle(m,n).$$

The degree of the linear map $\mu_{m,n}$ is equal to the degree of the generating (m,n)-corolla, i.e. to $2-m$. We can assemble the collection of linear maps, $\{\mu_{m,n}\}$, into one geometric object — a degree 1 polyvector field γ on V — as follows. Let $\{e_a\}$ be an arbitrary basis in V and $\{x^a\}$ the associated linear coordinates. Then

$$\begin{aligned} \mu_{m,n}(e_{b_1},\ldots,e_{b_n}) &= \sum_{a_1,\ldots,a_m} \mu_{b_1\ldots b_n}^{\alpha_1\ldots\alpha_m} e_{a_1} \wedge \ldots e_{a_m}, \\ d(e_b) &= \sum_b d_b^a e_a, \end{aligned}$$

for some $\mu_{b_1\dots b_n}^{a_1\dots a_m}\in\mathbb{K}$, $d_b^a\in\mathbb{K}$, and we set

$$\pi := \sum_{a,b} d_b^a x^b \frac{\partial}{\partial x^a} + \sum_{m+n\geq 3}\frac{1}{m!n!}\sum_{\substack{a_1,\dots,a_m\\ b_1,\dots,b_n}} \mu_{b_1\dots b_n}^{a_1\dots a_m} x^{b_1}\cdots x^{b_n}\frac{\partial}{\partial x^{a_1}}\wedge\dots\wedge\frac{\partial}{\partial x^{a_m}} \in \wedge^\bullet\mathcal{T}_V,$$

where $\mathcal{T}_V$ stands for the tangent sheaf to V. It is not hard to check that the compatibility of the morphism ρ with the differentials is equivalent to the equation,

$$[\pi,\pi]_S = 0,$$

where $[\ ,\ ]_S$ stand for the Schouten Lie brackets on $\wedge^\bullet\mathcal{T}_V$. Thus we get the following

3.4.1. Proposition. *There is a one-to-one correspondence between representations of the dg properad* $(\mathsf{Lie^1B}_\infty,\delta)$ *in a vector space* V *and degree 1 elements,* π, *in the graded Schouten Lie algebra of polyvector fields which satisfy the equations,* $[\pi,\pi]=0$ *and* $\pi|_{x=0}=0$.

3.4.2. Corollary. *There is a one-to-one correspondence between representations of the dg properad* $(\mathsf{Lie^1B}_\infty,\delta)$ *in the vector space* $\mathbb{R}^n$ *and formal Poisson structures,* π, *on* $\mathbb{R}^n$ *which satisfy the condition* $\pi|_{x=0}=0$.

Proof. As the vector space $V=\mathbb{R}^n$ is concentrated in degree 0, a representation of $(\mathsf{Lie^1B}_\infty,\delta)$ in $\mathbb{R}^n$ must have all the associated linear maps, $\mu_{m,n}$ of *non*-zero degree vanishing. Hence only $\mu_{2,n}$ can be non-vanishing, and, by Proposition 3.4.1, these assemble into Taylor components of a formal smooth Poisson structure.

Conversely, given a formal germ, $\pi=\sum_{a_1,a_2}\pi^{a_1a_2}(x)\partial_{a_1}\wedge\partial_{a_2}$, of Poisson structure on $\mathbb{R}^n$ which vanishes at the point $0\in\mathbb{R}^n$, its Taylor components, $\frac{\partial^n\pi^{a_1a_2}(x)}{\partial x^{b_1}\cdots\partial x^{b_n}}|_{x=0}\rightsquigarrow\mu_{2,n}$, define a representation of the properad $(\mathsf{Lie^1B}_\infty,\delta)$. □

3.4.3. Remark. The condition $\pi|_{x=0}=0$ above is not a serious restriction: given an arbitrary Poisson structure π on $\mathbb{R}^n$ (not necessary vanishing at $0\in\mathbb{R}^n$), then, for any parameter $\hbar$ viewed as a coordinate on $\mathbb{R}$, the product $\hbar\pi$ is a Poisson structure on $\mathbb{R}^{n+1}=\mathbb{R}^n\times\mathbb{R}$ vanishing at zero $0\in\mathbb{R}^{n+1}$ and hence is a representation of the prop $\mathsf{Lie^1B}_\infty$.

3.4.4. Remark. Notice that in our approach a formal Poisson structure on $\mathbb{R}^n$ is a *morphism* in the category of properads. This "arrow" type interpretation of local Poisson geometry is crucial for our approach to deformation quantization.

§4 Properads, props and wheeled props

4.1. Graphs. Let $\mathfrak{G}^\uparrow(m,n)$, $m,n\geq 0$, be the space of *directed* (m,n)*-graphs*, G, that is, 1-dimensional *CW* complexes satisfying the following conditions:

(i) each edge (that is, 1-dimensional cell) is equipped with a direction;

(ii) if we split the set of all vertices (that is, 0-dimensional cells) which have exactly one adjacent edge into a disjoint union, $V_{in}\sqcup V_{out}$,

– with V_{in} being the subset of vertices with the adjacent edge directed from the vertex,

– and V_{out} the subset of vertices with the adjacent edge directed towards the vertex,

then $|V_{in}|\geq n$ and $|V_{out}|\geq m$;

(iii) precisely n of vertices from V_{in} are labelled by $\{1,\dots,n\}$ and are called *inputs*;

(iv) precisely m of vertices from V_{out} are labelled by $\{1,\dots,m\}$ and are called *outputs*;

(v) there are no oriented *wheels*, i.e. directed paths which begin and end at the same vertex; in particular, there are no *loops* (oriented wheels consisting of one internal edge). Put another way, directed edges generate a continuous flow on the graph which we always assume in our pictures to go from bottom to the top.

Note that $G \in \mathfrak{G}^{\uparrow}(m,n)$ may not be connected. Vertices in the complement,

$$v(G) := \overline{inputs \sqcup outputs},$$

are called *internal vertices*. For each internal vertex v we denote by in_v (resp., by out_v) the set of those adjacent half-edges whose orientation is directed towards (resp., from) the vertex. Input (resp., output) vertices together with adjacent edges are called *input* (resp., *output*) *legs*. The graph with one internal vertex, n input legs and m output legs is called the (m,n)-*corolla*.

We set $\mathfrak{G}^{\uparrow} := \sqcup_{m,n}\mathfrak{G}^{\uparrow}(m,n)$.

4.2. $\mathbb{S}$-bimodules. An $\mathbb{S}$-*bimodule*, E, is, by definition, a collection of graded vector spaces, $\{\mathsf{E}(m,n)\}_{m,n\geq 0}$, equipped with a left action of the group $\mathbb{S}_m$ and with right action of $\mathbb{S}_n$ which commute with each other. For any graded vector space V the collection, $\mathsf{End}\langle V\rangle = \{\mathsf{End}\langle V\rangle(m,n) := \mathrm{Hom}(V^{\otimes n}, V^{\otimes m})\}_{m,n\geq 0}$, is naturally an $\mathbb{S}$-bimodule. A *morphism* of $\mathbb{S}$-bimodules, $\phi : \mathsf{E}_1 \to \mathsf{E}_2$, is a collection of equivariant linear maps, $\{\phi(m,n) : \mathsf{E}_1(m,n) \to \mathsf{E}_2(m,n)\}_{m,n\geq 0}$. A morphism $\phi : \mathsf{E} \to \mathsf{End}\langle V\rangle$ is called a *representation* of an $\mathbb{S}$-bimodule E in a graded vector space V.

4.3. Decorated graphs. Let E be an $\mathbb{S}$-bimodule and $G \in \mathfrak{G}^{\uparrow}$ an arbitrary graph. We define a vector space,

$$G\langle \mathsf{E}\rangle := \left(\bigotimes_{v\in v(G)} \mathsf{E}(out_v, in_v)\right)_{AutG}$$

where

- $\mathsf{E}(out_v, in_v) := \mathrm{Bij}([m], out_v) \times_{\mathbb{S}_m} \mathsf{E}(m,n) \times_{\mathbb{S}_n} \mathrm{Bij}(in_v, [n])$ with Bij standing for the set of bijections,
- $Aut(G)$ stands for the automorphism group of the graph G.

An element of $G\langle \mathsf{E}\rangle$ is often called a *graph G with internal vertices decorated by elements of* E, or just a *decorated graph*. Thus a decorated graph is essentially a pair, $(G, [a_1 \otimes \ldots \otimes a_{|v(G)|}])$, consisting of a graph and an equivalence class of tensor products of elements $a_\bullet \in \mathsf{E}$.
If G is an (m,n)-graph with one internal vertex (i.e. an (m,n)-corolla), then $G\langle \mathsf{E}\rangle$ is canonically isomorphic to $\mathsf{E}(m,n)$.

4.4. Props [McL]. Let $H \in \mathfrak{G}^{\uparrow}(m,n)$ be a graph with more than two vertices and $G \subset H$ a subgraph with two vertices. We call G *admissible* if contracting the two internal vertices G together with internal edges into a corolla produces a graph, H/G, without oriented wheels and loops, i.e. $H/G \in \mathfrak{G}^{\uparrow}(m,n)$.

A *prop* (without identity) is, by definition, an $\mathbb{S}$-bimodule, $\mathsf{P} = \{\mathsf{P}(m,n)\}$, together with a family of equivariant linear maps,

$$\{\mu_G : G\langle \mathsf{P}\rangle \longrightarrow \mathsf{P}(m,n)\}_{G\in\mathfrak{G}^{\uparrow}:\,|v(G)|=2}\,, \tag{6}$$

parameterized by graphs $G \in \mathfrak{G}^{\uparrow}(m,n)$ with two internal vertices, such that, for arbitrary graph $H \in \mathfrak{G}^{\uparrow}(m,n)$ with three internal vertices the composition

$$\mu_{H/G} \circ \mu_G : H\langle \mathsf{P}\rangle \longrightarrow \mathsf{P}(m,n)$$

does not depend on the choice of an admissible 2-vertex subgraph $G \subset H$. This is a kind of associativity condition which "lives" in space rather on a plane.

In fact, for a prop P there is a contraction map,

$$\mu_H : H\langle \mathsf{P}\rangle \longrightarrow \mathsf{P}(m,n),$$

which is well-defined for a graph $H \in \mathfrak{G}^{\uparrow}(m,n)$ with *any* number of vertices. This map is given by a sequence of contractions along a family of admissible 2-vertex subgraphs, and it does not depend on the choice of such a sequence.

4.4.1. Endomorphism prop. For any vector space V the $\mathbb{S}$-bimodule $\mathsf{End}\langle V\rangle = \{\mathrm{Hom}(V^{\otimes n}, V^{\otimes m})\}$ is naturally a prop called the *endomorphism prop of* V.

4.4.2. Free prop. For an $\mathbb{S}$-bimodule, $\mathsf{E} = \{\mathsf{E}(m,n)\}_{m,n\geq 0}$, one can construct another $\mathbb{S}$-bimodule, $\mathsf{Free}\langle\mathsf{E}\rangle = \{\mathsf{Free}\langle\mathsf{E}\rangle(m,n)\}$ with

$$\mathsf{Free}\langle\mathsf{E}\rangle(m,n) := \bigoplus_{G\in\mathfrak{G}^{\uparrow}(m,n)} G\langle\mathsf{E}\rangle.$$

The $\mathbb{S}$-bimodule $\mathsf{Free}\langle\mathsf{E}\rangle$ has a natural prop structure with contraction maps μ_G being tautological. The prop $\mathsf{Free}\langle\mathsf{E}\rangle$ is called the *free prop generated by the* $\mathbb{S}$-*bimodule* E.

4.4.3. Morphisms of props. A morphisms of props, $F : \mathsf{P}_1 \to \mathsf{P}_2$, is a morphism of $\mathbb{S}$-bimodules such that, for any 2-vertex graph G one has $F \circ \mu_G = \mu_G \circ (F \boxtimes F)$, where $F \boxtimes F$ means a map, $G\langle\mathsf{P}_1\rangle \to G\langle\mathsf{P}_2\rangle$, which changes decorations of each vertex in G in accordance with F.

A morphism of props, $\mathsf{P} \to \mathsf{End}\langle V\rangle$, is called a *representation* of the prop P in a graded vector space V.

If P is a free prop, $\mathsf{P}_1 = \mathsf{Free}\langle\mathsf{E}\rangle$, generated by an $\mathbb{S}$-bimodule E, then the set of morphisms of props, $\{F : \mathsf{P}_1 \to \mathsf{P}_2\}$, is in one-to-one correspondence with the set of morphisms of $\mathbb{S}$-bimodules, $\{F|_{\mathsf{E}} : \mathsf{E} \to \mathsf{P}_2\}$, i.e. a prop morphism is completely determined by its values on the generators.

4.4.4. Dg props and resolutions. A *differential graded* (dg, for short) prop, (P, δ), is a prop, $\mathsf{P} = \{\mathsf{P}(m,n)\}$, such that every $\mathbb{S}_m \times \mathbb{S}_n$-module $\mathsf{P}(m,n)$ is a complex with differential δ commuting with the action of $\mathbb{S}_m \times \mathbb{S}_n$ and satisfying the equations, $\delta \circ \mu_G = \mu_G \circ (d \boxtimes \mathrm{Id} + \mathrm{Id} \boxtimes d)$ for every 2-vertex graph $G \in \mathfrak{G}^{\uparrow}$ (Leibnitz type identities).

If V is a dg vector space, then the associated endomorphism prop, $\mathsf{End}\langle V\rangle$, is naturally a dg prop.

A *free resolution* of a dg prop P is, by definition, a dg free prop, $(\mathsf{Free}\langle\mathsf{E}\rangle, \delta)$, generated by some $\mathbb{S}$-bimodule E together with a morphism of dg props, $\pi : (\mathsf{Free}\langle\mathsf{E}\rangle, \delta) \to \mathsf{P}$, which is a cohomology isomorphism.

If the differential δ in $\mathsf{Free}\langle\mathsf{E}\rangle$ is decomposable (with respect to prop's vertical and /or horizontal compositions), then $\pi : (\mathsf{Free}\langle\mathsf{E}\rangle, \delta) \to \mathsf{P}$ is called a *minimal model* of P. In this case the free prop $\mathsf{Free}\langle\mathsf{E}\rangle$ is often denoted by P_∞.

4.5. Properads [Va]. Let $\mathfrak{G}^{\uparrow}_c \subset \mathfrak{G}^{\uparrow}$ be the subset consisting of *connected* graphs. Then, repeating all the above definitions with $\mathfrak{G}^{\uparrow}$ replaced in all formulae by $\mathfrak{G}^{\uparrow}_c$, we get the notions of *properad*, *morphisms of properads*, *representations of properads*, etc.

Examples of properads are $\mathsf{End}\langle V\rangle$ and $\mathsf{Lie^1B}$ constructed in §3.3. If we repeat the constructions of §3.3 with arbitrary graphs (not necessarily connected ones) we would get props which we denote from now on by the same symbols $\mathsf{Lie^1B}$ and $\mathsf{Lie^1B}_\infty$. It is clear, that the latter is an example of a free prop generated by the $\mathbb{S}$-bimodule $\mathsf{E}_0 = \{\mathsf{E}_0(m,n) := sgn_m \otimes 1\!\!1_n[m-2]\}$, where sgn_m is the one-dimensional sign representation of $\mathbb{S}_m$ and $1\!\!1_n$ is the one-dimensional trivial representation of $\mathbb{S}_n$.

4.6. Wheeled props [**Me4, SMM**]. Let us now enlarge the set $\mathfrak{G}^\uparrow$ by allowing wheels: $\mathfrak{G}^\circlearrowright(m,n)$ is, by definition, the set of all directed (m,n)-graphs which satisfy conditions (i)-(iv) in §4.1. We set $\mathfrak{G}^\circlearrowright := \sqcup_{m,n}\mathfrak{G}(m,n)$. Then, repeating all the definitions in §4.4 with the set $\mathfrak{G}^\uparrow$ replaced in all formulae by the set $\mathfrak{G}^\circlearrowright$ and with the family of basic compositions (6) enlarged to contain maps μ_G associated with 1-vertex graphs with wheels, we get the notions of *wheeled prop*, *morphisms of wheeled props*, *representations of wheeled props*, etc.

For any *finite-dimensional* vector space V the associated $\mathbb{S}$-bimodule $\mathsf{End}\langle V\rangle$ is naturally a wheeled prop. Finite dimensionality is required because the maps μ_G associated with graphs G containing wheels involve traces of (compositions of) linear maps.

If we repeated all the constructions of §3.3 with graphs which may have wheels, we would get wheeled props which we denote from now as $\mathsf{Lie^1B}^\circlearrowright$ and, respectively, $\mathsf{Lie^1B}^\circlearrowright_\infty$. In fact any prop P has an associated wheeled extension, $\mathsf{P}^\circlearrowright$. For example, if P is a nondifferential prop with a minimal resolution $(\mathsf{P}_\infty = \bigoplus_{G\in\mathfrak{g}^\uparrow} G\langle\mathsf{E}\rangle, \delta)$ generated by an $\mathbb{S}$-bimodule E, then

$$\mathsf{P}^\circlearrowright_\infty := \bigoplus_{G\in\mathfrak{g}^\circlearrowright} G\langle\mathsf{E}\rangle$$

and $\mathsf{P}^\circlearrowright = \mathsf{H}^0(\mathsf{P}^\circlearrowright_\infty, \delta)$ where we used a so called Tate-Jozefiak grading of P_∞ in which all the generators of P have, by definition, degree 0. In this grading, for example, the cohomology $\mathsf{H}^\bullet(\mathsf{P}_\infty,\delta)$ is concentrated in degree zero and equals P. In general, however, the cohomology the wheeled completion, $\mathsf{H}^\bullet(\mathsf{P}^\circlearrowright_\infty,\delta)$, is *not* concentrated in Tate-Jozefiak degree 0, i.e. the quasi-isomorphism,

$$\pi : (\mathsf{P}_\infty, \delta) \to (\mathsf{P}, 0).$$

fails to be quasi-isomorphism for the natural extension of the monomorphism π,

$$\pi^\circlearrowright : (\mathsf{P}^\circlearrowright_\infty, \delta) \to (\mathsf{P}^\circlearrowright, 0).$$

to the wheeled completions.

4.6.1. Example. The natural projection

$$\pi^\circlearrowright : (\mathsf{Ass}^\circlearrowright_\infty, \delta) \longrightarrow (\mathsf{Ass}^\circlearrowright, 0)$$

is *not* a quasi-isomorphism. Indeed, it follows from (5) that

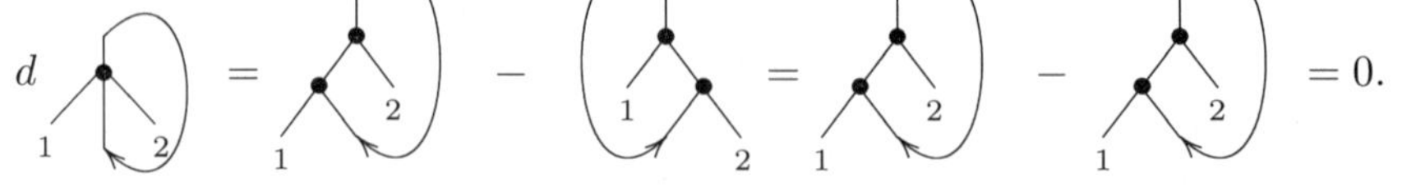

$$d \; [\text{graph}] = [\text{graph}] - [\text{graph}] = [\text{graph}] - [\text{graph}] = 0.$$

giving thereby a non-trivial cohomology class,

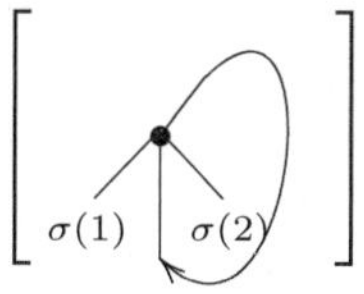

$$\in \mathsf{H}^{-1}\left(\mathsf{Ass}_\infty^{\circlearrowright}, d\right), \quad \sigma \in \Sigma_2,$$

which lies in the kernel of the projection $\mathsf{Ass}_\infty^{\circlearrowright} \to \mathsf{Ass}^{\circlearrowright}$.

4.6.2. Example. By contrast to the case of Ass_∞-algebras, the monomorphism

$$\pi^{\circlearrowright} : (\mathsf{Lie}_\infty^{\circlearrowright}, \delta) \longrightarrow (\mathsf{Lie}^{\circlearrowright}, 0)$$

of wheeled completions of props Lie_∞ and Lie remains a quasi-isomorphism (see [**Me4**] for a proof). Put another way, wheelification of the operad of Lie_∞-algebras does not create new cohomology classes.

4.6.3. Example. The natural projection,

$$\pi^{\circlearrowright} : (\mathsf{Lie^1B}_\infty^{\circlearrowright}, \delta) \longrightarrow (\mathsf{Lie^1B}^{\circlearrowright}, 0),$$

which sends to zero all generators of $\mathsf{Lie^1B}_\infty^{\circlearrowright}$ except corollas with three legs, is *not* a quasi-isomorphism[3]. Indeed, it is not hard to check that the graph [**Me4**],

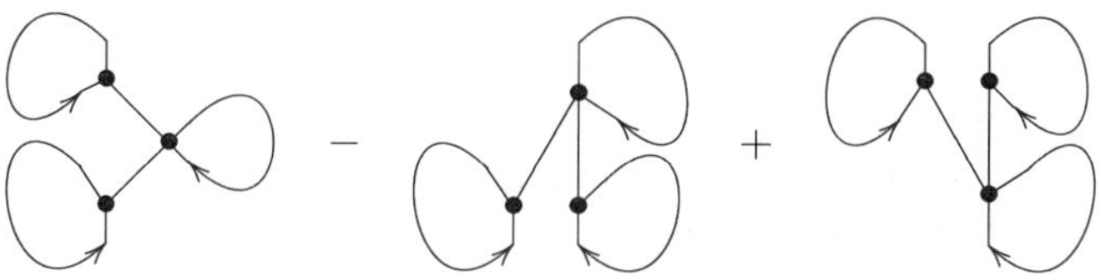

represents a non-trivial cohomology class in $\mathsf{H}^{-1}(\mathsf{Lie^1B}_\infty^{\circlearrowright}, \delta)$.

Note that the dg prop, $(\mathsf{Lie^1B}_\infty^{\circlearrowright}, \delta)$, is a *free* prop

$$\mathsf{Lie^1B}_\infty^{\circlearrowright} := \bigoplus_{G \in \mathfrak{G}^{\uparrow}} G\langle \mathsf{E}_0^{\circlearrowright} \rangle$$

on the $\mathbb{S}$-bimodule,

$$\mathsf{E}_0^{\circlearrowright} = \left\{ \mathsf{E}_0(m,n) := \bigoplus_{G \in \mathfrak{G}^{\circlearrowright}_{indec}(m,n)} G\langle \mathsf{E}_0 \rangle \right\}_{m,n \geq 0},$$

generated by indecomposable (with respect to grafting and disjoint union) decorated wheeled graphs. Note also that the induced differential is *not* quadratic with respect to the generating set $\mathsf{E}^{\circlearrowright}$.

4.6.4. Theorem. *There exists a dg free prop, $([\mathsf{Lie^1B}^{\circlearrowright}]_\infty, \delta)$, which fits into a commutative diagram,*

$$\begin{array}{ccc} [\mathsf{Lie^1B}^{\circlearrowright}]_\infty & \xrightarrow{\ \alpha\ } & \mathsf{Lie^1B}_\infty^{\circlearrowright} \\ & \searrow^{qis} & \downarrow \pi^{\circlearrowright} \\ & & \mathsf{Lie^1B}^{\circlearrowright} \end{array}$$

---

[3]However, this $\pi^{\circlearrowright}$ remains quasi-isomorphism on the subcomplex with purely operadic wheeels and, more generally, with wheels of genus 1, see [**Me4**] for proofs.

where α is an epimorphism of non-differential props, and qis a quasi-isomorphism of differential props. The prop $[\mathsf{Lie^1B}^{\circlearrowright}]_\infty$ *is called a* quasi-minimal prop resolution *of* $\mathsf{Lie^1B}^{\circlearrowright}$.

Proof. We shall use in the proof the Tate-Jozefiak grading of the prop $\mathsf{Lie^1B}^{\circlearrowright}_\infty$ which assigns to generating (m,n)-corollas the degree $3-m-n$. In this grading the cohomology of $\mathsf{Lie^1B}_\infty$ is concentrated in degree 0 and equals $\mathsf{Lie^1B}$, while $\mathsf{H}^0(\mathsf{Lie^1B}^{\circlearrowright}_\infty) = \mathsf{Lie^1B}^{\circlearrowright}$.

Let $s_1 : \mathsf{H}^{-1}(\mathsf{Lie^1B}^{\circlearrowright}_\infty) \to \mathsf{Lie^1B}^{\circlearrowright}_\infty$ be any representation of degree -1 cohomology classes (if there are any) as cycles. Set $\mathsf{E}_1 := \mathsf{H}^{-1}(\mathsf{P}^{\circlearrowright}_\infty)[1]$ and define a differential graded prop,

$$\mathsf{Q}_1 := \mathsf{Free}\langle \mathsf{E}^{\circlearrowright} \oplus \mathsf{E}_1 \rangle$$

with the differential δ extended to new generators as $s_1[1]$. By construction, $\mathsf{H}^0(\mathsf{Q}_1) = \mathsf{Lie^1B}^{\circlearrowright}$, and $\mathsf{H}^{-1}(\mathsf{Q}_1) = 0$.

Let $s_2 : \mathsf{H}^{-2}(\mathsf{Q}_1) \to \mathsf{Q}_1$ be any representation of degree -2 cohomology classes (if there are any) as cycles. Set $\mathsf{E}_2 := \mathsf{H}^{-2}(\mathsf{Q}_1)[1]$ and define a differential graded prop,

$$\mathsf{Q}_2 := \mathsf{Free}\langle \mathsf{E}^{\circlearrowright} \oplus \mathsf{E}_1 \oplus \mathsf{E}_2 \rangle$$

with the differential δ extended to new generators as $s_2[1]$. By construction, $\mathsf{H}^0(\mathsf{Q}_2) = \mathsf{Lie^1B}^{\circlearrowright}$, and $\mathsf{H}^{-1}(\mathsf{Q}_2) = \mathsf{H}^{-2}(\mathsf{Q}_2) = 0$.

Continuing by induction we construct a dg free prop, $[\mathsf{Lie^1B}^{\circlearrowright}]_\infty := \lim_{n\to\infty} \mathsf{Q}_n = \mathsf{Free}\langle \mathsf{E}^{\circlearrowright} \oplus \mathsf{E}_1 \oplus \mathsf{E}_2 \oplus \mathsf{E}_3 \oplus \ldots \rangle$ with all the cohomology concentrated in Tate-Jozefiak degree 0 and equal to $\mathsf{Lie^1B}^{\circlearrowright}$. □

4.6.5. Example. The above theorem generalizes to an arbitrary pair, P and P_∞, consisting of a prop P (with zero differential)) and its minimal resolution P_∞. The same arguments as in the prove of Theorem 4.6.3 show existence of a *quasi-minimal prop resolution* of $\mathsf{P}^{\circlearrowright}$ which fits the commutative diagram,

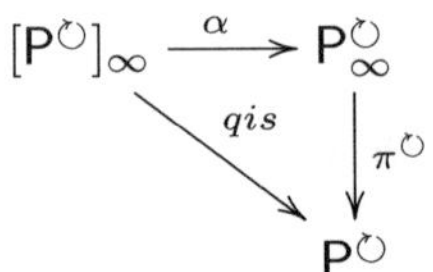

where α is an epimorphism of nondifferential props and qis a quasi-isomorphism.

In the case when P is an operad, Ass, of associative algebras such a quasi-minimal resolution of $\mathsf{Ass}^{\circlearrowright}$ has been found in [**SMM**]: this is a dg free prop $[\mathsf{Ass}^{\circlearrowright}]_\infty$

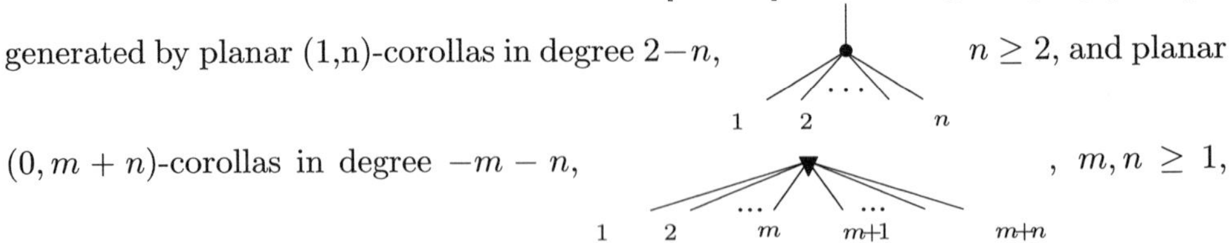

generated by planar (1,n)-corollas in degree $2-n$, $n \geq 2$, and planar $(0, m+n)$-corollas in degree $-m-n$, $m,n \geq 1$,

having the cyclic skew-symmetry

$$= (-1)^{\mathrm{sgn}(\zeta)} \qquad = (-1)^{\mathrm{sgn}(\xi)}$$

$1\ 2\ \ldots\ m\ \ m{+}1\ \ldots\ m{+}n \qquad \zeta(1)\ \zeta(2)\ \ldots\ \zeta(m)\ \ m{+}1\ \ldots\ m{+}n \qquad 1\ 2\ \ldots\ m\ \ \xi(m{+}1)\ \ldots\ \xi(m{+}n)$

with respect to the cyclic permutations $\zeta = (12\dots m)$ and $\xi = ((m+1)(m+2)\dots(m+n))$. The differential is given on generators as

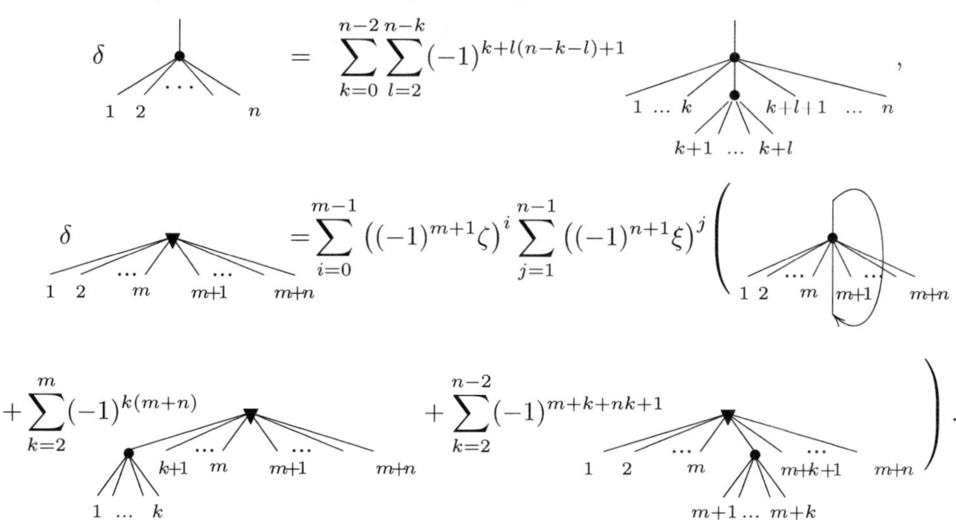

4.6.6. Wheeled Poisson structures. It follows from Proposition 3.4.1 that representations of the dg prop $\mathsf{Lie^1B}^{\circlearrowright}_\infty$ in a finite-dimensional dg space V are the same as Poisson structures on the formal manifold V. This fact prompts us to make the following

Definition. Representations of the dg prop $[\mathsf{Lie^1B}^{\circlearrowright}]_\infty$ in a finite-dimensional dg space V are called *wheeled Poisson structures* on the formal graded manifold M.

As generators of $[\mathsf{Lie^1B}^{\circlearrowright}]_\infty$ contain the generators of $\mathsf{Lie^1B}^{\circlearrowright}_\infty$, a wheeled Poisson structure includes a degree one polyvector field $\gamma \in \wedge^\bullet \mathcal{T}_V$ satisfying the Poisson equations, $[\gamma,\gamma] = 0$. However, there are other generators of $[\mathsf{Lie^1B}^{\circlearrowright}]_\infty]$, hence other *new* tensor fields on V enter, in general, the content list of a wheeled Poisson structure. One can get some insight into that content list from the above example of the prop $[\mathsf{Ass}^{\circlearrowright}]_\infty$ where new fields are cyclically invariant "functions" on the noncommutative manifold V — representations of the type $(0, m+n)$ corollas for all $n \geq 1, m \geq 1$.

The new fields of a wheeled Poisson structure satisfy systems of differential equations which involve *traces* of tensors formed from polyvector fields and their partial derivatives. Again the above example §4.6.5 (more precisely the value of the differential on $(0, m+n)$-corollas) gives some intuition into the possible structure of the equations but not that much: it is proven in [**Me4**] that the terms in that equations which involve traces of polyvector fields π can be neither linear in π nor contain only wheels of genus one (i.e. "wheeled" genus must be at least 2).

The geometric meaning of a wheeled Poisson structure is not clear to the author at present. One can only be sure that this notion is a *canonical* generalization of ordinary Poisson structure in *finite* dimensions. Clearer picture might emerge from computation of the cohomology of the complex $(\mathsf{Lie^1B}^{\circlearrowright}_\infty, \delta)$ which is a highly non-trivial problem comparable in complexity with the problem of computing the homology of the directed versions of famous Kontsevich's graph complexes [**Ko1**].

4.7. Example: prop profile of polydifferential operators. Let V be a graded vector space and $\mathcal{O}_V := \odot^\bullet V^\bullet$ the free graded commutative algebra generated by

V^*. We interpret the latter as formal smooth functions on the space V. It is well-known that the graded vector space,

$$\bigoplus_{k\geq 0} \mathrm{Hom}(\mathcal{O}_V^{\otimes k}, \mathcal{O}_V)[1-k],$$

has a natural structure of dg Lie algebra called the *Hochschild* dg Lie algebra. Let

$$\mathcal{D}_V \subset \bigoplus_{k\geq 0} \mathrm{Hom}(\mathcal{O}_V^{\otimes k}, \mathcal{O}_V)[1-k],$$

be its Lie subalgebra consisting of smooth formal polydifferential operators on $\mathcal{O}_V$ which, for $k \geq 1$, vanish on every element $f_1 \otimes \ldots \otimes f_k \in \mathcal{O}_V^{\otimes k}$ with at least one function, f_i, $i = 1, \ldots, k$, constant. A degree one element, $\Gamma \in \mathcal{D}_V$, decomposes into a sum, $\sum_{k\geq 0} \Gamma_k$, with Γ_k being polydifferential maps $\mathcal{O}_V^{\otimes k} \to \mathcal{O}_V$ of degree $2-k$. In a local coordinate system, $\{x^a\}$ on V, Γ can be represented as a Taylor series,

$$\Gamma = \sum_{k\geq 0} \sum_{I_1,\ldots,I_k,J} \Gamma_J^{I_1,\ldots,I_k} x^J \otimes \partial_{I_1} \otimes \ldots \otimes \partial_{I_k}, \tag{7}$$

with $\Gamma_J^{I_1,\ldots,I_k} \in \mathbb{K}$ is non-zero. The summation runs over the multi-indices $I = a_1 a_2 \ldots a_{|I|}$, and

$$\partial_I := \frac{\partial^{|I|}}{\partial x^{a_1} \cdots \partial x^{a_{|I|}}}, \quad x^I := x^{a_1} x^{a_2} \cdots x^{a_{|I|}}.$$

Such an element is called a *Maurer-Cartan* element if it satisfies the equation,

$$d_H \Gamma + \frac{1}{2}[\Gamma, \Gamma]_H = 0, \tag{8}$$

where d_H and $[\ ,\ ]_H$ stand for the differential and Lie bracket in the Hochschild dg Lie algebra. If all the components Γ_k of Γ are zero except $k = 2$, then such a Maurer-Cartan element defines an associative but, in general, non-commutative product, $\cdot + \Gamma_2 : \mathcal{O}_V^{\otimes 2} \to \mathcal{O}_V$, in $\mathcal{O}_V$, where $\cdot$ stands for the ordinary product in $\mathcal{O}_V$; this product is often called a *star product* and denoted by $*$. If the component Γ_0 is zero, then all other components, $\{\Gamma_k\}$ define on $\mathcal{O}_V$ the structure of an A_∞-algebra.

Remarkably enough, there exists a dg free prop whose representations in a vector space V are in one-to-one correspondence with Maurer-Cartan elements of the Hochschild dg Lie algebra on $\mathcal{O}_V$. Let us describe it.

4.7.1. Prop profile of "star products". Let us consider an $\mathbb{S}$-bimodule $\mathsf{D} = \{\mathsf{D}(m,n) := \mathsf{D}(m) \otimes \mathbf{1}_n[m-2]\}_{m,n\geq 0}$, where

$$\begin{aligned} \mathsf{D}(0) &:= \mathbb{R}[-2], \\ \mathsf{D}(m\geq 1) &:= \bigoplus_{k\geq 1} \bigoplus_{\substack{[m]=I_1\sqcup\ldots\sqcup I_k \\ |I_1|,\ldots,|I_k|\geq 1}} \mathrm{Ind}_{\mathbb{S}_{|I_1|}\times\ldots\times\mathbb{S}_{|I_k|}}^{\mathbb{S}_m} \mathbf{1}_{|I_1|} \otimes \ldots \otimes \mathbf{1}_{|I_k|}[k-2], \end{aligned}$$

and let $\mathsf{DefQ} := \mathsf{Free}\langle \mathsf{D} \rangle$ be the associated free prop. The generators of DefQ can be pictorially represented by degree $2-k$ directed planar corollas of the form,

$I_1 \quad I_i \; I_{i+1} \quad I_k$

$1 \; 2 \; 3 \quad \cdots \quad n$

where

- the input legs are labeled by the set $[n] := \{1, 2, \ldots, n\}$ and are symmetric (so that it does not matter how labels from $[n]$ are distributed over them),
- the output legs (if there are any) are labeled by the set $[m]$ partitioned into k disjoint non-empty subsets,

$$[m] = I_1 \sqcup \ldots \sqcup I_i \sqcup I_{i+1} \sqcup \ldots \sqcup I_k,$$

  and legs in each I_i-bunch are symmetric (so that it does not matter how labels from the set I_i are distributed over legs in I_ith bunch).

We define a differential, δ, in DefQ by its value on the generators,

$\delta\Big($ 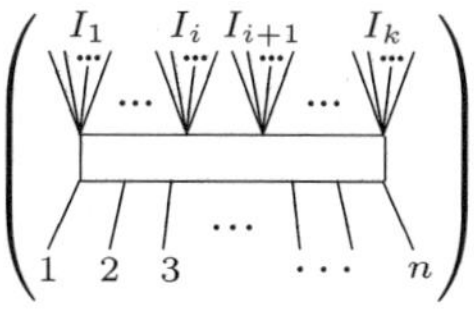 $\Big) = \sum_{i=1}^{k}(-1)^{i+1}$ 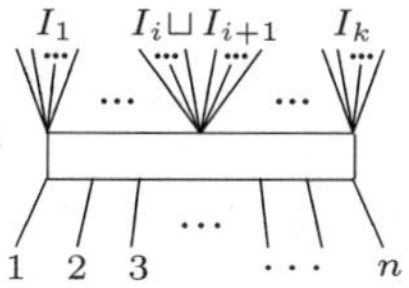

$$+ \sum_{\substack{p+q=k+1\\ p\geq 1, q\geq 0}} \sum_{i=0}^{p-1} \sum_{\substack{I_{i+1}=I'_{i+1}\sqcup I''_{i+1}\\ \cdots\cdots\\ I_{i+q}=I'_{i+q}\sqcup I''_{i+q}}} \sum_{[n]=J_1\sqcup J_2} \sum_{s\geq 0} (-1)^{(p+1)q+i(q-1)}$$

$\frac{1}{s!}$ 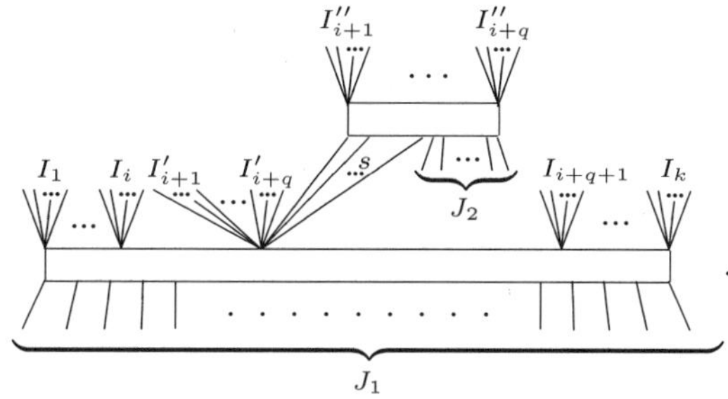 .

The s-summation runs over the number, s, of edges connecting the two internal vertices. As s can be zero, the r.h.s. above contains disconnected graphs (more precisely, disjoint unions of two corollas). As the summation over s is not bounded above, we have to assume from now on that the prop DefQ is completed with respect to the number of vertices. It is not hard to show [**Me4**] that $\delta^2 = 0$.

The main motivation behind the above definition of (DefQ, δ) is the following result.

4.7.2. Proposition [**Me4**]. *There is a one-to-one correspondence between representations,*

$$\phi : (\mathsf{DefQ}, \delta) \longrightarrow (\mathsf{End}\langle V\rangle, d),$$

and Maurer-Cartan elements, Γ*, in the Hochschild dg Lie algebra* $\mathcal{D}_V$*, that is, degree one polydifferential operators on* $\mathcal{O}_V$ *satisfying the equation,* $d_{\mathrm{H}}\Gamma + \frac{1}{2}[\Gamma, \Gamma]_{\mathrm{H}} = 0$.

Proof is similar to the proof of Proposition 3.4.1: the key to the correspondence between the generating corollas of DefQ and the Taylor series (7) is given by

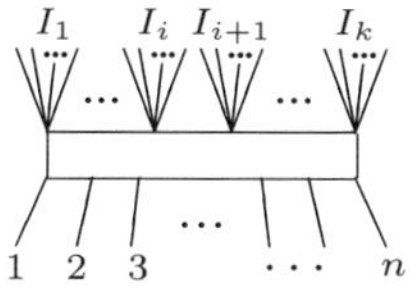 $\simeq \Gamma_J^{I_1,\ldots,I_k}.$

The differential δ in DefQ was specially designed so as to ensure that equation (8) is satisfied — the first sum in the formula for δ encodes the Hochschild differential d_{H} and the second sum is the graph encoding of the Hochschild brackets $[\ ,\ ]_{\mathrm{H}}$.

4.7.3. Remark. Notice that a "star product" Γ on $\mathcal{O}_V$ is interpreted in our approach as a *morphism* in the category of dg props (cf. §3.4.4).

4.7.4. Remark. Kontsevich's universal quantization [**Ko2**] gives rise to a morphism of dg props, $F_\infty : (\mathsf{DefQ}, \delta) \longrightarrow (\widehat{\mathsf{Lie}^1\mathsf{B}^{\circlearrowright}_\infty}, \delta)$. Vice versa, any morphism of the above dg props gives rise to a *universal* quantizations of formal Poisson structures.

4.8. Prop profile of perturbative "star products". Let $\mathbb{K}[[\hbar]]$ be the formal power series in a formal parameter $\hbar$, and let $\hbar\mathbb{K}[[\hbar]]$ be its (maximal) ideal spanned by series which vanish at $\hbar = 0$. Then $\mathcal{D}_V^\hbar := \mathcal{D}_V \otimes \hbar\mathbb{K}[[\hbar]]$ is a dg Lie algebra of polydifferential operators on $\mathcal{O}_V[[\hbar]]$ which vanish at $\hbar = 0$. A solution, Γ, of Maurer-Cartan equations in $\mathcal{D}_V^\hbar$ gives a (generalized) A_∞-structure on $\mathcal{O}_V[[\hbar]]$ which at $\hbar = 0$ reduces to the ordinary graded commutative multiplication in $\mathcal{O}_V$. Thus Maurer-Cartan elements in this Lie algebra describe perturbative deformations of the ordinary product in $\mathcal{O}_V$. Again, the set of all possible Maurer-Cartan elements in $\mathcal{D}_V^\hbar$ can be understood as the set of all possible representations of a certain dg free prop, $\mathsf{DefQ}^\hbar := \mathsf{Free}\langle \mathsf{D}^\hbar \rangle$, defined as follows:

- the $\mathbb{S}$-bimodule of generators, $\mathsf{D}^\hbar = \{\mathsf{D}^\hbar(m,n)\}$, is a direct sum, $\mathsf{D}^\hbar(m,n) = \bigoplus_{a=1}^\infty \mathsf{D}^a(m,n)$, $m,n \geq 0$, of labeled by $a \in \mathbb{N}^*$ copies, $\mathsf{D}^a(m,n) := \mathsf{D}(m) \otimes \mathbf{1}_n[2-m]$, of the $\mathbb{S}$-module corresponding to D; each copy corresponds to the "$\hbar^a$" coefficient in the formal power series;
- generators of $\mathsf{DefQ}^\hbar$ can be identified with planar corollas, 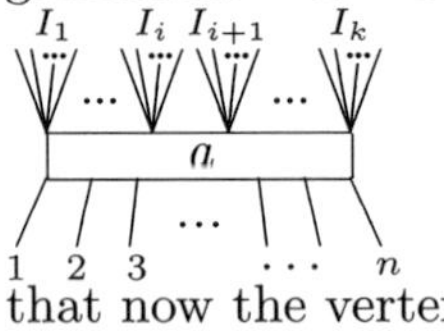, which are exactly the same as in §4.7 except that now the vertex gets a numerical label $a \in \mathbb{N}^*$;
- the differential δ is given on generators by

δ 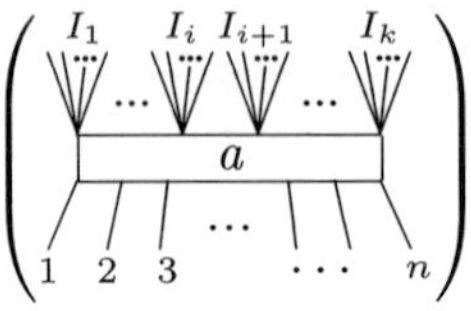 $= \sum_{i=1}^{k} (-1)^{i+1}$ 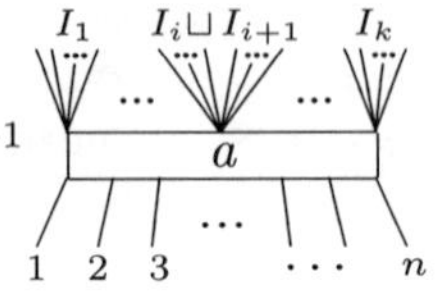

$$+ \sum_{\substack{b+c=a\\ b,c\geq 1}} \sum_{\substack{p+q=k+1\\ p\geq 1, q\geq 0}} \sum_{i=0}^{p-1} \sum_{\substack{I_{i+1}=I'_{i+1}\sqcup I''_{i+1}\\ \cdots\cdots\\ I_{i+q}=I'_{i+q}\sqcup I''_{i+q}}} \sum_{[n]=J_1\sqcup J_2} \sum_{s\geq 0} (-1)^{(p+1)q+i(q-1)}$$

(9) $\frac{1}{s!}$ 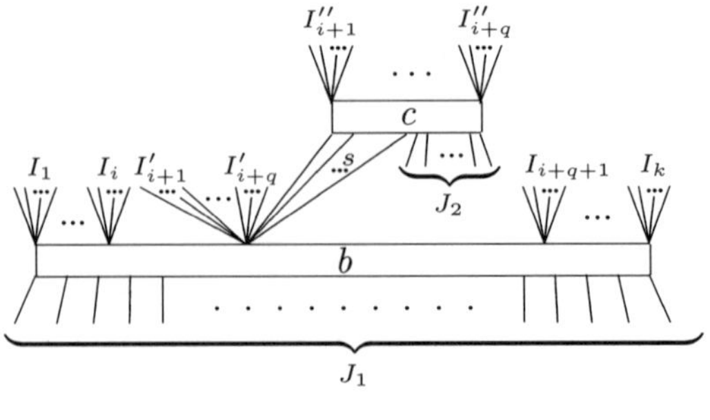

Note that $\mathsf{DefQ}^\hbar$ is spanned by $\mathsf{D}^\hbar$-decorated graphs over $\mathbb{K}$, *not* over $\mathbb{K}[[\hbar]]$. At the prop level the only remnant of the presence of the Planck constant $\hbar$ in the input geometry is in the decoration of vertices by a natural number $a \in \mathbb{N}^*$. In this respect the notation DefQ could be misleading.

4.8.1. Proposition. *There is a one-to-one correspondence between representations,*

$$\phi : \left(\mathsf{DefQ}^\hbar, \delta\right) \longrightarrow \left(\mathsf{End}_{\mathbb{R}[[\hbar]]}\langle V[[\hbar]]\rangle, d\right),$$

of $\mathsf{DefQ}^\hbar$ *in an* $\mathbb{K}[[\hbar]]$*-extension of a dg vector space* (V, d)*, and Maurer-Cartan elements,* Γ*, in* $\mathsf{D}_V[[\hbar]]$*, that is, degree one elements satisfying the equation,* $d_{\mathrm{H}}\Gamma + \frac{1}{2}[\Gamma, \Gamma]_{\mathrm{H}} = 0$ *and the condition* $\Gamma|_{\hbar=0} = d$*.*

Proof is similar to the proof of Proposition 3.4.1.

4.8.2. Remark. By constructions of (DefQ, δ) and $(\mathsf{DefQ}^\hbar, \delta)$, there is a canonical morphism of dg props,

$$\chi_\hbar : \quad (\mathsf{Free}\langle \mathsf{D}\rangle, \delta) \longrightarrow \left(\mathsf{Free}\langle \mathsf{D}^\hbar\rangle \otimes \mathbb{K}[[\hbar]], \delta\right)$$

$$\underbrace{\overset{I_1\quad I_i\ I_{i+1}\quad I_k}{\boxed{\qquad\qquad}}}_{1\ 2\ 3\ \ldots\ n} \longrightarrow \bigoplus_{a\geq 1} \hbar^a \underbrace{\overset{I_1\quad I_i\ I_{i+1}\quad I_k}{\boxed{\qquad a \qquad}}}_{1\ 2\ 3\ \ldots\ n}$$

If we assume that the prop $\mathsf{DefQ}^\hbar$ is completed with respect to the number of vertices, we can set $\hbar = 1$ above and get a well-defined morphism of dg props, $\chi_{\hbar=1} : \mathsf{DefQ} \to \mathsf{DefQ}^\hbar$. However, in general a morphism of dg props $\phi : \mathsf{DefQ}^\hbar \to \mathsf{P}$ can not be composed with $\chi_{\hbar=1}$ while $\phi \circ \chi_\hbar : \mathsf{DefQ} \to \mathsf{P}[[\hbar]]$ is always well defined.

§5. Deformation quantization via dg props

5.1. Reminder. Recall that in §3.3 we introduced the dg prop of polyvector fields, $\mathsf{Lie^1B}_\infty$, and then in §4.6 we studied its wheeled completion $\mathsf{Lie^1B}_\infty^\circlearrowright$ and proved that the latter can be further extended into a dg free prop, $[\mathsf{Lie^1B}^\circlearrowright]_\infty$, which fits into the commutative diagram,

$$\begin{array}{ccc} [\mathsf{Lie^1B}^\circlearrowright]_\infty & \xrightarrow{\ \alpha\ } & \mathsf{Lie^1B}_\infty^\circlearrowright \\ & {\scriptstyle qis}\searrow & \downarrow {\scriptstyle \pi^\circlearrowright} \\ & & \mathsf{Lie^1B}^\circlearrowright \end{array}$$

with α being an epimorphism of non-differential props and *qis* a quasi-isomorphism of differential props. Representations of $[\mathsf{Lie^1B}^\circlearrowright]_\infty$ in a dg vector space V are called wheeled Poisson structures.

In §4.7 we introduced the dg prop, DefQ, of "star products".

Let us denote by $\widehat{[\mathsf{Lie^1B}^\circlearrowright]}_\infty$ the completion of $[\mathsf{Lie^1B}^\circlearrowright]_\infty$ with respect to the number of vertices. The main purpose of this section is to prove Theorem 5.2 which says that wheeled Poisson structures can be deformation quantized.

5.2. Theorem. *There exists a morphism of dg props,*

$$\hat{Q} : \quad \mathsf{DefQ} \longrightarrow \widehat{[\mathsf{Lie^1B}^\circlearrowright]}_\infty,$$

such that

$$\pi_1 \circ \alpha \circ \hat{Q}\left(\text{graph: one box with outgoing corollas } I_1, \ldots, I_i, I_{i+1}, \ldots, I_k \text{ and incoming legs } 1, 2, 3, \ldots, n\right) = \begin{cases} \text{one-vertex corolla with outputs } 1, 2, \ldots, k-1, k \text{ and inputs } 1, 2, \ldots, n-1, n & \textit{for } |I_1| = \ldots = |I_k| = 1 \\ 0 & \textit{otherwise.} \end{cases}$$

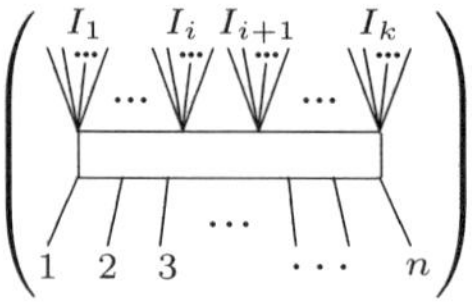

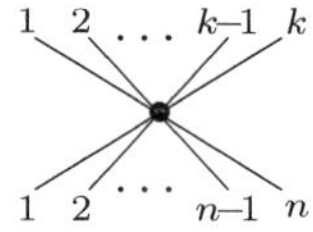

where π_1 *is the projection to the subspace of* $\mathsf{Lie^1B}^{\circlearrowright}_\infty$ *spanned by one-vertex graphs.*

Proof. We prove it below in three movements:

STEP I: we construct a morphism of dg props, $q : (\mathsf{DefQ}^\hbar, \delta) \longrightarrow (\mathsf{Lie^1B}^{\circlearrowright}, 0)$.

STEP II: we prove cofibrancy of $(\mathsf{DefQ}^\hbar, \delta)$ and then use this property to show that q can be lifted to a morphism Q making the diagram

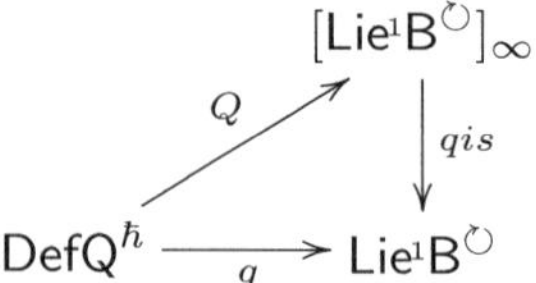

commutative;

STEP III: we set $\hat{Q}$ to be the composition,

$$\mathsf{DefQ} \xrightarrow{\chi_\hbar} \mathsf{DefQ}[[\hbar]] \xrightarrow{Q} [\mathsf{Lie^1B}^{\circlearrowright}]_\infty[[\hbar]]$$

at $\hbar = 1$ (which makes sense as a morphism between completed props).

Step I. To construct a morphism $q : (\mathsf{DefQ}^\hbar, \delta) \longrightarrow (\mathsf{Lie^1B}^{\circlearrowright}, 0)$ is the same as to deformation quantize an arbitrary finite-dimensional Lie 1-bialgebra, that is, a pair (ν, ξ), consisting of a linear Poisson structure ν and a quadratic homological vector field ξ such that $[\xi, \nu]_S = 0$.

As a first approximation to q we discuss first a well-known Poincare-Birkhoff-Witt quantization of linear Poisson structures, which, in the language of props, translates into existence of a morphism,

$$\mathcal{PBW} : (\mathsf{DefQ}^\hbar, \delta) \longrightarrow (\mathsf{CoLie}, 0),$$

where CoLie is the prop of coLie algebras. The morphism q we shall construct below in II.2 will make the diagram

$$\begin{array}{ccc} & & \mathsf{Lie^1B}^{\circlearrowright} \\ & \overset{q}{\nearrow} & \downarrow s \\ \mathsf{DefQ}^\hbar & \xrightarrow[\mathcal{PBW}]{} & \mathsf{CoLie}^{\circlearrowright} \end{array}$$

commutative, where s is the natural surjection,

$$\begin{array}{rccc} s: & \mathsf{Lie^1B} & \longrightarrow & \mathsf{CoLie} \\ & (\nu, \xi) & \longrightarrow & \nu, \end{array}$$

"forgetting" the quadratic homological vector field ξ.

I.1. Poincare-Birkhoff-Witt quantization. A representation of CoLie in a vector space V is the same as a linear Poisson structure on V viewed as a graded manifold. This Poisson structure makes the graded commutative algebra $\mathcal{O}_V[[\hbar]] := \widehat{\odot^\bullet} V^* \otimes \mathbb{K}[[\hbar]]$ into a Lie algebra with respect to the Poisson brackets, $\{\ ,\ \}$. Clearly,

$V^*[[\hbar]] \subset \mathcal{O}_V[[\hbar]]$ is a Lie subalgebra consisting of linear functions. Let $\mathcal{U}_\hbar$ be the associated universal enveloping algebra defined as a quotient,

$$\mathcal{U}_\hbar := \widehat{\otimes^\bullet} V^*[[\hbar]]/\mathcal{I},$$

where the ideal $\mathcal{I}$ is generated by all expressions of the form

$$X_1 \otimes X_2 - (-1)^{|X_1||X_2|} X_2 \otimes X_1 - \hbar\{X_1, X_2\}, \quad X_i \in V^*.$$

To construct a morphism of props $\mathcal{PBW}$ is the same as to deformation quantize an arbitrary linear Poisson structure ν. Which is a well-known trick: the Poincare-Birkhoff-Witt theorem says that the natural morphism,

$$s : \mathcal{O}_V[[\hbar]] \longrightarrow \mathcal{U}_\hbar,$$

is an isomorphism of vector spaces so that one can quantize ν by the formula,

$$f \star_\hbar g := s^{-1}(s(f) \circ s(g)), \quad \forall f, g \in \mathcal{O}_V,$$

where $\circ$ is the product in $\mathcal{U}(V_\hbar)$. This proves the existence of the required prop morphism $\mathcal{PBW}$.

I.2. Deformation quantization of $\mathsf{Lie^1B}$ **algebras.** A representation of $\mathsf{Lie^1B}$ in a graded vector space V is the same as a degree 0 linear Poisson structure,

$$\nu = \hbar \sum_{\alpha,\beta,\gamma} \Phi_\gamma^{\alpha\beta} x^\gamma \frac{\partial}{\partial x^\alpha} \wedge \frac{\partial}{\partial x^\beta},$$

together with a degree 1 quadratic vector field,

$$\xi = \hbar \sum_{\alpha,\beta,\gamma} C_{\alpha\beta}^\gamma x^\alpha x^\beta \frac{\partial}{\partial x^\gamma},$$

on V satisfying $[\xi, \xi]_S = 0$ and $Lie_\xi \nu = 0$. Here $\{x^a\}$ are arbitrary linear coordinates on V and Lie_ξ stands for the Lie derivative along the vector field ξ. The association,

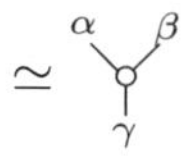

$$\Phi_\gamma^{\alpha\beta} \simeq \begin{smallmatrix} \alpha \quad \beta \\ \curlyvee \\ \gamma \end{smallmatrix}, \quad C_{\alpha\beta}^\gamma \simeq \begin{smallmatrix} \gamma \\ \curlywedge \\ \alpha \quad \beta \end{smallmatrix}$$

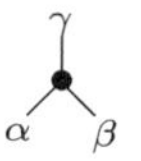

translates the equations,

$$[\nu, \nu]_S = 0, \quad [\xi, \xi]_S = 0, \quad Lie_\xi \nu = 0,$$

precisely into the graph relations, R, in §3.3.

To prove existence of a morphism $q : \mathsf{DefQ}^\hbar \to \mathsf{Lie^1B}^\circlearrowright$ is the same as to deformation quantize such a pair (ν, ξ), that is, to construct from (ν, ξ) a degree 2 function $\Gamma_0 \in \mathrm{Hom}_2(\mathbb{K}, \mathcal{O}_V)[[\hbar]] = \mathcal{O}_V[2][[\hbar]]$, a differential operator $\Gamma_1 \in \mathrm{Hom}_1(\mathcal{O}_V, \mathcal{O}_V)[[\hbar]]$, and a bi-differential operator $\Gamma_2 \in \mathrm{Hom}_0(\mathcal{O}_V^{\otimes 2}, \mathcal{O}_V)[[\hbar]]$ such that the following equations are satisfied[4],

$$\Gamma_1\Gamma_0 = 0, \qquad \Gamma_1^2 + [\Gamma_0, \Gamma_2]_{\mathrm{H}} = 0,$$
$$d_{\mathrm{H}}\Gamma_1 + [\Gamma_1, \Gamma_2]_{\mathrm{H}} = 0, \qquad d_{\mathrm{H}}\Gamma_2 + \frac{1}{2}[\Gamma_2, \Gamma_2]_{\mathrm{H}} = 0.$$

We can solve the last equation by setting Γ_2 to be related to $\Phi_\gamma^{\alpha\beta}$ via the $\mathcal{PBW}$ quantization, i.e. we choose the star product,

$$\star_\hbar = \text{usual product of functions} + \Gamma_2$$

---

[4]There can *not* be non-vanishing terms $\Gamma_k \in \mathrm{Hom}_{2-k}(\mathcal{O}_V^{\otimes k}, \mathcal{O}_V)[[\hbar]]$ with $k \geq 3$ for degree reasons.

to be given as before, $f \star_\hbar g := s^{-1}(s(f)\cdot s(g))$, $\forall f,g \in \mathcal{O}_V$.

Our next task is to find a degree 1 differential operator Γ_1 such that $d_{\mathrm{H}}\Gamma_1 + [\Gamma_1,\Gamma_2]_{\mathrm{H}} = 0$ which is equivalent to saying that Γ_1 is a derivation of the star product,

$$\Gamma_1(f\star_\hbar g) = (\Gamma_1 f)\star_\hbar g + (-1)^{|f|} f\star_\hbar (\Gamma_1 g), \qquad \forall f,g\in\mathcal{O}_V.$$

Consider first a derivation of the tensor algebra $\otimes^\bullet V[[\hbar]]$ given on the generators, $\{t^\gamma\}$, by

$$\hat{\xi}(t^\gamma) := \hbar \sum_{\alpha,\beta} C^\gamma_{\alpha\beta} x^\alpha \otimes x^\beta.$$

It is straightforward to check using equations $[\xi,\xi]_S = 0$ and $Lie_\xi \nu = 0$ that

$$\hat{\xi}\left(x^\alpha\otimes x^\beta - (-1)^{|\alpha||\beta|}x^\beta\otimes x^\alpha - \hbar\sum_\gamma \Phi^{\alpha\beta}_\gamma x^\gamma\right) = 0 \bmod \mathcal{I}$$

so that $\hat{\xi}$ descends to a derivation of the star product. Hence setting $\Gamma_1 = \hat{\xi}$ we solve the equation $d_{\mathrm{H}}\Gamma_1 + [\Gamma_1,\Gamma_2]_{\mathrm{H}} = 0$. However, $\hat{\xi}^2 \neq 0$, which is enough to check on a generator t^α,

$$\hat{\xi}^2(x^\alpha) = -\sum \frac{1}{3} C^\alpha_{\beta\gamma} C^\beta_{\mu\nu} \Phi^{\nu\gamma}_\tau \Phi^{\mu\tau}_\varepsilon x^\varepsilon,$$

or, in terms of graphs,

$$-\frac{1}{3}$$

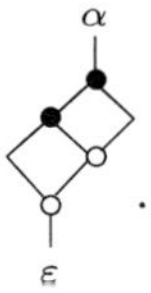

The data, $(\Gamma_2, \Gamma_1 = \hat{\xi})$, are given by directed graphs without cycles (in particular, these data make sense for *infinite* dimensional representations of the prop $\mathsf{Lie^1B}$). However, to solve the next equation, $\Gamma_1^2 + [\Gamma_0,\Gamma_2]_{\mathrm{H}} = 0$, one has to construct from the generators of $\mathsf{Lie^1B}$ a non-vanishing graph, Γ_0, with no output legs which is impossible to do without using graphs with oriented wheels. This is the reason why in Theorem 5.2 we attempt to construct a morphism into a *wheeled extension* of the prop $\mathsf{Lie^1B}_\infty$: there already does *not* exists a morphism, $\mathsf{DefQ}^\hbar \to \mathsf{Lie^1B}$, into the original unwheeled version of the prop of Lie 1-bialgebras which satisfies quasi-classical limit condition that terms linear in $\hbar$ are $\nu + \xi$.

In fact, one has to modify the naive choice, $\Gamma_1 = \hat{\xi}$, to get a solution of $\Gamma_1^2 + [\Gamma_0,\Gamma_2]_{\mathrm{H}} = 0$. Consider first a derivation of the tensor algebra $\otimes^\bullet V[[\hbar]]$ given on the generators, $\{x^\gamma\}$, by

$$\breve{\xi}(t^\gamma) := a\hbar^3\Theta^\gamma + \hbar\sum_{\alpha,\beta} C^\gamma_{\alpha\beta} x^\alpha\otimes x^\beta,$$

where

$$\Theta^\gamma := \sum_{\alpha,\beta,\mu,\nu} C^\gamma_{\alpha\beta}\Phi^{\alpha\mu}_\nu \Phi^{\beta\nu}_\mu = $$

and a is a constant.

I.2.1. Lemma. $\Phi_{\gamma}^{\alpha\beta}\Theta^{\gamma} = 0$, *i.e.,* [diagram] $= 0$.

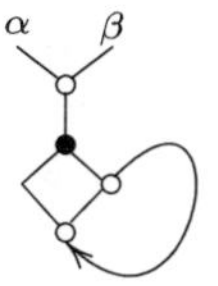

Proof. Using relations

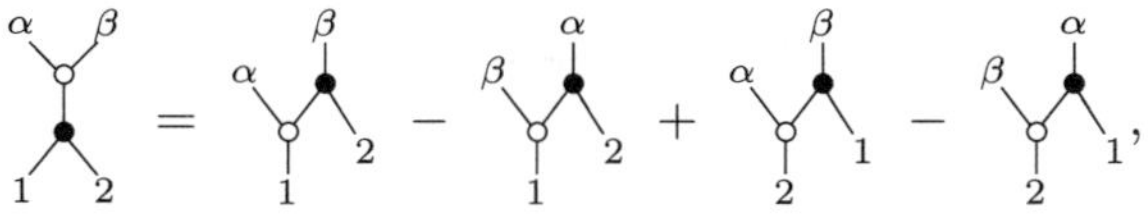

[diagram] $=$ [diagram] $-$ [diagram] $+$ [diagram] $-$ [diagram],

we first obtain,

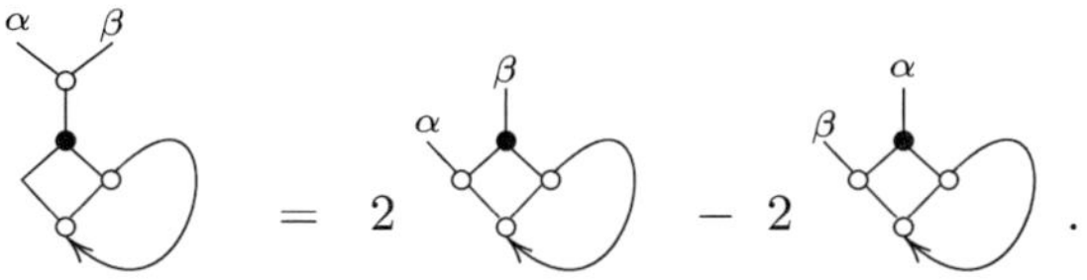

[diagram] $= 2$ [diagram] $- 2$ [diagram].

Next, using relations,

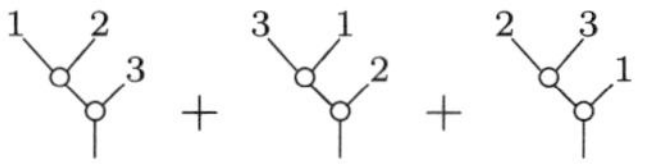

[diagram] $+$ [diagram] $+$ [diagram] $= 0$,

we finally obtain the desired result,

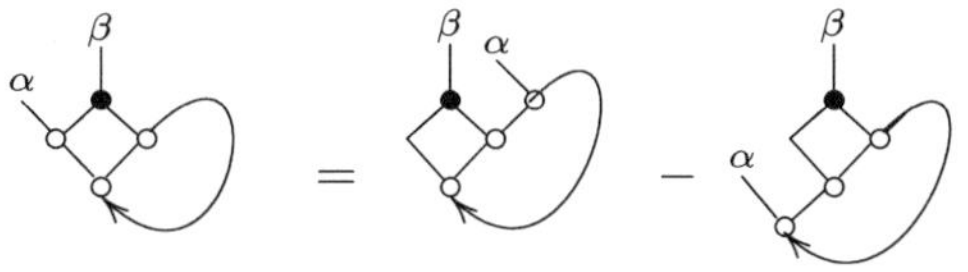

[diagram] $=$ [diagram] $-$ [diagram] $= 0$. □

I.2.2. Corollary-definition. *For any constant a the operator $\breve{\xi}$ descends to a derivation of the star product $\star_\hbar$ which we denote from now on by Γ_1.*

Proof.
$$\breve{\xi}\left(x^{\alpha}\otimes x^{\beta} - (-1)^{|\alpha||\beta|}x^{\beta}\otimes x^{\alpha} - \hbar\sum_{\gamma}\Phi_{\gamma}^{\alpha\beta}x^{\gamma}\right) = -a\hbar\sum_{\gamma}\Phi_{\gamma}^{\alpha\beta}\Theta^{\gamma} \bmod \mathcal{I}$$
$$= 0 \bmod \mathcal{I}. \qquad \square$$

Let us next define a linear function, $\Gamma_0 := b\hbar^3\sum_{\gamma}\Xi_{\gamma}x^{\gamma}$, where b is a constant and

$$\Xi_{\gamma} := \sum_{\alpha,\beta,\mu,\nu} C_{\alpha\beta}^{\mu} C_{\mu\nu}^{\beta}\Phi_{\gamma}^{\nu\alpha} =$$ [diagram]

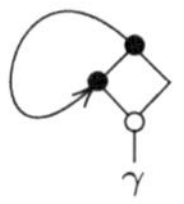

I.2.3. Lemma. $C_{\alpha\beta}^{\gamma}\Xi_{\gamma} = 0$, i.e. [diagram] $= 0$.

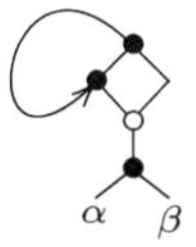

Proof is very similar to the proof of Lemma I.2.1. We omit the details.

I.2.4. Corollary. $\Gamma_1\Gamma_0 = 0$.

Therefore, for any Lie 1-bialgebra, $(C_{\alpha\beta}^{\gamma}, \Phi_{\varepsilon}^{\mu\nu})$, we constructed differential operators, $\Gamma_0 \in \mathrm{Hom}_2(\mathbb{R}, \mathcal{O}_V)[[\hbar]]$, $\Gamma_1 \in \mathrm{Hom}_1(\mathcal{O}_V, \mathcal{O}_V)[[\hbar]]$ and $\Gamma_2 \in$

$\mathrm{Hom}_0(\mathcal{O}_V^{\otimes 2}, \mathcal{O}_V)[[\hbar]]$ such that the equations, $\Gamma_1\Gamma_0 = 0$, $d_{\mathrm{H}}\Gamma_1 + [\Gamma_1, \Gamma_2]_{\mathrm{H}} = 0$ and $d_{\mathrm{H}}\Gamma_2 + \frac{1}{2}[\Gamma_2, \Gamma_2]_{\mathrm{H}} = 0$, are satisfied. It remains to check whether or not we can adjust the free parameters a and b in such a way that the last equation,

$$\Gamma_1^2 + [\Gamma_0, \Gamma_2]_{\mathrm{H}} = 0,$$

holds. As the l.h.s. of this equation is obviously a derivation of the star product, it is enough to check the latter only on generators x^α,

$$\Gamma_1^2(x^\alpha) + \Gamma_0 \star_\hbar x^\alpha - x^\alpha \star_\hbar \Gamma_0 = 0.$$

We have

$$\hbar^{-4}(\Gamma_0 \star_\hbar x^\alpha - x^\alpha \star_\hbar \Gamma_0) = b \sum \Xi_\gamma \Phi_\beta^{\gamma\alpha} x^\beta$$

$\simeq b$ 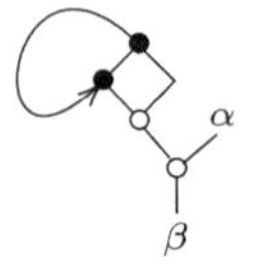 $= 2b$ 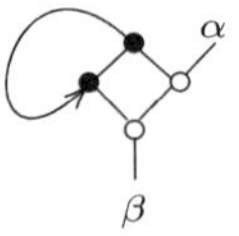

We also have,

$\hbar^{-4}\Gamma_1^2(x^\alpha) = -\frac{1}{3}$ 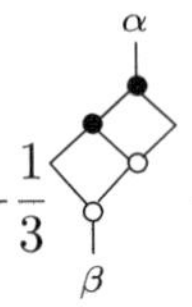 $- 2a$ 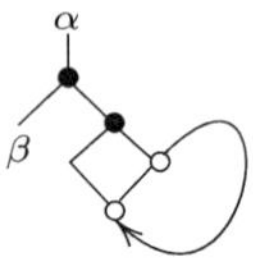 $= -\frac{1}{3}$ 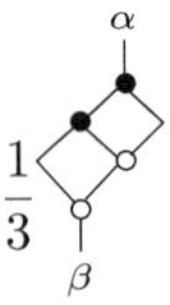 $+ 4a$ 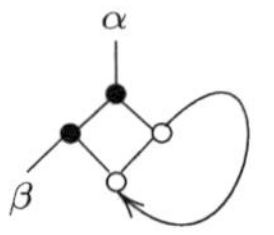 .

Consider a graph, 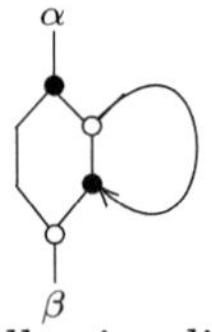 . Replacing its lower two vertices together with their half-edges by the following linear combination of four graphs,

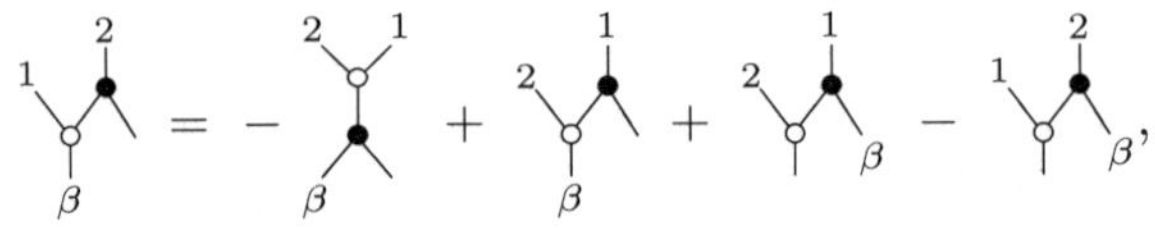

one gets, after a cancelation of two terms, the identity,

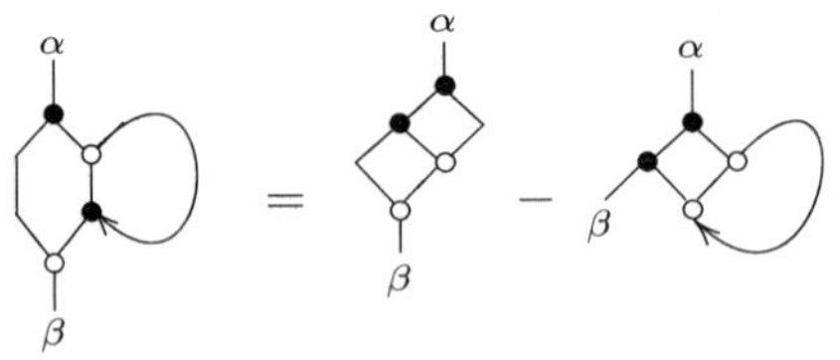

Playing a similar trick with upper two vertices one gets another identity,

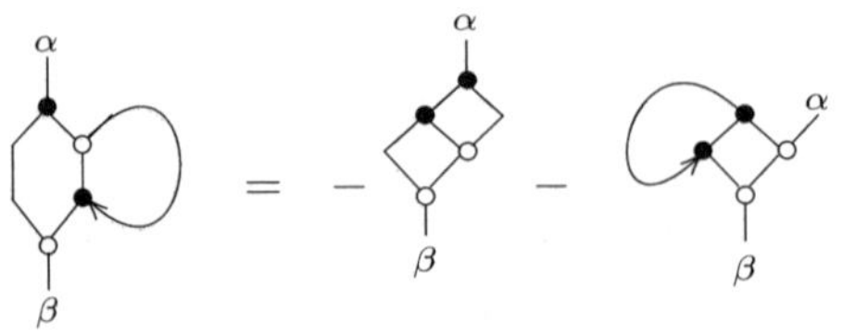

These two identities imply

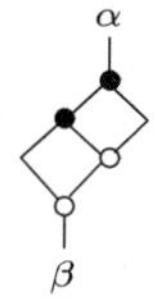

$$= \frac{1}{2} \qquad - \frac{1}{2}$$

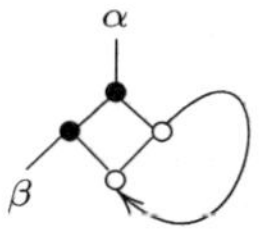

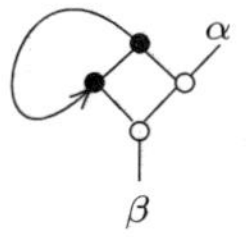

which in turn implies,

$$\hbar^{-4}\left(\Gamma_1^2(x^\alpha) + \Gamma_0 \star_\hbar x^\alpha - x^\alpha \star_\hbar \Gamma_0\right) = -\frac{1}{3} \qquad + 4a \qquad + 2b$$

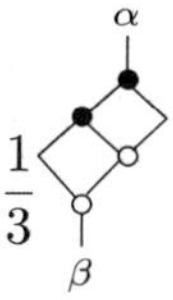

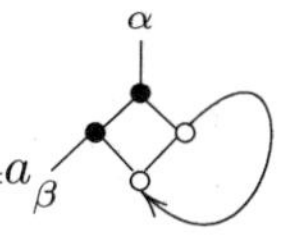

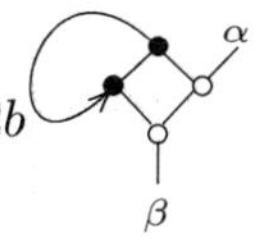

$$= (4a - \frac{1}{6}) \qquad + (2b + \frac{1}{6})$$

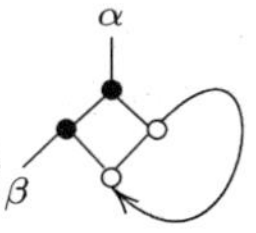

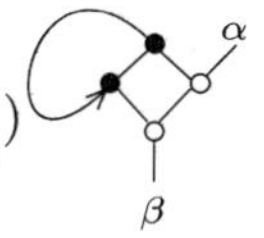

Hence setting $a = 1/24$ and $b = -1/12$, i.e. adding to the PBW star product, Γ_2, the operators,

$$\Gamma_0 = -\hbar^3 \frac{1}{12} \sum_{\alpha,\beta,\gamma,\mu,\nu} C^\mu_{\alpha\beta} C^\beta_{\mu\nu} \Phi^{\alpha\nu}_\gamma x^\gamma = -\hbar^3 \frac{1}{12} \sum_\gamma \qquad x^\gamma$$

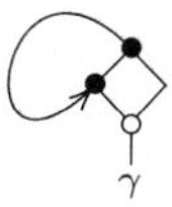

$$\Gamma_1(x^\gamma) = \hbar \sum_{\alpha,\beta} C^\gamma_{\alpha\beta} x^\alpha x^\beta + \frac{1}{24} \hbar^3 \sum_{\alpha,\beta,\mu,\nu} C^\gamma_{\alpha\beta} \Phi^{\alpha\mu}_\nu \Phi^{\beta\nu}_\mu = \hbar \sum_{\alpha,\beta} \qquad x^\alpha x^\beta + \frac{\hbar^3}{24}$$

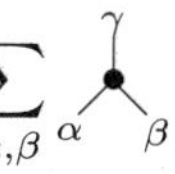

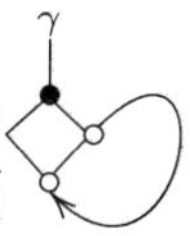

we complete deformation quantization of Lie 1-bialgebras. Thus we proved the following,

I.2.5. Proposition. *There exists a morphism of dg props,* $q : \mathsf{DefQ}^\hbar \to \mathsf{Lie^1B}^{\circlearrowright}$, *making the diagram*

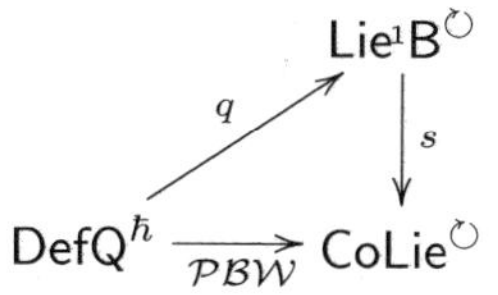

$$\begin{array}{ccc} & & \mathsf{Lie^1B}^{\circlearrowright} \\ & \overset{q}{\nearrow} & \downarrow s \\ \mathsf{DefQ}^\hbar & \xrightarrow[\mathcal{PBW}]{} & \mathsf{CoLie}^{\circlearrowright} \end{array}$$

commutative.

Hence Step 1 is done.

Step II. First we explain the iterative procedure behind our construction of a morphism Q fitting the commutative diagram

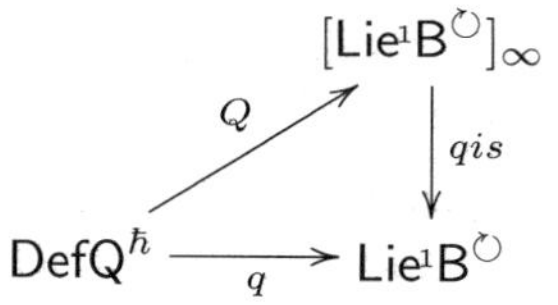

$$\begin{array}{ccc} & & [\mathsf{Lie^1B}^{\circlearrowright}]_\infty \\ & \overset{Q}{\nearrow} & \downarrow qis \\ \mathsf{DefQ}^\hbar & \xrightarrow[q]{} & \mathsf{Lie^1B}^{\circlearrowright} \end{array}$$

and then illustrate it with concrete examples. Define E_s to be zero for negative s and, for $s \geq 0$,

$$E_s \quad := \quad \mathrm{span}\left\{ \begin{array}{c} I_1 \quad I_i \; I_{i+1} \quad I_k \\ \boxed{\quad a \quad} \\ 1 \; 2 \; 3 \quad \cdots \quad n \end{array} \in \mathsf{DefQ}^\hbar \right\}_{\substack{2a+k-2=s \\ k\geq 0, a\geq 1, n\geq 0}}$$

For example,

$$E_0 = \mathrm{span}\left\{ \begin{array}{c} \boxed{1} \\ 1\,2\,\cdots n \end{array} \right\}, \; E_1 = \mathrm{span}\left\{ \begin{array}{c} \cdots \\ \boxed{1} \\ 1\,2\,\cdots n \end{array} \right\}, \; E_2 = \mathrm{span}\left\{ \begin{array}{c} \boxed{2} \\ 1\,2\,\cdots n \end{array}, \begin{array}{c} \cdots \;\; \cdots \\ \boxed{1} \\ 1\,2\,\cdots n \end{array} \right\}.$$

Let $\mathsf{DefQ}^\hbar_s \subset \mathsf{DefQ}^\hbar$ be the free prop generated by $\oplus_{i=0}^s E_i$. Thereby we get an increasing filtration, $0 \subset \mathsf{DefQ}^\hbar_0 \subset \ldots \subset \mathsf{DefQ}^\hbar_s \subset \mathsf{DefQ}^\hbar_{s+1} \ldots$ with

$$\lim_{s\to\infty} \mathsf{DefQ}^\hbar_s = \mathsf{DefQ}^\hbar.$$

A straightforward inspection of the formula for the differential δ in (9) implies

$$\delta E_{s+1} \subset \mathsf{DefQ}^\hbar_s,$$

i.e. that the dg prop $(\mathsf{DefQ}^\hbar, \delta)$ has a "cell" structure analogous to that of a CW complex. This simple but crucial for our purposes observation permits us to apply the well-known in algebraic topology Whitehead lifting trick and construct a morphism $Q : \mathsf{DefQ}^\hbar \to [\mathsf{Lie^1B}^{\circlearrowright}]_\infty$ by defining its values on the generators E_s via an induction on s. We begin the induction by specifying its values on $E_0 \oplus E_1 \oplus E_2$ as follows

$$Q\left(\begin{array}{c} \boxed{1} \\ 1\,2\,\cdots n \end{array} \right) = 0, \qquad Q\left(\begin{array}{c} I_1 \\ \boxed{1} \\ 1\,2\,\cdots n \end{array} \right) = \begin{cases} \curlywedge & \text{for } |I_1| = 1, n = 2 \\ 0 & \text{otherwise.} \end{cases}$$

$$Q\left(\begin{array}{c} \boxed{2} \\ 1\,2\,\cdots n \end{array} \right) = 0, \qquad Q\left(\begin{array}{c} I_1 \quad I_2 \\ \boxed{1} \\ 1\,2\,\cdots n \end{array} \right) = \begin{cases} \curlyvee & \text{for } |I_1| = 1, |I_2| = 1, n = 1 \\ 0 & \text{otherwise.} \end{cases}$$

Note that this choice respects both commutativity of the diagram and the condition on $\pi_1 \circ Q$ in the Proposition. Thus we constructed a morphism of dg props,

$$Q_3 : \mathsf{DefQ}^\hbar_3 \longrightarrow [\mathsf{Lie^1B}^{\circlearrowright}]_\infty$$

satisfying the required conditions.

Assume now that a morphism

$$Q_s : \mathsf{DefQ}^\hbar_s \longrightarrow [\mathsf{Lie^1B}^{\circlearrowright}]_\infty$$

satisfying the required conditions is already constructed for some $s \geq 3$. We want to extend Q_s to a morphism of dg props,

$$\begin{array}{rccc} Q_{s+1} : & \mathsf{DefQ}^\hbar_{s+1} & \longrightarrow & [\mathsf{Lie^1B}^{\circlearrowright}]_\infty \\ & e & \longrightarrow & Q_{s+1}(e) \end{array}$$

such that $qis \circ Q_{s+1} = q$ and the condition on $\pi_1 \circ Q_{s+1}$ is fulfilled. Let e' be a lift of $q(e)$ along the surjection qis. Then $Q_s(\delta e) - \delta e'$ is a cycle in $[\mathsf{Lie^1B}^{\circlearrowright}]_\infty$ which projects under qis to zero. As qis is a homology isomorphism, this element is exact, $Q_s(\delta e) - \delta e' = \delta e''$, for some $e'' \in [\mathsf{Lie^1B}^{\circlearrowright}]_\infty$. We set $Q_{s+1}(e) := e' + e''$ completing thereby the inductive construction of Q as a morphism of dg props. The condition $qis \circ Q = q$ is automatically satisfied. Another condition on $\pi_1 \circ \alpha \circ Q$ is also satisfied because the Hochschild brackets, $[\gamma, \gamma]_H$, of a degree 1 polyvector field viewed as a polydifferential operator contain the Schouten brackets, $[\gamma, \gamma]_S$, as one of the irreducible (in the $\mathbb{S}$-bimodule sense) summands. Using this fact it is easy to check by induction on $2a + k - 2$ that

$\pi_1 \circ \alpha \circ Q$ 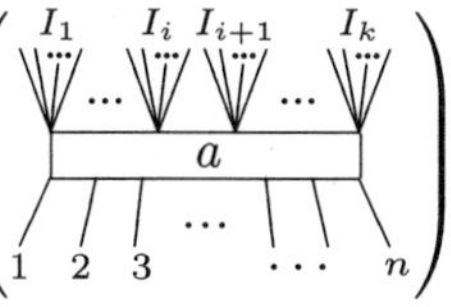 $=$ 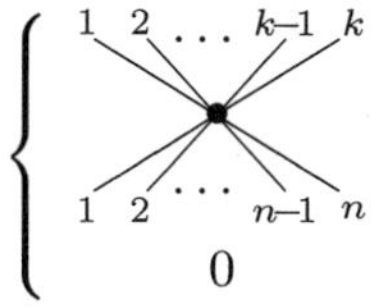 for $a = n + k - 2$ and $|I_1| = \ldots = |I_k| = 1$; 0 otherwise,

where $\pi_1 : \mathsf{Lie^1B}^{\circlearrowright}_\infty \to (\mathsf{Lie^1B}^{\circlearrowright}_\infty)_1$ is the projection to the subspace, $(\mathsf{Lie^1B}^{\circlearrowright}_\infty)_1$, consisting of graphs with precisely 1 internal vertex.

II.2. Corollary. (i) If $a_s \in E_s$ is such that $Q_{s-1}(\delta a_s) = 0$ then we can set $Q_s(a_s) := qis^{-1} \circ q(a_s)$, an arbitrary lifting of $q(a_s) \in \mathsf{Lie^1B}^{\circlearrowright}$ to $\mathsf{Lie^1B}^{\circlearrowright}_\infty \subset [\mathsf{Lie^1B}^{\circlearrowright}]_\infty$.

(ii) If $a_s \in E_s$ is such that $Q_{s-1}(\delta a_s) = 0$ and $q(a_s) = 0$, then we can set $Q(a_s) = 0$.

II.3. Iteration. Iteration (with respect to the "weight" parameter $s = 2a+k-2$) formula has the form,

Q 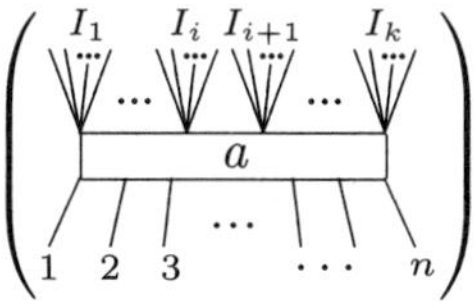 $= e' + e''$

where e' is an arbitrary lift of q 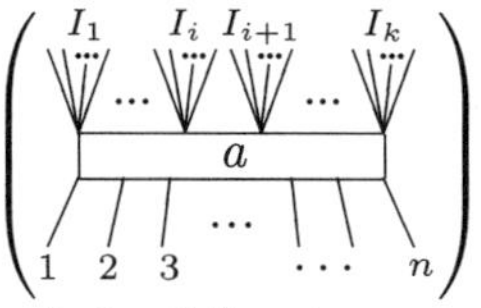 $\in \mathsf{Lie^1B}^{\circlearrowright}$ to a cycle in $\mathsf{Lie^1B}^{\circlearrowright}_\infty \subset [\mathsf{Lie^1B}^{\circlearrowright}]_\infty$, and e'' is a solution of the following equation in $[\mathsf{Lie^1B}^{\circlearrowright}]_\infty$,

$\delta e'' = Q\Big(\delta$ 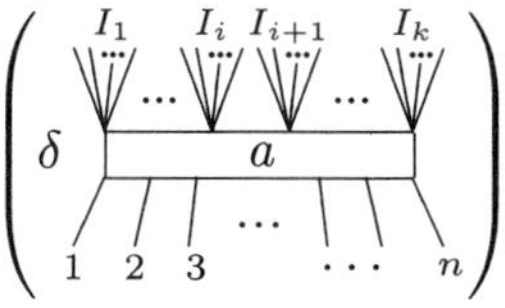 $\Big) = \sum_{i=1}^{k} (-1)^{i+1} Q$ 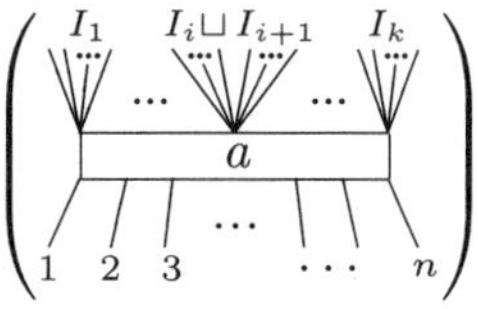

$+ \sum (-1)^{(p+1)q+i(q-1)} \frac{1}{s!}$ 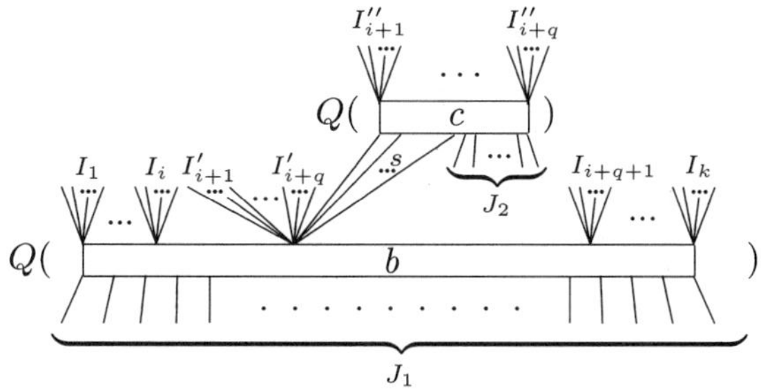 .

Note that the r.h.s. of the above equation contains values of Q on $[a', k', n']$-corollas with the weight $2a' + k' - 2 < 2a + k - 2$ which are, by the induction assumption, already known.

II.4. How it works. We shall illustrate the above construction of Q in a few examples.

Iteration level s=3: We have

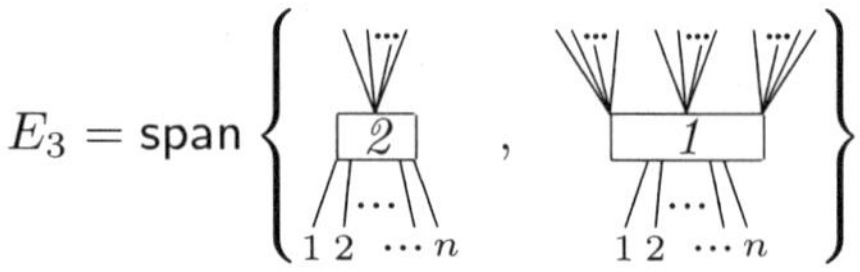

$$E_3 = \mathsf{span}\left\{ \text{[corolla 2; inputs } 1\,2\cdots n], \quad \text{[corolla 1; inputs } 1\,2\cdots n] \right\}$$

The equation in **II.3** for the first generator takes the form,

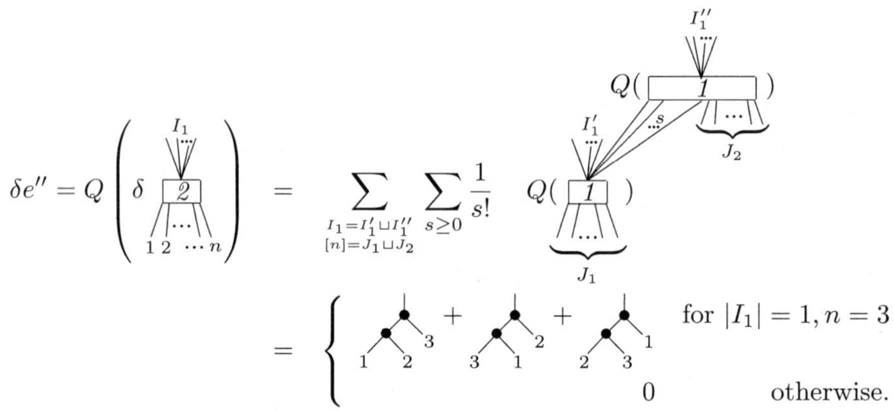

$$\delta e'' = Q\left(\delta \; [\text{corolla } 2;\ I_1;\ 1\,2\cdots n]\right) = \sum_{\substack{I_1 = I_1' \sqcup I_1'' \\ [n] = J_1 \sqcup J_2}} \sum_{s\geq 0} \frac{1}{s!} \; Q([1;\ I_1';\ J_1]) \circ_s Q([1;\ I_1'';\ J_2])$$

$$= \begin{cases} \text{[tree } (1,2),3] + \text{[tree } (3,1),2] + \text{[tree } (2,3),1] & \text{for } |I_1| = 1, n = 3 \\ 0 & \text{otherwise.} \end{cases}$$

implying

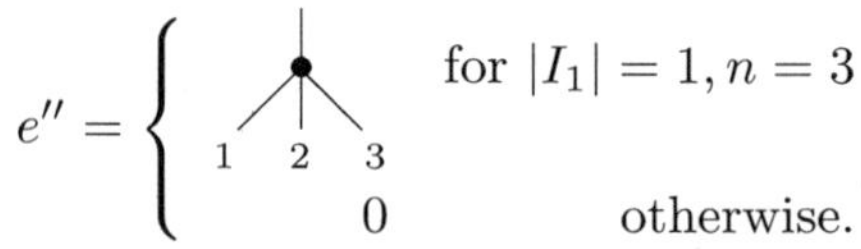

$$e'' = \begin{cases} \text{[corolla } 1\ 2\ 3] & \text{for } |I_1| = 1, n = 3 \\ 0 & \text{otherwise.} \end{cases}$$

As

$$q\left([\text{corolla } 2;\ I_1;\ 1\,2\cdots n]\right) = \begin{cases} -\text{[graph } 1,2;2;1] - \text{[graph } 2,1;2;1] - \text{[graph } 1,2;1;2] - \text{[graph } 2,1;1;2] & \text{for } |I_1| = 2, n = 2 \\ 0 & \text{otherwise.} \end{cases}$$

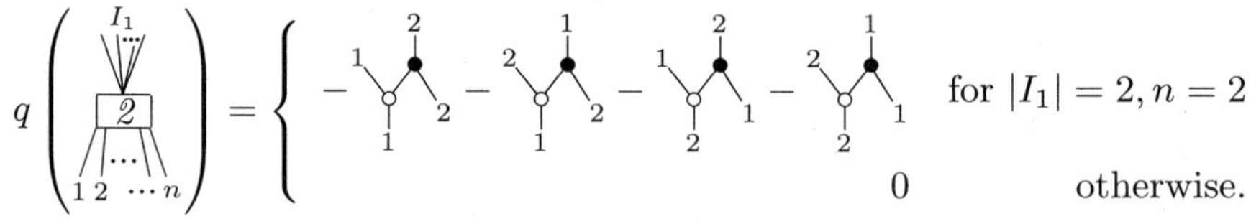

and identifying the r.h.s. with its natural lift, e', into $\mathsf{Lie^1B}_\infty$, we finally obtain

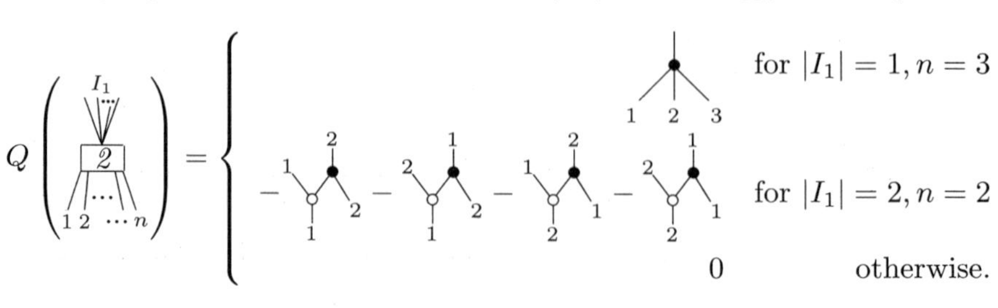

$$Q\left([\text{corolla } 2;\ I_1;\ 1\,2\cdots n]\right) = \begin{cases} \text{[corolla } 1\ 2\ 3] & \text{for } |I_1| = 1, n = 3 \\ -\text{[graph } 1,2;2;1] - \text{[graph } 2,1;2;1] - \text{[graph } 1,2;1;2] - \text{[graph } 2,1;1;2] & \text{for } |I_1| = 2, n = 2 \\ 0 & \text{otherwise.} \end{cases}$$

By Corollary II.2(ii), we can set

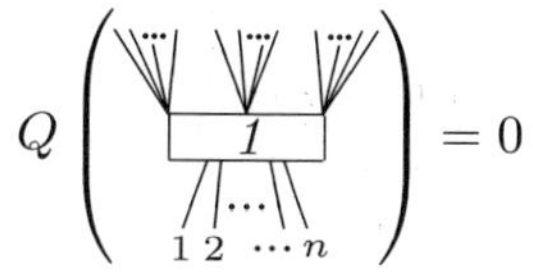
Q
1
1 2 ⋯ n
= 0

completing thereby the $s = 3$ iteration step.

Iteration level s=4: We have

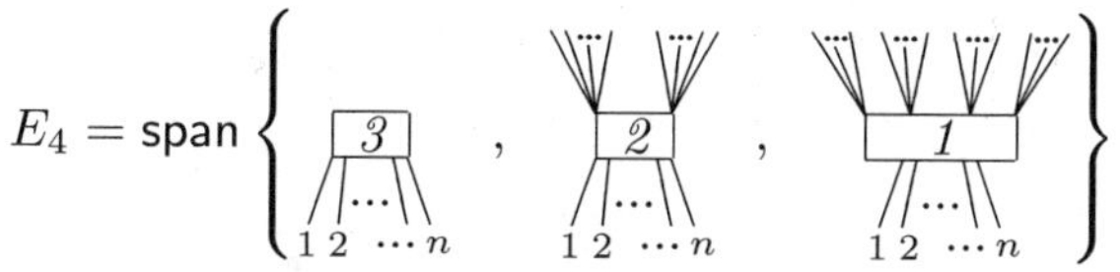
E_4 = span
3
2
1
1 2 ⋯ n

By Corollary II.1(i) the values of the morphism Q on corollas of the first type are completely determined by lifts of the values of the morphism q, i.e.

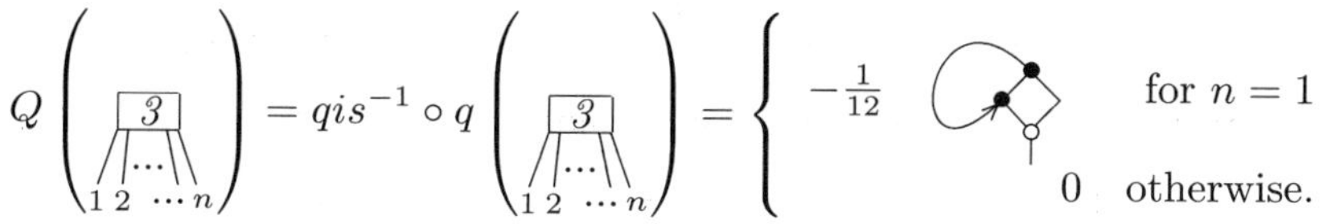
Q
3
1 2 ⋯ n
= qis^{-1} ∘ q
−1/12
for n = 1
0 otherwise.

It is also not hard to check that

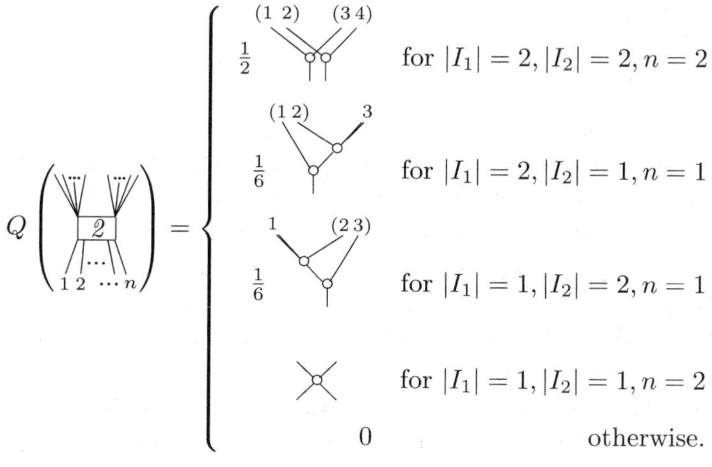
Q
2
1 2 ⋯ n
1/2
(1 2) (3 4)
for |I_1| = 2, |I_2| = 2, n = 2
1/6
(1 2) 3
for |I_1| = 2, |I_2| = 1, n = 1
1/6
1 (2 3)
for |I_1| = 1, |I_2| = 2, n = 1
for |I_1| = 1, |I_2| = 1, n = 2
0 otherwise.

Here round brackets stand for symmetrization of the labels.

By Corollary II.2(ii), we can set Q 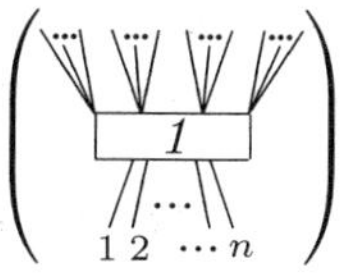 $= 0$ completing thereby the $s = 4$ iteration step.

Iteration level s=6: as a last but least trivial example we compute the value of the morphism Q on the generator

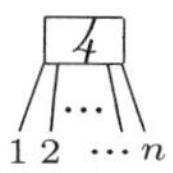
4
1 2 ⋯ n

$\in E_6$.

As the value of the morphism q on such a generator is zero, we have

$$Q\left(\begin{array}{c}\boxed{4}\\ 1\,2\,\cdots\,n\end{array}\right)=e'',$$

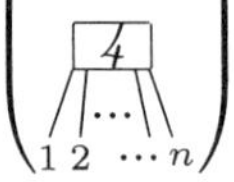

where e'' is a solution of the equation in $[\mathsf{Lie^1B}^{\circlearrowright}]_\infty$,

$$\delta e''=Q\left(\delta\begin{array}{c}\boxed{4}\\ 1\,2\,\cdots\,n\end{array}\right)=\sum_{b+c=4}\sum_{s\geq 0}\frac{1}{s!}\begin{array}{c}Q(\boxed{c})\\ s\\ Q(\boxed{b})\\ 1\,2\,\cdots\,n\end{array}$$

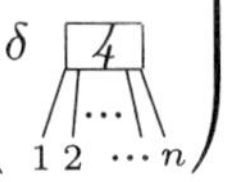

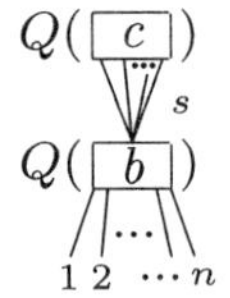

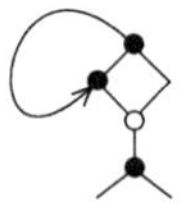

$$=\begin{cases}-\frac{1}{12}\,(\text{graph}) & \text{for } n=2\\ 0 & \text{otherwise.}\end{cases}$$

It is not hard to solve the latter for e'' and finally get

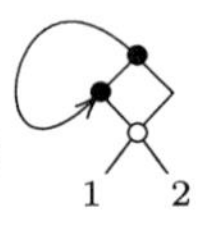

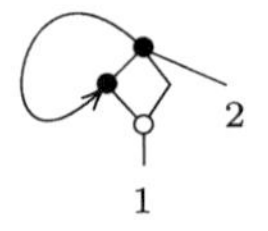

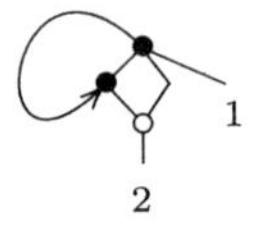

$$Q\left(\begin{array}{c}\boxed{4}\\ 1\,2\,\cdots\,n\end{array}\right)=\begin{cases}-\frac{1}{12}\,(\text{graph})_{1\,2}-\frac{1}{6}\,(\text{graph})_{1,2}-\frac{1}{6}\,(\text{graph})_{2,1} & \text{for } n=2\\ 0 & \text{otherwise.}\end{cases}$$

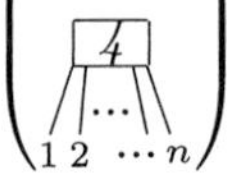

Step III. One can show by induction on the parameter $s=2a+k-2$ (we omit these details) that the map Q can be chosen so that the composition $Q\circ\chi_{\hbar=1}$ makes as a morphism into a completion of $[\mathsf{Lie^1B}^{\circlearrowright}]_\infty$ with respect to the genus and the number of vertices. We finally set $\hat{Q}:=Q\circ\chi_{\hbar=1}$ completing the construction. $\square$

5.3. Remark. There exists a canonical monomorphism of dg props, $i:\mathsf{Lie^1B}^{\circlearrowright}_\infty\to[\mathsf{Lie^1B}^{\circlearrowright}]_\infty$. Hence any morphism $\hat{Q}$ which factors through i gives rise to a universal quantization of ordinary Poisson structures. We shall discuss an analogous purely propic construction of such morphisms elsewhere.

5.4. Remark. The condition on the projection $\pi_1\circ\alpha\circ\hat{Q}$ in Theorem 5.2 implies high non-triviality of the quantization morphism $\hat{Q}$ in the sense that $\hat{Q}$ involves *all* possible jets of the polyvector part of the input wheeled Poisson structure.

Acknowledgement. I would like to thank Giuseppe Dito and an anonymous referee for careful reading of the manuscript and helpful remarks.

References

[D] V. Drinfeld, *On some unsolved problems in quantum group theory.* In: Lecture Notes in Math., Springer, **1510** (1992), 1-8.

[Ko1] M. Kontsevich. *Formal (non)commutative symplectic geometry.* In: The Gel'fand mathematics seminars 1990–1992, Birkhäuser, 1993.

[Ko2] M. Kontsevich, *Deformation quantization of Poisson manifolds*, Lett. Math. Phys. **66** (2003), 157-216

[McL] S. Mac Lane, Natural associativity and commutativity, Rice Univ. Stud. **49** (1963), 28-46.

[M] J.P. May, *The geometry if iterated loop spaces*, Lect. Notes in Math. **271**, Sprinder-Verlag, 1972.

[Me1] S.A. Merkulov, *Strong homotopy algebras of a Kähler manifold*, math.AG/9809172, Intern. Math. Res. Notices (1999),

[Me2] S.A. Merkulov, *Prop profile of Poisson geometry*, math.DG/0401034, Commun.Math.Phys. **262** (2006), 117-135.

[Me3] S.A. Merkulov, *Nijenhuis infinity and contractible dg manifolds*, math.AG/0403244, Compositio Mathematica **141** (2005), 1238-1254.

[Me4] S.A. Merkulov, *Prop profile of deformation quantization and graph complexes with loops and wheels*, unpublished preprint math.QA/0412257.

[Me5] S.A. Merkulov, *Deformation quantization of strongly homotopy Lie bialgebras*, preprint of Max-Planck Institute for Mathematics in Bonn, MPIM2006-151, 2006.

[Ta] D.E. Tamarkin, *Another proof of M. Kontsevich formality theorem*, math.QA/9803025.

[Va] B. Vallette, *A Koszul duality for props*, math.AT/0411542, to appear in Trans. AMS.

[SMM] S. Shadrin, S. Merkulov and M. Markl, Wheeled props and the master equation, math.AG/0610683.

[St] J.D. Stasheff, *On the homotopy associativity of H-spaces, I II*, Trans. Amer. Math. Soc. **108** (1963), 272-292 & 293-312.

Sergei A. Merkulov: Department of Mathematics, Stockholm University, 10691 Stockholm, Sweden

E-mail address: sm@math.su.se

Contemporary Mathematics
Volume **450**, 2008

Deformation quantization modules on complex symplectic manifolds

Pierre Schapira

Abstract. We study modules over the algebroid stack $\mathcal{W}_{\mathfrak{X}}$ of deformation quantization on a complex symplectic manifold $\mathfrak{X}$ and recall some results: construction of an algebra for $\star$-products, existence of (twisted) simple modules along smooth Lagrangian submanifolds, perversity of the complex of solutions for regular holonomic $\mathcal{W}_{\mathfrak{X}}$-modules, finiteness and duality for the composition of "good" kernels. As a corollary, we get that the derived category of good $\mathcal{W}_{\mathfrak{X}}$-modules with compact support is a Calabi-Yau category. We also give a conjectural Riemann-Roch type formula in this framework.

Introduction

Let X be a complex manifold, T^*X its cotangent bundle. The conic sheaf of $\mathbb{C}$-algebras $\mathcal{E}_{T^*X}$ of microdifferential operators on T^*X has been constructed functorially by Sato-Kashiwara-Kawai in [**25**]. This algebra is associated with the homogeneous symplectic structure and is also naturally defined on the projective cotangent bundle P^*X.

Another (no more conic) algebra on T^*X, denoted here by $\mathcal{W}_{T^*X}$ and defined over a subfield $\mathbf{k}$ of $\mathbb{C}[[\tau^{-1}, \tau]$ has been constructed in [**24**] (see [**5**] for related constructions). Its formal version has been considered by many authors after [**1**] and extended to Poisson manifolds in [**22**].

In general, neither the algebras $\mathcal{E}_{P^*X}$ glue on a complex contact manifold, nor the algebras $\mathcal{W}_{T^*X}$ glue on a complex symplectic manifold, although the categories of modules on these non existing algebras make sense. Indeed, one has to replace the notion of a sheaf of algebras by that of an algebroid stack, similarly as one replaces the notion of a sheaf by that of a stack. These constructions are performed in [**15**], [**21**], [**24**] (see also [**7**] for recent developments and [**4, 30, 31**] for an algebraic approach).

Here, we start by briefly recalling the constructions of the sheaves $\mathcal{E}_{T^*X}$ and $\mathcal{W}_{T^*X}$ as well as a new sheaf of algebras on T^*X containing $\mathcal{W}_{T^*X}$, invariant by quantized symplectic transformations, in which the $\star$-exponential is well defined (see [**10**]). Then we consider a complex symplectic manifold $\mathfrak{X}$, introduce the algebroid

---

2000 *Mathematics Subject Classification.* 46L65, 14A20, 32C38, 53D55.
Key words and phrases. Microdifferential operators, stacks, index theorem, deformation quantization.

stack $\mathcal{W}_{\mathfrak{X}}$ of deformation quantization on $\mathfrak{X}$ and discuss some recent results on $\mathcal{W}_{\mathfrak{X}}$-modules:

- If Λ is a smooth Lagrangian submanifold of $\mathfrak{X}$, there exist twisted simple $\mathcal{W}_{\mathfrak{X}}$-modules along Λ, the twist being associated with a square root of the line bundle Ω_Λ (see [**9**]).
- Let $\mathcal{L}_0$ and $\mathcal{L}_1$ be two regular holonomic modules supported by smooth Lagrangian submanifolds Λ_0 and Λ_1. Then the complex $R\mathcal{H}om_{\mathcal{W}_{\mathfrak{X}}}(\mathcal{L}_0, \mathcal{L}_1)$ is a perverse sheaf over the field $\mathbf{k}$ (see [**20**]).
- Let $\mathfrak{X}_i$ $(i = 1, 2, 3)$ be complex symplectic manifolds and denote by $\mathfrak{X}_i^a$ the symplectic manifold deduced from $\mathfrak{X}_i$ by taking the opposite symplectic form. Let $\mathcal{K}_i$ be a good $\mathcal{W}_{\mathfrak{X}_{i+1}\times\mathfrak{X}_i^a}$-module $(i = 1, 2)$ (good means coherent and endowed with a good filtration on each compact subset of $\mathfrak{X}$) and assume that a properness condition is satisfied by the supports of these modules. Then their composition $\mathcal{K}_2 \circ \mathcal{K}_1$ is a good $\mathcal{W}_{\mathfrak{X}_3\times\mathfrak{X}_1^a}$-module. Moreover, composition of kernels commutes with duality (see [**28**]). As a particular case, we obtain that the triangulated category consisting of good $\mathcal{W}_{\mathfrak{X}}$-modules with compact supports is Ext-finite over the field $\mathbf{k}$ and admits a Serre functor, namely the shift by $d_{\mathfrak{X}} := \dim_{\mathbb{C}} \mathfrak{X}$.
- The Hochschild homology of the algebroid stack $\mathcal{W}_{\mathfrak{X}}$ is concentrated in degree $-\dim_{\mathbb{C}} \mathfrak{X}$ and is isomorphic to $\mathbf{k}_{\mathfrak{X}}$. This allows us to construct the Euler class $\mathrm{Eu}(\mathcal{M}) \in H^{d_{\mathfrak{X}}}_{\mathrm{supp}\,\mathcal{M}}(\mathfrak{X}; \mathbf{k}_{\mathfrak{X}})$ of a coherent $\mathcal{W}_{\mathfrak{X}}$-module. We conjecture that in the situation above, $\mathrm{Eu}(\mathcal{K}_2 \circ \mathcal{K}_1)$ is the relative integral of the cup products $\mathrm{Eu}(\mathcal{K}_1) \cup \mathrm{Eu}(\mathcal{K}_2)$ (see [**28**]).

This paper summarizes various joint works with A. D'Agnolo [**9**], G. Dito [**10**], M. Kashiwara [**20**], P. Polesello [**24**] and J-P. Schneiders [**28**].

1. Microdifferential operators on cotangent bundles

Let X be a complex manifold, $\pi \colon T^*X \to X$ its cotangent bundle.

The ring $\mathcal{E}_{T^*X}$. The manifold T^*X is a complex *homogeneous* symplectic manifold, *i.e.*, T^*X is endowed with a canonical 1-form α_X such that $d\alpha_X$ is symplectic. On T^*X there exists a conic (*i.e.*, constant on the orbits of the action of $\mathbb{C}^\times$) sheaf $\mathcal{E}_{T^*X}$ constructed functorially by Sato-Kashiwara-Kawai [**25**] (see also [**16, 26**] for an exposition) which plays the role of a noncommutative localization of the ring $\mathcal{D}_X$ of differential operators. The sheaf $\mathcal{E}_{T^*X}$ enjoys the following properties:

- $\mathcal{E}_{T^*X}$ is a filtered sheaf of central $\mathbb{C}$-algebras and
$$\mathrm{gr}\, \mathcal{E}_{T^*X} \simeq \bigoplus_{j\in\mathbb{Z}} \mathcal{O}_{T^*X}(j).$$
($\mathcal{O}_{T^*X}(j)$ is the subsheaf of $\mathcal{O}_{T^*X}$ consisting of homogeneous functions of degree j in the fibers of π.)
- There is a flat monomorphism of filtered rings $\pi^{-1}\mathcal{D}_X \hookrightarrow \mathcal{E}_{T^*X}$.
- Denote by $\Omega_X^{\frac{1}{2}}$ the twisted sheaf of holomorphic half-forms of maximal degree on X (the notion of twisted sheaves will be recalled below). One defines the sheaf of algebras:

$$\mathcal{E}^{\sqrt{v}}_{T^*X} := \pi^{-1}\Omega_X^{\frac{1}{2}} \otimes_{\pi^{-1}\mathcal{O}_X} \mathcal{E}_{T^*X} \otimes_{\pi^{-1}\mathcal{O}_X} \pi^{-1}\Omega_X^{-\frac{1}{2}}. \tag{1.1}$$

(Note that $\mathcal{E}^{\sqrt{v}}_{T^*X}$ is a sheaf although $\Omega_X^{\frac{1}{2}}$ is not a sheaf but a twisted sheaf.) Denote by $a\colon T^*X \to T^*X$ the antipodal map, $(x;\xi) \mapsto (x;-\xi)$. There exists a $\mathbb{C}$-linear anti-isomorphism of sheaves of algebras[1] $a_*\mathcal{E}^{\sqrt{v}}_{T^*X} \xrightarrow{\sim} \mathcal{E}^{\sqrt{v}}_{T^*X}$ called the transposition and denoted $P \mapsto {}^tP$.

- Consider a $\mathbb{C}^\times$-homogeneous symplectic isomorphism $\varphi\colon T^*X \supset U \xrightarrow{\sim} V \subset T^*Y$. Then φ can be *locally* quantized as an isomorphism of filtered sheaf of rings commuting with the transposition
$$\Phi\colon \varphi_*\mathcal{E}^{\sqrt{v}}_{T^*X} \xrightarrow{\sim} \mathcal{E}^{\sqrt{v}}_{T^*Y}.$$
Denote by V^a the image of V by the antipodal map a on T^*Y and by $\Lambda_\varphi \subset U \times V^a$ the image of the graph of φ. This is a Lagrangian submanifold of $U \times V^a$. Locally, we may assume that Ω_X is trivial and there exists an ideal $\mathcal{I}_\varphi$ of $\mathcal{E}_{T^*(X\times Y)}$ whose associated graded ideal is reduced and coincides with the defining ideal of Λ_φ. Then, for each $\mathcal{E}_{T^*X} \ni P$ there exists a unique $Q \in \mathcal{E}_{T^*Y}$ such that $P - Q \in \mathcal{I}_\varphi$. The correspondence $P \mapsto Q$ is an anti-isomorphism of $\mathbb{C}$-algebras $\varphi_*\mathcal{E}_{T^*X} \xrightarrow{\sim} a_*\mathcal{E}_{T^*Y}$. One gets the isomorphism Φ by composing with the transposition in $\mathcal{E}_{T^*Y}$. One shall be aware that this isomorphism exists *only locally and is not unique* in general.

Moreover, when X is affine (*i.e.,* open in some n-dimensional complex vector space), $\mathcal{E}_{T^*X}$ satisfies:

- any section $P \in \mathcal{E}_{T^*X}(U)$ on an open subset $U \subset T^*X$ admits a total symbol
$$\sigma_{\rm tot}(P)(x;\xi) = \sum_{-\infty<j\le m} p_j(x;\xi),\ m\in\mathbb{Z} \quad p_j \in \mathcal{O}_{T^*X}(j)(U), \tag{1.2}$$
with the condition:
$$\begin{cases}\text{for any compact subset } K \text{ of } U \text{ there exists a positive constant } C_K\\ \text{such that } \sup\limits_K |p_j| \le C_K^{-j}(-j)! \text{ for all } j\le 0.\end{cases} \tag{1.3}$$
- The total symbol of the product is given by the Leibniz rule:
$$\sigma_{\rm tot}(P\circ Q) \;=\; \sum_{\alpha\in\mathbb{N}^n} \frac{1}{\alpha!}\partial_\xi^\alpha\sigma_{\rm tot}(P)\partial_x^\alpha\sigma_{\rm tot}(Q). \tag{1.4}$$
- The total symbol of the transposition is given by
$$\sigma_{\rm tot}({}^tP)(x;\xi) \;=\; \sum_{\alpha\in\mathbb{N}^n} \frac{(-1)^{|\alpha|}}{\alpha!}\partial_\xi^\alpha\partial_x^\alpha\sigma_{\rm tot}(P)(x;-\xi). \tag{1.5}$$

The field k. Let $\widehat{\mathbf{k}} := \mathbb{C}[[\tau^{-1},\tau]$ be the field of formal Laurent series in τ^{-1}. We consider the filtered subfield $\mathbf{k}$ of $\widehat{\mathbf{k}}$ consisting of series $a = \sum_{-\infty<j\le m} a_j\tau^j$ ($a_j\in\mathbb{C}$, $m\in\mathbb{Z}$) satisfying:

$$\text{there exists } C>0 \text{ such that } |a_j| \le C^{-j}(-j)! \text{ for all } j\le 0. \tag{1.6}$$

We denote by $\mathbf{k}_0$ the subring of $\mathbf{k}$ consisting of elements of order ≤ 0 and by $\mathbf{k}(r)$ the $\mathbf{k}_0$-module consisting of elements of order $\le r$.

[1] An anti-isomorphism of algebras $A \xrightarrow{\sim} B$ is an isomorphism of algebras $A \xrightarrow{\sim} B^{\rm op}$ where $B^{\rm op}$ is the opposite algebra.

We denote by ${}^t(\bullet)\colon \mathbf{k} \to \mathbf{k}$ the $\mathbb{C}$-linear automorphism of $\mathbf{k}$ induced by ${}^t(\tau) = -\tau$ and call it the transposition. We say that a $\mathbb{C}$-linear map $u\colon E \to F$ of $\mathbf{k}$-vector spaces is anti-$\mathbf{k}$-linear if it satisfies $u(a \cdot x) = {}^t a \cdot u(x)$ for any $x \in E$, $a \in \mathbf{k}$.

The ring $\mathcal{W}_{T^*X}$. On T^*X there exists a *no more conic* sheaf $\mathcal{W}_{T^*X}$ which enjoys the following properties:

- $\mathcal{W}_{T^*X}$ is a filtered sheaf of central $\mathbf{k}$-algebras and
$$\operatorname{gr} \mathcal{W}_{T^*X} \simeq \mathcal{O}_{T^*X}[\tau^{-1}, \tau].$$
- There is a faithful and flat monomorphism of filtered $\mathbb{C}$-algebras $\mathcal{E}_{T^*X} \hookrightarrow \mathcal{W}_{T^*X}$.
- Set

$$\mathcal{W}^{\sqrt{v}}_{T^*X} = \pi^{-1}\Omega_X^{\frac{1}{2}} \otimes_{\pi^{-1}\mathcal{O}_X} \mathcal{W}_{T^*X} \otimes_{\pi^{-1}\mathcal{O}_X} \pi^{-1}\Omega_X^{-\frac{1}{2}}. \tag{1.7}$$

  The sheaf of algebras $\mathcal{W}^{\sqrt{v}}_{T^*X}$ is endowed with an anti-$\mathbf{k}$-linear anti-automorphism $P \mapsto {}^tP$.
- Any symplectic isomorphism $\psi\colon T^*X \supset U \xrightarrow{\sim} V \subset T^*Y$ can be *locally* quantized as an isomorphism of filtered sheaves of $\mathbf{k}$-algebras commuting with the anti-$\mathbf{k}$-linear anti-isomorphism $P \mapsto {}^tP$:
$$\Psi\colon \psi_*\mathcal{W}^{\sqrt{v}}_{T^*X} \xrightarrow{\sim} \mathcal{W}^{\sqrt{v}}_{T^*Y}.$$
  (Again, this isomorphism Ψ exists *only locally and is not unique.*)

Moreover, when X is affine, $\mathcal{W}_{T^*X}$ satisfies:

- any section $P \in \mathcal{W}_{T^*X}(U)$ on an open subset $U \subset T^*X$ admits a total symbol

$$\sigma_{\mathrm{tot}}(P)(x;u,\tau) = \sum_{-\infty<j\leq m} p_j(x;u)\tau^j,\ m \in \mathbb{Z} \quad p_j \in \mathcal{O}_{T^*X}(U), \tag{1.8}$$

  with the condition:

$$\begin{cases} \text{for any compact subset } K \text{ of } U \text{ there exists a positive constant } C_K \\ \text{such that } \sup_K |p_j| \leq C_K^{-j}(-j)! \text{ for all } j \leq 0. \end{cases} \tag{1.9}$$

  Note that $\mathbf{k} = \mathcal{W}_{\mathrm{pt}}$.
- The total symbol of the product is given by the Leibniz rule:
$$\sigma_{\mathrm{tot}}(P \circ Q) \;=\; \sum_{\alpha\in\mathbb{N}^n} \frac{\tau^{-|\alpha|}}{\alpha!} \partial_u^\alpha \sigma_{\mathrm{tot}}(P) \partial_x^\alpha \sigma_{\mathrm{tot}}(Q).$$
- The total symbol of the transposition is given by

$$\sigma_{\mathrm{tot}}({}^tP)(x;u,\tau) \;=\; \sum_{\alpha\in\mathbb{N}^n} \frac{(-\tau)^{-|\alpha|}}{\alpha!} \partial_u^\alpha \partial_x^\alpha \sigma_{\mathrm{tot}}(P)(x;u,-\tau). \tag{1.10}$$

We denote by $\mathcal{W}_{T^*X}(0)$ the subsheaf of $\mathcal{W}_{T^*X}$ consisting of sections of order ≤ 0. Then $\mathcal{W}_{T^*X}(0)$ is a $\mathbf{k}_0$-algebra and there is a $\mathbf{k}$-linear isomorphism

$$\mathcal{W}_{T^*X}(0) \otimes_{\mathbf{k}_{0\,T^*X}} \mathbf{k}_{T^*X} \xrightarrow{\sim} \mathcal{W}_{T^*X}.$$

From $\mathcal{E}$ to $\mathcal{W}$. One can deduce the algebra $\mathcal{W}_{T^*X}$ from the algebra $\mathcal{E}_{T^*X}$. Let $t \in \mathbb{C}$ be the coordinate and set

$$\mathcal{E}_{T^*(X\times\mathbb{C}),\hat{t}} = \{P \in \mathcal{E}_{T^*(X\times\mathbb{C})};\ [P,\partial_t]=0\}.$$

Set $T^*_{\tau\neq 0}(X\times\mathbb{C}) = \{(x,t;\xi,\tau);\tau\neq 0\}$, and consider the map

$$\begin{array}{rcl}\rho\colon T^*_{\tau\neq 0}(X\times\mathbb{C}) & \to & T^*X, \\ (x,t;\xi,\tau) & \mapsto & (x;\xi/\tau).\end{array}$$

The ring $\mathcal{W}_{T^*X}$ on T^*X may be defined by setting (see [**24**]):

$$\mathcal{W}_{T^*X} := \rho_*(\mathcal{E}_{T^*(X\times\mathbb{C}),\hat{t}}|_{T^*_{\tau\neq 0}(X\times\mathbb{C})}).$$

REMARK 1.1. (i) Many authors use the parameter $\hbar$ instead of τ^{-1}.
(ii) There exist formal versions $\widehat{\mathcal{E}}_{T^*X}$ and $\widehat{\mathcal{W}}_{T^*X}$ of the sheaves $\mathcal{E}_{T^*X}$ and $\mathcal{W}_{T^*X}$, respectively, and most of the authors work with $\widehat{\mathcal{W}}_{T^*X}$.

Deformation quantization and exponential star products. If $P \in \mathcal{W}_{T^*X}$ has order 0, the operator $\exp\tau P$ does not exist in $\mathcal{W}_{T^*X}$. Using an extra central parameter t, a new $\mathbf{k}$-algebra $\mathcal{W}^t_{T^*X}$ on T^*X was constructed in [**10**]. This algebra enjoys the following properties:

(i) there is a monomorphism of $\mathbf{k}$-algebras $\iota\colon \mathcal{W}_{T^*X} \hookrightarrow \mathcal{W}^t_{T^*X}$ and a morphism res: $\mathcal{W}^t_{T^*X} \to \mathcal{W}_{T^*X}$ such that the composition $\mathcal{W}_{T^*X} \to \mathcal{W}^t_{T^*X} \to \mathcal{W}_{T^*X}$ is the identity,
(ii) any symplectic isomorphism $\psi : T^*X \supset U_X \xrightarrow{\sim} U_Y \subset T^*Y$ can be locally quantized as an isomorphism of $\mathbf{k}$-algebras $\Psi\colon \mathcal{W}^t_{T^*X} \xrightarrow{\sim} \mathcal{W}^t_{T^*Y}$,
(iii) for $P \in \mathcal{W}_{T^*X}(0)$, the section $\exp(t\tau P) = \sum_{n\geq 0}\dfrac{(t\tau P)^n}{n!}$ is well defined in $\mathcal{W}^t_{T^*X}$.

The algebra $\mathcal{W}^t_{T^*X}$ is constructed as follows.

Let s be a holomorphic coordinate on $\mathbb{C}$ and denote by $\mathcal{W}_{T^*(\mathbb{C}\times X),\widehat{\partial}_s}$ the subalgebra of $\mathcal{W}_{T^*(\mathbb{C}\times X)}$ consisting of sections which do not depend on ∂_s, *i.e.*, which commute with s. We look at $\mathcal{W}_{T^*(\mathbb{C}\times X),\widehat{\partial}_s}$ as a sheaf on $\mathbb{C}\times T^*X$ and we denote by $p\colon \mathbb{C}\times T^*X \to T^*X$ the projection. Set

$$\mathcal{W}^s_{T^*X} := R^1p_!(\mathcal{W}_{T^*(\mathbb{C}\times X),\widehat{\partial}_s}). \tag{1.11}$$

Hence, the sections of $\mathcal{W}^s_{T^*X}$ are sections of $\mathcal{W}_{T^*X}$ depending of an extra holomorphic parameter s defined for $|s| \gg 0$ modulo sections defined for all s. The convolution product in the s variable allows us to endow $\mathcal{W}^s_{T^*X}$ with a structure of an algebra.

The algebra $\mathcal{W}^t_{T^*X}$ is constructed as the Laplace transform of $\mathcal{W}^s_{T^*X}$ which interchanges s^{-n-1} and $\frac{(t\tau)^n}{n!}$. (Here, t and τ commute.)

Note that, if P has order 0, the section $s-P$ is invertible on each compact subset K of T^*X for $|s| \gg 0$. Therefore, $1/(s-P)$ is a well-defined section of $\mathcal{W}^s_{T^*X}$ and its Laplace transform $\exp(t\tau P)$ belongs to $\mathcal{W}^t_{T^*X}$.

2. Algebroid stacks

A local model for a complex symplectic manifold is an open subset of T^*X. Hence, it is natural to ask whether the construction of the sheaf of algebras $\mathcal{W}_{T^*X}$ still makes sense on complex symplectic manifolds. However, since the quantization of a symplectic isomorphism is not unique, one has to replace the notion of a sheaf of algebras by that of an algebroid stack, a notion introduced in [**21**]. We refer to [**8**] for a more systematic study and to [**19**] for an introduction to stacks.

In this section, $\mathbb{K}$ denotes a commutative unital algebra and X a topological space.

If A is a $\mathbb{K}$-algebra, we denote by A^+ the category with one object and having A as morphisms of this object. Let $\mathcal{A}$ be a sheaf of $\mathbb{K}$-algebras on X and consider the prestack $U \mapsto \mathcal{A}(U)^+$ (U open in X). We denote by $\mathcal{A}^+$ the associated stack and call $\mathcal{A}^+$ the algebroid stack associated with $\mathcal{A}$.

Consider an open covering $\mathcal{U} = \{U_i\}_{i\in I}$ of X, sheaves of $\mathbb{K}$-algebras $\mathcal{A}_i$ on U_i ($i \in I$) and isomorphisms $f_{ij} \colon \mathcal{A}_j|_{U_{ij}} \xrightarrow{\sim} \mathcal{A}_i|_{U_{ij}}$ ($i, j \in I$). The existence of a sheaf of $\mathbb{K}$-algebras $\mathcal{A}$ locally isomorphic to $\mathcal{A}_i$ requires the condition $f_{ij}f_{jk} = f_{ik}$ on triple intersections. Let us weaken this last condition by assuming that there exist invertible sections $a_{ijk} \in \mathcal{A}_i(U_{ijk})$ satisfying

$$\begin{cases} f_{ij}f_{jk} = \mathrm{Ad}(a_{ijk})f_{ik} \text{ on } U_{ijk}, \\ a_{ijk}a_{ikl} = f_{ij}(a_{jkl})a_{ijl} \text{ on } U_{ijkl}. \end{cases} \tag{2.1}$$

(Recall that $\mathrm{Ad}(a)(b) = a \cdot b \cdot a^{-1}$.) One calls

$$(\{\mathcal{A}_i\}_{i\in I}, \{f_{ij}\}_{i,j\in I}, \{a_{ijk}\}_{i,j,k\in I}) \tag{2.2}$$

a descent datum for $\mathbb{K}$-algebroid stacks on $\mathcal{U}$. For such a descent datum, we shall denote by $f_{ij}^+ \colon \mathcal{A}_j^+ \xrightarrow{\sim} \mathcal{A}_i^+$ the equivalences of stacks associated with the isomorphisms f_{ij}. The following result is stated (in a different form) in [**15**] and goes back to [**13**].

Theorem 2.1. *Consider a descent datum* (2.2) *on* $\mathcal{U}$. *Then there exist a stack* $\mathcal{A}^+$ *on* X, *equivalences of stacks* $\varphi_i \colon \mathcal{A}^+|_{U_i} \xrightarrow{\sim} \mathcal{A}_i^+$ *and isomorphisms of functors* $c_{ij} \colon f_{ij}^+ \xrightarrow{\sim} \varphi_i \circ \varphi_j^{-1}$ *satisfying* $c_{ij} \circ c_{jk} \circ a_{ijk} = c_{ik}$. *Moreover, the data* $(\mathcal{A}^+, \{\varphi_i\}_i, \{c_{ij}\}_{ij})$ *are unique up to equivalence of stacks, this equivalence being unique up to a unique isomorphism.*

One calls $\mathcal{A}^+$ an algebroid stack. Although $\mathcal{A}^+$ is not a sheaf of algebras, modules over $\mathcal{A}^+$ are well defined. They are described by pairs

$$\mathcal{M} = (\{\mathcal{M}_i\}_{i\in I}, \{\xi_{ij}\}_{i,j\in I}),$$

where $\mathcal{M}_i$ is an $\mathcal{A}_i$-module and $\xi_{ij} \colon {}_{f_{ji}}\mathcal{M}_j|_{U_{ij}} \to \mathcal{M}_i|_{U_{ij}}$ is an isomorphism of $\mathcal{A}_i$-modules such that for any $u_k \in \mathcal{M}_k$ one has

$$\xi_{ij}({}_{f_{ji}}\xi_{jk}(u_k)) = \xi_{ik}(a_{kji}^{-1}u_k). \tag{2.3}$$

Here, ${}_{f_{ji}}\mathcal{M}_j$ is the $\mathcal{A}_i$-module deduced from the $\mathcal{A}_j$-module $\mathcal{M}_j|_{U_{ij}}$ by the isomorphism f_{ji}.

One gets a Grothendieck category $\mathrm{Mod}(\mathcal{A}^+)$ and the prestack $\mathfrak{Mod}(\mathcal{A}^+)$ given by $U \mapsto \mathrm{Mod}(\mathcal{A}^+|_U)$ is a stack equivalent on U_i to the stack $\mathfrak{Mod}(\mathcal{A}_i)$.

Twisted sheaves. As a particular case of a module over an algebroid stack, one has the notion of a twisted sheaf. Assume that $\mathbb{K}$ is a field and denote by $\mathbb{K}^\times$ the group of its invertible elements.

Let X be a manifold and let $\mathbf{c} \in H^2(X; \mathbb{K}^\times)$. Represent $\mathbf{c}$ by a Čech cocycle $\{c_{ijk}\}_{i,j,k\in I}$ associated to an open covering $\mathcal{U} = \{U_i\}_{i\in I}$ of X. We thus get a descent datum for $\mathbb{K}$-algebroid stacks

$$\mathbb{K}_{X,\mathbf{c}} := (\{\mathbb{K}_{U_i}\}_{i\in I}, \{\mathrm{id}_{\mathbb{K}_{U_{ij}}}\}_{i,j\in I}, \{c_{ijk}\}_{i,j,k\in I}),$$

and cohomologous cocycles give equivalent stacks.

EXAMPLE 2.2. Assume now that X is a complex manifold. Consider the short exact sequence

$$1 \to \mathbb{C}_X^\times \to \mathcal{O}_X^\times \xrightarrow{d\log} d\mathcal{O}_X \to 0$$

which gives rise to the long exact sequence

$$H^1(X;\mathbb{C}_X^\times) \xrightarrow{\alpha} H^1(X;\mathcal{O}_X^\times) \xrightarrow{\beta} H^1(X;d\mathcal{O}_X) \xrightarrow{\gamma} H^2(X;\mathbb{C}_X^\times).$$

If $\mathcal{L}$ is a line bundle, it defines a class $[\mathcal{L}] \in H^1(X;\mathcal{O}_X^\times)$. For $\lambda \in \mathbb{C}$, one sets

$$\mathbf{c}_{\mathcal{L}}^\lambda = \gamma(\lambda \cdot \beta([\mathcal{L}])) \in H^2(X;\mathbb{C}_X^\times).$$

We shall apply this construction when $\mathcal{L} = \Omega_X$ and $\lambda = \frac{1}{2}$ and set for short:

$$\mathrm{Mod}(\mathbb{C}_{X,\frac{1}{2}}) = \mathrm{Mod}(\mathbb{C}_{X,\mathbf{c}_{\Omega_X}^{\frac{1}{2}}}).$$

3. Quantization of symplectic manifolds

On any complex contact manifold, the existence of a canonical $\mathbb{C}$-algebroid stack locally equivalent to the algebroid stack associated with the sheaf of algebras of microdifferential operators of [**25**] has been obtained by M. Kashiwara in [**15**].

On any complex Poisson manifold, the existence of a $\widehat{\mathbf{k}}$-algebroid stack of formal deformation quantization has been obtained by M. Kontsevich [**21**]. The analytic case on symplectic manifolds has been obtained in [**24**] by a different method, making a link with Kashiwara's construction. The classification of these algebroid stacks is discussed in [**23**].

In particular, for a complex symplectic manifold $\mathfrak{X}$, there is a canonical $\mathbf{k}$-algebroid stack $\mathcal{W}_{\mathfrak{X}}^{\sqrt{v},+}$ locally equivalent to the algebroid stack $\mathcal{W}_{T^*X}^{\sqrt{v},+}$ associated with the sheaf of algebras $\mathcal{W}_{T^*X}^{\sqrt{v}}$. The same result holds with $\mathcal{W}_{\mathfrak{X}}$ replaced by $\mathcal{W}_{\mathfrak{X}}(0)$ and $\mathbf{k}$ by $\mathbf{k}_0$.

NOTATION 3.1. For short, as far as there is no risk of confusion, we shall write $\mathcal{W}_{\mathfrak{X}}$ instead of $\mathcal{W}_{\mathfrak{X}}^{\sqrt{v},+}$.

Let $\mathfrak{X}$ be a complex symplectic manifold. Then $\mathrm{Mod}(\mathcal{W}_{\mathfrak{X}})$ is a Grothendieck category. We denote by $\mathrm{D^b}(\mathcal{W}_{\mathfrak{X}})$ its bounded derived category and call an object of this derived category a $\mathcal{W}_{\mathfrak{X}}$-module. One proves as usual that the sheaf of algebras $\mathcal{W}_{T^*X}$ is coherent and the support of a coherent $\mathcal{W}_{T^*X}$-module is a closed complex analytic subvariety of T^*X. This support is involutive in view of Gabber's theorem (see [**16**, Th. 7.33]). Hence, the (local) notions of a coherent or holonomic $\mathcal{W}_{\mathfrak{X}}$-module make sense.

Similarly as for $\mathcal{D}$-modules (see [**16**]), one says that a coherent $\mathcal{W}_{\mathfrak{X}}$-module $\mathcal{M}$ is good if, for any open relatively compact subset U of $\mathfrak{X}$, there exists a coherent $\mathcal{W}_{\mathfrak{X}}(0)|_U$-module $\mathcal{M}_0$ contained in $\mathcal{M}|_U$ which generates $\mathcal{M}|_U$.

Let us denote by:

- $\mathrm{D^b_{coh}}(\mathcal{W}_{\mathfrak{X}})$ the full triangulated subcategory of $\mathrm{D^b}(\mathcal{W}_{\mathfrak{X}})$ consisting of objects with coherent cohomologies.
- $\mathrm{D^b_{gd}}(\mathcal{W}_{\mathfrak{X}})$ the full triangulated subcategory of $\mathrm{D^b_{coh}}(\mathcal{W}_{\mathfrak{X}})$ consisting of objects with good cohomologies.

- $D^b_{gd,c}(\mathcal{W}_{\mathfrak{X}})$ the full triangulated subcategory of $D^b_{gd}(\mathcal{W}_{\mathfrak{X}})$ consisting of objects with compact supports.
- $D^b_{hol}(\mathcal{W}_{\mathfrak{X}})$ the full triangulated subcategory of $D^b_{coh}(\mathcal{W}_{\mathfrak{X}})$ consisting of objects with Lagrangian supports in $\mathfrak{X}$. (One calls such an object an holonomic $\mathcal{W}_{\mathfrak{X}}$-module.)
- $D^b_{rh}(\mathcal{W}_{\mathfrak{X}})$ the full triangulated subcategory of $D^b_{hol}(\mathcal{W}_{\mathfrak{X}})$ consisting of objects with regular holonomic cohomologies (to be defined below).
- Let $\mathfrak{X}$ and $\mathfrak{Y}$ be two complex symplectic manifolds and let $\mathcal{M} \in D^b(\mathcal{W}_{\mathfrak{X}})$, $\mathcal{N} \in D^b(\mathcal{W}_{\mathfrak{Y}})$. Their exterior product is given by $\mathcal{M}\underline{\boxtimes}\mathcal{N} := \mathcal{W}_{\mathfrak{X}\times\mathfrak{Y}} \boxtimes_{\mathcal{W}_{\mathfrak{X}}\boxtimes\mathcal{W}_{\mathfrak{Y}}} (\mathcal{M}\boxtimes\mathcal{N})$.

Simple $\mathcal{W}_{\mathfrak{X}}$-modules.

DEFINITION 3.2. Let Λ be a smooth Lagrangian submanifold of $\mathfrak{X}$.

(a) Let $\mathcal{L}(0)$ be a coherent $\mathcal{W}_{\mathfrak{X}}(0)$-module supported by Λ. One says that $\mathcal{L}(0)$ is simple along Λ if $\mathcal{L}(0)/\mathcal{L}(-1)$ is an invertible $\mathcal{O}_{\Lambda}$-module. Here, $\mathcal{L}(-1) = \mathbf{k}_{\mathfrak{X}}(-1)\mathcal{L}(0)$.
(b) Let $\mathcal{L}$ be a coherent $\mathcal{W}_{\mathfrak{X}}$-module supported by Λ. One says that $\mathcal{L}$ is simple along Λ if there locally exists a coherent $\mathcal{W}_{\mathfrak{X}}(0)$-submodule $\mathcal{L}(0)$ of $\mathcal{L}$ such that $\mathcal{L}(0)$ generates $\mathcal{L}$ over $\mathcal{W}_{\mathfrak{X}}$ and is simple along Λ.
(c) Let $\mathcal{L}$ be a coherent $\mathcal{W}_{\mathfrak{X}}$-module supported by Λ. One says that $\mathcal{L}$ is regular if, locally, it is a finite direct sum of simple modules.
(d) Let Λ be a, not necessarily smooth, Lagrangian subvariety of $\mathfrak{X}$. A coherent $\mathcal{W}_{\mathfrak{X}}$-module supported by Λ is regular if it is regular at generic points of Λ. One calls such an object a regular holonomic $\mathcal{W}_{\mathfrak{X}}$-module.

It follows from Gabber's theorem that when Λ is smooth, Definitions 3.2 (c) and (d) coincide (see [**16**, Th. 8.34]).

One proves easily that any two $\mathcal{W}_{\mathfrak{X}}$-modules simple along Λ are locally isomorphic and that if $\mathcal{L}_i$ $(i = 0, 1)$ are simple along Λ, then $R\mathcal{H}om_{\mathcal{W}_{\mathfrak{X}}}(\mathcal{L}_0, \mathcal{L}_1)$ is concentrated in degree 0 and is a $\mathbf{k}$-local system of rank one on Λ.

EXAMPLE 3.3. Let X be a complex manifold. We denote by $\mathcal{O}^{\tau}_X$ the $\mathcal{W}_{T^*X}$-module supported by the zero-section T^*_XX defined by $\mathcal{O}^{\tau}_X = \mathcal{W}_{T^*X}/\mathcal{I}$, where $\mathcal{I}$ is the left ideal generated by the vector fields which annihilate the section $1 \in \mathcal{O}_X$. A section $f(x,\tau)$ of this module may be written as a series:

$$f(x,\tau) = \sum_{-\infty<j\leq m} f_j(x)\tau^j, \quad m \in \mathbb{Z}, \tag{3.1}$$

the f_j's satisfying Condition (1.9). Then $\mathcal{O}^{\tau}_X$ is a simple $\mathcal{W}_{T^*X}$-module along T^*_XX.

The next result asserts that, up to a twist, there exist globally defined simple $\mathcal{W}_{\mathfrak{X}}$-modules.

THEOREM 3.4. [**9**] *Let Λ be a smooth Lagrangian submanifold of $\mathfrak{X}$. There is an equivalence of* $\mathbf{k}$*-additive stacks:*

$$\mathfrak{Mod}_{reg\text{-}\Lambda}(\mathcal{W}_{\mathfrak{X}})|_{\Lambda} \simeq \mathfrak{Mod}_{loc\text{-}sys}(\mathbf{k}_{\Lambda} \otimes_{\mathbb{C}} \mathbb{C}_{\Lambda,1/2}). \tag{3.2}$$

Here, the left-hand side is the substack of $\mathfrak{Mod}(\mathcal{W}_{\mathfrak{X}})|_{\Lambda}$ consisting of regular holonomic modules along Λ and the right-hand side is the substack of the stack of twisted sheaves of $\mathbf{k}_{\Lambda}$-modules with twist $\mathbb{C}_{\Lambda,1/2}$ consisting of objects locally isomorphic to local systems over $\mathbf{k}$. The proof uses the corresponding theorem for contact manifolds due to Kashiwara.

$\mathcal{W}_{\mathfrak{X}}$-module associated with the diagonal. Let $\mathfrak{X}$ be a complex symplectic manifold. We denote by $\mathfrak{X}^a$ the complex manifold $\mathfrak{X}$ endowed with the symplectic form $-\omega$, where ω is the symplectic form on $\mathfrak{X}$. There is a natural equivalence of algebroid stacks $\mathcal{W}_{\mathfrak{X}}^{\mathrm{op}} \simeq \mathcal{W}_{\mathfrak{X}^a}$.

We denote by $\Delta_{\mathfrak{X}}$ the diagonal of $\mathfrak{X} \times \mathfrak{X}^a$ and by $d_{\mathfrak{X}}$ the complex dimension of $\mathfrak{X}$.

THEOREM 3.5. (i) *There exists a simple $\mathcal{W}_{\mathfrak{X}\times\mathfrak{X}^a}$-module $\mathcal{C}_{\Delta_{\mathfrak{X}}}$ supported by the diagonal $\Delta_{\mathfrak{X}}$ of $\mathfrak{X} \times \mathfrak{X}^a$ with the property that if U is open in $\mathfrak{X}$ and isomorphic to an open subset V of a cotangent bundle T^*X, then $\mathcal{C}_{\Delta_{\mathfrak{X}}}|_U$ is isomorphic to $\mathcal{W}_{T^*X}|_V$ as a $\mathcal{W}_{T^*X} \otimes \mathcal{W}_{T^*X}^{\mathrm{op}}$-module.*

(ii) *There is a natural isomorphism $\mathcal{C}_{\Delta_{\mathfrak{X}^a}} \overset{\mathrm{L}}{\otimes}_{\mathcal{W}_{\mathfrak{X}\times\mathfrak{X}^a}} \mathcal{C}_{\Delta_{\mathfrak{X}}} \simeq \mathbf{k}_{\Delta_{\mathfrak{X}}}\,[d_{\mathfrak{X}}]$.*

(i) follows from general considerations on algebroid stacks. (ii) follows from a construction of Feigin and Tsygan [**12**] (see also [**11**]).

Let $\mathcal{M} \in \mathrm{D}^{\mathrm{b}}(\mathcal{W}_{\mathfrak{X}})$. We set

$$\mathrm{D}'_{\mathrm{w}}\mathcal{M} := R\mathcal{H}om_{\mathcal{W}_{\mathfrak{X}}}(\mathcal{M}, \mathcal{C}_{\Delta_{\mathfrak{X}}}), \quad \mathrm{D}_{\mathrm{w}}\mathcal{M} := \mathrm{D}'_{\mathrm{w}}\mathcal{M}\,[\frac{1}{2}d_{\mathfrak{X}}]. \tag{3.3}$$

These objects are well-defined in $\mathrm{D}^{\mathrm{b}}(\mathcal{W}_{\mathfrak{X}^a})$. Using Theorem 3.5 and the fact that $\mathcal{C}_{\Delta_{\mathfrak{X}}}$ is simple, one gets an isomorphism in $\mathrm{D}^{\mathrm{b}}_{\mathrm{gd}}(\mathcal{W}_{\mathfrak{X}^a\times\mathfrak{X}})$:

$$\mathrm{D}_{\mathrm{w}}(\mathcal{C}_{\Delta_{\mathfrak{X}}}) \simeq \mathcal{C}_{\Delta_{\mathfrak{X}^a}}. \tag{3.4}$$

REMARK 3.6. By extending the definition of Van den Bergh [**29**] (see also [**32**]) to algebroid stacks, the isomorphism (3.4) could be translated by saying that $\mathcal{C}_{\Delta_{\mathfrak{X}}}\,[d_{\mathfrak{X}}]$ is a rigid dualizing complex over $\mathcal{W}_{\mathfrak{X}}$.

Let $\mathcal{M}, \mathcal{N}$ be two objects of $\mathrm{D}^{\mathrm{b}}_{\mathrm{coh}}(\mathcal{W}_{\mathfrak{X}})$. Then, after identifying $\Delta_{\mathfrak{X}}$ with $\mathfrak{X}$ by the first projection, there is a natural isomorphism in $\mathrm{D}^{\mathrm{b}}(\mathbf{k}_{\mathfrak{X}})$:

$$R\mathcal{H}om_{\mathcal{W}_{\mathfrak{X}}}(\mathcal{M}, \mathcal{N}) \simeq R\mathcal{H}om_{\mathcal{W}_{\mathfrak{X}\times\mathfrak{X}^a}}(\mathcal{M}\underline{\boxtimes}\mathrm{D}'_{\mathrm{w}}\mathcal{N}, \mathcal{C}_{\Delta_{\mathfrak{X}}}).$$

4. Constructibility and perversity

We refer to [**18**] for basic notions on sheaves. Let Z be a real analytic manifold. Recall that one denotes by:

- $C(S_1, S_2)$ the normal cone of two subsets S_1 and S_2 of Z, a closed conic subset of the tangent space TZ, identified with a subset of T^*Z in case Z is symplectic,
- $\mathrm{D}^{\mathrm{b}}(\mathbf{k}_Z)$ the bounded derived category of sheaves of $\mathbf{k}$-modules on Z,
- $\mathrm{SS}(F)$ the microsupport of an object $F \in \mathrm{D}^{\mathrm{b}}(\mathbf{k}_Z)$, a closed conic involutive subset of T^*Z,
- or_Z the orientation sheaf on Z, ω_Z the dualizing complex (hence, $\omega_Z \simeq \mathrm{or}_Z\,[\dim_{\mathbb{R}} Z]$), $\mathrm{D}_Z := R\mathcal{H}om_{\mathbf{k}_Z}(\,\bullet\,, \omega_Z)$ the duality functor for sheaves,
- $\mathrm{D}^{\mathrm{b}}_{\mathbb{R}\mathrm{c}}(\mathbf{k}_Z)$ the full triangulated subcategory of $\mathrm{D}^{\mathrm{b}}(\mathbf{k}_Z)$ consisting of objects with $\mathbb{R}$-constructible cohomology and, in case Z is a complex manifold, $\mathrm{D}^{\mathrm{b}}_{\mathbb{C}\mathrm{c}}(\mathbf{k}_Z)$ the full triangulated subcategory of $\mathrm{D}^{\mathrm{b}}(\mathbf{k}_Z)$ consisting of objects with $\mathbb{C}$-constructible cohomology.

THEOREM 4.1. [**20**] *Let $\mathfrak{X}$ be a complex symplectic manifold and let $\mathcal{L}_i$ $(i = 0, 1)$ be two objects of $\mathrm{D}^{\mathrm{b}}_{\mathrm{rh}}(\mathcal{W}_{\mathfrak{X}})$ supported by smooth Lagrangian manifolds Λ_i. Then*

(i) *the object* $R\mathcal{H}om_{\mathcal{W}_{\mathfrak{X}}}(\mathcal{L}_1, \mathcal{L}_0)$ *belongs to* $\mathrm{D}^{\mathrm{b}}_{\mathbb{C}\mathrm{c}}(\mathbf{k}_{\mathfrak{X}})$ *and its microsupport is contained in the normal cone* $C(\Lambda_0, \Lambda_1)$,

(ii) *the natural morphism*

$$R\mathcal{H}om_{\mathcal{W}_{\mathfrak{X}}}(\mathcal{L}_1, \mathcal{L}_0) \to \mathrm{D}_{\mathfrak{X}}\big(R\mathcal{H}om_{\mathcal{W}_{\mathfrak{X}}}(\mathcal{L}_0, \mathcal{L}_1\,[d_{\mathfrak{X}}])\big)$$

is an isomorphism.

The proof makes use of tools from the theory of holonomic $\mathcal{D}$-modules and uses some functional analysis, namely Houzel's theorem [**14**, 2,§ 4 Th. 1'].

REMARK 4.2. In [**2**], K. Behrend and B. Fantecci construct complexes (over $\mathbb{C}$) naturally associated with the data of two smooth Lagrangian submanifolds. Their result should have some relations with Theorem 4.1.

COROLLARY 4.3. *Let* $\mathcal{L}_0$ *and* $\mathcal{L}_1$ *be two regular holonomic* $\mathcal{W}_{\mathfrak{X}}$*-modules supported by smooth Lagrangian manifolds. Then the object* $R\mathcal{H}om_{\mathcal{W}_{\mathfrak{X}}}(\mathcal{L}_1, \mathcal{L}_0)$ *of* $\mathrm{D}^{\mathrm{b}}_{\mathbb{C}\mathrm{c}}(\mathbf{k}_{\mathfrak{X}})$ *is perverse.*

CONJECTURE 4.4. [**20**] Theorem 4.1 remains true without assuming that the Λ_i's are smooth.

5. Composition of kernels and Calabi-Yau categories

Consider three complex symplectic manifolds $\mathfrak{X}_i$ $(i = 1, 2, 3)$ and denote as usual by p_i and p_{ji} the projections defined on $\mathfrak{X}_3 \times \mathfrak{X}_2 \times \mathfrak{X}_1$.

For Λ_i a closed subset of $\mathfrak{X}_{i+1} \times \mathfrak{X}_i$ $(i = 1, 2)$, we set

$$\Lambda_2 \circ \Lambda_1 := p_{31}(p_{32}^{-1}\Lambda_2 \cap p_{21}^{-1}\Lambda_1). \tag{5.1}$$

For $\mathcal{K}_i \in \mathrm{D}^{\mathrm{b}}_{\mathrm{gd}}(\mathcal{W}_{\mathfrak{X}_{i+1}\times\mathfrak{X}_i^a})$ $(1 \le i \le 2)$, we set

$$\mathcal{K}_2 \circ \mathcal{K}_1 := Rp_{31!}(p_{32}^{-1}\mathcal{K}_2 \overset{\mathrm{L}}{\otimes}_{p_2^{-1}\mathcal{W}_{\mathfrak{X}_2}} p_{21}^{-1}\mathcal{K}_1). \tag{5.2}$$

THEOREM 5.1. [**28**] *Assume that* p_{31} *is proper on* $p_{32}^{-1}\operatorname{supp}(\mathcal{K}_2) \cap p_{12}^{-1}\operatorname{supp}(\mathcal{K}_1)$. *Then*

(i) *the object* $\mathcal{K}_2 \circ \mathcal{K}_1$ *belongs to* $\mathrm{D}^{\mathrm{b}}_{\mathrm{gd}}(\mathcal{W}_{\mathfrak{X}_3\times\mathfrak{X}_1^a})$,

(ii) *there is a natural isomorphism in* $\mathrm{D}^{\mathrm{b}}_{\mathrm{gd}}(\mathcal{W}_{\mathfrak{X}_3^a\times\mathfrak{X}_1})$:

$$\mathrm{D_w}\mathcal{K}_2 \circ \mathrm{D_w}\mathcal{K}_1 \xrightarrow{\sim} \mathrm{D_w}(\mathcal{K}_2 \circ \mathcal{K}_1). \tag{5.3}$$

The proof of (i) uses again [**14**, 2,§ 4 Th. 1']. The construction of the duality morphism in (ii) uses the isomorphism (3.4).

Choosing $\mathfrak{X}_3 = \mathfrak{X}_1 = \{\mathrm{pt}\}$ in Theorem 5.1, we get:

COROLLARY 5.2. *Let* $\mathfrak{X}$ *be a complex symplectic manifold and let* $\mathcal{M}$ *and* $\mathcal{N}$ *be two objects of* $\mathrm{D}^{\mathrm{b}}_{\mathrm{gd}}(\mathcal{W}_{\mathfrak{X}})$. *Assume that* $\operatorname{supp}(\mathcal{M}) \cap \operatorname{supp}(\mathcal{N})$ *is compact. Then*

(i) *the object* $\mathrm{RHom}_{\mathcal{W}_{\mathfrak{X}}}(\mathcal{M}, \mathcal{N})$ *has* $\mathbf{k}$*-finite dimensional cohomology,*

(ii) *there is a natural isomorphism, functorial with respect to* $\mathcal{M}$ *and* $\mathcal{N}$:

$$\mathrm{RHom}_{\mathcal{W}_{\mathfrak{X}}}(\mathcal{M}, \mathcal{N}) \simeq \big(\mathrm{RHom}_{\mathcal{W}_{\mathfrak{X}}}(\mathcal{N}, \mathcal{M}\,[d_{\mathfrak{X}}])\big)^{\star},$$

where * *is the duality functor for* $\mathbf{k}$*-vector spaces.*

Recall [**3**] that a $\mathbf{k}$-triangulated category $\mathcal{T}$ is *Ext*-finite if for any two objects F, G of $\mathcal{T}$, the $\mathbf{k}$-vector space $\bigoplus_{i\in\mathbb{Z}} \mathrm{Hom}_{\mathcal{T}}(F, G[i])$ is finite dimensional. In this situation, a Serre functor $S\colon \mathcal{T} \to \mathcal{T}$ is an equivalence of $\mathbf{k}$-triangulated categories such that

$$(\mathrm{Hom}_{\mathcal{T}}(F,G))^\star \simeq \mathrm{Hom}_{\mathcal{T}}(G, S(F))$$

functorially in F and G. If the Serre functor is a shift by an integer d, one says that $\mathcal{T}$ is a Calabi-Yau category of dimension d.

COROLLARY 5.3. *Let $\mathfrak{X}$ be a complex symplectic manifold. Then $\mathrm{D}^{\mathrm{b}}_{\mathrm{gd,c}}(\mathcal{W}_{\mathfrak{X}})$ is a Calabi-Yau category of dimension $d_{\mathfrak{X}}$.*

REMARK 5.4. (i) The analogue of Corollary 5.2 on complex contact manifolds over the field $\mathbb{C}$ is false. Note that Corollary 5.2 may be considered as a direct image theorem over one point, and a manifold of dimension 0 has a complex symplectic structure, not a complex contact structure.

Nevertheless, the analogue of Corollary 5.2 is true over the field $\mathbb{C}$ for complex contact manifolds when restricting to the category of regular holonomic modules. This follows from results obtained in [**17**].

(ii) On a complex compact manifold X, the bounded derived category of sheaves with $\mathbb{C}$-constructible cohomology is an *Ext*-finite triangulated category, as well as the equivalent category, the bounded derived category of $\mathcal{D}_X$-modules with regular holonomic cohomology. Both categories do not seem to have a Serre functor.

6. Index theorem

In this section, we announce works in progress with J-P. Schneiders [**28**].

Euler class. Let $\mathfrak{X}$ be complex symplectic manifold and let $\mathcal{M} \in \mathrm{D}^{\mathrm{b}}_{\mathrm{coh}}(\mathcal{W}_{\mathfrak{X}})$. We have the chain of morphisms

$$\begin{aligned} R\mathcal{H}om_{\mathcal{W}_{\mathfrak{X}}}(\mathcal{M},\mathcal{M}) &\xleftarrow{\sim} R\mathcal{H}om_{\mathcal{W}_{\mathfrak{X}}}(\mathcal{M},\mathcal{C}_{\Delta_{\mathfrak{X}}})\overset{\mathrm{L}}{\otimes}_{\mathcal{W}_{\mathfrak{X}}}\mathcal{M} \\ &\simeq (R\mathcal{H}om_{\mathcal{W}_{\mathfrak{X}}}(\mathcal{M},\mathcal{C}_{\Delta_{\mathfrak{X}}}) \otimes_{\mathbf{k}_{\mathfrak{X}}} \mathcal{M})\overset{\mathrm{L}}{\otimes}_{\mathcal{W}_{\mathfrak{X}}\otimes\mathcal{W}_{\mathfrak{X}^a}}\mathcal{C}_{\Delta_{\mathfrak{X}}} \\ &\to \mathcal{C}_{\Delta_{\mathfrak{X}^a}}\overset{\mathrm{L}}{\otimes}_{\mathcal{W}_{\mathfrak{X}\times\mathfrak{X}^a}}\mathcal{C}_{\Delta_{\mathfrak{X}}} \simeq \mathbf{k}_{\mathfrak{X}}\,[d_{\mathfrak{X}}]. \end{aligned}$$

Here, we have used Theorem 3.5 (ii). We get a map

$$\mathrm{Hom}_{\mathcal{W}_{\mathfrak{X}}}(\mathcal{M},\mathcal{M}) \to H^{d_{\mathfrak{X}}}_{\mathrm{supp}(\mathcal{M})}(\mathfrak{X};\mathbf{k}_{\mathfrak{X}}).$$

The image of $\mathrm{id}_{\mathcal{M}}$ gives an element

$$\mathrm{Eu}(\mathcal{M}) \in H^{d_{\mathfrak{X}}}_{\mathrm{supp}(\mathcal{M})}(\mathfrak{X};\mathbf{k}_{\mathfrak{X}}). \tag{6.1}$$

Symplectic Riemann-Roch theorem. We consider the situation of Theorem 5.1. Hence, we have three complex symplectic manifolds $\mathfrak{X}_i$ $(i=1,2,3)$ and we have closed subsets Λ_i of $\mathfrak{X}_{i+1}\times\mathfrak{X}_i$ $(i=1,2)$. We set for short $d_i := d_{\mathfrak{X}_i}$ $(i=1,2,3)$ and consider cohomology classes $\lambda_i \in H^{d_{i+1}+d_i}_{\Lambda_{i+1}\times\Lambda_i}(\mathfrak{X}_{i+1}\times\mathfrak{X}_i;\mathbf{k}_{\mathfrak{X}_{i+1}\times\mathfrak{X}_i})$ $(i=1,2)$. Assuming that p_{31} is proper on $p_{32}^{-1}(\Lambda_2)\cap p_{12}^{-1}(\Lambda_1)$, we set

$$\lambda_2\circ\lambda_1 \;:=\; \int_{\mathfrak{X}_2}(p_{32}^{-1}\lambda_2 \cup p_{21}^{-1}\lambda_1) \in H^{d_3+d_1}_{\Lambda_2\circ\Lambda_1}(\mathfrak{X}_3\times\mathfrak{X}_1;\mathbf{k}_{\mathfrak{X}_3\times\mathfrak{X}_1}).$$

Here, $\cup$ is the cup product and

$$\int_{\mathfrak{X}_2}: H^{d_3+2d_2+d_1}_{p_{32}^{-1}\Lambda_2\cap p_{21}^{-1}\Lambda_1}(\mathfrak{X}_3\times\mathfrak{X}_2\times\mathfrak{X}_1;\mathbf{k}_{\mathfrak{X}_3\times\mathfrak{X}_2\times\mathfrak{X}_1})\to H^{d_3+d_1}_{\Lambda_2\circ\Lambda_1}(\mathfrak{X}_3\times\mathfrak{X}_1;\mathbf{k}_{\mathfrak{X}_3\times\mathfrak{X}_1})$$

is the Poincaré integration morphism.

Let $\mathcal{K}_i \in \mathrm{D}^{\mathrm{b}}_{\mathrm{gd}}(\mathcal{W}_{\mathfrak{X}_{i+1}\times\mathfrak{X}_i^a})$ $(1\le i\le 2)$ and assume that p_{31} is proper on $p_{32}^{-1}\operatorname{supp}(\mathcal{K}_2)\cap p_{12}^{-1}\operatorname{supp}(\mathcal{K}_1)$. The proof of the formula

$$\mathrm{Eu}(\mathcal{K}_2\circ\mathcal{K}_1)=\mathrm{Eu}(\mathcal{K}_2)\circ\mathrm{Eu}(\mathcal{K}_1) \tag{6.2}$$

is in progress. It would partly generalize the index theorems for coherent $\mathcal{D}_X$-modules proved in [**27**].

As a particular case of (6.2), one finds that for two objects $\mathcal{L}$ and $\mathcal{M}$ in $\mathrm{D}^{\mathrm{b}}_{\mathrm{gd}}(\mathcal{W}_{\mathfrak{X}})$ such that $\operatorname{supp}\mathcal{L}\cap\operatorname{supp}\mathcal{M}$ is compact, we have

$$\chi(\mathrm{RHom}_{\mathcal{W}_{\mathfrak{X}}}(\mathcal{L},\mathcal{M}))=\int_{\mathfrak{X}}\mathrm{Eu}(\mathrm{D}'_{\mathrm{w}}\mathcal{L})\cup\mathrm{Eu}(\mathcal{M}). \tag{6.3}$$

In the case of coherent $\mathcal{D}_X$-modules on a complex manifold X, the formula

$$\mathrm{Eu}(\mathcal{L})=[\mathrm{Ch}(\mathrm{gr}\,\mathcal{L})\cup\mathrm{Td}(TX)]^{d_{T^*X}}$$

had been conjectured in [**27**] and proved by [**6**]. On a complex symplectic manifold, these authors give a more general formula calculating $\mathrm{Eu}(\mathcal{L})$, a formula in which the class of the deformation quantization appears.

References

[1] F. Bayen, M. Flato, C. Fronsdal, A. Lichnerowicz and D. Sternheimer, *Deformation theory and quantization I,II,* Ann. Physics **111** (1978) 61–110, 111–151.

[2] K. Behrend and B. Fantechi, *Gerstenhaber and Batalin-Vilkovisky structures on Lagrangian intersections,* to appear in "Arithmetic and Geometry, Manin Festschrift".

[3] A. I. Bondal and M. Kapranov, *Representable functors, Serre functors and mutations,* (English translation) Math USSR Izv., **35** (1990) 519–541.

[4] R. Bezrukavnikov and D. Kaledin, *Fedosov quantization in positive characteristic,* `arXiv:math.AG/0501247`.

[5] L. Boutet de Monvel, *Related semi-classical and Toeplitz algebras,* in *Deformation Quantization,* IRMA Lectures in Mathematics and Theoretical Physics **1**, de Gruyter, (2002) 163–191.

[6] P. Bressler, R. Nest and B. Tsygan, *Riemann-Roch theorems via deformation quantization. I, II,* Advances in Math. **167** (2002) 1–25, 26–73.

[7] P. Bressler, A. Gorokhovsky, R. Nest and B. Tsygan, *Deformation quantization of gerbes,* `arXiv:math.QA/0512136`.

[8] A. D'Agnolo and P. Polesello, *Deformation quantization of complex involutive submanifolds,* in: Noncommutative geometry and physics, World Sci. Publ., Hackensack, NJ (2005) 127-137.

[9] A. D'Agnolo and P. Schapira, *Quantization of complex Lagrangian submanifolds,* Advances in Mathematics, **213** (2007) 358-379. (See also `arXiv:math.AG/0506064`.)

[10] G. Dito and P. Schapira, *An algebra of deformation quantization for star-exponentials on complex symplectic manifolds,* Commun. Math. Phys. **273** (2007) 395–414. (See also `arXiv:math.QA/0607235`.)

[11] M. Engeli and G. Felder, *A Riemann-Roch-Hirzebruch formula for traces of differential operators,* `arXiv:math.QA/0702461`.

[12] B. Feigin and B. Tsygan, *Riemann-Roch theorem and Lie algebra cohomology. I,* Proceedings of the Winter School on Geometry and Physics (Srní, 1988). Rend. Circ. Mat. Palermo (2) Suppl. **21**2 (1989) 15–5.

[13] J. Giraud, *Cohomologie non abélienne,* Grundlheren der Math. Wiss. **179** Springer-Verlag (1971).

[14] C. Houzel, *Espaces analytiques relatifs et théorèmes de finitude,* Math. Annalen **205** (1973) 13–54.

[15] M. Kashiwara, *Quantization of contact manifolds,* Publ. RIMS, Kyoto Univ. **32** (1996) 1–5.

[16] M. Kashiwara, *D-modules and Microlocal Calculus,* Translations of Mathematical Monographs, **217** American Math. Soc. (2003).
[17] M. Kashiwara and T. Kawai, *On holonomic systems of microdifferential equations III,* Publ. RIMS, Kyoto Univ. **17** (1981) 813–979.
[18] M. Kashiwara and P. Schapira, *Sheaves on Manifolds,* Grundlehren der Math. Wiss. **292** Springer-Verlag (1990).
[19] M. Kashiwara and P. Schapira, *Categories and Sheaves,* Grundlehren der Math. Wiss. Springer-Verlag (2005).
[20] M. Kashiwara and P. Schapira, *Constructibility and duality for simple holonomic modules on complex symplectic manifolds,* Amer. Journ. of Math. to appear. (See also `arXiv:math.QA/0512047`.)
[21] M. Kontsevich, *Deformation quantization of algebraic varieties,* in: EuroConférence Moshé Flato, Part III (Dijon, 2000) Lett. Math. Phys. **56** (3) (2001) 271–294.
[22] M. Kontsevich, *Deformation quantization of Poisson manifolds,* in: EuroConférence Moshé Flato, Part VI (Dijon, 2000) Lett. Math. Phys. **66** (2003) 157–216.
[23] P. Polesello, *Classification of deformation quantization algebroids on complex symplectic manifolds,* `arXiv:math.AG/0503400`.
[24] P. Polesello and P. Schapira, *Stacks of quantization-deformation modules over complex symplectic manifolds,* Int. Math. Res. Notices **49** (2004) 2637–2664.
[25] M. Sato, T. Kawai, and M. Kashiwara, *Microfunctions and pseudo-differential equations,* in Komatsu (ed.), *Hyperfunctions and pseudo-differential equations,* Proceedings Katata 1971, Lecture Notes in Math. Springer-Verlag **287** (1973) 265–529.
[26] P. Schapira, *Microdifferential systems in the complex domain,* Grundlehren der Math. Wiss. **269** Springer-Verlag (1985).
[27] P. Schapira and J-P. Schneiders, *Index theorem for elliptic pairs,* Astérisque Soc. Math. France 224 (1994).
[28] P. Schapira and J-P. Schneiders, *Finiteness and duality on complex symplectic manifolds,* In preparation (2006).
[29] M. Van den Bergh, *Existence theorems for dualizing complexes over non-commutative graded and filtered rings,* J. of Algebra, **195** (1997) 662–679.
[30] M. Van den Bergh, *On global deformation quantization in the algebraic case,* `arXiv:math.AG/0603200`.
[31] A. Yekutieli, *Deformation quantization in algebraic geometry,* Advances in Math. **198** (2005) 383–432.
[32] A. Yekutieli and J. Zhang, *Dualizing complexes and perverse modules over differential algebras,* Compos. Math. **141** (2005) 620–654.

Institut de Mathématiques, Université Pierre et Marie Curie, 175, rue du Chevaleret, 75013 Paris, France

E-mail address: `schapira@math.jussieu.fr`

URL: `http://www.math.jussieu.fr/~schapira`

Contemporary Mathematics
Volume **450**, 2008

Non-commutative quasi-Hamiltonian spaces

Michel Van den Bergh

ABSTRACT. In this paper we introduce non-commutative analogues for the quasi-Hamiltonian G-spaces introduced by Alekseev, Malkin and Meinrenken. We outline the connection with the non-commutative analogues of quasi-Poisson algebras which the author had introduced earlier.

CONTENTS

1. Introduction

There has been recent interest in developing a non-commutative version of differential geometry based on Kontsevich's philosophy [**10, 12**] that for a property of a non-commutative k-algebra A to have geometric meaning it should induce standard geometric properties on all representation spaces $\operatorname{Rep}(A, N) = \operatorname{Hom}(A, M_N(k))$. Non-commutative symplectic geometry was developed in [**3, 6, 8, 11**] and a non-commutative version of (quasi-)Poisson geometry was introduced in [**14**].

In this paper we introduce so-called *quasi-bisymplectic algebras* (see §6). These are a multiplicative analogue of algebras equipped with a bisymplectic form (see [**6**]). Our definition is such that the representation spaces of quasi-bisymplectic algebras are quasi-Hamiltonian G-spaces, as introduced in [**2**]. We develop non-commutative analogues for some aspects of the commutative theory [**1, 2**]. In particular we show that there is a one-one correspondence between quasi-bisymplectic algebras and

The author is a director of Research at the FWO.

Hamiltonian double quasi-Poisson algebras (introduced in [**14**]) which satisfy a suitable non-degeneracy condition (see Theorem 7.1 below).

As a side result we show that double quasi-Poisson algebras give rise to something we call a "double Lie algebroid" (see Theorem 5.3 below). This is a non-commutative version of [**4**, Thm. 2.5].

In the final section of the paper we show that the Hamiltonian double quasi-Poisson algebras derived from quivers which were introduced in [**14**] are non-degenerate. Hence these algebras are also in a natural way quasi-bisymplectic.

The main result of this paper was announced in [**14**, App A] where we discussed the relation between *ordinary* double Poisson brackets and bisymplectic forms (i.e. the "non-quasi" case). It should be said however that our proof for the equivalence between integrability of double quasi-Poisson brackets and quasi-bisymplectic forms is based on a brute force computation and is less satisfactory than the corresponding proof in [**14**, App A].

This paper depends rather heavily on [**6, 14**]. For the convenience of the reader we have included some preliminary sections explaining the relevant concepts and results. To simplify the exposition we have chosen to write out all our computations over a base ring which is a field, although that is not sufficient for the application to quivers. Therefore in the short section §8.1 we outline the modifications necessary to handle more general situations.

A change in presentation with respect to [**14**] is that throughout the paper we have emphasized a certain functor

$$(-)_N : \mathrm{Bimod}(A) \to \mathrm{Mod}(\mathcal{O}(\mathrm{Rep}(A, N)))$$

which connect an algebra with its N'th representation space. When applied to a non-commutative object this functor yields the corresponding classical object. For example if L is a double Lie algebroid over A (see above) then L_N is a classical Lie algebroid on $\mathrm{Rep}(A, N)$.

The author wishes to thank Victor Ginzburg for explaining some aspects of [**7**].

2. Preliminaries

2.1. Representation spaces. We assume that k is a field of characteristic zero although this hypotheses is often too strong. Throughout A is a finitely generated k-algebra. For $N \in \mathbb{N}$ the associated representation space of A is defined as

$$\mathrm{Rep}(A, N) = \mathrm{Hom}(A, M_N(k))$$

The group Gl_N acts on $\mathrm{Rep}(A, N)$ by conjugation on $M_N(k)$.

A natural point of view in non-commutative algebraic geometry is that for a property of a non-commutative ring A to have geometric meaning it should induce standard geometric properties on all $\mathrm{Rep}(A, N)$.

It is easy to see that $\mathrm{Rep}(A, N)$ is an affine variety and its coordinate ring has a very convenient description. The ring $A_N \overset{\mathrm{def}}{=} \mathcal{O}(\mathrm{Rep}(A, N))$ is generated by symbols $(a_{ij})_{ij=1,\ldots,N}$ for all $a \in A$, subject to the relations

$$(ab)_{ij} = a_{il} b_{lj}$$

together with additivity in a and $1_{ij} = \delta_{ij}$. Here and below we sum over repeated indices. While being convenient this description is of course also very uneconomical. For example if $A = k\langle (x_l)_{l=1,\ldots,n} \rangle$ then $\mathcal{O}(\mathrm{Rep}(A, N)) = k[(x_{l,ij})_{l=1,\ldots,n,i,j=1,\ldots,N}]$.

It is easy to verify that our assumption that A is finitely generated implies that $\mathcal{O}(\mathrm{Rep}(A, N))$ is finitely generated.

If $a \in A$ then $(a_{ij})_{ij}$ defines a matrix values function on $\mathrm{Rep}(A, N)$ which we sometimes denote by $X(a)$. Concretely this is the function which associates to every $\phi : A \to M_N(k) \in \mathrm{Rep}(A, N)$ the matrix $(\phi(a))_{ij}$.

We define the trace $\mathrm{tr}(a)$ of $a \in A$ as a_{ii}. This defines a Gl_n-invariant function on $\mathrm{Rep}(A, N)$. As $\mathrm{tr}([A, A]) = 0$ we see that elements of $A/[A, A]$ correspond to invariant functions on $\mathrm{Rep}(A, N)$. In fact this will be true in all cases we consider below: the non-commutative version of working with Gl_N invariants objects is working modulo commutators.

2.2. Differential forms. We now consider differential forms. The bimodule Ω_A of differentials is generated as an A-bimodule by the symbols da subject to the relations $d(ab) = a(db) + (da)b$ and linearity. As usual one puts $\Omega A = T_A\Omega_A$. Defining $d(da) = 0$, $d(a) = da$ makes Ω_A into a differential graded algebra. Every homogeneous element ω in ΩA has a representation $a_0 da_1 \cdots da_n$. To such an element one associates a matrix valued differential form $(\omega_{ij})_{ij}$ on $\mathrm{Rep}(Q, \alpha)$:

$$\omega_{ij} = a_{1,ii_1} da_{2,i_1 i_2} \cdots da_{n,i_{n-1}j} \tag{2.1}$$

If we write it as $X(\omega)$ then (2.1) may be rewritten as

$$X(\omega) = X(a_1)dX(a_2) \cdots dX(a_n)$$

It would be tempting to define the non-commutative de Rham complex of A as ΩA. However, quite remarkably (see [**6**, §2.5]), ΩA is acyclic. That is

$$H^m(\Omega A) = \begin{cases} k & m = 0 \\ 0 & \text{otherwise} \end{cases}$$

Nevertheless ΩA can be used as the basis for a new construction of the cyclic homology of A (see [**7**]).

A different non-commutative analogue of the de Rham complex is the *Karoubi-de Rham complex* which is defined by

$$\mathrm{DR}(A) = \Omega A/[\Omega A, \Omega A]$$

The de Rham cohomology of A is defined as the cohomology of the Karoubi complex. It is closely related to cyclic homology and to equivariant de Rham cohomology of representation spaces (see [**7**, Thm 4.2.3]). According to [**13**, 2.6.7] we have a short exact sequence of reduced (co)homology

$$0 \to \bar{H}^n(\mathrm{DR}(A)) \to \overline{\mathrm{HC}}_n(A) \to \overline{\mathrm{HH}}_{n+1}(A) \to 0$$

Below we will mostly deal with *smooth algebras*, i.e. algebras whose category of bimodules has homological dimension one. In that case $\overline{\mathrm{HH}}_n(A) = 0$ for $n > 1$ and hence $\bar{H}^n(\mathrm{DR}(A)) = \overline{\mathrm{HC}}_n(A)$ for $n \geq 1$.

To any element ω we associate a Gl_n invariant differential form $\mathrm{tr}(\omega) = \omega_{ii}$ on $\mathrm{Rep}(A, N)$. This yields a map

$$\mathrm{tr} : DR(A) \to \Omega(\mathrm{Rep}(A, N))^{\mathrm{Gl}(n)}$$

which is compatible with differentials and hence descends to cohomology.

EXAMPLE 2.1. Let $C = k[t, t^{-1}]$. Then

$$H^i(\mathrm{DR}(C)) = \begin{cases} k & \text{if } i = 0 \\ k(t^{-1}dt)^i & \text{if } i \text{ is odd} \\ 0 & \text{otherwise} \end{cases}$$

That the cohomology groups have the indicated dimension follows from [**13**, Cor 3.4.15]. It is easy to see that the generators for the cohomology groups are as claimed.

We have $\mathrm{Rep}(A, N) = \mathrm{Gl}_N$. The elements $\mathrm{tr}(t^{-1}dt)^{2i+1}$ are precisely the generators of the de Rham cohomology of Gl_N.

2.3. Vector fields. Now we discuss vector fields. Again there are two possible points of view.

If we insist that a vector field on A induces vector fields on all $\mathrm{Rep}(A, N)$ then a vector field on A should simply be a derivation $\Delta : A \to A$. The induced derivation δ on $\mathcal{O}(\mathrm{Rep}(A, N))$ is then given by

$$\delta(a_{ij}) = \Delta(a)_{ij}$$

A second point of view is that a vector field Δ on A should induce *matrix valued vector fields* $(\Delta_{ij})_{i,j=1,\ldots,n}$ on all $\mathrm{Rep}(A, N)$. Since now $\Delta_{ij}(a_{uv})$ depends on four indices $\Delta(a)$ should be an element of $A \otimes A$. It was first of observed by Crawley-Boevey that the second point of view is often more useful. Put

$$D_A \overset{\text{def}}{=} \mathrm{Der}(A, A \otimes A) = \mathrm{Hom}_{A \otimes A^\circ}(\Omega_A, A \otimes A)$$

where as usual we put the outer bimodule structure on $A \otimes A$. For $\Delta \in D_A$ the corresponding matrix valued vector field on $\mathrm{Rep}(A, N)$ is then given by

$$\Delta_{ij}(a_{uv}) = \Delta(a)'_{uj}\Delta(a)''_{iv} \tag{2.2}$$

where by convention we write an element x of $A \otimes A$ as $x' \otimes x''$ (i.e. we drop the summation sign). We will call the peculiar arrangement of indices in (2.2) the *standard index convention.* It will reappear often below.

Starting with D_A we define *the algebra of poly-vector fields* DA on A as the tensor algebra $T_A D_A$ of D_A where we make D_A into an A-bimodule by using the inner bimodule structure on $A \otimes A$. Any homogeneous element δ of DA induces polyvector fields $X(\delta)$ on all representation spaces using a formula similar to (2.1).

The counterpart to the differential on ΩA is the "double Schouten Nijenhuys" bracket on DA which was defined in [**14**]. A double bracket on an ordinary algebra A is a bilinear map

$$\{\!\{-,-\}\!\} : A \times A \to A \otimes A$$

which is a derivation in its second argument (for the outer bimodule structure on A) and which satisfies

$$\{\!\{a, b\}\!\} = -\{\!\{b, a\}\!\}^\circ$$

where $(u \otimes v)^\circ = v \otimes u$. If $\{\!\{-,-\}\!\}$ satisfies the following analogue of the Jacobi identity

$$0 = \{\!\{a, b, c\}\!\} \overset{\text{def}}{=} \{\!\{a, \{\!\{b, c\}\!\}\}\!\}_L + \tau_{(123)}\{\!\{b, \{\!\{c, a\}\!\}\}\!\}_L + \tau_{(132)}\{\!\{c, \{\!\{a, b\}\!\}\}\!\}_L$$

where for $\tau \in S_n$ we define

$$\tau(a_1 \otimes \cdots \otimes a_n) = a_{\tau^{-1}(1)} \otimes \cdots \otimes a_{\tau^{-1}(n)} \tag{2.3}$$

(with an appropriate sign in the graded case) then we call $\{\!\{-,-\}\!\}$ a double Poisson bracket. A double Poisson bracket induces a Lie bracket $\{-,-\}$ on $A/[A,A]$ via the formula

$$\{a,b\} = \{\!\{a,b\}\!\}'\{\!\{a,b\}\!\}''$$

The motivation for introducing double Poisson brackets is that they induce ordinary Poisson brackets on $\operatorname{Rep}(A,N)$ via the standard index convention. The precise formula is

$$\{a_{ij}, b_{uv}\} = \{\!\{a,b\}\!\}'_{uj}\{\!\{a,b\}\!\}''_{iv}$$

One of the main results of [**14**] is the following.

Proposition 2.2. *The graded algebra DA has the structure of a* double Gerstenhaber algebra *i.e. a (super) double Poisson algebra with a double Poisson bracket $\{\!\{-,-\}\!\}$ of degree -1.*

For the convenience of the reader we give the construction of the double Schouten-Nijenhuys bracket on D_A.

If $\delta, \Delta \in D_A$ then it is easy to see that

$$\begin{aligned}\{\!\{\delta,\Delta\}\!\}\tilde{}_l &= (\delta\otimes 1)\Delta - (1\otimes\Delta)\delta\\ \{\!\{\delta,\Delta\}\!\}\tilde{}_r &= (1\otimes\delta)\Delta - (\Delta\otimes 1)\delta = -\{\!\{\Delta,\delta\}\!\}\tilde{}_l\end{aligned}$$

define derivations $A \to A^{\otimes 3}$ for the outer bimodule structure on $A^{\otimes 3}$. Since Ω_A is finitely generated we obtain

$$\operatorname{Der}_B(A, A^{\otimes 3}) \cong \operatorname{Hom}_{A^e}(\Omega_{A/B}, A\otimes A)\otimes A$$

We view $\{\!\{\delta,\Delta\}\!\}\tilde{}_l$ and $\{\!\{\delta,\Delta\}\!\}\tilde{}_r$ as elements of $D_A \otimes_k A$ and $A \otimes_k D_A$ respectively. To this end we define

$$\begin{aligned}\{\!\{\delta,\Delta\}\!\}_l &= \tau_{(23)} \circ \{\!\{\delta,\Delta\}\!\}\tilde{}_l\\ \{\!\{\delta,\Delta\}\!\}_r &= \tau_{(12)} \circ \{\!\{\delta,\Delta\}\!\}\tilde{}_r\end{aligned}$$

and we write

$$\begin{aligned}\{\!\{\delta,\Delta\}\!\}_l &= \{\!\{\delta,\Delta\}\!\}'_l \otimes \{\!\{\delta,\Delta\}\!\}''_l\\ \{\!\{\delta,\Delta\}\!\}_r &= \{\!\{\delta,\Delta\}\!\}'_r \otimes \{\!\{\delta,\Delta\}\!\}''_r\end{aligned}$$

with $\{\!\{\delta,\Delta\}\!\}''_l, \{\!\{\delta,\Delta\}\!\}'_r \in A$, $\{\!\{\delta,\Delta\}\!\}'_l, \{\!\{\delta,\Delta\}\!\}''_r$ in D_A.

An easy verification shows that

$$\{\!\{\delta,\Delta\}\!\}_r = -\{\!\{\Delta,\delta\}\!\}^\circ_l$$

The double Schouten-Nijenhuys bracket is defined on generators by

$$\begin{aligned}\{\!\{a,b\}\!\} &= 0\\ \{\!\{\delta,a\}\!\} &= \delta(a)\\ \{\!\{\delta,\Delta\}\!\} &= \{\!\{\delta,\Delta\}\!\}_l + \{\!\{\delta,\Delta\}\!\}_r\end{aligned}$$

for $a, b \in A$, $\delta, \Delta \in D_A$, where we regard the righthand sides in the previous display as elements of $DA \otimes DA$.

One may show again that the double Schouten-Nijenhuys bracket on D_A induces the standard Schouten-Nijenhuys bracket on the algebra of polyvector fields on $\operatorname{Rep}(A,N)$ using the standard index convention.

In [**14**] it was shown how to associate a double bracket $\{\!\{-,-\}\!\}_P$ to an element $P \in D^2A$. The formula is obtained by linear extension from the following formula with $\delta, \Delta \in D_A$:

$$\{\!\{a,b\}\!\}_{\delta\Delta} = \Delta(b)'\delta(a)'' \otimes \delta(a)'\Delta(b)'' - \delta(b)'\Delta(a)'' \otimes \Delta(a)'\delta(b)'' \tag{2.4}$$

If A is smooth then the double Jacobi identity for $\{\!\{-,-\}\!\}$ is equivalent to $\{P,P\} = 0$.

The algebra DA has a remarkable element E which has no commutative counter part. It is the double derivation which sends a to $a \otimes 1 - 1 \otimes a$. It is intimately connected to the Gl_N action on $\mathrm{Rep}(A,N)$. More precisely the following result was proved in [**6, 14**]

PROPOSITION 2.3.1. *Let $f_{ij} \in M_\alpha = \mathrm{Lie}(\mathrm{Gl}_N)$ be the elementary matrix which is 1 in the (i,j)-entry and zero everywhere else. Then $(E_p)_{ij}$ acts as f_{ji} on $\mathcal{O}(\mathrm{Rep}(A,N))$.*

The element E appears in several pleasing formulas. For example for any poly-vector field Δ in DA we have

$$\{\!\{E, \Delta\}\!\} = \Delta \otimes 1 - 1 \otimes \Delta$$

A vector field of the form

$$H_a = \{\!\{a, -\}\!\}$$

is called *Hamiltonian.* An element $\Phi \in A$ is a *moment map* if it realizes E as a Hamiltonian vector field, i.e. if $E = H_\Phi$. We should think of Φ as defining a matrix valued map

$$\Phi_{ij} : \mathrm{Rep}(A,N) \to \mathrm{Rep}(k[t],N) = M_N(k)$$

If Φ is a moment map for $\{\!\{-,-\}\!\}$ then Φ_{ij} is a moment map for the induced Poisson bracket $\{-,-\}$.

2.4. Bisymplectic geometry. In the commutative case symplectic geometry is a special case of Poisson geometry. A first version of non-commutative symplectic geometry was introduced by Kontsevich in [**11**]. See also [**3, 8**]. A version of Poisson geometry following a similar philosophy was introduced by Crawley-Boevey in [**5**].

A different version of non-commutative symplectic geometry which follows a similar philosophy as the Poisson geometry outlined in the previous sections was introduced in [**6**] and baptized "bisymplectic geometry". We recall the definition. If $\delta \in D_A$ then we may define a double derivation

$$i_\delta : \Omega A \to \Omega A \otimes \Omega A$$

in the usual way. For $a \in A$ put

$$i_\delta(a) = 0 \qquad i_\delta(da) = \delta(a)$$

If C is a graded k-algebra and $c = c_1 \otimes \cdots \otimes c_n$ then we put

$$^\circ c = \tau_{(1\cdots n)}(c) = (-1)^{|c_n|(|c_1|+\cdots+|c_{n-1}|)} c_n c_1 \cdots c_{n-1} \tag{2.5}$$

and if $\phi : C \to C^{\otimes 2}$ is a linear map then we define

$$^\circ\phi : C/[C,C] \to C : c \mapsto {}^\circ(\phi(c))$$

We apply this with $C = \Omega A$. Following [**6**] we put

$$\imath_\delta = {}^\circ i_\delta$$

Also following [**6**] we say that an element $\omega \in \mathrm{DR}^2(A)$ is a *bi-non-degenerate* if the map of A-bimodules

$$\imath(\omega) : D_A \to \Omega_A : \delta \mapsto \imath_\delta \omega$$

is an isomorphism (in fact it is sufficient to assume surjectivity, see Corollary 3.1.3 below). If in addition ω is closed in $\mathrm{DR}(A)$ then we say that ω is *bisymplectic.*

It was shown in [**6**] that if ω is bisymplectic the $\mathrm{tr}(\omega)$ defines a symplectic form on representation spaces.

3. Generalities about bimodules

In the framework of this paper the non-commutative version of a coherent sheaf is a bimodule. In this short section we discuss some elementary aspects of bimodules.

3.1. Pairings. Let A be an arbitrary k-algebra. A pairing (or bilinear map) between $A-A$ bimodules P, Q is a map

$$\langle -, - \rangle : P \times Q \to A \otimes A$$

such that $\langle p, - \rangle$ is linear for the outer bimodule structure on $A \otimes A$ and $\langle -, q \rangle$ is linear for the inner bimodule structure on $A \otimes A$. The obvious example is of course $P = Q^*$ and $\langle -, - \rangle$ is the evaluation pairing. We say that the pairing is non-degenerate if P, Q are finitely generated projective bimodules and the pairing induces an isomorphism $Q \cong P^*$.

If $\langle -, - \rangle$ is a pairing between P and Q then the opposite pairing between Q and P is given by

$$\langle q, p \rangle^\circ = \langle p, q \rangle'' \otimes \langle p, q \rangle'$$

If we have pairings between P and Q and between P' and Q' the morphisms $\alpha : P \to P'$, $\beta : Q' \to Q$ are said to be *adjoint* if

$$\langle \alpha(p), q' \rangle = \langle p, \beta(q') \rangle$$

We say that α is left adjoint to β and that β is a right adjoint of α. Note that if we change the pairings into the opposite ones then a left adjoint becomes a right adjoint and vice versa. Therefore we usually drop the left/right adjectives if what is meant is clear from the context. If the pairing are non-degenerate then we denote an adjoint often by $(-)^*$.

If we have a morphism $\alpha : P \to Q$ then then we say that α is anti-symmetric (or anti-self adjoint) if $-\alpha$ is left adjoint to α for the appropriate pairings, i.e. if

$$\langle p, \alpha(p') \rangle = -\langle \alpha(p), p' \rangle^\circ$$

for $p, p' \in P$.

Or written out explicitly

$$\langle p, \alpha(p') \rangle' \otimes \langle p, \alpha(p') \rangle'' = -\langle p', \alpha(p) \rangle'' \otimes \langle p', \alpha(p) \rangle' \tag{3.1}$$

If we have a pairing between P and Q as above then $p \in P$ defines a double derivation of degree -1

$$i_p : T_A(Q) \to T_A(Q) \otimes T_A(Q)$$

such that $i_p(q) = \langle p, q \rangle$. Since Q and P are related by the opposite pairing we may define i_q for $q \in Q$ in the same way. I.e.

$$i_q : T_A(P) \to T_A(P) \otimes T_A(P)$$

is the double derivation of degree -1 such that $i_q(p) = \langle q, p\rangle^\circ$.

We define $\imath_p = {}^\circ i_p$ and $\imath_q = {}^\circ i_q$ as in §2.4. If $\omega \in T^2_A Q$ then we put

$$\imath(\omega) : P \to Q : p \mapsto \imath_p(\omega)$$

This is a bimodule morphism between P and Q. The following property of $\imath(\omega)$ will be used below.

PROPOSITION 3.1.1. *The bimodule morphism $\imath(\omega)$ is anti-symmetric.*

PROOF. We need to prove that $\langle p, \imath(\omega)(p')\rangle = \langle p, \imath_{p'}\omega\rangle$ satisfies (3.1). We write ω as $\omega' \otimes \omega''$ with $\omega', \omega'' \in Q$. Then we have

$$i_{p'}(\omega) = i_{p'}(\omega')\omega'' - \omega' i_{p'}(\omega'')$$

so that we get

$$\begin{aligned}\imath_{p'}(\omega) &= i_{p'}(\omega')''\omega'' i_{p'}(\omega')' - i_{p'}(\omega'')''\omega' i_{p'}(\omega'')' \\ &= \langle p', \omega'\rangle''\omega''\langle p', \omega'\rangle' - \langle p', \omega''\rangle''\omega'\langle p', \omega''\rangle'\end{aligned}$$

and hence

$$\begin{aligned}\langle p, \imath_{p'}\omega\rangle &= \langle p', \omega'\rangle''\langle p, \omega''\rangle\langle p', \omega'\rangle' - \langle p', \omega''\rangle''\langle p, \omega'\rangle\langle p', \omega''\rangle' \\ &= \langle p', \omega'\rangle''\langle p, \omega''\rangle' \otimes \langle p, \omega''\rangle''\langle p', \omega'\rangle' - \langle p', \omega''\rangle''\langle p, \omega'\rangle' \otimes \langle p, \omega'\rangle''\langle p', \omega''\rangle'\end{aligned}$$

which is clearly satisfies (3.1). □

COROLLARY 3.1.2. *Assume that the pairing $\langle -, -\rangle$ is non-degenerate. If $\imath(\omega)$ is surjective then it is an isomorphism.*

PROOF. Let $\alpha = \imath(\omega)$. If α is surjective then its left adjoint $\alpha^* : Q^* \to P^*$ is injective. Since the pairing is non-degenerate we have $Q^* = P$ and $P^* = Q$. Hence we may identify α^* with $-\alpha$. So α is both injective and surjective and hence it is an isomorphism. □

We can now state the following corollary which was asserted in §2.4.

COROLLARY 3.1.3. *Assume that $\omega \in \mathrm{DR}^2(A)$ is such that the map of bimodules*

$$\imath(\omega) : D_A \to \Omega_A : \delta \mapsto \imath_\delta\omega$$

is surjective. Then ω is bi-non-degenerate.

If $\langle -, -\rangle : P \times Q \to A \otimes A$ is non-degenerate then we call a *dual bases* sets of elements $p_\alpha \in P$, $q_\alpha \in Q$ such that $\langle p_\alpha, -\rangle' q_\alpha \langle p_\alpha, -\rangle''$ is the identity map on Q. It is easy to see that this equivalent to $\langle -, q_\alpha\rangle'' p_\alpha \langle -, q_\alpha\rangle'$ being the identity map on P.

PROPOSITION 3.1.4. *Assume that the pairing $\langle -, -\rangle : P \times Q \to A \otimes A$ is non-degenerate. Then the map*

$$\imath : T^2_A Q \to \mathrm{Hom}_{A^e}(P, Q)$$

defines an isomorphism between $(T_A Q/[T_A Q, T_A Q])_2$ and the anti-symmetric elements of $\mathrm{Hom}_{A^e}(P, Q)$.

PROOF. It is easy to write down an explicit inverse to $\imath$. Let $p_\alpha \in P$, $q_\alpha \in Q$ be dual bases. Then the element of $T^2_A Q$ corresponding to $\alpha \in \mathrm{Hom}_{A^e}(P, Q)^{\mathrm{asym}}$ is given by $-\frac{1}{2}\alpha(p_\alpha)q_\alpha$. □

We will also have occasion to use the following result

PROPOSITION 3.1.5. *Assume* $\langle -,-\rangle$ *is a non-degenerate pairing between* P *and* Q.

(1) *Let* $\omega \in (T_A Q/[T_A Q, T_A Q])_n$. *If* $\imath_p(\omega) = 0$ *for all* $p \in P$ *then* $\omega = 0$.
(2) *Let* $\eta \in T^n Q$. *If the projection of* $i_p(\eta)$ *on* $A \otimes T^{n-2} Q$ *is zero for al* $p \in P$ *then* $\eta = 0$.

PROOF. We select dual bases p_α, q_α. The following formulas are easily verified for $\omega \in T_A^n Q$

$$q_\alpha \imath_{p_\alpha}(\omega) = n\omega \qquad \operatorname{mod}[-,-]$$
$$(\mathrm{pr}_1(i_{p_\alpha}(\eta)))' q_\alpha (\mathrm{pr}_1(i_{p_\alpha}(\eta)))'' = \eta$$

From this the stated results follow. □

3.2. Double Lie algebroids. It will be convenient to make the following definition.

DEFINITION 3.2.1. A *double Lie algebroid* over A is an A-bimodule L together with a (graded) double Poisson bracket of degree -1 on $T_A L$.

The archetypical example of a double Lie algebroid is D_A where we equip $T_A D_A = DA$ with its double Schouten bracket (see §2.3).

Assume that L is a double Lie algebroid with double bracket $\{\!\{-,-\}\!\}$. As for DA we have associated operations $\{\!\{-,-\}\!\}_l$, $\{\!\{-,-\}\!\}_r$ which are homomorphisms $L \times L \to L \otimes A$ and $L \times L \to A \otimes L$ respectively that are defined by

$$\{\!\{l_1, l_2\}\!\} = \{\!\{l_1, l_2\}\!\}_l + \{\!\{l_1, l_2\}\!\}_r$$

$\{\!\{-,-\}\!\}_l$ and $\{\!\{-,-\}\!\}_r$ determine each other via

$$\{\!\{l_1, l_2\}\!\}_r = -\{\!\{l_2, l_1\}\!\}_l^\circ$$

It follows that the minimal data necessary to specify a double Lie algebroid is given by

$$\{\!\{-,-\}\!\}_l : L \times L \to L \otimes A$$
$$\{\!\{-,-\}\!\} : L \times A \to A$$

One can write down a minimal set of axioms these operations have to satisfy (see [**14**, (3.4-1)-(3.8-1)]). However it is often more straightforward to use Definition 3.2.1 directly.

3.3. Representation spaces. Let's now discuss how this plays out with representation spaces. if P is an A-bimodule then we define P_N as the A_N (see §2.1)-module generated by symbols p_{ij} which are linear in $p \in P$ and which satisfy for $a \in A$:

$$(ap)_{ij} = a_{iu} p_{uj}$$
$$(pa)_{ij} = a_{uj} p_{iu}$$

In this way we obtain an additive functor

$$(-)_N : \operatorname{Mod}(A^e) \to \operatorname{Mod}(A_N)$$

which sends finitely generated bimodules to finitely generated modules.

This functor has a more intrinsic description as follows.

LEMMA 3.3.1. [**6**] *Consider* $M_N(A_N)$ *as an* A*-bimodule via the* k*-algebra morphism* $A \to M_N(A_N) : a \mapsto: a_{ij}$. *Consider* $M_N(A_N)$ *in addition as an* A_N*-module via the diagonal embedding* $A_N \to M_N(A_N)$. *Then there is a natural isomorphism*

$$P_N \cong P \otimes_{A^e} M_N(A_N)$$

In particular $(-)_N$ *is right exact and sends projective bimodules to projective modules.*

PROOF. We will content ourselves by giving the maps. Let f_{ij} be the usual elementary matrix in $M_N(A_N)$. Then one isomorphism is given by

$$P_N \to P \otimes_{A^e} M_N(A_N) : p_{ij} \to p \otimes f_{ji}$$

with the obvious definition for the inverse isomorphism. □

LEMMA 3.3.2. *We have* $(T_A P)_N = S_{A_N} P_N$ *(where one the left hand side the symbol* $(-)_N$ *should be interpreted as acting on the* ring $T_A P$ *and not on the* A*-bimodule* $T_A P$*).*

PROOF. It is easy to see that both sides have the same generators and relations. □

LEMMA 3.3.3. *If* $\langle -, - \rangle : P \times Q \to A \otimes A$ *is non-degenerate then the corresponding pairing between* P_N *and* Q_N *(obtained by applying the standard index convention from §2.3)*

$$\langle -, - \rangle : P_N \otimes Q_N \to A : (p_{ij}, q_{uv}) \mapsto \langle p, q \rangle'_{uj} \langle p, q \rangle''_{iv}$$

is non-degenerate as well.

PROOF. It follows from non-degeneracy that we may select $p_\alpha \in P$, $q_\alpha \in Q$ such that

$$q = \langle p_\alpha, q \rangle' q_\alpha \langle p_\alpha, q \rangle''$$

for all $q \in Q$. It follows

$$\begin{aligned} \langle p_{\alpha,ij}, q_{uv} \rangle q_{\alpha,ji} &= \langle p_\alpha, q \rangle'_{uj} \langle p, q_\alpha \rangle''_{iv} q_{\alpha,ji} \\ &= (\langle p_\alpha, q \rangle' q_\alpha \langle p_\alpha, q \rangle'')_{uv} \\ &= q_{uv} \end{aligned}$$

and hence the induced map $Q_N \to P_N^* : q_{uv} \mapsto \langle -, q_{uv} \rangle$ is injective and split by $\phi \mapsto q_{\alpha,ji}\phi(p_{\alpha,ij})$. Employing the dual argument we find that it is an isomorphism. □

PROPOSITION 3.3.4. *Assume that* A *is smooth. Then the map* $D_A \to \operatorname{Der}(A)$ *defined by* (2.2) *yields an isomorphism* $(D_A)_N \to \operatorname{Der}(A)$. *In particular the tangent space to* $\operatorname{Rep}(A, N)$ *is generated by the vector fields* δ_{ij} *for* $\delta \in \operatorname{Der}(A, N)$.

PROOF. It is easy to see that $(\Omega_A)_N = \Omega_{A_N}$. By the previous lemma we find

$$\operatorname{Der}(A_N) \cong \Omega^*_{A_N} \cong ((\Omega_A)_N)^* \cong (D_A)_N$$

It is easy to check that that the actual isomorphism is the asserted one. □

COROLLARY 3.3.5. *We have* $(\Omega A)_N = \bigwedge_{A_N} \Omega_{A_N}$ *(where* Ω_{A_N} *refers to the ordinary commutative differentials) and if* A *is smooth then* $(DA)_N = \bigwedge^*_{A_N} \operatorname{Der}(A_N)$.

PROOF. The statement about ΩA is easy. The statement about DA follows from Proposition 3.3.4 and Lemma 3.3.2, where we interprete the latter in a graded way, taking into account that D_A and Ω_A have odd degree. □

Below we sometimes use the map

$$X : P \to M_N(P_N) : p \mapsto (p_{ij})_{ij}$$

For later use we mention some additional result.

LEMMA 3.3.6. *Let $P, Q, \langle -,- \rangle$ be as above and let $p \in P$. Then the standard index convention (see §2.3) applies to the operator $\imath_p$. I.e. if $\omega \in T_A^2 Q$ then we have*

$$i_{p_{ij}}(\omega_{uv}) = i_p(\omega)'_{uj} i_p(\omega)''_{iv}$$

In particular we obtain $i_{p_{ij}}(\mathrm{tr}(\omega)) = (\imath_p\omega)_{ij}$ or in more suggestive notation

$$\mathrm{tr}(\omega)(X(p)) = X(\imath(\omega)(p)) \tag{3.2}$$

where we view $\mathrm{tr}(\omega)$ as map from P_N to Q_N.

LEMMA 3.3.7. *If L is a double Lie algebroid over A then L_N is a Lie algebroid over A_N with Lie bracket and anchor map given by the standard index convention. I.e. for $l, l_1, l_2 \in L$, $a \in A$*

$$[l_{1,ij}, l_{2,uv}] = \{\!\{l_1, l_2\}\!\}'_{uj} \{\!\{l_1, l_2\}\!\}''_{iv}$$
$$\rho(l_{ij})(a_{uv}) = \{\!\{l, a\}\!\}'_{uj} \{\!\{l, a\}\!\}''_{iv}$$

4. Quasi-Poisson and quasi-Hamiltonian G-spaces

Here we summarize some definitions from [**1, 2**]. Let G be a linear algebraic group and put $\mathfrak{g} = \mathrm{Lie}(G)$. We assume that $\mathfrak{g}$ carries G-invariant symmetric bilinear form $(-,-)$.

Let $(f_a)_a$, $(f^a)_a$ be dual bases of $\mathfrak{g}$. Then there is a canonical invariant element $\phi \in \wedge^3\mathfrak{g}$ given by

$$\phi = \frac{1}{12} c^{abc} f_a \wedge f_b \wedge f_c$$

where

$$c^{abc} = (f^a, [f^b, f^c])$$

If $\xi \in \mathfrak{g}$ we have left and right invariant vector fields on G defined by

$$(\xi^L(f))(x) = \frac{d}{dt} f(xe^{\xi t})$$
$$(\xi^R(f))(x) = \frac{d}{dt} f(e^{\xi t}x)$$

Assume now that G acts on a smooth affine variety X. If $\xi \in \mathfrak{g}$ then $(\xi_x)_{x\in X}$ is the vector field on X defined by

$$(v_{\xi_x}(f))(x) = \frac{d}{dt} f(e^{-t\xi}x)$$

This convection is such that $\mathfrak{g} \to TX : \xi \mapsto (\xi_x)_x$ is a morphism of Lie algebras.

The element $\phi \in \wedge^3\mathfrak{g}$ induces a three vector field ϕ_X on X. Following [**1**] an element $P \in \bigwedge^2_{\mathcal{O}(X)} \mathrm{Der}(\mathcal{O}(X))$ is said to be a *quasi-Poisson bracket* if

$$\{P, P\} = \phi_X$$

A *Hamiltonian quasi-Poisson G-space* is a triple (X, P, Φ) such that (X, P) is quasi-Poisson and such that Φ is a so-called *multiplicative moment map*. I.e.

$$\{h \circ \Phi, -\} = \frac{1}{2} f_X^a \left((f_a^L + f_a^R)(h) \circ \Phi\right)$$

for all $f \in \mathcal{O}(G)$.

A Hamiltonian quasi-Poisson G-space is said to be *non-degenerate* if for $x \in X$ the map

$$T_x^* \oplus \mathfrak{g} \mapsto T_x : (\eta, \xi) \mapsto P_x(\eta) + \xi_x$$

is surjective.

Now we let θ and $\bar{\theta}$ be respectively the left and right invariant $\mathfrak{g}$-valued Maurer-Cartan forms:

$$\theta = g^{-1}dg \qquad \bar{\theta} = dg \cdot g^{-1}$$

We let χ be the canonical G-invariant three form on G:

$$\chi = \frac{1}{12}(\theta, [\theta, \theta])$$

A quasi-Hamiltonian G-variety is a triple (X, ω, Φ) where M is a smooth G-variety, $\omega \in (T^{*,2})^G$ and $\Phi : X \to G$ is a G-equivariant map (for the given action on X and the adjoint action of G on G) such that the following axioms are satisfied:

(B1) $d\omega = \Phi^*\chi$.[1]

(B2) $\forall \xi \in \mathfrak{g} : i_{\xi_X}(\omega) = \frac{1}{2}\Phi^*(\theta + \bar{\theta}, \xi)$.

(B3) For all x in X we have

$$\ker \omega_x = \{\xi_x \mid \xi \in \ker(\operatorname{Ad}_{\Phi(x)} + 1) \subset \mathfrak{g}\}$$

Let us explain how to read (B3). By definition ω_x is the map

$$T_X \to T_X^* : \delta \mapsto i_\delta(\omega)$$

evaluated in x. For $\xi \in \mathfrak{g}$, ξ_x is the vector field on X given by ξ evaluated in x. (B3) states that $\ker \omega_x$ is a specific part of the image of the map $\mathfrak{g} \to T_{X,x} : \xi \mapsto \xi_x$. As the form of condition (B3) is not so convenient for us so we state an equivalent version.

(B3') The map

$$T_x \oplus \mathfrak{g} \to T_x^* : (\delta, \xi) \mapsto \omega_x(\delta) + (\xi, \Phi^*(\theta)_x)$$

is surjective.

LEMMA 4.1. *(B3) and (B3') are equivalent.*

PROOF. (Sketch) We first dualize (B3'). Let $\beta : T_x \to \mathfrak{g} : \delta \mapsto i_\delta(\Phi^*(\theta))$. Then (B3') is equivalent to the condition

(B3") The map $(\omega_x, \beta) : T_x \to T_x^* \oplus \mathfrak{g}$ is injective.

It is a straightforward verification using (B2) that the following diagram

$$(4.1) \qquad \begin{array}{ccc} \mathfrak{g} & \xrightarrow{\alpha} & \mathfrak{g} \\ {\scriptstyle\beta}\uparrow & & \uparrow{\scriptstyle\gamma} \\ T_x & \xrightarrow[\omega_x]{} & \Omega_x \end{array}$$

with $\alpha(\xi) = -\frac{1}{2}(1 + \operatorname{Ad}(\Phi(x)))(\xi)$ and $(\gamma(\eta), \xi) = i_{\xi_x}(\eta)$ is commutative, Furthermore using the properties of θ, $\bar{\theta}$ (see e.g [**9**, Ch II]) we find $\beta(\xi_x) = (1 - \operatorname{Ad}(\Phi(x))^{-1})(\xi)$.

Let us show that (B3) implies (B3"). Assume that δ is such that $i_\delta(\omega_x) = 0$ and $i_\delta(\Phi^*(\theta)) = 0$. By (B3) we have that $\delta = \xi_x$ where $(1 + \operatorname{Ad}(\Phi(x)))(\xi) = 0$.

[1] We follow the convention from [**1**, Def 10.1]. In [**2**] this formula is given with a minus sign

Since $\beta(\xi_x) = 0$ we also have $(1 - \mathrm{Ad}(\Phi(x)))(\xi) = 0$ by diagram (4.1). These two facts together imply that $\xi = 0$.

Now we prove the converse. Assume that $\delta \in \ker(\omega_x)$. Then $\xi = \beta(\delta) \in \ker \alpha$. I.e.

$$(1 + \mathrm{Ad}(\Phi(x)))(\xi) = 0 \tag{4.2}$$

Then

$$\beta(\xi_x) = (1 - \mathrm{Ad}(\Phi(x)^{-1})(\xi) = 2\xi$$

Since $(1 + \mathrm{Ad}(\Phi(x)))(\xi) = 0$ we also have $\omega_x(\xi_x) = 0$.

It follows that $= \delta - \frac{1}{2}\xi_x$ is both in the kernel of β an ω_x. Hence by (B3") $\delta = \frac{1}{2}\xi_x$. (B3) now follows from (4.2). □

We now state the main theorem of [**1**, §10]. Write $\theta = f^a\theta_a$ where $\theta_a \in T_X^*$ and similarly for $\bar{\theta}$.

THEOREM 4.2. [**1**, Thm 10.3] *Every non-degenerate Hamiltonian quasi-Poisson space (X, P, Φ) carries a unique 2-form ω such that (X, ω, Φ) is a quasi-Hamiltonian G-space and such that ω and P satisfy the following compatibility condition*

$$P \circ \omega = 1 - \frac{1}{4} f_X^a \otimes \Phi^*(\theta_a - \bar{\theta}_a) \tag{4.3}$$

(as maps $T_X^* \to T_X$). Conversely on every quasi-Hamiltonian G-space (X, ω, Φ) there is a unique bivector field P such that (M, X, Φ) is a non-degenerate quasi-Hamiltonian G-manifold and (4.3) is satisfied.

We will say that (P, ω, Φ) are compatible if (4.3) holds.

5. Hamiltonian double quasi-Poisson algebras

The non-commutative version of quasi-Poisson algebras was worked out in [**14**].

CONVENTION 5.1. *From now on our non-commutative algebras are always smooth.*

Let $P \in (DA/[DA, DA])_2$. We say that P is a *double quasi-Poisson bracket* if the following condition holds

($\mathbb{P}$1) $\{P, P\} = \frac{1}{12}E^3 \quad \mathrm{mod}[-,-]$.

In addition an invertible element $\Phi \in A$ is said to be a *multiplicative moment map* if the following condition holds.

($\mathbb{P}$2) $\{\Phi, -\} = \frac{1}{2}(\Phi E + E\Phi)$.

Finally we say that P is *non-degenerate* if the following condition holds

($\mathbb{P}$3) The map $\Omega_A \oplus AEA \to D_A : (\eta, \delta) \mapsto \imath(P)(\eta) + \delta$ is surjective.

It was proved in [**14**] that ($\mathbb{P}$1) and ($\mathbb{P}$2) imply the corresponding properties on representation spaces. The same is also true for ($\mathbb{P}$3) as the following proposition shows.

PROPOSITION 5.2. *Assume that ($\mathbb{P}$1)($\mathbb{P}$3) hold. Then $\mathrm{tr}(P)$ defines a non-degenerate quasi-Poisson bracket on $\mathrm{Rep}(A, N)$*

PROOF. Applying the right exact functor $(-)_N$ (lemma 3.3.1) and using Proposition 2.3.1 together with Corollary 3.3.5 we find that the map

$$\Omega_{A_N} \oplus A_N \otimes \mathfrak{gl}_N \to \operatorname{Der}(A_N)$$

is surjective. This finishes the proof. □

The following result is a non-commutative version of [**4**, Thm. 2.5].

THEOREM 5.3. *Assume that (A, P) is a double quasi-Poisson algebra. Then $\tilde{\Omega}_A = \Omega_A \oplus AEA$ has the structure of a double Lie algebroid where the double bracket is defined as follows.*

$$\begin{aligned}
\{\!\{da, b\}\!\}_{\tilde{\Omega}A} &= \{\!\{a, b\}\!\} \\
\{\!\{da, db\}\!\}_{\tilde{\Omega}A} &= d\{\!\{a, b\}\!\} + \frac{1}{4}[b, [a, E \otimes 1 - 1 \otimes E]_*] \\
\{\!\{E, X\}\!\}_{\tilde{\Omega}A} &= X \otimes 1 - 1 \otimes X
\end{aligned}$$

for $a, b \in A$, $X \in T_A\tilde{\Omega}_A$ and where $[-,-]_*$ denotes the commutator for the inner A-bimodule structure on $AEA \otimes AEA$. Furthermore the map

$$\Omega_A \oplus AEA \to D_A$$

defined in ($\mathbb{P}3$) is a morphism of double Lie algebroids.

Our proof of this theorem is a rather painful direct computation. We omit the details. The fact that ($\mathbb{P}$3) defines a morphism of double Lie algebroids translates into the following proposition which is an analogue of [**14**, Prop. 3.5.1].

PROPOSITION 5.4. *The following are equivalent*

(1) *$\{\!\{-,-\}\!\}$ is a double quasi-Poisson bracket.*
(2) *The following identity holds for all $a, b \in A$:*

$$\{\!\{H_a, H_b\}\!\}_l - H'_{\{\!\{a,b\}\!\}} \otimes \{\!\{a, b\}\!\}'' = \frac{1}{4}[b, [a, E \otimes 1]_*]$$

(3) *The following identity holds for all $a, b \in A$:*

$$\{\!\{H_a, H_b\}\!\}_r - \{\!\{a, b\}\!\}' \otimes H_{\{\!\{a,b\}\!\}''} = -\frac{1}{4}[b, [a, 1 \otimes E]_*]$$

(4) *The following identity holds for all $a, b \in A$:*

$$\{\!\{H_a, H_b\}\!\} - H_{\{\!\{a,b\}\!\}} = \frac{1}{4}[b, [a, E \otimes 1 - 1 \otimes E]_*]$$

where we use the convention $H_{x' \otimes x''} = H_{x'} \otimes x'' + x' \otimes H_{x''}$.

PROOF. Below it will be convenient to use the notation $^\circ(-)$ introduced in (2.5) as well as the convention

$$(a_1 \otimes \cdots \otimes a_m)(b_1 \otimes \cdots \otimes b_n) = a_1 \otimes \cdots \otimes a_m b_1 \otimes \cdots b_n$$

and similarly for longer products. According to the proof of [**14**, Prop. 3.5.1] we have for $a, b, c \in A$

$$\{\!\{a, b, c\}\!\} = \tau_{23}((\{\!\{H_a, H_b\}\!\}_l - H_{\{\!\{a,b\}\!\}'}(-) \otimes \{\!\{a, b\}\!\}'')(c))$$

Also according to [**14**, §5] a bracket is quasi-double Poisson if it satisfies

$$\{\!\{-,-,-\}\!\} = \frac{1}{12}\{\!\{-,-,-\}\!\}_{E^3}$$

By the formulas in [**14**, §4] it follows

$$\{\!\{-,-,-\}\!\}_{E^3} = 3^\circ(E(a)^\circ E(b)^\circ E(c)^\circ)$$

Hence

$$(\{\!\{H_a, H_b\}\!\}_l - H_{\{\!\{a,b\}\!\}'}(-) \otimes \{\!\{a,b\}\!\}'')(c) = \frac{1}{4}\tau_{23}{}^\circ(E(a)^\circ E(b)^\circ E(c)^\circ) \tag{5.1}$$

A straightforward verification yields

$$(E(a)^\circ E(b)^\circ E(-)^\circ) = [b, [a, E \otimes 1]_*](c)$$

This proves the equivence between (1) and (2). The equivalence between (2) and (3) follows easily from the fact that $\{\!\{a,b\}\!\}_r = -\{\!\{b,a\}\!\}_l^\circ$. The sum of (2) and (3) yields (4). To go back we use projection. □

6. Quasi-bisymplectic algebras

The algebras we will introduce are a non-commutative analogue of quasi-Hamiltonian G-spaces introduced in [**2**] (see §4). By definition a *quasi-bisymplectic algebra* will be a triple (A, ω, Φ) where $\omega \in \mathrm{DR}^2(A)$ and $\Phi \in A^*$ satisfying the following conditions

($\mathbb{B}1$) $d\omega = \frac{1}{6}(\Phi^{-1}d\Phi)^3 \quad \mathrm{mod}[-,-]$.
($\mathbb{B}2$) $\imath_E\omega = \frac{1}{2}(\Phi^{-1}d\Phi + d\Phi\cdot\Phi^{-1})$
($\mathbb{B}3$) The map

$$D_A \oplus Ad\Phi A \to \Omega_A : (\delta,\eta) \mapsto \imath(\omega)(\delta) + \eta$$

is surjective.

Proposition 6.1. *If* (A,ω,Φ) *is a quasi-bisymplectic algebra then* $(\mathrm{Rep}(A,N), \mathrm{tr}(\omega), X(\Phi))$ *is a quasi-Hamiltonian* Gl_N*-space.*

Proof. It is easy to see that we have

$$X(\Phi^{-1}d\Phi) = X(\Phi)^*(\theta) \qquad X(d\Phi\cdot\Phi^{-1}) = X(\Phi)^*(\bar\theta) \tag{6.1}$$

and

$$X(\Phi)^*(\chi) = \frac{1}{6}\,\mathrm{tr}(\Phi^{-1}d\Phi)^3 \tag{6.2}$$

This implies (B1). To check (B2) we test it with $\xi = f_{ji}$. Then according to lemma 3.3.6 and Prop. 2.3.1 we have

$$i_{f_{ji}}(\mathrm{tr}(\omega)) = (\imath_E\omega)_{ij}$$

We also have

$$X(\Phi)^*(\theta+\bar\theta, f_{ji}) = (\Phi^{-1}d\Phi + d\Phi\cdot\Phi^{-1})_{ij}$$

so that we see that ($\mathbb{B}2$) implies (B2).

Now we consider (B3) (or rather its variant (B3')). Apply the functor $(-)_N$ to ($\mathbb{B}3$). Using Corollary 3.3.5 we find that the map

$$\mathrm{Der}(A_N) \oplus \sum_{ij} A_N(d\Phi)_{ij} \to \Omega_{A_N}$$

is surjective. But the forms $(d\Phi)_{ij}$ generate the same A_N-module as $X(\Phi)^*(\theta)_{ij} = (\Phi^{-1}d\Phi)_{ij}$, finishing the proof. □

REMARK 6.2. Note that the appearance of the element $(\Phi^{-1}d\Phi)^3 \in \mathrm{DR}^3(A)$ is rather natural in view of Example 2.1. This is the only non-zero element in $H^3(\mathrm{DR}(A))$ that can be constructed from the single element Φ.

Let us say that $P \in D^2A$, $\omega \in \Omega^2 A$, $\Phi \in A^*$ are *compatible* if the following identity holds for all $\delta \in D_A$:

$$(\imath(P)\circ i(\omega))(\delta) = \delta - \frac{1}{4}\delta(\Phi)''(E\Phi^{-1} - \Phi^{-1}E)\delta(\Phi)' \tag{$\mathbb{C}$}$$

PROPOSITION 6.3. *If P, ω, Φ are compatible then* $(\mathrm{tr}(P), \mathrm{tr}(\omega), X(\Phi))$ *are compatible as well (see* (4.3)*).*

PROOF. We apply the map $X(-)$. Using Cor. 3.3.5, (3.2) and Prop. 2.3.1 we find

$$\begin{aligned}
(\mathrm{tr}(P)\circ\mathrm{tr}(\omega))(\delta_{ij}) &= \delta_{ij} - \frac{1}{4}\delta(\Phi)''_{iu}(f_{vu}(\Phi^{-1})_{vw} - (\Phi^{-1})_{uv}f_{wv})\delta(\Phi)'_{wj}\\
&= \delta_{ij} - \frac{1}{4}(f_{vu}(\Phi^{-1})_{vw} - (\Phi^{-1})_{uv}f_{wv})\delta(\Phi)'_{wj}\delta(\Phi)''_{iu}\\
&= \delta_{ij} - \frac{1}{4}(f_{vu}(\Phi^{-1})_{vw} - (\Phi^{-1})_{uv}f_{wv})\delta_{ij}(\Phi_{wu})\\
&= \delta_{ij} - \frac{1}{4}(f_{vu}(\Phi^{-1})_{vw}i_{\delta_{ij}}(d\Phi_{wu}) - i_{\delta_{ij}}(d\Phi_{wu})(\Phi^{-1})_{uv}f_{wv})\\
&= (\mathrm{id} - \frac{1}{4}f_{vu}((\Phi^{-1}d\Phi)_{vu} - (d\Phi\cdot\Phi^{-1})_{vu}))(\delta_{ij})
\end{aligned}$$

where we have viewed the f_{ij} as vector fields on $\mathrm{Rep}(A,N)$. We are now done by (6.1). □

7. Compatibility

Our aim is to prove a non-commutative analogue of Theorem 4.2.

THEOREM 7.1. *Fix $\Phi \in A^*$.*

1. *For every $P \in (DA/[DA,DA])_2$ satisfying ($\mathbb{P}2$)($\mathbb{P}3$) there exists a unique $\omega \in (\Omega A/[\Omega A,\Omega A])_2$ satisfying ($\mathbb{B}2$)($\mathbb{B}3$) and ($\mathbb{C}$).*
2. *For every $\omega \in (\Omega A/[\Omega A,\Omega A])_2$ satisfying ($\mathbb{B}2$)($\mathbb{B}3$) there exists a unique $P \in (DA/[DA,DA])_2$ satisfying ($\mathbb{P}2$)($\mathbb{P}3$) and ($\mathbb{C}$).*
3. *If ω, P correspond to one another as in (1)(2) then the integrability conditions ($\mathbb{B}1$) and ($\mathbb{P}1$) are equivalent.*

We will prove this theorem in this section. Throughout we fix $\Phi \in A^*$ and we let $\omega \in (\Omega A/[\Omega A,\Omega A]))_2$, $P \in (DA/[DA,DA]))_2$ be such that ($\mathbb{B}2$) and ($\mathbb{P}2$) are satisfied. First we consider the following diagram which summarizes a number of relevant maps which we will use below.

$$\begin{array}{ccccccc}
\Omega_A & \xrightarrow{e} & AE^*A & \xrightarrow{T^0} & Ad\Phi A & \xrightarrow{c} & \Omega_A\\
\downarrow{\scriptstyle \imath(P)} & & \downarrow{\scriptstyle \jmath} & & \downarrow{\scriptstyle \imath} & & \downarrow{\scriptstyle \imath(P)}\\
D_A & \xrightarrow[e]{} & A(d\Phi)^*A & \xrightarrow[S^0]{} & AEA & \xrightarrow[c]{} & D_A\\
\downarrow{\scriptstyle \imath(\omega)} & & \downarrow{\scriptstyle \jmath} & & \downarrow{\scriptstyle \imath} & & \downarrow{\scriptstyle \imath(\omega)}\\
\Omega_A & \xrightarrow[e]{} & AE^*A & \xrightarrow[T^0]{} & Ad\Phi A & \xrightarrow[c]{} & \Omega_A
\end{array} \tag{7.1}$$

The following conventions are used: first of all we view AE^*A, $A(d\Phi)^*A$, AEA and $Ad\Phi A$ as free bimodules with one generator. Furthermore we regard the diagram as being doubly infinite with period $(3,2)$. The bimodules occur in pairs related by an obvious non-degenerate pairing: (D_A, Ω_A), (AE^*A, AEA) and $(A(d\Phi)^*A, A(d\Phi)A)$. We now define the maps.

(1) c stands for "canonical map".

(2) e is adjoint to c. The formulas for the two variants are as follows.

$$e(d\phi) = \phi E^* - E^*\phi$$
$$e(\delta) = \delta(\Phi)''(d\Phi)^*\delta(\Phi)'$$

(3) $\imath$ is the restricted version of $\imath(P)$ and $\imath(\omega)$. The two variants are given by ($\mathbb{P}2$) and ($\mathbb{B}2$). I.e. explicitly

$$\imath(d\Phi) = \frac{1}{2}(E\Phi + \Phi E)$$
$$\imath(E) = \frac{1}{2}(\Phi^{-1}d\Phi + d\Phi \cdot \Phi^{-1})$$

(4) $\jmath$ is adjoint to $-\imath$. The two variants are given by the following formulas

$$\jmath(d\Phi^*) = -\frac{1}{2}(\Phi^{-1}E^* + E^*\Phi^{-1})$$
$$\jmath(E^*) = -\frac{1}{2}(\Phi(d\Phi)^* + (d\Phi)^*\Phi)$$

(5) The maps S^0 and T^0 are adjoint to one another. They are respectively given by the following formulas.

$$S^0((d\Phi)^*) = E\Phi^{-1} - \Phi^{-1}E$$
$$T^0(E^*) = \Phi^{-1}d\Phi - d\Phi\Phi^{-1}$$

From the stated adjointness properties it follows that the (7.1) is self dual, up to sign. We have the following fact.

LEMMA 7.2. (1) *The diagram (7.1) is commutative.*

(2) *Consider the diagram as doubly infinite. For any configuration of arrows*

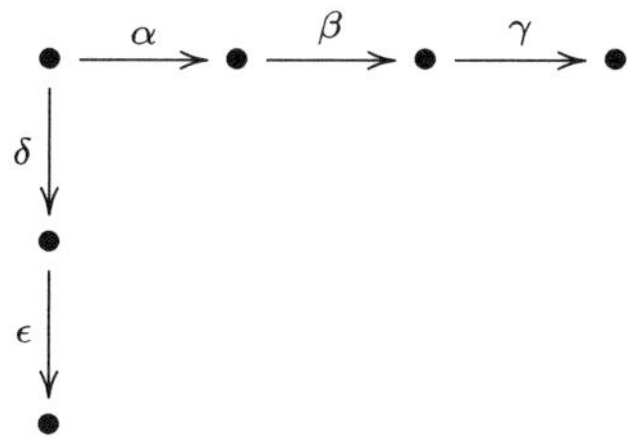

such that the vertical arrows do not involve $\imath(\omega), \imath(P)$ we have

$$\frac{1}{4}\gamma\beta\alpha + \epsilon\delta = 1 \tag{7.2}$$

(3) *If we contract in (7.1) the pairs of horizontal consecutive arrows that have Ω_A or D_A in the middle then the 2×2 subdiagrams in the resulting diagram are bicartesian.*

(4) *If* ($\mathbb{B}3$) *holds then any subdiagram*

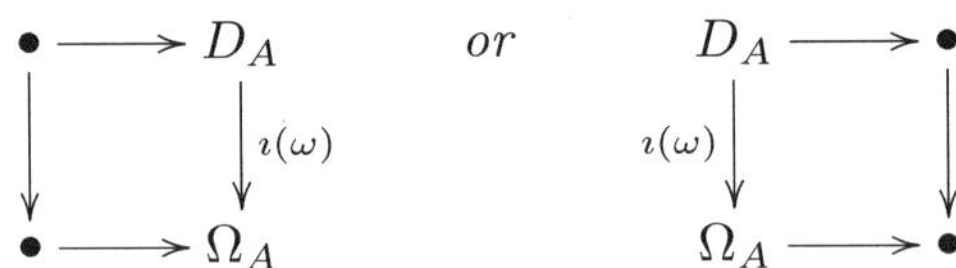

of (7.1) *is bicartesian.*

(5) *Similarly if* ($\mathbb{P}3$) *holds then any subdiagram*

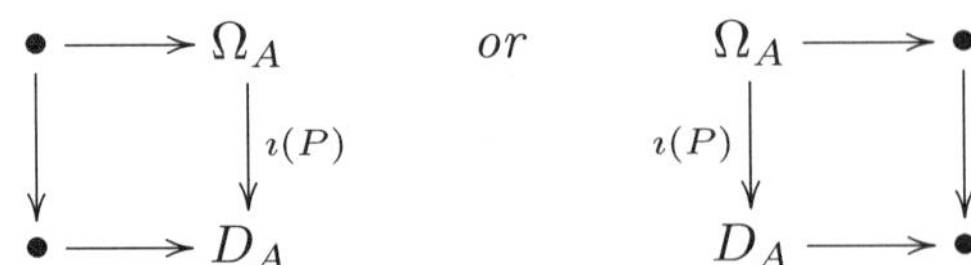

of (7.1) *is bicartesian.*

REMARK 7.3. To avoid confusion: statements in the above lemma which do not refer to $\imath(\omega)$ or $\imath(P)$ remain true if these maps are not present in the diagram.

PROOF. We leave the verification of (1) and (2) to the reader. For (3) will only give an example. E.g. we need to check in particular that the following diagram is bicartesian

$$\begin{array}{ccc} Ad\Phi A & \xrightarrow{ec} & AE^*A \\ {\scriptstyle \imath}\downarrow & & \downarrow{\scriptstyle \jmath} \\ AEA & \xrightarrow[ec]{} & A(d\Phi)^*A \end{array}$$

Put $R = A^e$, $\Phi_1 = \Phi \otimes 1$, $\Phi_2 = 1 \otimes \Phi$. The latter are two commuting invertible elements of R. Removing all confusing decoration the above diagram basically looks like

$$\begin{array}{ccc} R & \xrightarrow{\Phi_1-\Phi_2} & R \\ {\scriptstyle \Phi_1+\Phi_2}\downarrow & & \downarrow{\scriptstyle \Phi_1+\Phi_2} \\ R & \xrightarrow[\Phi_1-\Phi_2]{} & R \end{array}$$

It is now clear that this diagram is bicartesian (using the fact that Φ_1 and Φ_2 are invertible). The other 2×2-diagrams in (7.1) are similar.

(4) and (5) are similar. To prove (4) we only need to consider the first diagram as the second follows by duality. Assume that there are $\delta \in D_A$, $\eta \in Ad\Phi A$ such that

$$\imath(\omega)(\delta) = c(\eta) \tag{7.3}$$

We need to prove that there is a unique $\Delta \in AEA$ such that $c(\Delta) = \delta$, $\imath(\Delta) = \eta$. We need to worry only about existence since (2) implies that

$$(ec, \imath) : AEA \to A(d\Phi)^*A \oplus Ad\Phi A$$

is injective and hence $(c, \imath)$ is also injective.

Applying e to (7.3) we find $\jmath e(\delta) = e\imath(\omega)(c) = ec(\eta)$. By (3) there is an element $\delta' \in AEA$ such that $\imath(\delta') = \eta$, $ec(\delta') = e(\delta)$. Replacing η by $\eta - \imath(\delta')$, δ by $\delta - c(\delta')$ we reduce to the case $\eta = 0$, $e(\delta) = 0$, $\imath(\omega)(\delta) = 0$.

Now we note that ($\mathbb{B}3$) implies by duality that the following map is injective

$$(\imath(\omega), e) : D_A \to \Omega_A \oplus A(d\Phi)^* A$$

Hence $\delta = 0$ and we can take $\Delta = 0$. □

Using the maps in (7.1) the compatibility condition ($\mathbb{C}$) can be reformulated as

$$\imath(P)\imath(\omega) = 1 - \frac{1}{4}S \tag{7.4}$$

where $S = cS^0 e$. In fact it will be convenient to give some other equivalent formulations of (7.4). Consider the following matrices

$$\bar{\omega} = \begin{pmatrix} \imath(\omega) & c \\ \frac{1}{4}S^0 e & -\imath \end{pmatrix} \qquad \bar{P} = \begin{pmatrix} \imath(P) & c \\ \frac{1}{4}T^0 e & -\imath \end{pmatrix}$$

and view them as maps between $D_A \oplus Ad\Phi A$ and $\Omega_A \oplus AEA$.

Furthermore consider the matrices

$$\tilde{\omega} = \begin{pmatrix} \imath(\omega) & 1 \\ \frac{1}{4}S & -\imath(P) \end{pmatrix} \qquad \tilde{P} = \begin{pmatrix} \imath(P) & 1 \\ \frac{1}{4}T & -\imath(\omega) \end{pmatrix}$$

and view them as maps between $D_A \oplus \Omega_A$ and $\Omega_A \oplus D_A$.

PROPOSITION 7.4. *The following conditions are all equivalent.*

$$\imath(P)\imath(\omega) = 1 - \frac{1}{4}S \tag{7.5}$$

$$\imath(\omega)\imath(P) = 1 - \frac{1}{4}T \tag{7.6}$$

$$\bar{P}\bar{\omega} = \mathrm{id} \tag{7.7}$$

$$\bar{\omega}\bar{P} = \mathrm{id} \tag{7.8}$$

$$\tilde{P}\tilde{\omega} = \mathrm{id} \tag{7.9}$$

$$\tilde{\omega}\tilde{P} = \mathrm{id} \tag{7.10}$$

PROOF. (7.5) and ($\mathbb{B}$.6) are equivalent since they are adjoint. To understand (7.7) we compute the product

$$\bar{P}\bar{\omega} = \begin{pmatrix} \imath(P) & c \\ \frac{1}{4}T^0 e & -\imath \end{pmatrix} \begin{pmatrix} \imath(\omega) & c \\ \frac{1}{4}S^0 e & -\imath \end{pmatrix} = \begin{pmatrix} \imath(P)\imath(\omega) + \frac{1}{4}S & \imath(P)c - c\imath \\ \frac{1}{4}(T^0 e\imath(\omega) - \imath S^0 e) & \frac{1}{4}T^0 ec + \imath\imath \end{pmatrix} = \begin{pmatrix} \imath(P)\imath(\omega) + \frac{1}{4}S & 0 \\ 0 & 1 \end{pmatrix}$$

where we have used lemma 7.2. It is now clear that (7.5) and (7.7) are equivalent. Similarly (7.6) and (7.8) are equivalent. To understand (7.9) we write the product out again explicitly.

$$\tilde{P}\tilde{\omega} = \begin{pmatrix} \imath(P) & 1 \\ \frac{1}{4}T & -\imath(\omega) \end{pmatrix} \begin{pmatrix} \imath(\omega) & 1 \\ \frac{1}{4}S & -\imath(P) \end{pmatrix} = \begin{pmatrix} \imath(P)\imath(\omega) + \frac{1}{4}S & 0 \\ 0 & \imath(\omega)\imath(P) + \frac{1}{4}T \end{pmatrix}$$

where we have used lemma 7.2 again. It follows that (7.9) is equivalent to (7.5)(7.6) simultaneously. The same holds for (7.9). □

PROOF OF THEOREM 7.1(1). If $\imath(\omega)$ exists it will have the following two properties for $\eta \in \Omega_A$, $\delta \in AEA$

$$\imath(\omega)\imath(P)(\eta) = \eta - \frac{1}{4}T(\eta) \tag{7.11}$$

$$\imath(\omega)c(\delta) = c\imath(\delta) \tag{7.12}$$

where we have used lemmas 7.2 and Prop. 7.4. Since we are assuming ($\mathbb{P}3$) every element of D_A can be written as a sum $\imath(P)(\eta) + c(\delta)$ for $\eta \in \Omega_A$ and $\delta \in AEA$. Hence the properties (7.11)(7.12) characterize $\imath(\omega)$ uniquely (and ω as well by Prop. 3.1.4).

We still need to prove two things.

(1) The equations (7.11)(7.12) are non-contradictory.
(2) The resulting $\imath(\omega)$ is actually anti-symmetric (so that it genuinely comes from an element $\omega \in \Omega_A^2$).
(3) $\imath(\omega)$ satisfies ($\mathbb{B}3$).

We discuss (1) first. Suppose that $\imath(P)(\eta) = c(\delta)$. Then by lemma 7.1 there exist $\eta_0 \in Ad\Phi_p A$ such that $\eta = c(\eta_0)$, $\delta = \imath(\eta_0)$. Then

$$\begin{aligned} \eta - \frac{1}{4}T(\eta) &= c(\eta_0) - \frac{1}{4}cT^0 ec(\eta_0) \\ &= c(\eta_0 - \frac{1}{4}T^0 ec(\eta_0)) \\ &= c\imath\imath(\eta_0) \\ &= c\imath(\delta) \end{aligned}$$

where we have used lemma 7.1.

Now we prove (2). To avoid confusion we write X for the map $\imath(\omega)$ we have constructed. Thus we have

$$\begin{aligned} X\imath(P) &= 1 - \frac{1}{4}T \\ Xc &= c\imath \end{aligned} \tag{7.13}$$

Dualizing these equations we get

$$\begin{aligned} -\imath(P)X^* &= 1 - \frac{1}{4}S \\ eX^* &= -\jmath e \end{aligned} \tag{7.14}$$

We write (7.13) in matrix form.

$$X(\imath(P)\; c) = (1 - \frac{1}{4}T\; c\imath)$$

Left multiplication by $(i(P) \;-e)^t$ yields

$$\begin{aligned} \begin{pmatrix} i(P) \\ -e \end{pmatrix} X(\imath(P)\; c) &= \begin{pmatrix} \imath(P) \\ -e \end{pmatrix} (1 - \frac{1}{4}T\; c\imath) \\ &= \begin{pmatrix} 1 - \frac{1}{4}S \\ -\jmath e \end{pmatrix} (\imath(P)\; c) \end{aligned}$$

Since $(\imath(P)\; c)$ is injective we conclude

$$\begin{aligned} \imath(P)(X) &= 1 - \frac{1}{4}S \\ -eX &= -\jmath e \end{aligned}$$

Comparing with (7.14), and using that $(\imath(P) \;-e)$ is surjective we conclude that indeed $X^* = -X$.

We finish the proof by noting that (3) follows immediately from (7.6). $\square$

REMARK 7.5. Assume that P and ω are compatible. We have unearthed quite a bit of structure in the diagram (7.1). I have not seen this type of structure at other places.

(1) The diagram is $(3, 2)$ periodic.
(2) Denote the horizontal and vertical maps by h and v respectively. h and v satisfy the following equations:

$$hv = vh$$
$$v^2 + \frac{1}{4}h^3 = 1$$

(3) Every 2×2 subdiagram is bicartesian.
(4) The diagram is self dual up to a sign which must be applied to the vertical maps.

REMARK 7.6. If we view the maps $\tilde{\omega}$ and $\tilde{P}$ as endomorphisms of $D_A \oplus \Omega_A$ then in matrix form they look like

$$\tilde{\omega} = \begin{pmatrix} \frac{1}{4}S & -\imath(P) \\ \imath(\omega) & 1 \end{pmatrix} \qquad \tilde{P} = \begin{pmatrix} 1 & \imath(P) \\ -\imath(\omega) & \frac{1}{4}T \end{pmatrix}$$

We define a symmetric non-degenerate pairing between $D_A \oplus \Omega_A$ and itself using the formula

$$\langle(\delta, \eta), (\delta', \eta')\rangle = \langle\delta, \eta'\rangle + \langle\delta', \eta\rangle$$

For this pairing it is easy to check that $\tilde{\omega}$ is adjoint to $\tilde{P}$. In other words $\tilde{\omega}$ is a unitary transformation of $D_A \oplus \Omega_A$. This suggest a connection with a non-commutative version of Dirac geometry (yet to be created). In the commutative case this connection is well understood [**4**].

Part (2) or Theorem 7.1 is proved in the same way as (1). It remains to prove (3), i.e. the equivalence between the integrability conditions on P and ω.

To do this we may follow a similar method as in [**14**, Appendix]. We start by expressing the integrability condition in terms of Hamiltonian vector fields (see §2.3). Throughout we assume that P, ω, Φ satisfy $(\mathbb{P}2)(\mathbb{P}3)(\mathbb{B}2)(\mathbb{B}3)$ as well as the compatibility condition $(\mathbb{C})$. We let $\{\!\{-,-\}\!\}$ be the double bracket on A induced by P (see (2.4)). We recall that by Proposition 5.4 P is quasi-Poisson if and only if for all $a, b \in A$ we have the following identity in $D_A \otimes A$

$$\{\!\{H_a, H_b\}\!\}_l - H_{\{\!\{a,b\}\!\}'}(-) \otimes \{\!\{a, b\}\!\}'' = \frac{1}{4}[b, [a, E \otimes 1]_*] \tag{7.15}$$

In order to work with Hamiltonian vector fields one needs some formulas which are straightforward verifications

LEMMA 7.7. *One has*

$$i_{H_a}(\omega) = da - \frac{1}{4}[a, \Phi^{-1}d\Phi - d\Phi\Phi^{-1}]$$
$$i_{H_a}(d\Phi) = -\frac{1}{2}[a, \Phi \otimes 1 + 1 \otimes \Phi]_*$$

PROOF. The first formula follows from the following computation

$$\begin{aligned}\imath(\omega)(H_a) &= \imath(\omega)\imath(P)(da)\\ &= (1-\frac{1}{4}T)(da)\\ &= da - \frac{1}{4}\sum_p [a, \Phi^{-1}d\Phi - d\Phi\Phi^{-1}]\end{aligned}$$

Now we prove the second formula We have

$$i_{H_a}(d\Phi) = H_a(\Phi) = \{\!\{a, \Phi\}\!\} = -\{\!\{\Phi, a\}\!\}^\circ$$

and according to [**14**, Def 5.1.4]

$$-\{\!\{\Phi, a\}\!\}^\circ = -\frac{1}{2}((\Phi E + E\Phi)(a))^\circ$$

We have

$$(uEv)(a) = av \otimes u - v \otimes ua$$

and thus

$$(uEv)(a)^\circ = u \otimes av - ua \otimes v$$

so that we get

$$i_{H_a}(d\Phi) = -\frac{1}{2}(\Phi \otimes a - \Phi a \otimes 1 + 1 \otimes a\Phi - a \otimes \Phi)$$

as asserted. □

Using these formulas and a tedious computation one verifies the following lemma

LEMMA 7.8. *The formula* (7.15) *always holds when evaluated on* $d\Phi \otimes 1$.

From this lemma we deduce

LEMMA 7.9. *The element* P *defines a quasi-Poisson algebra if and only if the following identity in* $\Omega_A \otimes A$ *holds*

$$\imath_{\{\!\{H_a, H_b\}\!\}_l'}(\omega) \otimes \{\!\{H_a, H_b\}\!\}_l'' - \imath_{H_{\{\!\{a,b\}\!\}'}}(\omega) \otimes \{\!\{a, b\}\!\}'' = \frac{1}{4}[b, [a, \imath_E(\omega) \otimes 1]_*] \tag{7.16}$$

PROOF. By lemma 7.2 we have that $(\imath(\omega), e) : D_A \to \Omega_A \oplus A(d\Phi)^* A$ is injective. Hence to check that $\delta \in D_A$ is zero it is sufficient to verify that $\imath_\delta(\omega) = 0$ and $e(\delta) = \delta(\Phi)''(d\Phi)^*\delta(\Phi)' = 0$. The latter holds if $\delta(\Phi) = 0$. The lemma now follows by writing (7.15) as $\delta' \otimes \delta'' = 0$ with $\delta' \in D_A$ and using Lemma 7.8. □

To compute the part of (7.16) involving $\{\!\{H_a, H_b\}\!\}_l$ we may use the formula [**14**, (A.6)].

$$i_{H_a}\mathcal{L}_{H_b} - \tau_{12}L_{H_b}\imath_{H_a} = \imath_{\{\!\{H_a, H_b\}\!\}_l'} \otimes \{\!\{H_a, H_b\}\!\}_l'' + \{\!\{H_a, H_b\}\!\}_r' \otimes \imath_{\{\!\{H_a, H_b\}\!\}_r''}$$

Here $L_\delta = di_\delta + i_\delta d$ and $\mathcal{L}_\delta = {}^\circ L_\delta$. Applying the left hand side of this formula to ω we see that everything is computable except the term $i_{H_a}\imath_{H_b}d\omega$ appearing as part of $i_{H_a}\mathcal{L}_{H_b}(\omega)$. After a long and tedious computation we arrive at the following lemma.

LEMMA 7.10. *The formula* (7.16) *holds if and only if*

$$pr_1\, i_{H_a} \imath_{H_b} d\omega = \frac{1}{6}\, pr_1\, i_{H_a} \imath_{H_b} (\Phi^{-1} d\Phi)^3$$

holds for al a, b.

PROOF OF THEOREM 7.1(3). It follows from lemma 7.10 that if ω is integrable then so is P. To prove the converse we first do some computations

$$\begin{aligned}
\imath_E d\omega &= -d(\imath_E \omega) \qquad [\mathbf{14}, \text{(A,7)}] \\
&= -\frac{1}{2} d(\Phi^{-1} d\Phi + d\Phi \cdot \Phi^{-1}) \\
&= \frac{1}{2} (\Phi^{-1} d\Phi \cdot \Phi^{-1} d\Phi - d\Phi \cdot \Phi^{-1} d\Phi \cdot \Phi^{-1})
\end{aligned}$$

On the other hand

$$\begin{aligned}
\frac{1}{6} \imath_E (\Phi^{-1} d\Phi)^3 &= \frac{1}{6}{}^\circ i_E (\Phi^{-1} d\Phi)^3 \\
&= \frac{1}{6}{}^\circ (\Phi^{-1} (\Phi \otimes 1 - 1 \otimes \Phi) \Phi^{-1} d\Phi \cdot \Phi^{-1} d\Phi + \cdots) \\
&= \frac{1}{2} (\Phi^{-1} d\Phi \cdot \Phi^{-1} d\Phi - d\Phi \cdot \Phi^{-1} d\Phi \cdot \Phi^{-1})
\end{aligned}$$

So that we have deduced

$$\imath_E d\omega = \frac{1}{6} \imath_E (\Phi^{-1} d\Phi)^3$$

Write

$$\eta = d\omega - \frac{1}{6} (\Phi^{-1} d\Phi)^3$$

Assuming that P is integrable we need to prove that $\eta = 0$ (modulo commutators). We already know by lemma 7.10 that $\mathrm{pr}_1\, i_{H_a} \imath_{H_b} \eta = 0$ and by the computation above we have $\imath_E \eta = 0$. By [**14**, Lemma A.5.2] we have $\imath(da) = H_a$. Hence by ($\mathbb{P}3$) we have $\sum_{a \in A} A H_a A + A E A = D_A$. We observe

$$\begin{aligned}
\mathrm{pr}_1\, i_E \imath_{H_b} \eta &= -\,\mathrm{pr}_1\, \tau_{12} i_{H_b} \imath_E \eta \qquad [\mathbf{14}, \text{(A.5)}] \\
&= 0
\end{aligned}$$

Hence by Prop. 3.1.5(2) it follows that $\imath_{H_b} \eta = 0$. Using $\imath_E \eta = 0$ once again we conclude by Prop. 3.1.5(1) that $\eta = 0$ (modulo commutators). □

8. More general base rings and quivers

8.1. Generalities. In this section we show that the double quasi-Poisson brackets constructed on (localized) path algebras of double quivers are non-degenerate and hence Theorem 7.1 applies to them.

However first we note that for quivers it is more natural to use as base ring not a field but a direct sum of fields indexed by the vertices. In [**6**] it was shown how to set up the theory over an arbitrary semi-simple base ring B. In [**14**] we worked over the base ring $B = ke_1 + \cdots + ke_n$ with the e_p's being commuting idempotents. Since we rely on results from [**14**] we will do the same in this section.

So assume that B is as in the previous paragraph. We use differentials and polyvector fields relative to B (relevant notations: $\Omega_{A/B}$, $\Omega_B A$, $D_{A/B}$, $D_B A$). In this setting the canonical element E is defined as $\sum_p E_p$ where

$$E_p(a) = a e_p \otimes e_p - e_p \otimes e_p a$$

A (multiplicative) moment map Φ is now of the form $\sum_p \Phi_p$ with $\Phi \in e_p A e_p$. With these conventions the definitions and results in this paper go through verbatim. We will accept this without further discussion. For use below we introduce the following convention. If $c = e_p A e_q$ then, changing standard terminology, an *inverse* of c is an element c^{-1} of $e_q A e_p$ such that $c \cdot c^{-1} = e_p$, $c^{-1}c = e_q$. It is easy to see that c^{-1} is uniquely determined.

8.2. Fusion. As in [**14**] our application to quivers depends on a process called fusion (introduced in the commutative case in [**1**]). In the case of quivers fusion amounts to gluing vertices but it is beneficial to work somewhat generally. We first construct and algebra $\bar{A}$ from A by formally adjoining two variables e_{12}, e_{21} satisfying the usual matrix relations $e_{uv}e_{wt} = \delta_{wv}e_{ut}$ (with $e_{ii} = e_i$). The fusion algebra of A along e_1, e_2 is defined as

$$A^f = \epsilon \bar{A} \epsilon$$

where $\epsilon = 1 - e_2$. Clearly $\bar{A}$ is a $\bar{B}$-algebra and A^f is a B^f-algebra. We will identify B^f with $ke_1 + ke_3 + \cdots + ke_n$.

If $a \in A$ then we consider it as an element of $\bar{A}$ and we write a^f for $\epsilon a \epsilon + e_{12} a e_{21}$. We extend this convention to forms and polyvector fields. Note that in [**14**, §5.3] it was shown than the operations $\bar{(-)}$ and $(-)^f$ are compatible with the formation of $D_B A$ and its Schouten bracket. A similar result is true for $\Omega_B A$. The following result was proved in [**14**]

THEOREM 8.2.1. [**14**, Thm 5.3.1, 5.3.2] *Assume that (A, P, Φ) is a Hamiltonian double quasi-Poisson algebra (smooth as always). Then the same is true for (A^f, P^{ff}, Φ^{ff}) where*

$$P^{ff} = P^f - \frac{1}{2}E_1^f E_2^f \qquad \textit{and} \qquad \Phi_i^{ff} = \begin{cases} \Phi_1^f \Phi_2^f & \textit{if } i = 1 \\ \Phi_i^f & \textit{if } i > 2 \end{cases}$$

In this section we prove the following result.

PROPOSITION 8.2.2. *Assume that (A, P, Φ) is a non-degenerate Hamiltonian double quasi-Poisson algebra. Then the same is true for (A^f, P^{ff}, Φ^{ff}).*

PROOF. In order to avoid confusing notations we define $F_p \in D_{B^f}(A^f)$ for $p \neq 2$ by $F_p(a) = ae_p \otimes e_p - e_p \otimes e_p a$. It follows from [**14**, (5.3)] that

$$F_p = \begin{cases} E_1^f + E_2^f & \text{if } p = 1 \\ E_p^f & \text{otherwise} \end{cases} \tag{8.1}$$

Since A is non-degenerate the following map is surjective

$$(\imath(P), c) : \Omega_{A/B} \oplus \sum_p AE_pA \to D_{A/B}$$

From this we deduce (by base extension) that the following map is also surjective.

$$(\imath(\bar{P}), c) : \Omega_{\bar{A}/\bar{B}} \oplus \sum_p \bar{A}\bar{E}_p\bar{A} \to D_{\bar{A}/\bar{B}}$$

By [**14**, Prop 4.2.1, lemma A.5.2] we have

$$\imath(P)(db) = -\{P, b\} \tag{8.2}$$

Compatibility of fusion with Schouten brackets yields that

$$(\imath(P^f), c) : \Omega_{A^f/B^f} \oplus \sum_p \epsilon \bar{A} \bar{E}_p \bar{A} \epsilon \to D_{A^f/B^f}$$

is also surjective. In addition we have $\epsilon \bar{A} \bar{E}_p \bar{A} \epsilon = A^f E_p^f A^f$.

It follows easily that

$$(\imath(P^{ff}), c) : \Omega_{A^f/B^f} \oplus \sum_p A^f E_p^f A^f \to D_{A^f/B^f}$$

is surjective as well. Using (8.1) we see that the current proposition is proved provided we can show that E_1^f is in the image of $\imath(P^{ff})$, modulo $(F_p)_p$. We do this next. We have

$$\imath(P^{ff})(d\Phi_2^f) = -\{P^f - \frac{1}{2} E_1^f E_2^f, \Phi_2^f\}$$

We use some formulas we have already proved. I.e. the formula before [**14**, (5.5)] yields

$$\begin{aligned} \{P^f, \Phi_2^f\} &= -\frac{1}{2}(E_2^f \Phi_2^f + \Phi_2^f E_2^f) \\ &= -\frac{1}{2}((F_1 - E_1^f)\Phi_2^f + \Phi_2^f(F_1 - E_1^f)) \\ &= -\frac{1}{2}(F_1 \Phi_2^f + \Phi_2^f F_1) + \frac{1}{2}(E_1^f \Phi_2^f + \Phi_2^f E_1^f) \end{aligned}$$

We also have (using the formulas after [**14**, (5.5)])

$$\begin{aligned} \{\!\{E_1^f E_2^f, \Phi_2^f\}\!\} &= E_1^f * \{\!\{E_2^f, \Phi_2^f\}\!\} - \{\!\{E_1^f, \Phi_2^f\}\!\} * E_2^f \\ &= E_1^f * (\Phi_2^f \otimes e_1 - e_1 \otimes \Phi_2^f) \\ &= \Phi_2^f \otimes E_1^f - E_1^f \otimes \Phi_2^f \end{aligned}$$

("$*$" represents the inner bimodule structure) so that we get

$$\{E_1^f E_2^f, \Phi_2^f\} = \Phi_2^f E_1^f - E_1^f \Phi_2^f$$

Hence

$$\begin{aligned} \{P^f - \frac{1}{2} E_1^f E_2^f, \Phi_2^f\} &= -\frac{1}{2}(F_1 \Phi_2^f + \Phi_2^f F_1) + \frac{1}{2}(E_1^f \Phi_2^f + \Phi_2^f E_1^f) - \frac{1}{2}(\Phi_2^f E_1^f - E_1^f \Phi_2^f) \\ &= -\frac{1}{2}(F_1 \Phi_2^f + \Phi_2^f F_1) + E_1^f \Phi_2^f \end{aligned}$$

So we are done. □

8.3. Quivers. Below we assume that Q is a finite quiver whose vertices are indexed from 1 to n. We also use Q to refer to the set of arrows of Q. The head and tail of an arrow a are denoted by $h(a)$, $t(a)$ respectively.

Associated to Q is the double quiver which has the same vertices as Q and arrows $\{a, a^* \mid a \in Q\}$ where $h(a^*) = t(a)$, $t(a^*) = h(a)$. It will be convenient to write $(a^*)^* = a$ and to define for $a \in \bar{Q}$, $\epsilon(a) = 1$ if $a \in Q$ and $\epsilon(a) = -1$ otherwise. We let A be the path algebra of $k\bar{Q}$ to which we adjoin the inverses of $1 + aa^*$ for all $a \in \bar{Q}$.

For $a \in \bar{Q}$ one has a corresponding "partial derivative" in $D_{A/B}$ defined by

$$\frac{\partial b}{\partial a} = \begin{cases} e_{t(a)} \otimes e_{h(a)} & \text{if } a = b \\ 0 & \text{otherwise} \end{cases}$$

It is easy to see that $D_{A/B}$ is a projective A-bimodule with generators $\partial/\partial a \in e_{h(a)} D_{A/B} e_{t(a)}$. Note that $\partial/\partial a$ goes in the "opposite direction" as a.

According to [**14**, Thm 6.7] A has a Hamiltonian double quasi-Poisson structure given by

$$P = \frac{1}{2}\left(\sum_{a\in\bar{Q}}\left(\epsilon(a)(1+a^*a)\frac{\partial}{\partial a}\frac{\partial}{\partial a^*}\right) - \sum_{a<b\in\bar{Q}}\left(\frac{\partial}{\partial a^*}a^* - a\frac{\partial}{\partial a}\right)\left(\frac{\partial}{\partial b^*}b^* - b\frac{\partial}{\partial b}\right)\right)$$

$$\Phi = \prod_{a\in\bar{Q}}(1+aa^*)^{\epsilon(a)} \tag{8.3}$$

where "<" refers to an arbitrary ordering on the edges of Q, which is also used to order the terms in the product (8.3).

PROPOSITION 8.3.1. *This Hamiltonian double quasi-Poisson structure is non-degenerate.*

PROOF. This Hamiltonian double quasi-Poisson structure was constructed in [**14**] using fusion starting from (multiple copies of) the following basic quiver.

$$1 \underset{a^*}{\overset{a}{\rightleftarrows}} 2 \tag{8.4}$$

By Proposition 8.2.2 we may assume that $\bar{Q}$ is equal to (8.4). The formula for P then simplifies to

$$P = \frac{1}{2}\left((1+a^*a)\frac{\partial}{\partial a}\frac{\partial}{\partial a^*} - (1+aa^*)\frac{\partial}{\partial a^*}\frac{\partial}{\partial a}\right)$$

and by [**14**, (6.2)]

$$E_1 = \frac{\partial}{\partial a^*}a^* - a\frac{\partial}{\partial a}$$
$$E_2 = \frac{\partial}{\partial a}a - a^*\frac{\partial}{\partial a^*}$$

To check non-degeneracy we compute $\imath(P)(da)$ and $\imath(P)(da^*)$. We find

$$\begin{aligned}\imath(P)(da) &= \frac{1}{2}{}^\circ\left((1+a^*a)i_{da}\frac{\partial}{\partial a}\frac{\partial}{\partial a^*} + (1+aa^*)\frac{\partial}{\partial a^*}i_{da}\frac{\partial}{\partial a}\right)\\ &= \frac{1}{2}\left(\frac{\partial}{\partial a^*}(1+a^*a) + (1+aa^*)\frac{\partial}{\partial a^*}\right)\end{aligned}$$

A similar computation yields

$$\imath(P)(da^*) = -\frac{1}{2}\left((1+a^*a)\frac{\partial}{\partial a} + \frac{\partial}{\partial a}(1+aa^*)\right)$$

We have

$$\begin{aligned} a^*E_1 + E_2a^* &= -a^*a\frac{\partial}{\partial a} + \frac{\partial}{\partial a}aa^* \\ &= -(1+a^*a)\frac{\partial}{\partial a} + \frac{\partial}{\partial a}(1+aa^*) \end{aligned}$$

$$\begin{aligned} E_1a + aE_2 &= \frac{\partial}{\partial a^*}a^*a + -aa^*\frac{\partial}{\partial a^*} \\ &= \frac{\partial}{\partial a^*}(1+a^*a) - (1+aa^*)\frac{\partial}{\partial a^*} \end{aligned}$$

Hence

$$\begin{aligned} \imath(P)(da) &= (1+aa^*)\frac{\partial}{\partial a^*} \qquad \mathrm{mod}(E_1,E_2) \\ \imath(P)(da^*) &= -(1+a^*a)\frac{\partial}{\partial a} \qquad \mathrm{mod}(E_1,E_2) \end{aligned}$$

Since both $(1+aa^*)$ and $(1+a^*a)$ are invertible we are done. □

References

[1] A. Alekseev, Y. Kosmann-Schwarzbach, and E. Meinrenken, *Quasi-Poisson manifolds*, Canad. J. Math. **54** (2002), no. 1, 3–29.

[2] A. Alekseev, A. Malkin, and E. Meinrenken, *Lie group valued moment maps*, J. Differential Geom. **48** (1998), no. 3, 445–495.

[3] R. Bocklandt and L. Le Bruyn, *Necklace Lie algebras and noncommutative symplectic geometry*, Math. Z. **240** (2002), no. 1, 141–167.

[4] H. Bursztyn and M. Crainic, *Dirac structures, momentum maps, and quasi-Poisson manifolds*, The breadth of symplectic and Poisson geometry (Boston, MA), Progr. Math., vol. 232, Birkhäuser Boston, Boston, MA, 2005, pp. 1–40.

[5] W. Crawley-Boevey, *A note on non-commutative Poisson structures*, math.QA/0506268.

[6] W. Crawley-Boevey, P. Etingof, and V. Ginzburg, *Noncommutative geometry and quiver algebras*, Adv. Math. **209** (2007), no. 1, 274–336.

[7] V. Ginzburg, *Double derivations and cyclic homology*, math.AG/050230.

[8] V. Ginzburg, *Non-commutative symplectic geometry, quiver varieties, and operads*, Math. Res. Lett. **8** (2001), no. 3, 377–400.

[9] S. Helgason, *Differential geometry, Lie groups, and symmetric spaces*, Graduate Studies in Mathematics, vol. 34, American Mathematical Society, Providence, RI, 2001, Corrected reprint of the 1978 original.

[10] M. Kontsevich and A. Rosenberg, *Noncommutative smooth spaces*, The Gelfand Mathematical Seminars, 1996–1999 (Boston, MA), Birkhäuser Boston, Boston, MA, 2000, pp. 85–108.

[11] M. Kontsevich, *Formal (non)commutative symplectic geometry*, The Gel′fand Mathematical Seminars, 1990–1992 (Boston, MA), Birkhäuser Boston, Boston, MA, 1993, pp. 173–187.

[12] L. Le Bruyn, *Noncommutative Geometry and Caley-smooth Orders*, Pure and Applied Mathematics, vol. 250, Chapman & Hall/CRC, 2008.

[13] J.-L. Loday, *Cyclic homology*, second ed., Grundlehren der Mathematischen Wissenschaften [Fundamental Principles of Mathematical Sciences], vol. 301, Springer-Verlag, Berlin, 1998, Appendix E by María O. Ronco, Chapter 13 by the author in collaboration with Teimuraz Pirashvili.

[14] M. Van den Bergh, *Double Poisson algebras*, to appear in Trans. Amer. Math. Soc.

Departement WINI, Universiteit Hasselt, 3090 Diepenbeek, Belgium
E-mail address: michel.vandenbergh@uhasselt.be

Contemporary Mathematics
Volume **450**, 2008

Locally noncommutative space-times

Stefan Waldmann

ABSTRACT. A new concept for noncommutative space-times is reviewd: the noncommutativity is only present for small distances in the product space of pairs of points. Using a diffeomorphism of a small neighborhood of the diagonal to a neighborhood of the zero section of the tangent bundle the non-commutative structure is encoded using a vertical Poisson structure on TM together with a formal star product quantizing it. Several consequences of the verticality requirement are analyzed. After a detailed discussion of states also a C^*-algebraic version of the deformation is presented, based on Rieffel's quantization.

1. Introduction and Motivation

In this work I shall discuss a recent approach to noncommutative space-times obtained from a joint work with Dorothea Bahns [**2**] and Jakob Heller and Nikolai Neumaier [**9**]. However, before doing so, there are some words to be said about the whole motivation for noncommutative models of space-time as they are presently discussed very actively in mathematical physics.

It is common to believe that on one hand, quantum theory is a fundamental theory which describes nature on microscopic scales very accurately. Most prominent examples for quantum theoretical models are the quantum field theories and in particular the standard model of particle physics. On the other hand, general relativity is the most sophisticated theory for the interactions between space-time geometry and matter content at macroscopic and cosmological scales. The conceptual problem arising for a mathematical physicist is the inconsistency of both of these 'fundamental' theories: a rather simple *Gedankenexperiment* shows that at very small scales, i.e. at the Planck length $\ell_{\text{Planck}} \cong 1.6 \times 10^{-35}\text{m}$, one encounters contradictory predictions if both theories are kept unchanged, see e.g. [**8**] for a discussion. Presently, nobody has a good intuition how the two theories can be combined to overcome these difficulties: unfortunately, there is no reasonable and convincing theory of 'quantum gravity' available.

A much more modest approach to a quantum gravity and hence quantum geometry is behind the idea of noncommutative space-times: roughly speaking, one

1991 *Mathematics Subject Classification.* Primary: 53D55 Secondary 81T75, 53D17, 46L87.
Key words and phrases. Star Products, Deformation Quantization, Poisson Geometry, Noncommutative Differential Geometry.

considers a noncommutative algebra of 'functions' as a replacement for the commutative space-time M, usually obtained by a deformation of the function algebra on M, very much in the spirit of Connes' noncommutative geometry [**6**]. More physically speaking, the quantum nature of gravity amounts of imposing uncertainty relations to the coordinate functions whence points can no longer be localized sharply. This has to be seen as an effective theory where the details of a (not yet available) theory for quantum gravity are subsumed into the background of noncommutativity of space-time. In particular, the usual models do not consider a back-reaction from the matter content to the noncommutative structures. There are of course many ways to implement a noncommutative structure on space-time, but common to all these approaches is the hope that by introducing this noncommutativity on one hand the singularities of general relativity will become more smooth. On the other hand, the singularities of quantum field theories on such a noncommutative space-time, when properly defined, are also hopefully less singular.

To understand the criticism on the existing models, let me present one example, the prominent noncommutative Minkowski space: one considers the classical Minkowski space-time $M = \mathbb{R}^{1+n}$, equipped with the usual Minkowski metric $\eta = \operatorname{diag}(+, -, \ldots, -)$. To encode a noncommutativity, one chooses a *constant* bivector field $\theta \in \Gamma^\infty(\Lambda^2 TM)$, which therefore is a Poisson structure. Then we have the usual Weyl-Moyal star product

$$f \star_\theta g = \mu \circ \exp\left(\frac{\mathrm{i}\lambda}{2}\theta^{k\ell}\partial_k \otimes \partial_\ell\right)(f \otimes g), \tag{1.1}$$

where $\mu(f \otimes g) = fg$ is the pointwise product. The coefficients $\theta^{k\ell} = -\theta^{\ell k}$ of θ are real constants and one usually assumes that θ has maximal rank, in order to make the model nontrivial. For dimensional reasons, the deformation parameter λ plays the role of the Planck area, i.e. ℓ^2_{Planck}. Of course, one has to be slightly more careful by specifying the meaning of (1.1): For simplicity and for the moment, I choose the framework of *formal* star products [**3**] (see also [**15**] for a pedagogical introduction), i.e. (1.1) is interpreted as a formal power series in λ and $f, g \in C^\infty(M)[[\lambda]]$. By specifying appropriate function spaces one can also use a convergent and even C^*-algebraic version of (1.1), typically involving integral formulas using oscillatory integrals, see e.g. [**8, 12, 13**].

The important conceptual problem with this model is, among others, the fact that θ is a global quantity and hence, in principle, the effects of noncommutativity should be visible globally. In fact, for $n = 3$, (1.1) is mathematically equivalent to non-relativistic quantum mechanics on a 4 dimensional phase space with symplectic structure θ^{-1}, and hence ordinary quantum mechanics in two spacial degrees of freedom. Since quantum mechanics is very much observable at certain large distances in *phase space*, namely in the momentum directions, we have to expect similar effects also in (1.1): for certain directions of Minkowski space (those which correspond to the momenta) the noncommutativity is visible at macroscopic scales. This contradicts of course the daily life experience. In fact, models of quantum field theories on this noncommutative Minkowski space support this simple argument: the so-called UV/IR mixing properties of these theories is just an effect of this type, namely, that the behavior of these quantum field theories intertwines small and large scales and results in absurd dispersion relations, see e.g. the discussion in [**1, 11**]. In conclusion, one should not be too surprised by this as the effect of θ is global from the very beginning. Hence one should better ask what this model has to do with the original motivation to change the geometry at small distances. To conclude, let me just point out that a nontrivial spacial dependence of θ, i.e. a non-constant Poisson structure, does not solve the problem as it is still a global object on space-time.

Acknowledgements: It is a pleasure for me to thank the organizers of this wonderful conference for all their work. Moreover, I would like to thank the participants for interesting discussions as well as the referee for helpful remarks.

2. Noncommutative Structures at Small Distances

In this section, I want to advocate for a more refined version of noncommutativity on space-times which will try to take care of the following issues:

- It should have a truly geometric formulation.
- It should work on generic space-times (M, g).
- It should use formal and/or C^*-algebraic deformation quantization as a tool.
- It should modify geometry only at small distances.

The main idea to achieve this, is actually very simple: M is not sufficient to encode the notion of 'distance', one needs *two* points to speak of distances and not one. Thus one has to consider $M \times M$ instead of M. So the idea is to make $M \times M$ noncommutative instead of M in such a way, that the noncommutativity is nontrivial only close to the *diagonal* $\Delta_M \subseteq M \times M$.

Before proceeding with the set-up of the model, I have to recall some basic geometric features of $M \times M$. As M is assumed to be a space-time, it is equipped with a pseudo Riemannian metric g and hence a Levi-Civita connection ∇. In the following, I shall only use properties of a torsion-free connection, whether it is the Levi-Civita connection of some metric or not is not important. I choose an open neighborhood $\mathcal{U} \subseteq TM$ of the zero section and a corresponding open neighborhood $\mathcal{V} \subseteq M \times M$ of the diagonal Δ_M such that the map

$$\Phi : \mathcal{U} \ni v_p \ \mapsto \ (\exp_p(-v_p), \exp_p(v_p)) \in \mathcal{V} \tag{2.1}$$

is a diffeomorphism. Here $v_p \in T_pM$ denotes a tangent vector at the point $p \in M$ and $\exp_p$ is the exponential map of ∇. Clearly, this can be arranged on any manifold with connection.

In a next step I choose a Poisson structure $\tilde{\theta} \in \Gamma^\infty(\Lambda^2 T(M \times M))$ on $M \times M$ whose support is within $\mathcal{V}$. As it should give raise to noncommutativity only close to the diagonal this seems to be appropriate. The idea is that $\mathcal{V}$ is still much larger than 'small distances'. If $\operatorname{supp} \tilde{\theta} \subseteq \mathcal{V}$, one can pull-back $\tilde{\theta}$ by the diffeomorphism Φ to a Poisson structure $\theta = \Phi^* \tilde{\theta} \in \Gamma^\infty(\Lambda^2 T(TM))$ on the tangent bundle with $\operatorname{supp} \theta \subseteq \mathcal{U}$. I have chosen the diffeomorphism Φ as it is most symmetric in the two copies in $M \times M$. In addition to Φ, I want the neighborhoods $\mathcal{U}$ and $\mathcal{V}$ to be symmetric under the flips $\tau : v_p \mapsto -v_p$ and $(q, q') \mapsto (q', q)$, respectively. Moreover, also θ and $\tilde{\theta}$ should be invariant under these two flip diffeomorphisms, simply because 'distance' is a symmetric concept. Of course, one can even start thinking about this feature and allow for more general and non-symmetric versions of the model.

In a next step, I want to incorporate the idea that the noncommutativity only becomes apparent at small distances. Clearly, the notion of metric distance is involved on a pseudo Riemannian manifold. Thus I choose the following, purely topological characterization: for all points $p \in M$, I require $\operatorname{supp}\theta \cap T_pM$ to be *compact.*

In a final step, the Poisson structures θ as well as $\tilde{\theta}$ are quantized by choosing a formal star product $\star$ and $\tilde{\star}$, respectively. Here I require that the support of the higher order cochains in $\star$ and $\tilde{\star}$ are contained in $\operatorname{supp}\theta$ and $\operatorname{supp}\tilde{\theta}$, respectively, in order to keep the idea of noncommutativity at small distances. Clearly, one has only to specify one of the two and use Φ to transport it back and forth. This way, Φ^* will become an isomorphism between the algebras $(C^\infty(\mathcal{V})[[\lambda]], \tilde{\star})$ and $(C^\infty(\mathcal{U})[[\lambda]], \star)$.

Clearly, this construction is in accordance with the four requirements I mentioned before. However, as there are many such Poisson structures and corresponding star products I want to restrict the model to some more specific Poisson structures. In fact, the requirements in the beginning can be sharpened as follows: *only* the relative distances should be responsible for noncommutativity, not their absolute position in M. The absolute position in M may nevertheless affect the 'amount' of noncommutativity. So in order to speak of relative coordinates on $M \times M$ one has to use some extra structure. In general this is not easy to justify, but here I have already used the diffeomorphism Φ anyway: this allows for a notion of relative coordinates at least within the neighborhood $\mathcal{V}$. Since the noncommutativity is located within $\mathcal{V}$, this is clearly sufficient. Thus the following requirement makes sense: if for $f \in C^\infty(M \times M)[[\lambda]]$ the restriction $f|_{\mathcal{V}}$ is constant along the relative coordinates defined by means of Φ, it should behave like a scalar with respect to $\tilde{\star}$, i.e.

$$f \,\tilde{\star}\, g = fg = g \,\tilde{\star}\, f \tag{2.2}$$

for all other functions $g \in C^\infty(M \times M)[[\lambda]]$. Note that (2.2) is fulfilled trivially outside $\mathcal{V}$. Equivalently, in terms of the product $\star$ on TM this yields the condition

$$\pi^*u \star g = \pi^*ug = g \star \pi^*u \tag{2.3}$$

for all $u \in C^\infty(M)[[\lambda]]$ and $g \in C^\infty(TM)[[\lambda]]$.

This last condition has now a very simple geometric interpretation: the cochains C_r of $\star$ have to be *vertical* bidifferential operators. Recall first that a k-differential operator D on a vector bundle $\pi : E \longrightarrow M$ is called vertical if for all functions $f_1, \ldots, f_k \in C^\infty(E)$ and $u \in C^\infty(M)$ one has

$$D(f_1, \ldots, \pi^*uf_r, \ldots, f_k) = \pi^*uD(f_1, \ldots, f_k) \tag{2.4}$$

for all $r = 1, \ldots, k$. This is equivalent to say that D differentiates only in vertical directions. Then clearly (2.3) implies that the cochains C_r of the star product differentiate *at least once* in vertical directions. The fact that they *only* differentiate in vertical directions follows from the associativity of $\star$.

3. Existence and Classification Results

Let me now focus on existence and classification results on such vertical Poisson structures and their corresponding vertical star products. The following lemma provides the basis for many vertical Poisson structures [**9**, Lem. 2.2]:

LEMMA 3.1. *Let $\pi : E \longrightarrow M$ be a vector bundle with a fiber metric h and let $e_i \in \Gamma^\infty(E)$ be a collection of sections for $i \in I$. Then there exist vector fields $X_i \in \Gamma^\infty(TE)$ on E with the following properties for all $i, j \in I$:*

(1) *$X_i \in \Gamma^\infty(TE)$ is vertical.*

(2) $[X_i, X_j] = 0$.
(3) X_i *coincides with the vertical lift* e_i^{ver} *on* $B^h_{1/2}(0)$.
(4) $\operatorname{supp} X_i \subseteq B^h_1(0)$.
(5) $\tau^* X_i = -X_i$.

Here $B^h_\epsilon(0) \subseteq E$ denotes the open neighborhood of the zero section of E built fiberwise out of the open ϵ-balls in each fiber with respect to the fiber metric. One can use these vector fields to show the following [**9**, Prop. 2.3]:

PROPOSITION 3.2. *Let* $\gamma \in \Gamma^\infty(\Lambda^2 E)$ *and let* $\mathfrak{U} \subseteq E$ *be an open neighborhood of the zero section. For d sufficiently large, there exist* $2d$ *vertical, pairwise commuting vector fields* $X_1, \ldots, X_{2d} \in \Gamma^\infty(TE)$ *and an open neighborhood* $\tilde{\mathfrak{U}} \subseteq \mathfrak{U}$ *of the zero section such that*

(1) $\operatorname{supp} X_i \subseteq \mathfrak{U}$ *and* $\operatorname{supp} X_i \cap E_p$ *is compact for all* $p \in M$ *and* $i = 1, \ldots, 2d$.
(2) $\theta = \sum_{i=1}^d X_i \wedge X_{i+d}$ *is a vertical Poisson structure which coincides with the vertical lift* γ^{ver} *on* $\tilde{\mathfrak{U}}$.

Thanks to the support condition it is clear that the vector fields X_i all have a complete flow. The fact that θ is a Poisson structure follows trivially from the fact that the vector fields all commute. The non-trivial part of the proposition is to achieve that θ coincides with the vertical lift γ^{ver} on $\tilde{\mathfrak{U}}$. From this point of view, the proposition says that there are as many vertical Poisson structures meeting the support conditions as there could possibly be. When it comes to C^*-algebraic quantization these Poisson structures play an even more important role as they arise from an action of $\mathbb{R}^{2d}$: indeed, the flows of the vector fields $X_1, \ldots, X_{2d}$ combine into such an action. Moreover, one can easily arrange the vector fields to satisfy $\tau^* X_i = -X_i$ whence $\tau^*\theta = \theta$ follows.

Let me now come to the quantized picture. Given a vertical Poisson structure θ fulfilling the support conditions Kontsevich's formality theorem [**10**] guarantees the existence of a star product also fulfilling the support conditions since the formality itself is local. However, it is not clear from the global formulations of the formality theorem [**5, 7**] whether this will yield a *vertical* star product. Instead of implementing this compatibility of the global formality with vertical Poisson structures by hand, it is much simpler to use a vertical formality from the beginning: This will not be a formality map on all multivector fields but only on the vertical ones. Using this, one arrives at the following statement [**2**, Thm. A.9]:

THEOREM 3.3. *For any vertical Poisson structure* $\theta \in \Gamma^\infty(\Lambda^2 TE)$ *there exists a vertical star product* $\star$ *such that the support of the cochains of* $\star$ *are contained in* $\operatorname{supp}\theta$. *In fact, the equivalence classes of formal vertical Poisson structures modulo formal vertical diffeomorphisms are in bijection to the equivalence classes of formal vertical star products modulo vertical equivalence.*

The proof of this theorem consists in two steps. First one shows a vertical Hochschild-Kostant-Rosenberg theorem relating the Gerstenhaber algebra of vertical multivector fields $\Gamma^\infty_{\mathsf{ver}}(\Lambda^\bullet TE)$ to the vertical differential Hochschild cohomology $\mathrm{HH}^\bullet_{\text{diff, ver}}(C^\infty(E))$. Applying the usual HKR morphism on each fiber yields an isomorphism of Gerstenhaber algebras

$$U^{(1)}_{\mathsf{ver}} : \Gamma^\infty_{\mathsf{ver}}(\Lambda^\bullet TE) \longrightarrow \mathrm{HH}^\bullet_{\text{diff, ver}}(C^\infty(E)). \tag{3.1}$$

In a second step one extends $U^{(1)}_{\mathrm{ver}}$ to a formality map, i.e. a L_∞-morphism $U^{(\bullet)}_{\mathrm{ver}}$ by applying Kontsevich's formula for the formality on $\mathbb{R}^d$ fiberwise on each fiber E_p of E. Since Kontsevich's formality is $\mathrm{GL}_d(\mathbb{R})$-invariant, this is well-defined. From this the theorem immediately follows. Note also, that if the Poisson structure θ is invariant under the flip τ, i.e. $\tau^*\theta = \theta$ then τ^* becomes an automorphism of $\star$. Again, this can be checked fiberwise and follows from the $\mathrm{GL}_d(\mathbb{R})$-invariance of Kontsevich's formality on $\mathbb{R}^d$.

4. Properties of Vertical Star Products

Suppose now that a vertical θ obeying all the conditions concerning support together with a corresponding star product $\star$ has been chosen. In order to get more familiar with the (noncommutative) geometry resulting from such choices, I shall now point out some of the consequences of the above construction.

First and most important, one can use the embedding $\iota_p : T_pM \longrightarrow TM$ of a tangent space of a point $p \in M$ into the tangent bundle to restrict the bivector field θ. This is possible thanks to the verticality property. It yields a bivector field $\theta_p \in \Gamma^\infty(\Lambda^2 T(T_pM))$ which is still a Poisson structure. Indeed, θ and θ_p are ι_p-related. Moreover, this implies that ι_p is a Poisson map whence

$$\iota_p^*\{f, g\}_\theta = \{\iota_p^* f, \iota_p^* g\}_{\theta_p} \tag{4.1}$$

for all $f, g \in C^\infty(TM)[[\lambda]]$. From the construction of the star product the same holds for $\star$. One has a restricted star product $\star_p$ quantizing θ_p such that ι_p^* is an algebra morphism, i.e.

$$\iota_p^*(f \star g) = \iota_p^* f \star_p \iota_p^* g \tag{4.2}$$

for all $f, g \in C^\infty(TM)[[\lambda]]$. The support condition on θ and $\star$ then imply that θ_p as well as $\star_p$ have their support inside a compact subset around $0_p \in T_pM$. For later use, let me denote the open neighborhood $\mathcal{U} \cap T_pM$ of 0_p in T_pM by $\mathcal{U}_p$. Then clearly $\operatorname{supp}\theta_p \subseteq \mathcal{U}_p$ and $\star_p$ is the pointwise product outside $\mathcal{U}_p$.

To interpret this in physical terms, recall that the starting point for the construction of θ and $\star$ was the diffeomorphism Φ: a point $v_p \in \mathcal{U}_p \subseteq T_pM$ corresponds to two points (q, q') in $M \times M$ given by $\Phi(v_p)$. In particular, p is the *geodesic midpoint* between (q, q'), being well-defined as $(q, q') \in \mathcal{V}$ are close enough. Thus, using the exponential map $\exp_p$ at p gives us a normal chart

$$\exp_p : \mathcal{U}_p \subseteq T_pM \longrightarrow \mathcal{V}_p \subseteq M \tag{4.3}$$

for an open neighborhood $\mathcal{V}_p$ of $p \in M$. Mapping $v_p \in \mathcal{U}_p$ to $\exp(v_p) = q$ in $\mathcal{V}_p$ comes along with the point $q' = \exp_p(-v_p)$. So one can now transport everything to the manifold M itself using $\exp_p$. This gives a star product $\tilde{\star}_p$ quantizing a Poisson structure $\tilde{\theta}_p = (\exp_p)_*\theta_p$ whose support is close around p. Thus every point in M obtains its own small *noncommutative neighborhood* encoded by the star products $\{\tilde{\star}_p\}_{p\in M}$. Again, the interpretation comes from the original picture on $M \times M$: the point p is the geodesic midpoint between points q and q': So if one is interested in noncommutative effects related to the small distance of the points q and q' then one has to determine their geodesic midpoint p and use $\tilde{\star}_p$ for the observables $C^\infty(M)[[\lambda]]$ which encode the relative coordinates of q and q'.

In a next step I want to discuss possible measurable effects coming from such a structure of local noncommutativity. To this end one needs to fix a notion of *states* first, then one can formulate expectation values and variances etc. In order to have a consistent notion for states, I shall assume from now on that all star products will be *Hermitian*, i.e. complex conjugation is a *-involution. Again, for real $\theta = \overline{\theta}$ and the construction of $\star$ according to the vertical formality theorem, this will follow automatically from symmetry properties of Kontsevich's formality theorem in $\mathbb{R}^d$.

In deformation quantization there is a well-established notion of states in complete analogy to the usual situation for *-algebras over $\mathbb{C}$ using positive functionals, see e.g. [**14**, **15**] for a details: The only difference is that the linear functionals take now values in $\mathbb{C}[[\lambda]]$ and, consequently, are $\mathbb{C}[[\lambda]]$-linear instead of only $\mathbb{C}$-linear. For positivity, one uses the notion of a positive (real) formal power series coming from the canonical ring ordering of $\mathbb{R}[[\lambda]]$, i.e. $a = \sum_{r=r_0}^{\infty} \lambda^r a_r \in \mathbb{R}[[\lambda]]$ is called positive, if $a_{r_0} > 0$. Then a *state* of a star product algebra $(C^\infty(N)[[\lambda]], \star)$ on some manifold N is a $\mathbb{C}[[\lambda]]$-linear functional $\omega : (C^\infty(N)[[\lambda]], \star) \longrightarrow \mathbb{C}[[\lambda]]$ such that

$$\omega(\overline{f} \star f) \geq 0 \quad \text{and} \quad \omega(\mathbb{1}) = 1 \tag{4.4}$$

for all $f \in C^\infty(N)[[\lambda]]$. The $\mathbb{C}[[\lambda]]$-linearity ensures that ω is actually of the form

$$\omega = \sum_{r=0}^{\infty} \lambda^r \omega_r \quad \text{with } \mathbb{C}\text{-linear maps} \quad \omega_r : C^\infty(N) \longrightarrow \mathbb{C}, \tag{4.5}$$

and ω_0 is a *classical* state, i.e. $\omega_0(\overline{f}f) \geq 0$ for all $f \in C^\infty(N)$. By Riesz' representation theorem one knows that ω_0 is the integration with respect to a compactly supported Borel measure on N. Conversely, a non-trivial theorem states that one can add higher order correction terms $\omega_1, \ldots$ to deform any classically positive functional into a positive functional with respect to $\star$, see [**4**]. In general, such correction terms are necessary and can be chosen to be local: they can be chosen to be distributional derivatives of the measure ω_0. Note however, that the correction terms are by no means unique.

Let me now apply this concept of states to the locally noncommutative space-time. A particular interest is *what geometry* one actually can observe in this model. Thus I want to discuss *distance measurements* as the notion of distance was the starting point of the whole analysis. Classically, the geodesic distance is encoded in the smooth function $d^2 \in C^\infty(TM)$ given by

$$d^2(v_p) = g_p(v_p, v_p) \tag{4.6}$$

where $v_p \in T_pM$ and g_p is the (pseudo-) Riemannian metric at p. Then the quantity $d^2(v_p)$ describes the square of the geodesic distance between the geodesic midpoint point p and $\exp_p(v_p)$ from which one obtains the geodesic distance between $\exp_p(v_p)$ and $\exp_p(-v_p)$. Note that d^2 is smooth, while the function d itself needs not to be smooth and, in the Lorentz situation, is not very meaningful anyway, as d^2 assumes both signs. This is the reason why the choice of d^2 seems to be more appropriate to encode the metric geometry. Note also, that in the Riemannian situation and for close enough points the square root d of d^2 indeed coincides with the metric distance defined as usual as the infimum over all lengths of connecting curves.

Now turning to the measurement in the noncommutative regime I re-interpret the quantity $d^2(v_p)$ as the evaluation $\delta_{v_p}(d^2)$ of the observable $d^2 \in C^\infty(TM)$ in the classical state δ_{v_p}. By the general philosophy of deformation quantization, the observables themselves are untouched when deforming, but the states have to be corrected in order to obtain some *positive* functionals again. Thus one has to chose corrections to the δ-functional at v_p in order to make it positive again. Denote by

$\omega_{v_p} = \delta_{v_p} + \lambda\omega_{v_p}^{(1)} + \cdots$ such a deformation. Then 1/4th of the *noncommutative distance square* between $\exp_p(v_p)$ and $\exp_p(-v_p)$ is given by

$$\omega_{v_p}(d^2) = d^2(v_p) + \lambda\omega_{v_p}^{(1)}(d^2) + \cdots . \tag{4.7}$$

Recall that for the noncommutative effects concerning the distance between $\exp_p(v_p)$ and $\exp_p(-v_p)$ only their relative coordinates should matter whence one can compute it using $\star_p$ where p is their geodesic midpoint.

In (4.7) the crucial observation is that the higher order corrections may depend on the choice of the deformation ω_{v_p} but typically can not be chosen to be zero. This is in some sense the effect of the uncertainty relations corresponding to the noncommutativity of $\star_p$. In particular, it typically will happen that the noncommutative distance is even non-zero for $v_p = 0$. This simply means that one is in a truly noncommutative geometry at these small distances, so the notion of a *point* does not make sense any longer. Let me finally mention that the possible deformations of δ_{v_p} into a positive functional ω_{v_p} are still not very well-understood. Here one certainly has to find more specific deformations and classifications. In [**2**], more explicit examples of such deformations are discussed.

5. Pro-C^* Algebraic Version

For a general Poisson structure θ meeting the above conditions it seems that only formal star products are available for quantizing. It is clearly one of the most challenging questions in deformation quantization to find general results on convergence of star products. However, for more particular choices of θ one can use techniques of Rieffel to find C^*-algebraic deformations: in the particular case where θ is build out of commuting vector fields as in Proposition 3.2 one obtains an action of $\mathbb{R}^d$ by integrating the commuting vector fields: since they have compact support in fiber directions, they clearly have complete flows. Thus one can use Rieffel's deformation quantization for actions of $\mathbb{R}^d$ as in [**12**].

Rieffel's quantization starts with the commutative C^*-algebra of bounded continuous functions $C_b^0(M \times M)$ as well as $C_b^0(TM)$, $C_b^0(T_pM)$ and $C_b^0(M)$, each of which is equipped with an action of $\mathbb{R}^d$ together with the corresponding Poisson structures. The choice for these function spaces is reasonable, but not the only possible one: in fact, I shall also make use of the spaces of bounded continuous functions defined only on the subsets $\mathfrak{U}$ and $\mathfrak{V}$ as well as on $\mathfrak{U}_p$ and $\mathfrak{V}_p$. While these are all C^*-algebras with unit, one can also consider the continuous functions vanishing at the boundaries of the above open subsets, i.e. $C_\infty^0(M \times M)$, $C_\infty^0(TM)$, $C_\infty^0(T_pM)$, $C_\infty^0(M)$ and similarly for $\mathfrak{U}$ etc.

For all these C^*-algebras, Rieffel's construction provides a continuous deformation in direction of the corresponding Poisson structures in the sense of continuous fields of C^*-algebras. The deformations are first defined on the dense subspaces of smooth vectors with respect to the actions and yield products $\star_\ell$ where $\ell > 0$ is now a real parameter. Asymptotically for $\ell \longrightarrow 0$ one obtains the corresponding Weyl-type deformations $\star$ with formal λ instead of ℓ. I shall not go into any details of his (by now well-known) construction but emphasize some particular aspects relevant for the present framework:

First, it is clear that transporting the commuting vector fields to $M \times M$ and restricting them to a tangent space T_pM and pushing them forward to M by $\exp_p$ makes all the involved maps Φ, ι_p and $\exp_p$ equivariant for the corresponding actions of $\mathbb{R}^d$. Then by general features of Rieffel's construction the maps Φ^*, ι_p^* and $\exp_p^*$ turn into *-homomorphisms of the deformed C^*-algebras, at least if we restrict to the open subsets $\mathcal{U}$, etc., where the pull-backs are well-defined.

Second, even though Rieffel's deformation yields *nonlocal* products, the support conditions on the vector fields still preserve some sort of weak locality: this is a general feature of Rieffel's construction which has not much been used up to now, as in the interesting examples the support of the fundamental vector fields of the action typically was the whole manifold. Here however, the support is compact in fiber directions. This leads to the following results [**9**, Prop. 3.2, 3.3, and 3.4] which I formulate for TM for simplicity:

PROPOSITION 5.1. *Let $f, g \in C_b^0(TM)^\infty$ be smooth vectors for the action of $\mathbb{R}^d$ and let $\star_\ell$ be the corresponding Rieffel star product for some $\ell > 0$. Denote by K the union of the supports of the fundamental vector fields X_i of the action.*

(1) $\operatorname{supp}(f \star_\ell g) \subseteq K \cup (\operatorname{supp} f \cap \operatorname{supp} g)$.
(2) *If* $\operatorname{supp} f \cap K = \emptyset$ *then* $f \star_\ell g = fg = g \star_\ell f$.
(3) *If* $X_i(v_p) = 0$ *for all* $i = 1, \ldots, d$, *then* $\delta_{v_p}(f \star_\ell g) = f(v_p)g(v_p)$.

In particular, the last part shows that the points far away from the noncommutativity still remain 'points' as the corresponding δ-functionals are still positive functionals (even characters) of the deformed algebra. This matches very well the original idea of changing the geometry from classical to noncommutative only for small distances. Clearly, the above results hold also for $M \times M$, T_pM and M after the obvious modifications.

The last part also raises the following question: why should one pose conditions on the functions at infinity, if the whole deformation takes place only in a compact subset of each tangent space. Indeed, in this situation, Rieffel's construction for C^*-algebras can be easily generalized to pro-C^*-algebras: recall that a pro-C^*-algebra is a projective limit of C^*-algebras in the category of locally convex algebras. The continuous functions $C^0(TM)$ will be the motivating example. Here one only has a directed set of C^*-seminorms, the usual sup-norms over compact subsets, which make $C^0(TM)$ into a pro-C^*-algebra. The crucial observation is now that all of Rieffel's construction can be carried through for pro-C^*-algebras as well, provided the action of $\mathbb{R}^d$ is *cofinally isometric*: this means that for every continuous C^*-seminorm $\|\cdot\|_\alpha$ on the pro-C^*-algebra $\mathcal{A}$ there exists another continuous C^*-seminorm $\|\cdot\|_\beta$ with $\|\cdot\|_\alpha \leq \|\cdot\|_\beta$ such that the action is isometric with respect to $\|\cdot\|_\beta$. In the geometric case I am considering, this is clearly the case, as for all compact subsets L which contain K, the action is isometric with respect to the sup-norm $\|\cdot\|_{\infty,L}$ over L. As result, there are also deformations $\star_\ell$ for $C^0(TM)$, $C^0(M \times M)$ etc.

6. Outlook and Open Questions

Let me conclude with some remarks and questions arising from this approach to locally noncommutative space-times.

From a physical point of view, the above construction can only be a first step to obtain reasonable models for space-times at small distances. Clearly, the next crucial step will be to endow the locally noncommutative space-times with some dynamics on it and study e.g. a (quantum) field theory, or certainly much easier, just some point-like particles moving in M. Here the hope is to make contact to formulations of noncommutative field theory as currently discussed. However, one should be aware that things might not be too easy as now the fields are defined on TM instead of M, at least if one wants to use $\star$. Otherwise, one can use $\star_p$ for fields living on M, but has a dependence on the point p. Here a reasonable physical interpretation is clearly missing.

A second and probably much harder question is to understand the role of the Poisson structure θ better: if one wants to interpret this locally noncommutative space-time as an effective theory of the true quantum gravity, the Poisson structure θ should have some *dynamical* content. With other words, θ should not be god-given but itself a field subject to some field equations. Clearly, this is a general question concerning all models of noncommutative space-times, it is not particular to this approach. It seems to be very hard to formulate some field equation in a geometric way which on one hand couples θ to other fields like the metric or matter and Yang-Mills fields. On the other hand one still has to take care of the Jacobi identity $[\![\theta, \theta]\!] = 0$, since otherwise θ can not be used as the first order term of an *associative* deformation. Probably, the Jacobi identity can be implemented as a constraint but it seems to be hard to encode this in a geometric fashion in some reasonable action principle.

The third aspect of further investigations is more mathematical in nature: one would like to understand better the role of the deformed states. Even though one has the statement that classical states can be deformed into states with respect to any formal star product, the proof in [**4**] is far from being explicit. Thus, a more explicit deformation with more specific features would be highly desirable. Moreover, one would like to understand better in how many essentially different ways a classical state can be deformed. The C^*- and pro-C^*-algebraic situation seems to be much more obscure. Can one always deform the classical states and if so, how explicit and how unique?

The last remark I want to make goes beyond the presented model: if one allows for noncommutativity arising from small distances of two points, why should one stop here? Clearly, it would be interesting to consider a whole hierarchy of noncommutative structures on higher Cartesian products $M \times M \times M$, M^4, $\ldots$, endowed with suitable compatibility conditions. Here the notion of 'vertical' has to be adapted in a suitable way. Succeeding with this, one possibility of using such a hierarchy would be to modify the short distance behavior of n-point functions in quantum field theory. So the above construction would be only the first step concerning two-point functions. Apart from the above geometric motivation of modifying the notions of distance, this would probably give a very different and yet related motivation: the introduction of θ and its higher analogs on M^n can serve as regulators for a quantum field theory. The question is then, whether this is their only purpose or if there is some intrinsic meaning, as I have argued above.

References

[1] Bahns, D., Doplicher, S., Fredenhagen, K., Piacitelli, G.: *Ultraviolet finite quantum field theory on quantum spacetime.* Commun. Math. Phys. **237** (2003), 221–241.

[2] Bahns, D., Waldmann, S.: *Locally Noncommutative Space-Times.* Preprint **math.QA/0607745** (2006), 28 pages. To appear in Rev. Math. Phys.

[3] Bayen, F., Flato, M., Frønsdal, C., Lichnerowicz, A., Sternheimer, D.: *Deformation Theory and Quantization.* Ann. Phys. **111** (1978), 61–151.

[4] Bursztyn, H., Waldmann, S.: *Hermitian star products are completely positive deformations.* Lett. Math. Phys. **72** (2005), 143–152.

[5] Cattaneo, A. S., Felder, G., Tomassini, L.: *From local to global deformation quantization of Poisson manifolds.* Duke Math. J. **115**.2 (2002), 329–352.

[6] Connes, A.: *Noncommutative Geometry.* Academic Press, San Diego, New York, London, 1994.

[7] Dolgushev, V. A.: *Covariant and equivariant formality theorems.* Adv. Math. **191** (2005), 147–177.

[8] Doplicher, S., Fredenhagen, K., Roberts, J. E.: *The Quantum Structure of Spacetime at the Planck Scale and Quantum Fields.* Commun. Math. Phys. **172** (1995), 187–220.

[9] Heller, J. G., Neumaier, N., Waldmann, S.: *A C^*-Algebraic Model for Locally Noncommutative Space-Times.* Preprint Freiburg FR-THEP 2006/14 **math.QA/0609850** (September 2006), 13 pages.

[10] Kontsevich, M.: *Deformation Quantization of Poisson manifolds.* Lett. Math. Phys. **66** (2003), 157–216.

[11] Liao, Y., Sibold, K.: *Spectral representation and dispersion relations in field theory on noncommutative space.* Phys. Lett. B **549** (2002), 352–361.

[12] Rieffel, M. A.: *Deformation quantization for actions of $\mathbb{R}^d$.* Mem. Amer. Math. Soc. **106**.506 (1993), 93 pages.

[13] Rieffel, M. A.: *On the operator algebra for the space-time uncertainty relations.* In: Doplicher, S., Longo, R., Roberts, J. E., Zsido, L. (eds.): *Operator algebras and quantum field theory,* 374–382. International Press, Cambridge, MA, 1997. Proceedings of the conference held in Rome, July 1–6, 1996.

[14] Waldmann, S.: *States and Representation Theory in Deformation Quantization.* Rev. Math. Phys. **17** (2005), 15–75.

[15] Waldmann, S.: *Poisson-Geometrie und Deformationsquantisierung. Eine Einführung.* Springer-Verlag, Heidelberg, Berlin, New York, 2007. To appear in September 2007.

Fakultät für Mathematik und Physik, Albert-Ludwigs-Universität Freiburg, Physikalisches Institut, Hermann Herder Strasse 3, D 79104 Freiburg, Germany

E-mail address: `Stefan.Waldmann@physik.uni-freiburg.de`

Titles in This Series

450 **Giuseppe Dito, Jiang-Hua Lu, Yoshiaki Maeda, and Alan Weinstein, Editors,** Poisson geometry in mathematics and physics, 2008

449 **Robert S. Doran, Calvin C. Moore, and Robert J. Zimmer, Editors,** Group representations, ergodic theory, and mathematical physics: A tribute to George W. Mackey, 2007

448 **Alberto Corso, Juan Migliore, and Claudia Polini, Editors,** Algebra, geometry and their interactions, 2007

447 **François Germinet and Peter Hislop, Editors,** Adventures in mathematical physics, 2007

446 **Henri Berestycki, Michiel Bertsch, Felix E. Browder, Louis Nirenberg, Lambertus A. Peletier, and Laurent Véron, Editors,** Perspectives in Nonliner Partial Differential Equations, 2007

445 **Laura De Carli and Mario Milman, Editors,** Interpolation Theory and Applications, 2007

444 **Joseph Rosenblatt, Alexander Stokolos, and Ahmed I. Zayed, Editors,** Topics in harmonic analysis and ergodic theory, 2007

443 **Joseph Stephen Verducci and Xiaotong Shen, Editors,** Prediction and discovery, 2007

442 **Yi-Zhi Huang and Kailash C Misra, Editors,** Lie algebras, vertex operator algbras and their applications, 2007

441 **Louis H. Kauffman, David E. Radford, and Fernando J. O. Souza, Editors,** Hopf algebras and generalizations, 2007

440 **Fernanda Botelho, Thomas Hagen, and James Jamison, Editors,** Fluids and Waves, 2007

439 **Donatella Danielli, Editor,** Recent developments in nonlinear partial differential equations, 2007

438 **Marc Burger, Michael Farber, Robert Ghrist, and Daniel Koditschek, Editors,** Topology and robotics, 2007

437 **José C. Mourão, Joào P. Nunes, Roger Picken, and Jean-Claude Zambrini, Editors,** Prospects in mathematical physics, 2007

436 **Luchezar L. Avramov, Daniel Christensen, William G Dwyer, Michael A Mandell, and Brooke E Shipley, Editors,** Interactions between homotopy theory and algebra, 2007

435 **Krzysztof Jarosz, Editor,** Function spaces, 2007

434 **S. Paycha and B. Uribe, Editors,** Geometric and topological methods for quantum field theory, 2007

433 **Pavel Etingof, Shlomo Gelaki, and Steven Shnider, Editors,** Quantum groups, 2007

432 **Dick Canery, Jane Gilman, Juha Heinoren, and Howard Masur, Editors,** In the tradition of Ahlfors-Bers, IV, 2007

431 **Michael Batanin, Alexei Davydov, Michael Johnson, Stephen Lack, and Amnon Neeman, Editors,** Categories in algebra, geometry and mathematical physics, 2007

430 **Idris Assani, Editor,** Ergodic theory and related fields, 2007

429 **Gui-Qiang Chen, Elton Hsu, and Mark Pinsky, Editors,** Stochastic analysis and partial differential equations, 2007

428 **Estela A. Gavosto, Marianne K. Korten, Charles N. Moore, and Rodolfo H. Torres, Editors,** Harmonic analysis, partial differential equations, and related topics, 2007

427 **Anastasios Mallios and Marina Haralampidou, Editors,** Topological algebras and applications, 2007

426 **Fabio Ancona, Irena Lasiecka, Walter Littman, and Roberto Triggiani, Editors,** Control methods in PDE-dynamical systems, 2007